Spontaneous Symmetry Breaking

Spontaneous symmetry breaking lies at the heart of modern physics, shaping our understanding of matter, forces, and even the universe itself. From condensed matter physics to particle physics and cosmology, spontaneous symmetry breaking unifies phenomena that at first seem worlds apart. This graduate-level text offers a comprehensive yet accessible guide to the conceptual theory and practical consequences of spontaneous symmetry breaking. It introduces topics ranging from Noether's theorem, thermodynamic limits, and gauge freedoms to Nambu–Goldstone modes, topological defects, effective field theory, the Mermin–Wagner–Hohenberg–Coleman theorem, and the Anderson–Higgs mechanism within the Standard Model. Packed with exercises, with solutions available online, in-depth projects, and a myth-busting frequently asked questions section addressing common pitfalls, this book equips readers to master both the fundamentals and modern frontiers of spontaneous symmetry breaking, making it an indispensable resource for students, teachers, and researchers.

Aron J. Beekman obtained a PhD in theoretical physics from Leiden University. He has held research positions in Japan at RIKEN, the National Institute for Materials Science, and Keio University, focussing on the symmetry structure of quantum liquid crystals. He currently leads the Computational Physics group at Sioux Technologies in Eindhoven, the Netherlands.

Louk Rademaker obtained a PhD in theoretical physics from Leiden University. He has worked at the Kavli Institute for Theoretical Physics in Santa Barbara, the Perimeter Institute in Waterloo, and the University of Geneva. He is currently Associate Professor at Leiden University.

Jasper van Wezel obtained a PhD in theoretical physics from Leiden University. He subsequently held prize fellowships at the University of Cambridge and Argonne National Laboratory and was a lecturer at the University of Bristol. He is currently Professor of Condensed Matter Theory at the University of Amsterdam and has won multiple teaching prizes.

Spontaneous Symmetry Breaking

ARON J. BEEKMAN
Sioux Technologies

LOUK RADEMAKER
University of Geneva and Leiden University

JASPER VAN WEZEL
University of Amsterdam

CAMBRIDGE
UNIVERSITY PRESS

CAMBRIDGE
UNIVERSITY PRESS

Shaftesbury Road, Cambridge CB2 8EA, United Kingdom

One Liberty Plaza, 20th Floor, New York, NY 10006, USA

477 Williamstown Road, Port Melbourne, VIC 3207, Australia

314–321, 3rd Floor, Plot 3, Splendor Forum, Jasola District Centre,
New Delhi – 110025, India

Cambridge University Press is part of Cambridge University Press & Assessment,
a department of the University of Cambridge.

We share the University's mission to contribute to society through the pursuit of
education, learning and research at the highest international levels of excellence.

www.cambridge.org
Information on this title: www.cambridge.org/9781009575768
DOI: 10.1017/9781009575782

When citing this work, please include a reference to the DOI 10.1017/9781009575782

First published 2026

Cover image: While any pencil can in principle be balanced, the difficulty of achieving balance becomes
insurmountable for sharper pencils. This is an example of the type of singular limit that underlies
spontaneous symmetry breaking. When the pencil falls, a symmetry is broken. Image credit: Xavier Ravinet,
University of Geneva.

A catalogue record for this publication is available from the British Library

A Cataloging-in-Publication data record for this book is available from the Library of Congress

ISBN 978-1-009-57576-8 Hardback

To Chikako, Lusine, and Annelies

Contents

Preface

Symmetry is one of the great unifying themes in physics. From cosmology to nuclear physics, from soft matter to quantum materials, symmetries determine which shapes, interactions, and evolutions occur in nature. Perhaps the most important aspect of symmetry in physics is the idea that the state of a system does not need to have the same symmetries as the laws of nature that describe the system. Such *spontaneous breakdown* of symmetries governs the dynamics of phase transitions, the emergence of new particles and excitations, the rigidity of collective states of matter, and is one of the main ways classical physics emerges in a quantum world.

The basic idea of spontaneous symmetry breaking (SSB) is well known, and repeated in different ways throughout all fields of physics. More specific aspects of SSB, and their physical ramifications, however, are dispersed among the specialised literature of multiple subfields, and not widely known or readily transferred to other areas. Quantum field theory textbooks, for example, often focus on the mathematical structure, while only leisurely borrowing and discussing physical examples and nomenclature from condensed matter physics. Most condensed matter-oriented texts, on the other hand, treat SSB on a case-by-case basis, without building a rigorous understanding of the mathematical and physical concepts underlying them.

In 2019, we published a set of lecture notes attempting to bridge these approaches [23]. The notes focussed on core physical concepts, placing less emphasis on any field-theoretical calculations describing particular instances of SSB. While the lecture notes give a solid introduction to SSB, they lack an in-depth treatment of its applications, in which the full power of its concepts becomes apparent. This book can be seen as a substantial expansion of the lecture notes into a general and comprehensive introduction to SSB and its applications throughout physics. The main focus is to provide a sound physical understanding of the core concepts of SSB, while at the same time being sufficiently mathematically rigorous.

The central perspective on SSB taken in this book, often neglected or even denied in other sources, is that the limit of infinite system size, often called the thermodynamic limit, is *not* a necessary condition for SSB to occur in practice. Stronger, for almost all realistic applications of the theory of symmetry breaking, it is a rather useless limit, in the sense that it is never strictly realised. Even in situations where the object of interest can be considered large, the coherence length of ordered phases is generically small, and a single domain cannot in good faith be considered to approximate any sort of infinite size. Fortunately, even relatively small objects can spontaneously break symmetries, and almost all of the generic physical consequences already appear in some form at small scales. The central message of SSB then is that objects of all sizes reside in thermodynamically stable states, rather than energy eigenstates. In this book, we explore the physical manifestations of this viewpoint.

This book is divided into four parts: Basics (Part I), Concepts (Part II), Applications (Part III), and Projects (Part IV). Parts I and II describe the basics and core concepts that apply to SSB in all fields of physics and are in large part based on the original lecture notes, with many additions and improvements.

Chapters 1–3 in Part I constitute the basis of this treatise, systematically explaining all principles underlying SSB. The most important part of this book is Chapter 3, which we believe will be insightful for readers from all backgrounds learning about SSB. Throughout that chapter, we show that the description of symmetry breaking is not just a mathematical trick, but a practical theory governing the behaviour of real, small, dirty, and non-isolated systems. The remaining chapters in Part I, namely Chapter 4 on Nambu–Goldstone modes and Chapter 5 on phase transitions, are heavily referenced in the other chapters, and complete a basic understanding of SSB.

In Part II, we discuss various consequences of symmetries being broken, and each chapter there could be taught more or less independently from the others. The concepts introduced in this part include corrections to the order parameter (Chapter 6), topological defects (Chapter 7), and gauge fields (Chapter 8). A new chapter, Chapter 9, concerning the dynamics of symmetry breaking has been added as compared to the lecture notes.

In-text exercises are dispersed throughout the book and provide an opportunity to deepen the understanding of concepts introduced in the text. They are intended primarily to help guide the reader through the main text, as opposed to the Projects in Part IV of the book, which provide more extensive and challenging problems.

Part III delves into some of the main applications of SSB: effective field theory, particle physics, and the Standard Model in Chapters 10 and 11, fermion condensates and superconductivity in Chapter 12, SSB of spacetime symmetries in Chapter 13, and applications in cosmology in Chapter 14. As each of these topics can and do have entire textbooks dedicated to them, we give concise overviews with a focus on their relations to SSB.

Part IV contains projects challenging motivated readers to do real calculations on specialised topics related to SSB at a level of detail beyond the in-text exercises. Besides the first, introductory project, each is intended to take between one and a few days to complete. Projects 16 and 18 involve writing numerical codes and are aimed at readers who appreciate a computational approach to physics.

Additional related topics we do not discuss in detail are collected in Appendix A, and suggestions for further reading are collected in Appendix B. As the exercises are intended to support physical understanding, the solutions to all exercises are published on the accompanying website: www.cambridge.org/9781009575768.

The main text is preceded by a list of frequently asked questions about SSB. Each of these questions, which address often-occurring uncertainties about core concepts as well as common misconceptions about SSB, is answered with references to the main text.

Parts I and II of this book are suitable for a one-semester graduate course, whereas topics in Parts III and IV can be added independently, according to the readers' interests. Undergraduate students may find that large parts of each chapter are fun and interesting to read. Knowledge of basic quantum mechanics and Hamiltonian classical mechanics is essential, while familiarity with (classical) field theory, relativity, basic group theory, as well as some condensed matter or solid state physics will be useful. Throughout the book, we regularly switch between the Hamiltonian and Lagrangian paradigms. Many aspects of symmetry breaking can be formulated in either formalism, but sometimes one provides a clearer understanding than the other, while at other times it is insightful to compare the two approaches.

We hope this book brings to the fore the ubiquitous effects of symmetry breaking across all realms of physics. It explains how to predict both the number and the nature of collective excitations arising in any physical system with a broken symmetry. It shows that an Anderson–Higgs mechanism may arise from the breakdown of a global symmetry whenever it is accompanied by a gauge freedom, regardless of whether this occurs in general relativity, elementary particle physics, or superconductors. It allows you to appreciate why sitting on a chair is essentially the same thing as levitating a piece of superconducting material in a magnetic field. And it prepares you for further exploring the many wonderful connections between fields of physics brought together by the unifying themes of symmetry and symmetry breaking.

Notation

We adopt relativistic notation for vector fields where $A^\nu = (A^t, A^n)$. Greek indices run over time and space, and Roman indices run over space only. The metric is mostly plus: $\eta_{\mu\nu} = \text{diag}(-1, +1, +1, +1)$ (note that this is opposite to the sign convention used in the lecture notes [23]). The shorthand notation for derivatives is $\partial_\nu = \partial/\partial x^\nu$. For arguments of fields, we use (x) to denote $(t, \mathbf{x})$.

Acknowledgements

We thank Tomas Brauner and Hal Tasaki for discussions concerning gapped partner modes and effective field theories. We further benefited from discussions with Haruki Watanabe, Yoshimasa Hidaka, and Felix Flicker. We are grateful to Wouter Bos, Cecile Dekker, José Dupont, Anna Efimova, Ivo Gabrovski, Pim Herbschleb, Chen Yang, and Yanis Wenger for helping us improve the projects, exercises, and early versions of the draft. Finally, we thank our late mentor Jan Zaanen for instilling upon us the idea that all condensates, ranging from crystals and magnets to superconductors and superfluids, are to be seen as classical matter.

Frequently Asked Questions

Spontaneous symmetry breaking (SSB) is a topic that is ubiquitous throughout all fields of physics. Regardless of specialisation, students, researchers, and interested readers learning about specific aspects of SSB or looking for a deeper understanding of them often ask similar questions. Here we collect and answer the most frequently asked ones, sorted by topic. We include in each answer a reference to where in the book you can find more details.

Symmetry

Q1: What is symmetry in physics?

A physical theory or model is expressed in terms of dynamical **degrees of freedom**. These degrees of freedom may change under 'transformations', such as translations, rotations, and so forth. A **model** is said to be symmetric when the Hamiltonian, Lagrangian, or action describing the model is invariant under such transformations (Section 1.2).

Moreover, physical states (e.g., the wave function in quantum mechanics) can also be invariant under transformations, regardless of the model defining its equations of motion. We then speak of a **symmetric state** (Section 1.1).

Q2: Why is symmetry so important?

Any symmetry in a model gives rise to a **conserved quantity** via Noether's theorem (Section 1.3.1).

This viewpoint is ingrained in modern physics, and looking for symmetries and conserved quantities is the first approach to understanding a new situation encountered in any of its specialisations.

We can also invert this approach: given a set of symmetries that we expect a model to possess, the exact form of its Lagrangian is constrained. One can then learn a lot of general or qualitative properties of the theory without having to know its detailed form, simply by writing down all terms that would be allowed in the Lagrangian by symmetry. This is known as the **effective field theory** approach (Section 10.1).

Q3: What is Noether's theorem?

Every **continuous symmetry** gives rise to a **locally conserved current** (Section 1.3).

Q4: Does Noether's theorem hold if the symmetry is spontaneously broken?

Yes, Noether's theorem holds regardless of whether or not the state of the system respects the symmetry (Chapter 4).

Q5: What is the difference between discrete and continuous symmetry?

The discrete or continuous nature of symmetries refers to whether the group of symmetry operations *G* is discrete (e.g., $\mathbb{Z}_2$) or continuous (e.g., $U(1)$). Simply put: a discrete transformation you can either enact completely or not at all; continuous transformations, on the other hand, you can do 'a little bit' (Section 1.5.1). There is more richness in continuous symmetries than in discrete ones. For example, only continuous symmetries satisfy a local continuity equation based on Noether's theorem (Section 1.3.2), and only breaking a continuous symmetry gives rise to gapless Nambu–Goldstone modes (Chapter 4).

Q6: What is an approximate symmetry?

When all terms in a Hamiltonian or Lagrangian are invariant under a certain symmetry, **except a small term** (as compared to the others), that symmetry is referred to as an **approximate symmetry**.
Strictly speaking, the model does not obey that symmetry. Nevertheless, many properties of the system can be understood by first assuming it is a real symmetry, and only then including small corrections.
A famous example is the **pion** in particle physics, which can be viewed as resulting from the spontaneous breaking of an approximate chiral symmetry (Section 11.2.2).

Q7: What is a local symmetry?

A **local symmetry** is a symmetry transformation that acts differently on different local degrees of freedom (Section 1.5.3). An example is translating each particle in an ideal gas differently. Elitzur's theorem guarantees that local symmetries can never be broken spontaneously (Exercise 2.3).
Unfortunately, in the literature 'gauge symmetry' is often confused with 'local symmetry'. This is incorrect: a **gauge freedom** can be global or local and is not a physical symmetry transformation, but reflects a redundancy in the theoretical description of a physical situation (Section 1.6). See also Q26.

Symmetry Breaking

Q8: What is symmetry breaking?

Symmetry breaking is said to occur when a physical theory or **model** possesses a certain symmetry, while a physically realised **state** in the system governed by that model does not possess the same symmetry (Section 2.1).

Q9: What ingredients are necessary for SSB?

1. Spontaneous symmetry breaking requires a system that has a **symmetry** in its effective (Hamiltonian or Lagrangian) description (Section 1.2).
2. This system should have a susceptibility towards symmetry-breaking perturbations that diverges in the thermodynamic limit. In other words, the symmetric state should be **unstable** (Section 3.7).
3. A sufficiently strong external symmetry-breaking **perturbation** should be present. Because of the diverging susceptibility, the required perturbation can be exceedingly weak for large systems (Section 2.1).

Q10: What is explicit symmetry breaking?

If a symmetric theory or model is modified to include terms that are not invariant under the symmetry transformations, the symmetry is said to be broken **explicitly**.
This phenomenon is therefore **subjective** and involves comparing one model (with symmetry) to another (without symmetry).
Nevertheless, it is a useful concept to argue what should happen in case the model is modified. Explicit symmetry breaking is usually invoked when considering **external fields** (that can be applied in an experiment, for instance), such as an external magnetic field. A field that explicitly breaks the symmetry is also used to understand SSB, in the limit where the strength of the symmetry-breaking field goes to zero (Section 2.2).

Q11: What is SSB?

Spontaneous symmetry breaking refers to the situation in which a system occurs in a **stable low-energy state** that is not symmetric under the transformations leaving its Hamiltonian, Lagrangian, or action invariant (Chapter 2).
The system may be in one of several **degenerate states** that are related to each other by the broken symmetry transformations (Section 3.1), and these states become true ground states in the thermodynamic limit (Section 2.1).

Q12: What does 'spontaneous' mean?

The 'spontaneous' in SSB means '**without an apparent external cause**'. A system undergoing SSB does so even in the limit of vanishing external perturbations, provided we take the thermodynamic limit first (Sections 2.1 and 2.2).
Finite-sized systems do require a weak but non-zero external perturbation to break a symmetry. We still call this symmetry breaking 'spontaneous' if it occurs in systems large enough for the required perturbation to be immeasurably weak.

Q13: What is a Goldstone mode?

The spectrum of systems in which a continuous symmetry is spontaneously broken in the absence of long-range interactions is guaranteed to contain modes known as **Nambu–Goldstone modes**, whose energy vanishes in the limit of infinite wave length (Section 4.1). They are excited by acting on the symmetry-breaking state with the local Noether charge operator.

Q14: Does SSB require an infinite volume?

In sufficiently large, finite-sized systems, **the symmetric ground state is highly unstable**, and any real system will occur in a stable, symmetry-broken state (Section 3.7).
The energy difference between the symmetric ground state and the stable symmetry-breaking states is much smaller than any other energy gaps, including those of the Nambu–Goldstone modes (Section 3.6). The point of view taken in this book is that the stability of states is at least as important as their energy, and that **SSB is a useful concept even in small systems**.

Q15: When can SSB not occur?

There are several known circumstances that preclude spontaneous breaking of symmetry. The most important condition is known as the **Mermin–Wagner–Hohenberg–Coleman theorem**, which indicates that the spontaneous breaking of a continuous symmetry in low spatial dimensions (usually 2 or lower) is prevented by strong quantum corrections or thermal fluctuations even at the lowest temperatures (Section 6.2).

Q16: Is every phase transition associated with a symmetry breaking?

No. Topological transitions like the Berezinskii–Kosterlitz–Thouless (**BKT**) phase transition are not captured by the language of symmetry breaking (Section 7.4). Other transitions, such as the **liquid–gas transition**, may seem unrelated to symmetry breaking, but can be understood in the framework of universality classes for symmetry-breaking transitions (Chapter 17).

Q17: Does SSB only occur at zero temperature?

No. The concepts of symmetry breaking also apply to low-energy states, including low-temperature thermal equilibrium states (Chapter 2). Whenever there is a broken symmetry at low energy, there is a **critical temperature** above which the symmetry is restored (Chapter 5).

Q18: What is dynamical symmetry breaking?

Dynamical symmetry breaking is a type of SSB arising from symmetric interactions between multiple systems (Chapter 20). Examples include **electron mass generation** in the Standard Model (Section 11.3) or electron pairing in **BCS superconductors** (Section 12.2).

Q19: Why do SSB systems not occur in a symmetric state?

In a system that undergoes SSB, the symmetric state is extremely **unstable** against local perturbations. This instability can be quantified in terms of the **cluster decomposition** condition on correlation functions, or in terms of the scaling of the variance of extensive observables (Section 3.7).

Q20: Are there signatures of SSB in the symmetric state?

The symmetric state has the **same connected correlation functions** as the symmetry-broken state. In addition, a set of excitations from the symmetric ground state appear as a **tower of states** with energy gaps vanishing as $1/N$, which indicates the type of symmetry breaking (Section 3.6). This tower can be calculated using exact diagonalisation (Chapter 18).

Q21: Does SSB occur in one or two dimensions?

Yes: Discrete symmetry breaking can occur at zero temperature in $d = 1$ (Section 7.3) and at finite temperature in $d = 2$. Breaking a continuous symmetry in $d \leq 2$, however, is limited by the **Mermin–Wagner–Hohenberg–Coleman theorem** (Section 6.2 and Table 6.3). Finally, a special case is realised by systems with **algebraic long-range order**, occurring for example for $U(1)$ symmetry in $d = 2$ dimensions (Section 7.4).

Order Parameters

Q22: What is an order parameter?

A minimal, often used definition of the order parameter is an observable that **vanishes in the symmetric state** but is **non-zero in symmetry-broken states**. As such, it signals when symmetry is broken. In this book, we use a stronger defintion and additionally demand that the order parameter takes different values for distinct symmetry-breaking states (Section 3.2). In other words, its values are in one-to-one correspondence with the so-called quotient space of broken symmetries (Section 3.1).
It is important to distinguish between the **order parameter operator**, and the **order parameter** itself, which is the expectation value of the order parameter operator in a state with broken symmetry (Section 3.2).

Q23: Is there an absolute reference for an order parameter?

While the occurrence of the phenomenology of SSB is absolute, and can be inferred, for instance, from the existence of the tower of states (Section 3.6), or long-range order (Section 3.4), the value of the order parameter is always relative to another system with the same broken symmetry. This is rooted in the fact that symmetries of states can only be defined with respect to other states (Sections 1.1).

Q24: Do all symmetry-broken phases have an order parameter?

Yes (Section 3.2). However, it may be difficult to find an experimentally accessible quantity that measures the order parameter directly.

Q25: Is the gap a good order parameter for superconductors?

The energy gap in the electronic spectrum of superconductors is an order parameter in the sense of **Landau theory**, meaning it is zero in the symmetric phase and non-zero in the symmetry-broken phase (Section 5.2). However, this superconducting gap does not distinguish between inequivalent symmetry-broken states. It is only the amplitude of the order parameter, but does not contain information about the value of the broken $U(1)$ phase. It is therefore **not** an order parameter in the modern sense (Section 3.2).
Because of the presence of **gauge freedom**, one needs to be careful in defining a proper **gauge-invariant order parameter** (Section 8.2). In a superconductor, the correct order parameter is therefore an expectation value of the form $\langle c^\dagger c^\dagger \rangle$ supplemented with a Dirac string that makes it gauge invariant.

Gauge Freedom

Q26: What is gauge symmetry?

Gauge symmetry is a **misnomer**. What is usually meant is **gauge freedom**: the existence of superfluous or redundant degrees of freedom in our mathematical expression of a physical model (Hamiltonian or Lagrangian). These additional degrees of freedom are **not real, physical symmetries** and do not correspond to conserved quantities (Section 1.6).
In models with gauge freedom, physical quantities are and must always be **gauge invariant**, and gauge freedom can never be broken, spontaneously or otherwise (Section 1.6.2). Gauge freedoms can be either global or local, but in practice, the term is used almost exclusively for local redundancies. Not all local transformations leaving a model invariant are gauge freedoms (Section 1.5.3).
To add to the confusion, the generic, global $U(1)$-phase symmetry of quantum mechanical wave functions, $\psi(x) \rightarrow \psi(x)e^{i\alpha}$, has been referred to in early literatures as a 'global gauge transformation' or a 'gauge symmetry of the first kind'. This transformation, however, is an example of a real symmetry, corresponding to the number conservation of quantised excitations, and can be broken spontaneously (Section 1.6.2).
Many field theories contain **gauge fields** (Chapter 8), which are physical fields, whose mathematical description contains redundant gauge degrees of freedom.

Q27: Why is gauge freedom so important?

In quantum field theory, forces between particles are expressed by **gauge fields** that couple to the matter fields, such as the electromagnetic field (Chapter 8). Gauge invariance must hold for all physical quantities. Taking this as a starting principle, conservation laws for the physical particles follow from global symmetries in the coupling of particle to gauge fields. They lead to additional constraints on physical observables through **Noether's second theorem** (Section 1.6.3) and **Ward identities** (Section 10.4). Using gauge invariance in this way has provided guidance, for example, to the development of elementary particle physics.

Q28: What is meant by 'broken gauge symmetry'?

A gauge freedom describes superfluous degrees of freedom. Since physical quantities must be gauge invariant, **gauge invariance can never be broken** (Section 1.6).
However, local gauge freedoms are often accompanied by global symmetries (Section 1.6.3). Those global symmetries can be broken spontaneously (Chapter 8). The accompanying gauge fields then usually become massive through **the Anderson–Higgs mechanism** (Section 8.3). It is important to remember that the gauge transformations are never symmetries, can never be broken, and every physical quantity is gauge invariant at all times.

Q29: What is the Higgs mechanism?

When a global continuous symmetry that is coupled to **gauge fields** is broken spontaneously, the corresponding Nambu–Goldstone mode (as well as the gauge fields themselves) will become massive. This is known as the Anderson–Higgs mechanism (Section 8.3). Sometimes, the phrase **Higgs mechanism** is used to refer to the specific case of the Anderson–Higgs mechanism applied to the $SU(2) \times U(1) \rightarrow U(1)$ electroweak symmetry breaking (Section 11.3). In this context, the scalar field that couples to the gauge fields is called the **Higgs field** and its amplitude mode is called the **Higgs boson** (Section 8.3.2).

Q30: Is a symmetry broken in superconductors?

Yes. A superconductor **breaks the global $U(1)$ symmetry** associated with conservation of the number of Cooper pairs, while it retains its local $U(1)$ gauge freedom. The order parameter in this case is a non-local quantity (Section 8.2). Superconductors can be viewed as (electrically) charged versions of superfluids (Section 8.1), which break the same symmetry.

Long-Range Order

Q31: What is long-range order?

We say there is **long-range order** when the two-point correlation function of the local order parameter $C(x,x') = \langle \hat{\mathcal{O}}^\dagger(x)\hat{\mathcal{O}}(x')\rangle$ approaches a non-zero constant value as the distance $|x - x'|$ goes to infinity (Section 3.4).

Q32: What is off-diagonal long-range order (ODLRO)?

This term was introduced in the 1950s in the context of **superfluid helium**, as a way of contrasting the superfluid order with ordering in solids. The concept is largely historical, and the distinction between ODLRO and other types of order has become obsolete with the modern definition of the order parameter (Section 3.4).

Q33: What is algebraic long-range order?

A situation in which the two-point correlation function of the local order parameter $C(x,x') = \langle \hat{\mathcal{O}}^\dagger(x)\hat{\mathcal{O}}(x')\rangle$ vanishes with a **power-law** at large distances is called **algebraic long-range order**. This phenomenon occurs, for example, in the XY-model in $d = 2$ dimensions (Section 7.4).

Q34: Can long-range order exist without broken symmetry?

No, not in practice. Superpositions of broken symmetry states can be symmetric and long-range correlated, but these superpositions are unstable and cannot be realised in practice for sufficiently large systems (Section 3.7). In systems without long-range order, the correlation function will decay, either as a power-law when there is algebraic long-range order, or exponentially, in the **disordered** phase (Section 3.4).

Superfluids

Q35: Can you have $U(1)$ SSB in a system with an exact number of particles?

No, but this is a subtle point. Eigenstates of the number operator do not have a well-defined value for the corresponding global $U(1)$ phase. However, all the phenomenology of SSB, including phase coherence, Goldstone modes, and dissipationless (super)flow may occur in systems with an exactly defined number of particles. Breaking the global $U(1)$ symmetry requires an external perturbation. Even if the perturbation is infinitesimally weak, it may break particle conservation and allows for SSB in a sufficiently large system. To measure the global $U(1)$ phase, you need a **Josephson coupling** (Section 3.5), which necessarily introduces the external perturbation required for SSB.

Q36: What is a 'macroscopic wave function'?

It is a popular way of referring to the **local order parameter** $\psi(x)$ in a **superfluid** or **superconductor**. In both cases, the order parameter $U(1)$-valued and therefore 'looks' like a standard complex quantum-mechanical wave function (Section 2.5).

Other Topics

Q37: What are quantum fluctuations?

Quantum 'fluctuations' are generally a **misnomer**: it almost never refers to something that is fluctuating in time (which, for example, the classical noisy bath in Section 9.1 does do). The phrase is most often used to describe **quantum corrections** to mean-field theories or semi-classical approximations. In the context of SSB, it usually refers to quantum corrections to the expectation value of the order parameter (Chapter 6).

Q38: What are first-order and second-order phase transitions?

First-order phase transitions refer to discontinuous transitions, in which the order parameter jumps discontinuously. **Second-order** (or higher-order) phase transitions instead refer to continuous transitions, in which the order parameter vanishes continuously (Section 5.1).

Q39: What is a topological defect?

A **topological defect** is an excitation in a symmetry-breaking phase that causes the local value of the order parameter to deviate from its average over large areas, and that cannot be created or removed by smooth transformations of the symmetry-broken state (Chapter 7).

Q40: What is the role of SSB in the Standard Model?

The Standard Model includes the unification of the electromagnetic and weak nuclear forces through the process of SSB. This is known as **electroweak symmetry breaking** (Section 11.3).

Q41: What is the relation between SSB and quantum measurements?

Several authors have pointed out similarities between the two. Most notably, if a system breaks a symmetry, it 'spontaneously' chooses one of the possible symmetry-broken states, in a way that is reminiscent of how a projective quantum measurement 'spontaneously' chooses one of the measurement outcomes following Born's rule. Based on this observation, it has been suggested that **quantum measurement** can be viewed as a **spontaneous breaking of the unitarity** of quantum mechanics (Appendix A.6).

PART I

BASICS

1 Symmetry

Before talking about the breaking of symmetry, we need to define and understand what is meant by symmetry itself. Here, we need to distinguish between, on the one hand, the symmetries of the laws of nature as described by equations of motion, the action, or a Hamiltonian, and on the other hand, the symmetries of states and objects given by solutions to the equations of motion. This distinction lies at the core of spontaneous symmetry breaking, which is said to occur whenever a physical state or object has less symmetry than the laws of nature that govern it.

1.1 Symmetries of States

Intuitively, we would say an object possesses a symmetry if it looks identical from different viewpoints. For example, a sphere looks identical from any angle and is therefore concluded to be rotationally symmetric. To put it more formally, the continuous rotational symmetry in this case implies that the physical description of the sphere does not depend on its orientation in space. In the same way, an equilateral triangle possesses threefold rotational symmetry, since it is unaffected by rotations over any multiple of 120°.

Within quantum mechanics, the formal definition of symmetric states closely mimics the intuitive one. Taking a 'different viewpoint' towards a given state is mathematically represented in the quantum formalism by applying a unitary transformation $\hat{U}$ to it.[1] As a consequence, the symmetry of quantum states can be defined as follows:

Definition 1.1 A state $|\psi\rangle$ is said to be symmetric under a unitary transformation U if the transformed state is identical to the original state, up to a phase factor:

$$\hat{U}|\psi\rangle = \mathrm{e}^{\mathrm{i}\varphi}|\psi\rangle. \tag{1.1}$$

The possible appearance of a phase factor in this definition is due to the usual axiom in quantum mechanics that the total phase of a quantum state is unmeasurable, and physical states correspond to rays, rather than vectors, in the Hilbert space (see any textbook on quantum mechanics, such as [174, 217]). For the moment, the phase φ may be ignored, and a symmetry may be understood to imply $\hat{U}|\psi\rangle = |\psi\rangle$.

This simple definition for the symmetry of a state in quantum mechanics brings to light a seemingly trivial, but actually consequential, issue. Acting on the symmetric state $|\psi\rangle$ with $\hat{U}$ does nothing. So how does $\hat{U}$ differ from the identity operator? In terms of our earlier

[1] Classical physics can also be formulated in terms of operators acting on a Hilbert space. The definition of the symmetries of a classical state is then similar to the quantum case.

example, $\hat{U}$ might represent a rotation of the sphere. Such a rotation can only exist (or only makes sense) when it is applied with respect to something else. If the observer holding the sphere rotates her hand to look at it from a different angle, and thus applies the operation $\hat{U}$, the sphere is rotated with respect to the observer, who remains stationary. The fact that a state $|\psi\rangle$ is symmetric under $\hat{U}$ can thus only be observed and is only relevant, when there exist other states that are not symmetric under the same transformation. Phrased more generally, we notice the following:

A state, object, or system can only be defined as symmetric with respect to a different, non-symmetric state, object, or system.

As we will see in Section 1.6, this seemingly obvious observation is paramount to understanding the difference between a symmetry and a gauge freedom.

1.2 Symmetries of Hamiltonians

Like states, also the laws of nature, or equivalently the equations of motion, may be said to possess symmetries. Just as for states, defining this second type of symmetry is most naturally done within quantum mechanics, where it again appears in a deceptively simple form. Since Schrödinger's equation and the dynamics of quantum states follow directly from the Hamiltonian, symmetries of the equations of motion correspond to symmetries of the Hamiltonian operator. In general, any quantum operator $\hat{A}$ is called invariant under the unitary transformation $\hat{U}$ if $\hat{U}^\dagger\hat{A}\hat{U} = \hat{A}$, or equivalently, if $[\hat{U},\hat{A}] = 0$. If the Hamiltonian $\hat{H}$ is invariant under some unitary transformation $\hat{U}$, then $\hat{U}$ is called a symmetry of the Hamiltonian. The rationale behind this definition is that the expectation value of a symmetric Hamiltonian $\hat{H}$ in any state $|\psi\rangle$, symmetric or not, is invariant under application of the transformation $\hat{U}$ to the state,

$$\left(\langle\psi|\hat{U}^\dagger\right)\hat{H}\left(\hat{U}|\psi\rangle\right) = \langle\psi|\left(\hat{U}^\dagger\hat{H}\hat{U}\right)|\psi\rangle = \langle\psi|\hat{H}|\psi\rangle. \tag{1.2}$$

The symmetry of Hamiltonians, and therefore the quantum mechanical equations of motion, can thus be defined by stating as follows:

Definition 1.2 A unitary transformation $\hat{U}$ is a symmetry of the Hamiltonian $\hat{H}$ if their commutator vanishes: $[\hat{U},\hat{H}] = 0$.

The vanishing commutator for a symmetric Hamiltonian, $[\hat{U},\hat{H}] = 0$, implies that both the eigenstates of $\hat{H}$ and those of $\hat{U}$ may be used as a complete basis of energy eigenstates in describing the dynamics of the system. Similarly, if a symmetric Hamiltonian $\hat{H}$ has an eigenstate $|\psi\rangle$ with eigenvalue E_ψ, then the transformed state $\hat{U}|\psi\rangle$ is also an eigenstate of $\hat{H}$, with the same eigenvalue:

$$\hat{H}(\hat{U}|\psi\rangle) = \hat{U}\hat{H}|\psi\rangle = \hat{U}E_\psi|\psi\rangle = E_\psi(\hat{U}|\psi\rangle). \tag{1.3}$$

In case the state $|\psi\rangle$ is itself symmetric under $\hat{U}$, the fact that its energy is invariant under the symmetry operation may seem obvious. The non-trivial implication of Eq. (1.3) is that for every energy eigenstate that is not itself symmetric, there exists a degenerate state, which can be reached by applying the symmetry transformation. Applying the operator $\hat{U}$ a second time will lead to the next degenerate state and so on.

Of course, symmetries can be defined within the Lagrangian formulation of physics just as well as within the Hamiltonian formalism. This definition will appear naturally in Section 1.3, when we discuss Noether's theorem.

1.3 Noether's Theorem

The dynamics of many physical systems may be described almost entirely in terms of conservation laws, like the conservation of energy, momentum, angular momentum, and so on. One of the most profound insights in all of physics is the fact that the existence of conserved quantities is always rooted in symmetries of the applicable laws of nature. Translational invariance, for instance, meaning that the equations of motion look the same at any spatial location, implies conservation of momentum. This holds true regardless of what the equations of motion happen to be.

1.3.1 Conserved Quantities

The relation between symmetries and conserved quantities can be readily understood by realising that time evolution is determined by the Hamiltonian. Explicitly, the **time evolution operator** for time-independent Hamiltonians is shown by Schrödinger's equation to be the exponentiation of the Hamiltonian: $\hat{\mathcal{U}}(t) = \mathrm{e}^{i\hat{H}t/\hbar}$. As we saw in Section 1.2, a transformation $\hat{U}$ defines a symmetry of the Hamiltonian if it commutes with $\hat{H}$. It then follows that symmetry transformations also commute with the time evolution operator $\hat{\mathcal{U}}(t)$. Because any symmetry transformation is unitary, it can be written as the exponential $\hat{U} = \mathrm{e}^{i\hat{Q}}$ of some Hermitian operator $\hat{Q}$. The fact that $\hat{U}$ commutes with the time evolution operator, then implies that also $[\hat{Q}, \hat{\mathcal{U}}(t)] = 0$. This in turn means that the expectation value of $\hat{Q}$ in any state $|\psi\rangle$ is conserved in time:

$$\langle\psi(t)|\hat{Q}|\psi(t)\rangle = \left(\langle\psi|\hat{\mathcal{U}}^\dagger(t)\right)\hat{Q}\left(\hat{\mathcal{U}}(t)|\psi\rangle\right) = \langle\psi|\left(\hat{\mathcal{U}}^\dagger(t)\hat{Q}\hat{\mathcal{U}}(t)\right)|\psi\rangle = \langle\psi|\hat{Q}|\psi\rangle. \quad (1.4)$$

Now recall that in quantum mechanics, Hermitian operators represent observables. We thus find that the existence of the symmetry transformation $\hat{U}$ directly implies a conservation law for the observable $\hat{Q}$, which is known as a conserved quantity or **constant of motion**. Following an argument similar to Eq. (1.3), it can furthermore be shown that an eigenstate of $\hat{Q}$ can only evolve in time to eigenstates of $\hat{Q}$ with the same eigenvalue.

So the fact that the Hamiltonian generates time translations leads immediately to the conceptually important result:

> Any unitary symmetry $\hat{U}$ corresponds to an observable $\hat{Q}$ such that $\hat{U} = \mathrm{e}^{i\hat{Q}}$. The observable $\hat{Q}$ is a **conserved quantity**.

Exercise 1.1 (Exponential of an operator).
The exponential of an operator $\hat{A}$ is defined by its power series $e^{\hat{A}} = \sum_{n=0}^{\infty} \frac{1}{n!}\hat{A}^n$. Show that any operator that commutes with the Hamiltonian also commutes with the time evolution operator. Also show that if $\hat{U} = e^{i\alpha\hat{Q}}$ commutes with the Hamiltonian for all values of α, then $\hat{Q}$ must commute with the Hamiltonian as well.

1.3.2 Continuity Equations

The existence of any symmetry transformation implies a conservation law. For continuous symmetry transformations parametrised by a continuous variable, such as translations over a continuous distance or rotations over a continuous angle, there is an even stronger result, known as **Noether's theorem**. This theorem states that each continuous symmetry is associated with a current $j^\nu(\mathbf{x}, t)$ obeying the local conservation law or continuity equation $\partial_\nu j^\nu = 0$. For non-continuous, or discrete, symmetries, we have only a global constant of motion Q, but no local continuity equation.

Before formally proving Noether's theorem, let us give an intuitive interpretation of its main result. Firstly, continuous symmetries can be parametrised by a continuous (real-valued) parameter α. In terms of the unitary symmetry transformations discussed earlier, this means the Hamiltonian commutes with a family of related transformations $\hat{U}_\alpha = e^{i\alpha\hat{Q}}$. Taking continuous translations, for example, $\hat{U}_\alpha$ would be the operator that translates a state over distance α. In general, the operator $\hat{Q}$ is called the **symmetry generator**, because any symmetry transformation $\hat{U}_\alpha$ for finite α can be obtained from the action of $\hat{Q}$. For transformations with an infinitesimally small value of the parameter α, the operator $\hat{U}_\alpha$ may be expanded in a Taylor series, and terms beyond first order may be neglected:

$$\hat{U}_\alpha = 1 + i\alpha\hat{Q} + \mathcal{O}(\alpha^2). \tag{1.5}$$

Noether's theorem now follows from the observation that the conserved quantity $\hat{Q}$ can always be written as an integral over purely local operators, $\hat{Q} = \int d^D x\, \hat{\rho}(\mathbf{x}, t)$, where $\hat{\rho}(\mathbf{x}, t)$ is defined to be the local 'density of $\hat{Q}$'. Even though the global observable $\hat{Q}$ is conserved, the local expectation values $\rho(\mathbf{x}, t) = \langle\hat{\rho}(\mathbf{x}, t)\rangle$ are generally not. If, say, the value of $\rho(\mathbf{x}, t)$ at position $\mathbf{x}$ decreases, conservation of the global $\langle\hat{Q}\rangle$ implies that the value of $\rho(\mathbf{x}', t)$ at some other position $\mathbf{x}'$ must simultaneously increase. This means there is a 'current of ρ' flowing from $\mathbf{x}$ to $\mathbf{x}'$. Assuming the flow to be continuous in space and time, and calling the current $j^n(\mathbf{x}, t)$, the conservation of the global $\langle\hat{Q}\rangle$ is seen to be equivalent to the density and current together satisfying a continuity equation:

$$\partial_\nu j^\nu(\mathbf{x}, t) = \partial_t \rho(\mathbf{x}, t) + \partial_n j^n(\mathbf{x}, t) = 0. \tag{1.6}$$

Here, the density $\rho(\mathbf{x}, t)$ is written as the time component of the four-vector $j^\nu(\mathbf{x}, t)$.

Argued the other way around, if a continuity equation holds, there must be a global quantity Q that is conserved in time. This can be seen by integrating the continuity equation over all space:

$$0 = \int \mathrm{d}^D x\, \partial_\nu j^\nu(x) = \int \mathrm{d}^D x\, \partial_t \rho(\mathbf{x}, t) + \int \mathrm{d}^D x\, \partial_n j^n(\mathbf{x}, t)$$
$$= \partial_t Q(t) + \oint \mathrm{d}S_n\, j^n(\mathbf{x}, t). \tag{1.7}$$

Here we used Gauss' divergence theorem in going to the last line. The second term is a boundary term at spatial infinity and assuming that j^n falls off sufficiently quickly, this term vanishes. Therefore, we find $\partial_t Q = 0$, or in other words, Q is a conserved quantity. The direct relation between the presence of a symmetry and a corresponding continuity equation constitutes Noether's theorem.

1.3.3 Proving Noether's Theorem

Noether's theorem applies to any theory of physics that has a continuous symmetry, and which can be described in terms of a Lagrangian, or minimum-action principle. It is equally valid in quantum physics, classical mechanics, gravity, and even as-yet unknown realms of physics. Here, we will present a proof for a general field theory containing several fields $\phi_a(x) = \phi_a(\mathbf{x}, t)$. The fields $\phi_a(x)$ need not take the form of a vector. The index a could also, for example, label the two real components of a complex scalar field, or even different types of fields. The Lagrangian (density) is a functional of both the fields and their derivatives, $\mathcal{L} = \mathcal{L}[\phi_a(x), \partial_\nu \phi_a(x)]$, while the action, $S = \int \mathrm{d}t \mathrm{d}^D x\, \mathcal{L}$, is the integral of the Lagrangian over space and time.

To prove Noether's theorem, we will compare the effects of infinitesimal symmetry transformations of the fields on the Lagrangian before and after imposing the equations of motion. In general, a transformation of the fields can be written as follows:

$$\phi_a(x) \to \phi'_a(x) = \phi_a(x) + \delta_s \phi_a(x). \tag{1.8}$$

This expression may be interpreted as the definition of the variation $\delta_s \phi_a = \phi'_a - \phi_a$. We will consider $\phi'_a(x)$ to be the field resulting from the action of a symmetry transformation on $\phi_a(x)$, and the variation is assumed to be infinitesimally small. We can then use variational calculus to write the effect of the transformation on the Lagrangian in terms of the transformations of the fields and their derivatives:

$$\mathcal{L} \to \mathcal{L}' = \mathcal{L} + \delta_s \mathcal{L},$$
$$\delta_s \mathcal{L} = \frac{\partial \mathcal{L}}{\partial \phi_a} \delta_s \phi_a + \frac{\partial \mathcal{L}}{\partial(\partial_\nu \phi_a)} \delta_s (\partial_\nu \phi_a). \tag{1.9}$$

Here $\partial \mathcal{L}/\partial \phi$ means taking a derivative of $\mathcal{L}$ as if it were an ordinary function of a variable ϕ, and sums over a and ν are implied by the Einstein summation convention. The expansion of $\delta_s \mathcal{L}$ in terms of infinitesimal variations of the fields is what makes Noether's theorem valid only for continuous, and not discrete, symmetries.

Equation (1.9) holds true for any infinitesimal variation $\delta_s \phi_a$ of the fields. It can be seen simply as the 'variational chain rule'. If the variation $\delta_s \mathcal{L}$ happens to be zero, the transformation that caused it can be called a symmetry of the action. In fact, even if $\delta_s \mathcal{L}$ is a total derivative $\partial_\nu K^\nu$ of some function K^ν, we will call the transformation a symmetry. In that case, $\delta_s \mathcal{L}$ will only add a boundary term to the action, which does not affect the equations of motion. For most symmetry transformations, the variation of the Lagrangian will simply be zero, but an important example of a non-vanishing boundary term is that of spacetime translations, which will be discussed in Section 1.4.4.

A transformation of the fields, $\phi_a(x) \to \phi_a(x) + \delta_s\phi_a(x)$, is a symmetry of the action if the corresponding change in the Lagrangian $\mathcal{L}[\phi_a(x), \partial_\nu\phi_a(x)] \to \mathcal{L} + \delta_s\mathcal{L}$ is at most a total derivative: $\delta_s\mathcal{L} = \partial_\nu K^\nu$.

So far, we have not specified whether or not the fields ϕ_a satisfy any equations of motion. The condition $\delta_s\mathcal{L} = \partial_\nu K^\nu$ defines what it means for a transformation to be a symmetry of the action, regardless of the type of field it acts on.

We now return to Eq. (1.9), and recall that it holds for any infinitesimal transformation of the fields, whether they constitute a symmetry or not. Using an elementary result from variational calculus, $\delta_s(\partial_\nu\phi_a) = \partial_\nu(\delta_s\phi_a)$, and using the chain rule allows us to write it in the following form:

$$\delta_s\mathcal{L} = \partial_\nu\left(\frac{\partial\mathcal{L}}{\partial(\partial_\nu\phi_a)}\delta_s\phi_a\right) + \left[\frac{\partial\mathcal{L}}{\partial\phi_a} - \partial_\nu\left(\frac{\partial\mathcal{L}}{\partial(\partial_\nu\phi_a)}\right)\right]\delta_s\phi_a. \tag{1.10}$$

The part in the square brackets should look familiar: this is precisely what the Euler–Lagrange equations of motion prescribe to be zero. If we thus restrict attention to field configurations ϕ_a that satisfy these equations of motion, we are left with only the first term on the right-hand side, which is a total derivative. Notice that this condition for $\delta_s\mathcal{L}$ being a total derivative holds for specific field configurations ϕ_a, without putting any requirements on the transformation $\delta_s\phi_a$ we consider. Conversely, the previous result of $\delta_s\mathcal{L} = \partial_\nu K^\nu$ constituting a symmetry of the action was a requirement on the transformation $\delta_s\phi_a$, for arbitrary ϕ_a [15].

Noether's theorem is now obtained by considering fields that obey the Euler–Lagrange equation of motion and transformations that are symmetries of the action. We can then subtract the two conditions on $\delta_s\mathcal{L}$ and obtain to the continuity equation:

$$\partial_\nu\left(\frac{\partial\mathcal{L}}{\partial(\partial_\nu\phi_a)}\delta_s\phi_a - K^\nu\right) \equiv \alpha\partial_\nu j^\nu = 0. \tag{1.11}$$

Here we introduced an infinitesimal parameter α for later convenience, and defined the **Noether current** j^ν related to the symmetry transformation $\phi_a(x) \to \phi_a(x) + \delta_s\phi_a(x)$, as follows:

$$j^\nu = \frac{1}{\alpha}\left(\frac{\partial\mathcal{L}}{\partial(\partial_\nu\phi_a)}\delta_s\phi_a - K^\nu\right) = \frac{\partial\mathcal{L}}{\partial(\partial_\nu\phi_a)}\Delta_s\phi_a - \frac{1}{\alpha}K^\nu. \tag{1.12}$$

In the final expression, we introduced the notation $\Delta_s\phi = \delta_s\phi/\alpha$. The presence of a continuous symmetry implying the existence of a Noether current that is locally conserved, $\partial_\nu j^\nu = 0$, is the main result of Noether's theorem.

Theorem 1.3 (Noether's theorem). To any continuous symmetry of a local action corresponds a current $j^\nu(\mathbf{x}, t)$ that is locally conserved. That is, it satisfies the continuity equation $\partial_\nu j^\nu(\mathbf{x}, t) = 0$.

1.3.4 Noether Charge

Recall that Eq. (1.7) showed any local continuity equation to imply the existence of a globally conserved quantity. In the context of Noether's theorem, this is called the **Noether charge** (not to be confused with electric charge, which is in first instance not related) and may be defined as follows:

$$Q(t) = \int \mathrm{d}^D x\, j^t(\mathbf{x}, t) = \int \mathrm{d}^D x \left(\frac{\partial \mathcal{L}}{\partial(\partial_t \phi)} \Delta_s \phi - \frac{1}{\alpha} K^t \right). \tag{1.13}$$

Similarly, $j^t = \rho$ is called the **Noether charge density**. Notice that because $\partial_\nu j^\nu = 0$, the Noether charge $Q(t)$ is in fact independent of time, or in other words, a constant of motion.

To see how the Noether charge is related to the symmetry transformation, consider the canonical momentum $\pi_a(x) = \partial\mathcal{L}/\partial(\partial_t\phi_a)$, conjugate to $\phi_a(x)$. The field values $\pi_a(x)$ and $\phi_a(x)$ are eigenvalues of corresponding canonically conjugate quantum mechanical operators $\hat{\pi}_a(x)$ and $\hat{\phi}_a(x)$, which obey the commutation relations $[\hat{\pi}_a(x), \hat{\phi}_b(y)] = -\mathrm{i}\delta_{ab}\delta(x - y)$. Focussing on the most commonly encountered case with $K^\nu = 0$ and $[\Delta_s\hat{\phi}_b, \hat{\phi}_a] = 0$, the commutator of the field and the Noether charge operators is found to be as follows:

$$\begin{aligned}
[\mathrm{i}\alpha\hat{Q}, \hat{\phi}_a(x)] &= \mathrm{i}\alpha \int \mathrm{d}^D y\, [\hat{\pi}_b(y)\Delta_s\hat{\phi}_b(y), \hat{\phi}_a(x)] \\
&= \mathrm{i}\alpha \int \mathrm{d}^D y\, [\hat{\pi}_b(y), \hat{\phi}_a(x)]\Delta_s\hat{\phi}_b(y) = \alpha\Delta_s\hat{\phi}_a(x) = \delta_s\hat{\phi}_a(x).
\end{aligned} \tag{1.14}$$

Because the commutator of the Noether charge and the field equals the variation of the field, the Noether charge is also called the **generator** of the symmetry. A symmetry transformation of the field operators can now be written as follows:

$$\begin{aligned}
\hat{\phi}_a(x) \to \hat{\phi}'_a(x) &= \hat{\phi}_a(x) + \mathrm{i}\alpha[\hat{Q}, \hat{\phi}_a(x)] \\
&= \mathrm{e}^{\mathrm{i}\alpha\hat{Q}}\hat{\phi}_a(x)\mathrm{e}^{-\mathrm{i}\alpha\hat{Q}} + \mathcal{O}(\alpha^2).
\end{aligned} \tag{1.15}$$

In the final line, we applied the Baker–Campbell–Hausdorff formula while assuming α to be infinitesimal. Since this expression for the symmetry transformation is of the same form as that implied by Eq. (1.5), we see that the Noether charge Q indeed corresponds to the observable $\hat{Q}$ obeying $[\hat{Q}, \hat{H}] = 0$ in the Hamiltonian formalism. This result also holds in the more general case with non-zero boundary terms K^ν, but the derivation is lengthier.

It should be emphasised that the local form of Noether's theorem $\partial_\nu j^\nu = 0$ depends only on the symmetry of the action, and is valid for any state that satisfies the equations of motion. In quantum mechanics, it is an operator identity that does not refer to any particular state. However, if the physical state of a system happens not to share the symmetry of the action—that is, if the symmetry of the action is spontaneously broken—one should be careful about what the continuity equation implies physically. We will come back to this point in Section 4.6, when discussing so-called Nambu–Goldstone modes.

1.4 Examples of Noether Currents and Charges

To put the formal definitions of Section 1.3 into context, we will consider some explicit examples of Noether currents and Noether charges.

1.4.1 Schrödinger Field

The Schrödinger equation for the wave function $\psi(x)$ of a quantum particle in an external potential can be interpreted as an equation of motion for the complex scalar field $\psi(x,t)$. The Schrödinger equation is then derived from an action of the following form:

$$S[\psi,\psi^*] = \int \mathrm{d}t\mathrm{d}^D x \left(\mathrm{i}\frac{\hbar}{2}(\psi^*(\partial_t\psi) - (\partial_t\psi^*)\psi) - \frac{\hbar^2}{2m}(\partial_n\psi^*)(\partial_n\psi) - V(x)\psi^*\psi \right). \tag{1.16}$$

Here the potential $V(x)$ is an ordinary function of space. It is easily seen that the Euler–Lagrange equation containing derivatives with respect to ψ^* indeed yields the Schrödinger equation,

$$\begin{aligned} 0 &= \partial_t \left(\frac{\partial \mathcal{L}}{\partial(\partial_t\psi^*)} \right) + \partial_n \left(\frac{\partial \mathcal{L}}{\partial(\partial_n\psi^*)} \right) - \frac{\partial \mathcal{L}}{\partial\psi^*} \\ &= -\mathrm{i}\hbar\partial_t\psi - \frac{\hbar^2}{2m}\partial_n^2\psi + V(x)\psi. \end{aligned} \tag{1.17}$$

One way to handle the two degrees of freedom contained in a complex scalar field is to treat ψ and ψ^* independently. The canonical momenta associated with the Schrödinger field and its complex conjugate can then be identified as follows:

$$\pi = \partial\mathcal{L}/\partial(\partial_t\psi) = \mathrm{i}\hbar\psi^*/2, \qquad \pi^* = \partial\mathcal{L}/\partial(\partial_t\psi^*) = -\mathrm{i}\hbar\psi/2. \tag{1.18}$$

In terms of these independent fields, the action in Eq. (1.16) has a continuous symmetry. It is invariant under phase rotations of the form

$$\psi(x) \to \mathrm{e}^{-\mathrm{i}\alpha}\psi(x), \qquad \psi^*(x) \to \mathrm{e}^{\mathrm{i}\alpha}\psi^*(x), \tag{1.19}$$

with α a real and continuous parameter. Notice that α does not depend on x, so that the phase rotation is a 'global' transformation, affecting all points in the system in the same way. The symmetry of a global phase rotation is often called a $U(1)$ symmetry, after the Lie group with the same name that describes such rotations.

Taking α to be infinitesimal, the exponent can be expanded, and the variations of the field under a phase rotation become $\Delta_S\psi(x) = -\mathrm{i}\psi(x)$ and $\Delta_S\psi^*(x) = \mathrm{i}\psi^*(x)$. Using Eq. (1.12), the Noether current and conserved Noether charge are found to equal

$$j^t = \pi\Delta_S\psi + \pi^*\Delta_S\psi^* = \hbar\psi^*\psi, \qquad Q = \int \mathrm{d}^D x\, \hbar\psi^*\psi, \tag{1.20}$$

$$j^n = \mathrm{i}\frac{\hbar^2}{2m}\left((\partial_n\psi^*)\psi - \psi^*(\partial_n\psi) \right). \tag{1.21}$$

The quantity $\int \psi^*\psi = \int |\psi|^2$ is of course just the total amplitude of the wave function which indeed should retain its normalisation in any well-defined quantum theory. In the Lagrangian treatment, the conservation of the norm can be interpreted as a consequence of the invariance of Eq. (1.16) under $U(1)$ global phase rotations. Similarly, the local amplitude of the wave function, $|\psi(x)|^2$, can only change when it flows elsewhere in the form of a probability current j^n. Noether's theorem can then be interpreted as a continuity equation for the probability current.

1.4.2 Relativistic Scalar Fields

Rather than starting from Schrödinger's equation, we can also consider a different form for the action of a complex scalar field:

$$S = \int \mathrm{d}t\mathrm{d}^D x \, \frac{\hbar^2}{m} \left(\frac{1}{c^2} (\partial_t \psi^*)(\partial_t \psi) - (\partial_n \psi^*)(\partial_n \psi) - \frac{m^2 c^2}{\hbar^2} \psi^* \psi \right). \tag{1.22}$$

This action is invariant under Lorentz transformations and therefore said to be 'relativistic'. Its equation of motion is known as the Klein–Gordon equation, and it is often written in natural units, absorbing all factors of $\hbar$ and c, as well as an overall prefactor.

Besides Lorentz invariance, the action also has the same $U(1)$ global phase rotation symmetry as Eq. (1.16). However, the canonical momentum becomes $\pi = \partial \mathcal{L}/\partial(\partial_t \psi) = \partial_t \psi^*$, written in natural units. The spatial part of the Noether current is the same as that in Eq. (1.21), up to a prefactor, but the relativistic Noether charge becomes

$$Q = \int \mathrm{d}^D x \, \left(\pi \, \Delta_\mathrm{s} \psi + \pi^* \Delta_\mathrm{s} \psi^* \right) = \int \mathrm{d}^D x \, \mathrm{i} \big(\psi^* (\partial_t \psi) - (\partial_t \psi^*) \psi \big). \tag{1.23}$$

This conserved charge plays the role of the conserved wave function normalisation in the relativistic Klein–Gordon theory.

The relativistic theory for a single complex field can alternatively be written in terms of a vector field with two real components, by defining $\vec{\psi} = (\mathrm{Re}(\psi), \mathrm{Im}(\psi))$. In natural units, the action is then written as follows:

$$S = \int \mathrm{d}t\mathrm{d}^D x \, \left(\eta^{\mu\nu} \partial_\mu \vec{\psi} \cdot \partial_\nu \vec{\psi} - m^2 \vec{\psi} \cdot \vec{\psi} \right). \tag{1.24}$$

Here, we used the metric $\eta_{\mu\nu} = \mathrm{diag}(-1, 1, 1, 1)$ to make the Lorentz invariance explicit. In this form, the global $U(1)$ phase symmetry of the complex field appears as a global $O(2)$ symmetry rotating the real vector components into one another. Equation (1.24) can in turn be generalised to describe a set of fields forming an N-component vector. In that case, the action of Eq. (1.24) has an $O(N)$ symmetry, describing global rotations of all vector components among each other.

Exercise 1.2 ($SO(2)$ and $U(1)$ symmetry).
Consider an action of two real fields $A(x)$ and $B(x)$, written in natural units:

$$S = \int \mathrm{d}t\mathrm{d}^D x \, (\partial_t A)^2 + (\partial_t B)^2 - (\partial_n A)^2 - (\partial_n B)^2 - V(x)(A^2 + B^2). \tag{1.25}$$

Show that the action is invariant under so-called $SO(2)$ rotations of the fields:

$$\begin{pmatrix} A \\ B \end{pmatrix} \to \begin{pmatrix} A' \\ B' \end{pmatrix} = \begin{pmatrix} \cos\alpha & -\sin\alpha \\ \sin\alpha & \cos\alpha \end{pmatrix} \begin{pmatrix} A \\ B \end{pmatrix}. \tag{1.26}$$

Determine the Noether current and Noether charge associated with this symmetry transformation. Compare your result with that of Eq. (1.23) by combining the two real fields into a single complex field $\psi = A + iB$. In group theory terms, the correspondence between phase rotations of a complex scalar variable and rotations within a vector of two real components is known as the isomorphism between the groups $U(1)$ and $SO(2)$.

Exercise 1.3 (Noether's trick).
There is a technique, sometimes called **Noether's trick**, which can be used to obtain the Noether current more directly than by following the derivation preceding Eq. (1.12). It starts from a transformation of the action based on the global symmetry, but in which the parameter of the symmetry transformation is taken to depend on space and time: $\alpha \to \alpha(x)$. The action is not invariant under the spacetime-dependent transformation. However, because it is invariant under the global transformation with constant α, all terms in the variation of the action must be proportional to derivatives of $\alpha(x)$. To lowest order, we can then write $\delta S = \int \mathrm{d}x\ j^\mu \partial_\mu \alpha + \mathcal{O}(\partial^2)$, with j^μ some expansion coefficients. Using a partial integration this equals $\delta S = -\int \mathrm{d}x\ \alpha(\partial_\mu j^\mu)$. Focussing on field configurations that satisfy the Euler–Lagrange equations of motion, we moreover know that $\delta S = 0$ for any infinitesimal local variation $\alpha(x)$, because the equations of motion are defined by extremising the action. For field configurations that satisfy the Euler–Lagrange equations as well as local transformations whose global component is a symmetry of the action, we thus find $\int \mathrm{d}x\ \alpha(\partial_\mu j^\mu) = 0$. This can only be true for arbitrary local variations of the global symmetry if $\partial_\mu j^\mu = 0$. In other words, the coefficients j^μ appearing in the lowest-order variation of the action when considering local versions of a global symmetry, turn out to be precisely the conserved Noether current associated with that global symmetry.

Use Noether's trick to derive the Noether current for the Schrödinger field and the relativistic complex scalar field, and compare your result to Eq. (1.21).

1.4.3 Many-Body Lattice Theories

Theories involving macroscopic numbers of quantum particles can often be conveniently formulated on a spatial or spacetime lattice, in terms of a Hamiltonian in the second-quantisation formalism. A (relativistic) field theory can then be obtained by taking the continuum limit, in which the lattice spacing vanishes in a controlled manner. A typical example describing free bosons on a lattice is given by

$$\hat{H} = -J \sum_{\langle ij \rangle} \hat{b}_i^\dagger \hat{b}_j - \mu \sum_i \hat{n}_i. \tag{1.27}$$

Here, $\hat{b}_i^\dagger$ and $\hat{b}_i$ are bosonic creation and annihilation operators obeying $[\hat{b}_i, \hat{b}_j^\dagger] = \delta_{ij}$, with the discrete index i labelling lattice sites. The bosonic number operator is given by $\hat{n}_i = \hat{b}_i^\dagger \hat{b}_i$, while μ is the chemical potential, J a hopping parameter, and the sum $\langle ij \rangle$ runs over nearest neighbours only. In the continuum limit, the second-quantised operators will be replaced by fields, $\hat{b}_i \to \phi(\vec{x}_i)$, while the discrete difference becomes a derivative, $\hat{b}_j - \hat{b}_i \to \partial_x \phi(\vec{x})$. This way, the Schrödinger action of Eq. (1.16) is recovered.

Just like the relativistic scalar field theory and the Schrödinger field theory, the lattice theory of Eq. (1.27) has a global $U(1)$ phase rotation symmetry:

$$\hat{b}_i \to \mathrm{e}^{-\mathrm{i}\alpha} \hat{b}_i, \qquad \hat{b}_i^\dagger \to \mathrm{e}^{\mathrm{i}\alpha} \hat{b}_i^\dagger. \tag{1.28}$$

The symmetry operator can be written as follows:

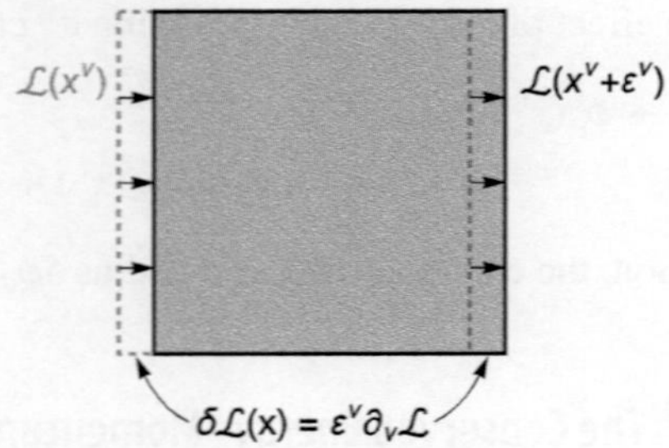

Figure 1.1

Under a spacetime translation $x^\nu \to x^\nu - \epsilon^\nu$ the value of the Lagrangian density $\mathcal{L}$ at position x^ν changes to the density that was originally found at position $x^\nu + \epsilon^\nu$. For an infinitesimal translation, the Lagrangian density thus changes with $\delta\mathcal{L} = \partial_\nu(\epsilon^\nu\mathcal{L})$.

$$\hat{U}_\alpha = e^{-i\alpha\sum_i \hat{n}_i}. \tag{1.29}$$

One can verify that this operator indeed defines a symmetry of the model, $[\hat{U},\hat{H}] = 0$, and that it acts as a phase rotation on the bosonic operators, $\hat{U}^\dagger \hat{b}_i \hat{U} = e^{-i\alpha}\hat{b}_i$. By explicitly writing the $U(1)$ symmetry operator in Eq. (1.29), we immediately see that the associated conserved quantity is the total number of bosons: $\hat{Q} = \sum_i \hat{n}_i$. In many-body quantum theories, a $U(1)$ symmetry is therefore a reflection of the conservation of particle number. If the particles have an electric charge, the $U(1)$ symmetry corresponds to charge conservation.

The fact that $U(1)$ symmetry implies particle number conservation is not restricted to bosonic theories. Also theories of fermions, such as the Fermi–Hubbard model or the Dirac field theory, have a $U(1)$ symmetry that enforces conservation of the total number of particles.

1.4.4 Spacetime Translations

One of the few examples in which a symmetry transformation does not leave the Lagrangian explicitly invariant, but rather changes it by a total derivative, $\delta\mathcal{L} = \partial_\nu K^\nu$, is that of **spacetime translations**, which are discussed more thoroughly in Section 11.1. Writing spacetime coordinates as usual, $x^\nu = (ct, \mathbf{x})$, a global spacetime translation can be written as

$$x^\nu \to x'^\nu = x^\nu - \epsilon^\nu. \tag{1.30}$$

Here ϵ^ν is a constant spacetime vector. Under the global translation, all fields transform as $\phi_a(x) \to \phi'_a(x) = \phi_a(x+\epsilon)$, since the transformation can equivalently be seen as a shift the of the coordinate system. This is visualised in Figure 1.1.

The Lagrangian transforms in the same way, so that $\mathcal{L}(x) \to \mathcal{L}'(x) = \mathcal{L}(x+\epsilon)$. The change in the local Lagrangian density can also be written as $\delta\mathcal{L} = \mathcal{L}' - \mathcal{L} = \epsilon^\nu\partial_\nu\mathcal{L} + \mathcal{O}(\epsilon^2)$, which for an infinitesimal translation becomes just $\delta\mathcal{L} = \epsilon^\nu\partial_\nu\mathcal{L}$. Using the fact that the global spacetime translation vector ϵ is constant, the effect of the transformation can be cast in the form of a total derivative:

$$\delta\mathcal{L} = \epsilon^\nu\partial_\nu\mathcal{L} = \partial_\nu\left(\epsilon^\nu\mathcal{L}\right). \tag{1.31}$$

This way of writing the transformation shows that indeed the Lagrangian changes under it by the total derivative $\partial_\nu K^\nu$, with $K^\nu = \epsilon^\nu\mathcal{L}$.

For the fields $\phi_a(x^\nu)$, the effect of a spacetime translation ϵ^ν can similarly be written as

$$\begin{aligned}\phi_a(x^\nu) \to \phi_a'(x^\nu) &= \phi_a(x^\nu + \epsilon^\nu) \\ &= \phi_a(x^\nu) + \epsilon^\mu \partial_\mu \phi_a(x^\nu) + \mathcal{O}(\epsilon^2).\end{aligned} \tag{1.32}$$

For an infinitesimal translation, the change in the field is thus $\delta\phi_a(x) = \partial_\nu \epsilon^\nu \phi_a$.

1.4.4.1 The Conserved Energy–Momentum Tensor

To derive the Noether current corresponding to spacetime translations, we use the variational calculus of Eq. (1.9). First, we express the change in the Lagrangian in terms of the change in fields and their derivatives:

$$\delta\mathcal{L} = \frac{\partial\mathcal{L}}{\partial\phi_a}\delta\phi_a + \frac{\partial\mathcal{L}}{\partial(\partial_\mu\phi_a)}\delta(\partial_\mu\phi_a). \tag{1.33}$$

Next, we use the Euler–Lagrange equations of motion and insert the definition of $\delta_\nu\phi_a$ to simplify the expression to

$$\delta\mathcal{L} = \partial_\mu\left(\frac{\partial\mathcal{L}}{\partial(\partial_\mu\phi_a)}\partial_\nu\epsilon^\nu\phi_a\right). \tag{1.34}$$

The change of the Lagrangian under the transformation has to equal the earlier result of Eq. (1.31). Combining these expressions, we find that

$$\partial_\mu\left(\frac{\partial\mathcal{L}}{\partial(\partial_\mu\phi_a)}\partial_\nu\epsilon^\nu\phi_a - \delta^\mu_\nu\epsilon^\nu\mathcal{L}\right) = 0. \tag{1.35}$$

Here, we introduced the Kronecker delta δ^μ_ν, which equals one if $\mu = \nu$ and zero otherwise. Comparing this result to Eq. (1.12), and treating the components of ϵ^ν as independent spacetime translations, the conserved Noether current associated with a translation in just the ν-direction is found to be

$$(j^\mu)_\nu = \frac{\partial\mathcal{L}}{\partial(\partial_\mu\phi_a)}\partial_\nu\phi_a - \delta^\mu_\nu\mathcal{L}. \tag{1.36}$$

The label ν on the components of the conserved Noether current, j^μ, indicates that they are associated with translations in the ν-direction. Since it is a spacetime index, however, we can also interpret $(j^\mu)_\nu$ as elements of a tensor. This tensor of Noether currents is traditionally written with the letter T:

$$T^\mu{}_\nu \equiv (j^\mu)_\nu. \tag{1.37}$$

Note that this so-called **energy–momentum tensor** or **stress–energy tensor** satisfies $\partial_\mu T^\mu{}_\nu = 0$ for every ν.

The energy–momentum tensor describes the local flow of energy ($T^0{}_0$), momentum ($T^0{}_i$) and stress ($T^i{}_j$). To see this, consider, for example, the Noether charge associated with time translations:

$$Q_0 = \int \mathrm{d}^D x\, T^0{}_0 = \int \mathrm{d}^D x \left(\pi^0\partial_0\phi - \mathcal{L}\right) = \int \mathrm{d}^D x\, \mathcal{H} = H. \tag{1.38}$$

Here, we used the expression for the canonical momentum $\pi^\mu = \partial\mathcal{L}/\partial(\partial_\mu\phi)$ conjugate to the field ϕ. The conserved charge associated with time translation symmetry is given by the Hamiltonian, or energy operator. Energy is thus conserved whenever the action has time translation symmetry. Similarly, the Noether charge for spatial translations is given by

$$Q_i = \int \mathrm{d}^D x \, T^0{}_i = \int \mathrm{d}^D x \, \pi^0 \partial_i \phi. \tag{1.39}$$

This expression can be recognised as the total momentum associated with the field ϕ. Spatial translation symmetry of the action thus ensures conservation of momentum.

The derivation of the energy–momentum tensor in Eq. (1.36) was done without specifying any particular form of the action. We therefore find that energy and momentum are conserved in any physical theory described by an action that is invariant under spacetime translations.

1.5 Types of Symmetry Transformations

In our definitions of symmetries and the discussion of their relation to conservation laws so far, we did not yet need to be very precise about the different types of symmetry transformations that may be encountered in various physical situations. However, not all symmetry transformations are susceptible to spontaneous symmetry breaking, and moreover, it is not obvious how some of the historically introduced classifications of symmetry transformations fit into the modern framework of symmetry breaking and its consequences. In this section, we therefore give a brief account of the different types of transformations that may be encountered in the literature, focussing on the physical relevance of each distinction to the phenomenon of spontaneous symmetry breaking.

1.5.1 Discrete versus Continuous Symmetries

Contrary to continuous symmetries, a discrete symmetry cannot be parametrised by a continuous, real variable. Simply put, you either do the discrete transformation or you do not. You cannot do it just a little bit. For some types of discrete symmetries, such as reflections, performing an arbitrary fraction of the symmetry transformation is simply not possible. In other cases, such fractional transformations are possible, but they are not symmetries. For example, a triangle is symmetric under rotations of 120°, but not under rotations over any smaller angle. This is unlike the continuous rotational symmetry of a circle, which looks the same after rotations over any arbitrary angle.

While there is no mathematical difference between continuous and discrete symmetries beyond their parametrisation and both can be spontaneously broken, there is an important physical difference: Noether's theorem applies only to continuous symmetries. The proof of Noether's theorem, and thus of the existence of a locally conserved current, requires the invocation of infinitesimal symmetry transformations. This is not possible for discrete symmetries.[2] Other concepts related to symmetry breaking, including the emergence of

[2] Notice that in some cases, a remnant of the conserved Noether charges may survive even in systems with only a discrete symmetry. The discrete translation symmetry of a crystalline lattice, for example, is responsible for the fact that lattice momentum (or crystal momentum) is conserved modulo reciprocal lattice vectors [13].

Nambu–Goldstone modes (Chapter 4) and the Mermin–Wagner–Hohenberg–Coleman theorem (Section 6.2), likewise only apply in the presence of spontaneously broken continuous symmetries.

In short, there is more richness in the breaking of continuous symmetries than in discrete ones. Some examples of discrete symmetry breaking are certainly interesting and instructive – it will for instance be worthwhile considering the Ising model when discussing the stability of broken symmetry states in Section 3.7 – but throughout this book we refer to discrete symmetries mostly in passing.

1.5.2 Anti-unitary Symmetries

In Section 1.1, the symmetry of a state was defined as a unitary transformation that leaves the state unaffected, up to a total phase. A different point of view on why symmetry transformations are required to be unitary, is that such transformations conserve the inner product between two states, $\left(\langle\psi|U^{\dagger}\right)\left(U|\psi'\rangle\right) = \langle\psi|\psi'\rangle$. Contrary to a single state, the amplitude of the overlap between two states is directly measurable, giving a physical reason to consider this definition of symmetry operators. In fact, the amplitude will be conserved also under somewhat relaxed conditions, and we could instead consider transformations that only leave inner products unaffected up to a phase factor. This then allows for so-called **anti-unitary symmetries**, whose transformations turn out to satisfy $\left(\langle\psi|U^{\dagger}\right)\left(U|\psi'\rangle\right) = \langle\psi|\psi'\rangle^*$.

One very important anti-unitary symmetry is **time reversal**, which reverses the time evolution of all observable expectation values. Like other symmetries, time-reversal symmetry can be spontaneously broken. Examples of systems with broken time-reversal symmetries are ferromagnets and the A-phase of superfluid Helium-3 (Section 12.3.4). However, we will largely ignore time-reversal and other anti-unitary symmetries, as they are necessarily discrete.

1.5.3 Global versus Local Symmetries

Spontaneous symmetry breaking occurs in physical systems with many microscopic degrees of freedom, such as a large collection of atoms, electrons, spins, or a continuous field spread out over all space and time. A symmetry of the full system is then defined by how it acts on the individual constituents (i.e., the atoms, the field amplitude at each location, and so on). A **global symmetry** is a symmetry that acts in the same way on each individual constituent. All of the examples in Section 1.4 concerned global symmetries, from the global phase rotation in Section 1.4.1 to the global shift of spacetime coordinates in Section 1.4.4.

As will become clear in Chapter 2, only global symmetries can be spontaneously broken. There also exist, however, many kinds of **local symmetry** defined as a physical symmetry transformation that acts differently on different local degrees of freedom. These are not to be confused with **gauge freedoms**, which are purely mathematical manipulations leaving a system's description invariant, but which do not correspond to any physical transformation of the actual system. Such gauge freedoms may have important consequences and will be discussed in detail in Section 1.6, but for now, we will focus on actual local symmetries.

The easiest example of a local symmetry appears in a classical ideal gas of N particles of mass m with position operators $\hat{\mathbf{X}}_i(t)$ and momenta $\hat{\mathbf{P}}_i(t)$. The index i labels the particles and runs from 1 to N. Since ideal particles do not interact, the Hamiltonian contains only the kinetic energy of the individual particles, and can be written as $\hat{H} = \sum_i |\hat{\mathbf{P}}_i|^2/(2m)$. Because the Hamiltonian does not depend on the position of any particle, it is not affected by translations of each particle individually:

$$\mathbf{X}_i(t) \rightarrow \mathbf{X}_i(t) + \mathbf{a}_i. \tag{1.40}$$

The local displacements $\mathbf{a}_i$ may be different for each i. We could even consider an extreme case in which the displacement is zero for all particles except one, making it obvious that the symmetry is local, rather than global.

You may argue that the example of the ideal gas is somewhat artificial, since the particles are really independent, and each come with their own symmetry. This example can be easily extended, however, to that of a free field theory [16]. Consider, for example, the relativistic complex field of Eq. (1.22) with no external potential, $V(x) = 0$. Local spacetime displacements of the fields are then described by the transformation:

$$\psi(x) \rightarrow \psi(x) + \alpha(x). \tag{1.41}$$

Here, $\alpha(x)$ is complex-valued and depends on the spacetime coordinate x. If $\alpha(x)$ is completely general, the local shift is not a symmetry of the action. However, if α satisfies the equations of motion, then using $\partial^\nu \psi^* \partial_\nu \alpha + \partial^\nu \psi \partial_\nu \alpha = \partial^\nu(\psi^* \partial_\nu \alpha + \psi \partial_\nu \alpha^*) - \psi^* \partial^2 \alpha - \psi \partial^2 \alpha^*$, we see that the transformation only adds a boundary term to the action:

$$\delta_s S = \int \mathrm{d}t\mathrm{d}^D x\ \partial_\nu(\psi \partial^\nu \alpha + \psi^* \partial^\nu \alpha). \tag{1.42}$$

Here, the equations of motion were imposed for α, but not for ψ.

Although local symmetries cannot be spontaneously broken, as we will discuss in Exercise 2.3, they do give rise to conserved charges. For the free complex scalar field, we could derive Noether currents related to $\delta_s \psi = \alpha(x)$ at each x individually. One interpretation of the corresponding Noether charges is that each of the components in the Fourier transform of $\psi(x)$ is individually conserved [16].

The examples of the ideal gas and free field both concern non-interacting systems, but local symmetries may exist in interacting systems as well. For example, interacting particles in a disordered potential in some cases undergo so-called many-body localisation (MBL), and the MBL-phase is characterised precisely by having a large number of emergent local symmetries (that are difficult to write down explicitly). Even in such interacting systems, spontaneous symmetry breaking does not occur for local symmetries.

1.5.4 Active versus Passive and Internal versus External Symmetries

In addition to the physically relevant distinctions between global versus local and continuous versus discrete symmetries, various other classifications of symmetry transformations may be encountered in the literature. These are, in our opinion, not useful for the clarification of any physical effects, and often lead to confusion. For the sake of completeness, we will briefly comment on some of these alternative notions here, but they have no role in the remainder of the book.

The first distinction drawn in the literature, by mainly mathematically inspired authors, is between so-called active and passive transformations. An active transformation is said to be 'an actual transformation of the coordinates and fields' as if to physically manipulate the system, whereas a passive transformation would be a coordinate transformation, or a 'relabelling of the numerical values assigned to coordinates and fields'. Taking the perspective of a physicist, we should note that the effect of both transformations on the description of the system is the same, and hence, that there is only a philosophical distinction between active and passive transformations. For an opposing viewpoint, see for instance Ref. [158].

Another distinction made by some authors, is between internal and external symmetry transformations. An external symmetry, which is also sometimes called a spacetime symmetry, is said to involve a transformation of spacetime coordinates, while internal symmetries concern properties of the fields other than its spacetime coordinate. For example, the phase-rotation symmetry of Section 1.4.1 would be an internal symmetry, whereas the translations in Section 1.4.4 are an example of external or spacetime symmetries. Notice that external symmetries encompass not just translations and rotations but also dilatations and boosts. From a practical physical point of view, there is no fundamental difference between breaking a global spacetime symmetry or a global internal symmetry: in either case, observable effects result from the transformation properties of the physical fields [15]. This observation notwithstanding, some specific physical effects may of course be special to certain types of symmetry. For instance, there are cases in which the dispersion relation of Nambu–Goldstone modes is fractional (i.e., $\omega \propto q^{\gamma}$ with γ non-integer), and these only occur for specific broken spatial symmetries [210]. The breaking of spacetime symmetries and its related subleties are discussed in Chapter 13.

1.6 Gauge Freedom

Some transformations that can be applied to a model description may leave its Lagrangian or Hamiltonian invariant, and yet have no physical consequence whatsoever. They do not give rise to Noether currents or charges, cannot be spontaneously broken and are not associated with any sort of Nambu–Goldstone modes. Instead, these transformations appear purely as a mathematical property of the models with which we choose to describe nature. Any such mathematical transformation that leaves the description of nature invariant and moreover does not correspond to any physical effect, may be called a **gauge freedom**. These freedoms can take many forms, and although they typically simplify the mathematical representation of a model, they also often complicate its physical interpretation.

1.6.1 Relabelling your Measuring Rod

A straightforward example of a gauge freedom is the fact that in any description of nature, the origin of the coordinate system may be freely chosen. Surely, the physics of any system, object, field theory, or anything else cannot depend upon this choice. More generally, we are free to choose any type of coordinate system that we like, be it Cartesian, spherical or something more exotic. We are even free to switch from using one type of coordinate system to another at any point in time. Nothing physical ever changes as a result of this choice. The equations we use may look different in different coordinate systems, but they represent the same physical object.

The freedom of choosing axes does not apply only to the coordinates of space and time. As soon as we set out to measure any physical quantity whatsoever, we must choose a scale, which must have a zero and a set distance between units. Consider temperature, for example. Nothing physical changes in the system whose temperature we take, when we replace the thermometer's Celsius scale with one using Fahrenheit. Both scales correspond to an arbitrary choice of zero, and an arbitrary distance between units, in the sense that significance is arbitrarily assigned to the freezing point of water or the average temperature of a human body. The zero of the Kelvin scale is less arbitrary, but still, no physical process will change the moment you express its temperature in Kelvins, rather than Celsius or Fahrenheit. In fact, the term gauge invariance was coined by Hermann Weyl in 1919 in German as 'eichinvarianz', where 'eich' (gauge) refers to the scale or standard dimension used by a measurement device [80]. It is the 'choice of tick marks', and we are free to choose the ticks on our measurement devices, without ever affecting the physical properties of the objects we measure.

Right now, this emphasis on the arbitrariness of the choice of coordinate system may sound a bit pedantic, but exactly the same arbitrariness underlies more sophisticated forms of gauge freedom. This is made more explicit in Project 15 using simple toy models. In general, it is therefore worthwhile keeping the simple picture of the choice of tick marks in mind whenever gauge freedoms appear in any model of physics.

1.6.2 Superfluous Degrees of Freedom

A more profound type of gauge freedom encountered in many theories of physics, appears when new, auxiliary degrees of freedom are introduced to simplify our mathematical description of nature.

The most familiar example of this type of gauge freedom can be found in Maxwell's equations of electromagnetism. Here, the physical observables are the electric and magnetic fields $\mathbf{E}$ and $\mathbf{B}$. These fields cannot take arbitrary forms, but are constrained by the Faraday–Maxwell equation and the requirement that the magnetic field is solenoidal:

$$\nabla \times \mathbf{E} + \partial_t \mathbf{B} = 0, \qquad \nabla \cdot \mathbf{B} = 0. \tag{1.43}$$

These equations fix three of the six components that together make up $\mathbf{E}$ and $\mathbf{B}$: the Faraday–Maxwell equation provides two constraints, and $\nabla \cdot \mathbf{B} = 0$ provides another one. The constraints can be explicitly enforced by writing $\mathbf{E}$ and $\mathbf{B}$ in terms of the scalar and vector potentials V and $\mathbf{A}$:

$$\mathbf{E} = -\nabla V - \partial_t \mathbf{A}, \qquad \mathbf{B} = \nabla \times \mathbf{A}. \tag{1.44}$$

Written in this way, it automatically follows that the divergence of $\mathbf{B}$ vanishes and the Faraday–Maxwell equation is satisfied, regardless of what form the fields V and $\mathbf{A}$ take.

Using the scalar and vector potentials, Maxwell's equations too take on a more convenient form. Introducing the four-potential $A_\mu = (\frac{1}{c}V, \mathbf{A})$ and the relativistic gradient $\partial_\mu = (\frac{1}{c}\partial_t, \partial_m)$, the Maxwell action can be written as

$$S_{\text{Maxw}} = \frac{1}{2\mu_0} \int \mathrm{d}t\mathrm{d}^3x \left(\frac{1}{c^2}\mathbf{E}^2 - \mathbf{B}^2 \right) = \frac{1}{4\mu_0} \int \mathrm{d}t\mathrm{d}^3x \, (\partial_\mu A_\nu - \partial_\nu A_\mu)^2. \tag{1.45}$$

Here, c is the speed of light and μ_0 is the vacuum permeability or magnetic constant. The equations of motion associated with this action yield precisely the two remaining Maxwell equations.

It can be easily checked that the action is left invariant when the relativistic gradient of an arbitrary scalar field $\alpha(x)$ is added to the vector potential:

$$A_\nu(x) \to A_\nu(x) + \partial_\nu \alpha(x). \tag{1.46}$$

This transformation does not affect the physical fields **E** and **B**. It is therefore not a (local) symmetry, but rather a **gauge transformation**. The four-potential $A_\nu(x)$ is often called a **gauge field**, or **gauge potential**, to reflect this role. Note that although the gauge field $A_\nu(x)$ may be changed by a gauge transformation without affecting any physical fields, it does contain all information required to construct the physical electric and magnetic fields. It is important to remember that almost all gauge fields you will encounter in physics, contain both 'real', physical, degrees of freedom, as well as superfluous, 'gauge', degrees of freedom.

One way to understand why a gauge freedom emerges from expressing the Maxwell equations in terms of the four-potential, is to note that A_μ has four components, while only three physical components are needed to determine **E** and **B** after implementing the constraints of Eq. (1.43). The fourth component of A_ν is redundant, and there is some freedom in choosing its value. Since the Maxwell equations take on a simple form in terms of A_ν, it is often convenient to do calculations in terms of the four-potential, rather than **E** and **B**. At the end of any calculation, however, the final results should not depend on the superfluous degree of freedom, which was introduced purely for mathematical convenience. That is, performing a gauge transformation of the type of Eq. (1.46) can never affect any predictions for physical observables.

At this point, it should be noted that gauge freedom is referred to as 'gauge symmetry' throughout much of the physics literature. Although the action is left invariant by gauge transformations, however, it is misleading to call them symmetries, because they do not correspond to measurable properties of any physical degree of freedom. There are no conserved currents or charges associated with gauge transformations. Gauge freedoms can also never be broken. Not only because they are local transformations, which cannot be spontaneously broken anyway (see Exercise 2.3), but also, and more fundamentally, because gauge freedoms do not correspond to any physical manipulation and therefore there cannot exist any measurable quantity that could conceivably be observed to vary under a gauge transformation.

Gauge transformations are not symmetries.

To make matters even more confusing, global transformations of internal symmetries, and in particular global phase rotations like Eq. (1.19), are sometimes called 'global gauge transformations' [9, 112, 168] or 'gauge symmetry of the first kind'. This terminology is simply outdated [28, 200]. A transformation is either a symmetry or a gauge freedom, and the two notions should not be mixed.

1.6.3 Coupling to Matter

Despite the fact that gauge freedoms are concerned with superfluous degrees of freedom, they are not just obnoxious complications arising from our ineptness in finding better mathematical representations of the laws of nature. In fact, our understanding of elementary particle physics relies heavily on the structure of gauge transformations, with the gauge fields connected to the W boson, the Z boson, the photon, and the gluon appearing as force fields mediating interactions between fermions. No matter how useful, however, gauge freedom is never a kind of symmetry, and much confusion can be avoided by taking that fact to heart.

Theories that include gauge fields in their description (fields containing both physical and superfluous degrees of freedom) are called **gauge theories**. In those cases, a gauge freedom can exist next to real, physical symmetries. Such a combination of physical symmetry and gauge freedom is very clear in Maxwell electromagnetism, for example. In the presence of electrically charged matter, the part of the action describing the interaction between charges and the electromagnetic fields can be written as follows:

$$S_{\text{int}} = \int \mathrm{d}t\mathrm{d}^3x\, eA_\nu j^\nu. \tag{1.47}$$

Here, j^ν is the Noether four-current of a global $U(1)$ symmetry, and e is the coupling constant, which in this case is the elementary electron charge. This form of the action, tying gauge fields to matter fields, is known as **minimal coupling** and appears more generally in theories that combine symmetries and gauge freedom. Now, if we perform a gauge transformation of the type of Eq. (1.46), we obtain an additional term in the action:

$$S_{\text{int}} \to S_{\text{int}} + \int \mathrm{d}t\mathrm{d}^3x\, e(\partial_\nu\alpha)j^\nu = S_{\text{int}} - \int \mathrm{d}t\mathrm{d}^3x\, e\alpha(\partial_\nu j^\nu). \tag{1.48}$$

This expression clearly shows that the gauge fields can only ever be minimally coupled to a conserved Noether current, satisfying $\partial_\nu j^\nu = 0$, because only then the action will be invariant under gauge transformations. The local $U(1)$ gauge freedom in the Maxwell action is therefore tightly connected to the simultaneously present global $U(1)$ symmetry: without global $U(1)$ symmetry, we cannot couple matter to the $U(1)$ gauge field.

There is another way to understand the link between global symmetries and gauge freedom, which uses **Noether's second theorem**. Briefly stated, it says that if the transformations that leave the Lagrangian invariant (including both symmetries and gauge freedoms) depend on some parameters $\alpha_n(x)$ and their derivatives $\partial_\mu\alpha_n(x)$, then there are constraints on the possible field configurations, regardless of whether or not the equations of motion are satisfied. We will not attempt to prove Noether's second theorem here, or even to state it in general form. Instead, we will illustrate its implications by considering the example of a gauge field A_μ that is coupled to a complex scalar field ψ. As we will discuss at length in Chapter 8, this is the main ingredient of the Ginzburg–Landau theory for $U(1)$ symmetry breaking in the presence of gauge fields, giving rise to superconductivity.

Assume that the theory is invariant under a local transformation that acts on the fields as

$$\delta_s\psi = -\mathrm{i}\alpha(x)\psi, \qquad \delta_s\psi^* = \mathrm{i}\alpha(x)\psi^*, \qquad \delta_s A_\mu = -\frac{\hbar}{e^*}\partial_\mu\alpha(x). \tag{1.49}$$

Here, e^* is the electric charge of the field ψ. We can now write the Euler–Lagrange equation of motion obtained by varying $\mathcal{L}$ with respect to ψ as $E_\psi = 0$, and similarly for variations with respect to ψ^* and A_μ. The expression E_ψ is the term between square brackets in Eq. (1.10). Noether's second theorem then states that

$$E_\psi(-\mathrm{i}\psi) + E_{\psi^*}(\mathrm{i}\psi^*) = \left(-\frac{\hbar}{e^*}\right)\partial_\mu E_{A_\mu}. \tag{1.50}$$

Again, this identity holds whether or not the equations of motion are satisfied. In classical physics, only field configurations that obey the equations of motion are usually of any importance, and for these Noether's second theorem reduces to a trivial equation. Within quantum field theory, however, the calculation of quantum corrections involves contributions from so called off-shell field configurations that do not satisfy the equations of motion. Noether's second theorem, and the closely related Ward–Takahashi identities, then impose important and influential constraints on the field configurations that need to be considered. This will be discussed further in Section 10.4.

Noether's second theorem gives an alternative way of understanding the fact that gauge fields can only couple to conserved currents. To see this, consider a situation in which the equations of motion for A_μ are satisfied, so that $E_{A_\mu} = 0$, and the right-hand side of Eq. (1.50) vanishes. Furthermore, the left-hand side can be rewritten as $E_\psi \Delta_s \psi + E_{\psi^*} \Delta_s \psi^*$. Since we also know that for complex scalar fields $\frac{\partial \mathcal{L}}{\partial \psi} \Delta_s \psi + \frac{\partial \mathcal{L}}{\partial \psi^*} \Delta_s \psi^* = 0$, the left-hand side is then precisely of the form $\partial_\mu j^\mu$, with j^μ the Noether current associated with the global, constant α, part of the symmetry transformation on the field ψ. Equating the left- and right-hand sides gives Noether's (first) theorem, $\partial_\mu j^\mu = 0$, which we can now interpret as saying that as long as the gauge field satisfies its equation of motion, the current it couples to must be a conserved one. This is then true regardless of whether or not the field configuration of ψ itself obeys its equations of motion. A more thorough account of this viewpoint is given in Ref. [30].

Notice that even if the global, constant α, $U(1)$ symmetry of the field ψ is spontaneously broken, the gauge freedom and gauge invariance of the theory persists unaffected [28, 200]. This fact is often under-appreciated in the literature, to the point that sometimes you read incorrect phrases like 'spontaneous symmetry breaking of gauge symmetries'. We emphasise that gauge freedom *cannot be broken*. However, the fact that there is a minimal coupling with gauge fields as in Eq. 1.47 leads to the so-called Anderson–Higgs effect when we break a global $U(1)$ symmetry. This will be discussed in detail in Section 8.3. The Anderson–Higgs mechanism is responsible for mass creation in the Standard Model (see Chapter 11) as well as for the Meissner effect describing the flux expulsion observed in superconductors.

1.6.4 Distinguishing Gauge Freedom from Symmetry

Looking only at the action, there does not seem to be much difference between the $\alpha(x)$ describing local spacetime translations in Eq. (1.41), and the $\alpha(x)$ describing local gauge transformations in Eq. (1.46). The former, however, is a symmetry transformation that yields conserved Noether currents, whereas the latter is a gauge transformation signifying the presence of a superfluous degree of freedom. It would thus be convenient if there were a way of telling these two physically distinct types of local transformation apart, even though both appear as local transformations leaving the action invariant. Fortunately, there is a method for identifying constraints that lead to local gauge freedoms, such as the Faraday–Maxwell equation and the requirement of the magnetic field being solenoidal in electromagnetism. The method may be referred to as the Dirac treatment of Hamiltonian constraints, and starts from the expression of the Hamiltonian in terms of canonical fields and their conjugate momenta. Any relations between these, such that linear combinations of the fields and momenta vanish, then constitute constraints. This in turn implies there are associated redundant degrees of freedom. A detailed discussion can be found in Refs. [49, 71].

The distinction between symmetries and gauge freedom becomes even more subtle for global transformations, which are not easily described in terms of constraints on the Hamiltonian. To illustrate this, consider a many-body spin system whose action is invariant under the global rotation of all spins simultaneously. Surely, the global spin rotation is a symmetry, which may be spontaneously broken into a ferromagnetic arrangement. Taking a different point of view, however, one could also argue that the global transformation that seemingly rotates the direction of all spins, really only describes the rotation of the coordinate system that we use to measure the spin direction with. Such a relabelling of coordinates is the archetype of a global gauge transformation, which should not have any physical implications whatsoever, and which cannot be spontaneously broken.

The way out of this paradox lies in the observation we made at the beginning of this chapter, that symmetry can only be defined with respect to a reference. As long as we exclude from the universe any objects that can measure the magnetisation of our material, it is impossible to tell whether the spins have aligned to a certain direction. Global rotations of all spins are then unobservable by construction, even for the ferromagnet. The magnetisation becomes measurable only if we allow some interaction to exist between the magnet and an external reference. For example, even if the spin-rotation symmetry is not spontaneously broken, the paramagnet can be subjected to an externally applied magnetic field which forces the spins to align in a given direction according to the coupling:

$$\mathcal{L}_{\text{coupling}} = -\mathbf{h} \cdot \mathbf{S}(x). \tag{1.51}$$

Here, the spins are described by $\mathbf{S}(x)$, while $\mathbf{h}$ is the uniform applied field. Since the external field is applied in some given direction, the total action including the coupling to the field is no longer invariant under global rotations of the spins, and the ground state may be magnetised. Incidentally, notice that this is an example of explicit symmetry breaking. In contrast, spontaneous symmetry breaking occurs when the ferromagnetic state survives in the limit of the field strength $|\mathbf{h}|$ going to zero, as will be discussed in detail in Chapter 2.

The crucial observation is now, that in the presence of the external field, the global spin-rotational symmetry and the global rotation of the coordinate system are different. Rotating the coordinate system that we use to define directions in space, implies a simultaneous transformation of both the spins $\mathbf{S}(x)$, and the applied field $\mathbf{h}$. After all, the directions of both are described within our arbitrarily chosen coordinate system. It is easily checked that the coupling of Eq. (1.51) is invariant under this global gauge transformation, which therefore applies equally to the ferromagnetic and paramagnetic states. On the other hand, the term $\mathcal{L}_{\text{coupling}}$ is not invariant under the global rotation of only the spins $\mathbf{S}$, keeping the reference field $\mathbf{h}$ fixed. This physical, global symmetry is broken in the ferromagnetic state, whether it be spontaneously or explicitly.

Even though the coupling strength between the applied field and the magnet may be taken to vanish, the fact that an external field may exist in principle is crucial. In other words, the global symmetry that may be broken in a magnet is not simply the rotation of all its spins, but rather the global spin-rotation relative to some reference. To put this in a more mathematically precise formulation, we can consider the applied magnetic field to be generated by a second, external ferromagnet, which itself is also invariant under global spin rotations. Each magnet in complete isolation, then has a global symmetry group denoted by $SU(2)$. As long as interactions between the two magnets are strictly forbidden, the total system of two magnets has a combined $SU(2)_S \times SU(2)_h$ symmetry, where the indices indicate the magnet with which each symmetry is associated. When interactions between the two magnets are allowed, the symmetry of the combined system is reduced to the so-called diagonal subgroup which contains only simultaneous rotations of **S** and **h**. The symmetry breaking that occurs in a ferromagnet is thus the reduction of $SU(2)_S \times SU(2)_h$ to its diagonal subgroup, rather than the breakdown of just $SU(2)_S$ that we might naively expect. As we will see in Section 2.2, this is the case even with spontaneously broken symmetries, for which the coupling to an external field may be infinitely weak, but must necessarily be allowed to exist.

1.7 Symmetry Groups and Lie Algebras

In the final section of this chapter, we give a brief overview of some of the mathematical structure underlying symmetry transformations. This is not intended as a complete treatment of any of the topics discussed, but rather as a reminder or entry point towards further study, for which many excellent books may be consulted [56, 81].

1.7.1 Symmetry Groups

Symmetry transformations correspond to manipulations of a physical system that leave a particular state or action invariant. This definition alone has three important implications that together determine how symmetry transformations are represented mathematically:

1. The combined effect of two consecutive symmetry transformations is also a symmetry transformation. This is obvious because if the first transformation does not affect the state, then the second transformation simply acts on the initial state. Likewise, if both transformations leave the action invariant regardless of the state, then acting consecutively will also have no effect on the action. Mathematically, the consecutive action of two symmetry transformations may be considered an (ordered) product of transformations.
2. There always exists a transformation that does nothing to any state or action it acts on, even non-symmetric ones. It is a symmetry transformation for any system, albeit a trivial one. We can call this trivial symmetry the **identity transformation** $\mathbb{I}$.
3. For each symmetry transformation, there exists an **inverse transformation**, such that the consecutive action of the transformation and its inverse yields the identity transformation. This follows from the fact that for each possible state, we can identify the state it transforms into under the action of the symmetry. The inverse transformation can then be defined as the operation that takes the transformed states back to the original ones. It is itself a symmetry transformation, because by definition, the symmetry-transformed and original states are either equal or possess the same action. Denoting the symmetry transformation by U, the inverse transformation can be written as U^{-1}, and it satisfies $UU^{-1} = U^{-1}U = \mathbb{I}$, with $\mathbb{I}$ the identity transformation.

These three properties are obeyed by the set of symmetry transformations $\{U\}$ for any given system and endow them with the mathematical structure of a **group**. The description of groups and their various properties and manipulations is a large and vibrant area of mathematics. If this is your first encounter with group theory, it is probably helpful to consult a more complete text, such as Refs. [56, 81]. Here, we will only remind you of some of the most important ingredients of group theory, with a focus on the parts relevant to the discussion of spontaneous symmetry breaking.

Let us start with some definitions. First, if a set of symmetry transformations can be parametrised by one or more continuous variables, the group they form is called **continuous**. If it cannot, the group is said to be **discrete**. Furthermore, if the number N of transformations within a discrete group is finite, it is called a finite group of order N, otherwise it is an infinite discrete group. Second, if all symmetry transformations in a group commute, that is, $U_1 U_2 = U_2 U_1$ for all possible pairs of transformations in the group, the group is called **Abelian** or **commutative**. If not all elements commute, the group is called **non-Abelian** or **non-commutative**.

Groups appear in many places throughout all of physics. As a reminder of just how common they are, consider the following examples:

- The trivial group consists of one element, the identity $\mathbb{I}$.
- The cyclic group $\mathbb{Z}_n$ or C_n is a finite and Abelian group of order n, which describes the rotation symmetries of a regular, n-sided polygon (triangle, square, etc.)
- The dihedral group D_n is a finite group of order $2n$ describing the rotations and reflections of a regular, n-sided polygon. It is non-Abelian if n is larger than two.
- The discrete translation group $\mathbb{Z}^D$ describes translations on a regular D-dimensional lattice. It is discrete, infinite and Abelian.
- The translation group $\mathbb{R}^D$ describes translations of D-dimensional continuous space. It is continuous and Abelian.
- The rotations of a circle are given by the group $SO(2)$ of real, orthogonal 2×2-matrices with determinant 1. It is continuous and Abelian.
- The rotations and reflections of a circle are given by the group $O(2)$ of real, orthogonal 2×2-matrices. It is continuous and non-Abelian.
- The rotations of a sphere are given by the group $SO(3)$ of real, orthogonal 3×3-matrices with determinant 1. It is continuous and non-Abelian.
- The rotations and reflections of a sphere are given by the group $O(3)$ of real, orthogonal 3×3-matrices. It is continuous and non-Abelian.
- The set of complex unitary 2×2-matrices with determinant 1 form a continuous and non-Abelian group called $SU(2)$. It describes spin rotations and the weak force.
- The continuous, non-Abelian group hosting complex unitary 3×3-matrices with determinant 1 is $SU(3)$. It underlies the strong nuclear force.

Exercise 1.4 (Dihedral groups).
Consider dihedral groups D_n. Denote the identity by e, the rotations over multiples of $2\pi/n$ by $r, r^2, \ldots, r^{n-1}$, and the reflections by $s, sr, sr^2, \ldots, sr^{n-1}$, where $s^2 = e$.
a. Make a table of all the multiplication rules (transformation g_1 labelling the columns, g_2 the rows, and the products $g_1 g_2$ as the entries of the table) for the group D_4.
b. Check explicitly that every element in D_4 has an inverse.
c. Give the multiplications for general n of (i) the two rotations r^k and r^l; (ii) a rotation r^k and the reflection s; (iii) two reflections sr^k and sr^l.

A subset H of the elements making up a group G may by itself also form a group, if it possesses all of the three properties that define a group. In particular, the multiplication of elements in the subset must be closed, so that $h_1 h_2 \in H$ for all h_1 and h_2 in H. If the subset is a group by itself, it is called a **subgroup** of G. The set of all rotations within the dihedral group D_n, for example, form a subgroup, known as the cyclic group C_n. Similarly, translations in the x-direction are a subgroup of all translations in three-dimensional space. Subgroups appear in the discussion of symmetry breaking because all transformations that still leave the system invariant after a system has gone through a symmetry-breaking transition, form a subgroup of the original symmetry group.

For a given group G and subgroup H, we may identify a (left) **coset** for each element $g \in G$, denoted by gH. It is defined as the set of all elements gh with $h \in H$.[3] The set of all distinct cosets associated with elements of G is denoted by G/H and called the **quotient set**. Since the set inherits topological structure from the Lie group, we usually refer to G/H as the **coset space**. Neither the individual cosets nor the quotient set are typically groups by themselves.

As an example, consider the group of rotations of a regular hexagon, C_6. It has six elements, which may be written as r^k, with $k = 0, \ldots, 5$. Here, r is a rotation over $2\pi/6$, and $r^0 = e$ is the identity transformation. This group has a subgroup C_2, consisting of the rotations of a line (a two-sided polygon). The subgroup has elements e and r^3. The unique cosets that can be generated from C_6 and its subgroup C_2 are $\{e, r^3\}$, $\{r, r^4\}$, and $\{r^2, r^5\}$. The quotient set C_6/C_2 is a set with three elements, each of which may be represented by a single element from its corresponding coset, here for instance $\{e, r, r^2\}$. Cosets play an important role in the classification of broken-symmetry states, as we will see in Section 3.1.

Exercise 1.5 (Equivalence and quotient sets).
Two elements in the same coset may be said to be equivalent. That is, $g_1 \sim g_2$ if $g_1 = g_2 h$ for some $h \in H$.
Show that this is indeed an equivalence relation in the mathematical sense, by showing that it satisfies the three properties: (i) $g \sim g$; (ii) if $g_1 \sim g_2$ then $g_2 \sim g_1$; (iii) if $g_1 \sim g_2$ and $g_2 \sim g_3$ then $g_1 \sim g_3$, for all $g, g_1, g_2, g_3 \in G$.
The set of cosets or quotient set G/H can alternatively be defined as the set of equivalence classes under this definition of equivalence.

Exercise 1.6 (Subgroups of dihedral groups).
In the dihedral group D_n, the set $\{e, s\}$ forms a subgroup.
a. Find all cosets with respect to this subgroup.
If n is even, the element $\{r^{n/2}\}$ commutes with all other elements. It is therefore called a central element, and the subgroup $\{e, r^{n/2}\}$ is called a normal subgroup. Normal subgroups H have the property that $ghg^{-1} \in H$ for all $g \in G$ and all $h \in H$.
b. Find all cosets with respect to the normal subgroup $\{e, r^{n/2}\}$.

[3] The right coset of g is defined as $\{hg | h \in H\}$. Left and right cosets are not generally identical, but there is a bijection between them. We will use only left cosets and refer to them as just cosets.

The set of cosets with respect to a normal subgroup has a group structure itself.
c. What is that group in this case?

1.7.2 Lie Groups and Algebras

If the parameters of continuous symmetry transformations in a group are smooth, the group is called a **Lie group**. To be precise, a Lie group is a continuous group with the structure of a differentiable manifold, but for our purposes, having a description in terms of smooth functions suffices. All continuous groups encountered in this book, including all examples we have already seen, are Lie groups. The number of variables needed to parametrise the full set of continuous transformations is called the dimension of the Lie group.

Unlike discrete groups, Lie groups always contain transformations that are arbitrarily close to the identity. The fact that the transformations can be written in terms of differentiable functions then allows for them to be expanded around the identity. For example, consider a quantum system that is symmetric under the continuous transformations $\hat{U}_\alpha = \exp(\mathrm{i}\alpha_a \hat{Q}_a)$, with α_a a set of small parameters and $\hat{Q}_a$ a set of Hermitian operators, both labelled by the index $a = 1, \ldots, N$. Here, we used the fact that because symmetry transformations must be unitary, they can always be written as the exponent of a Hermitian operator. The identity is the symmetry transformation with all α_a equal to zero, and a general continuous transformation can be expanded around the identity as $\hat{U}_\alpha = \hat{\mathbb{I}} + \mathrm{i}\alpha_a \hat{Q}_a + \mathcal{O}(\alpha^2)$. Because continuous symmetry transformations close to the identity are determined entirely by the action of the operators $\hat{Q}_a$, these are called the **generators** of the Lie group. We actually already used an expansion around the identity in our proof of Noether's theorem, when we considered infinitesimal symmetry transformations in Eq. (1.5) and found the expectation values of the symmetry generators $\hat{Q}_a$ to be conserved Noether charges.

The sum of two symmetry generators is itself a generator, which implies that the set of symmetry generators has the mathematical structure of a vector space. Generators can also be multiplied, but their product is generally not itself a generator. Instead their commutator, or more generally, the Lie bracket, is a linear combination of other generators:

$$[Q_a, Q_b] = \mathrm{i} \sum_c f_{ab}{}^c Q_c. \tag{1.52}$$

Here, the **structure constants** $f_{ab}{}^c$ are real numbers. The vector space of symmetry generators together with their commutation relations is called a **Lie algebra**. The Lie algebra is determined entirely by its structure constants, and all Lie algebras satisfy the so-called **Jacobi identity**:

$$[Q_a, [Q_b, Q_c]] + [Q_b, [Q_c, Q_a]] + [Q_c, [Q_a, Q_b]] = 0. \tag{1.53}$$

Finally, if a group is Abelian, all structure constants of the Lie algebra are zero.

While each Lie group gives rise to a single Lie algebra, a given Lie algebra does not uniquely define a corresponding Lie group. The reason for this is that the Lie algebra describes the structure of the symmetry generators, rather than the symmetry transformations themselves. The two are equivalent for continuous transformations close to the identity, but in addition to those, the complete group may also contain some discrete transformations, which cannot be expanded around the identity. For example, the groups $SO(3)$ and $O(3)$ have the same Lie algebra, describing continuous rotations in three dimensions, but the discrete reflections present in $O(3)$ are not captured by the Lie algebra. The Lie algebra is important for the structure of Nambu–Goldstone modes, discussed in Section 4.2.

Exercise 1.7 ($SU(2)$ Lie algebra).
The generators for the Lie algebra of the group $SU(2)$ are the Pauli matrices:

$$\sigma_x = \frac{1}{2}\begin{pmatrix} 0 & 1 \\ 1 & 0 \end{pmatrix}, \quad \sigma_y = \frac{1}{2}\begin{pmatrix} 0 & -\mathrm{i} \\ \mathrm{i} & 0 \end{pmatrix}, \quad \sigma_z = \frac{1}{2}\begin{pmatrix} 1 & 0 \\ 0 & -1 \end{pmatrix}. \tag{1.54}$$

a. Show that the structure constants are given by the Levi-Civita symbol ε_{abc}, which is defined by the element ε_{xyz} being one, and by being completely antisymmetric in a, b, and c.
b. Using the property of the Pauli matrices that $(\sigma_a)^2 = \frac{1}{4}\mathbb{I}$ for $a = x$, y, or z, and the operator expansion $\mathrm{e}^A = \sum_{n=0}^{\infty} \frac{1}{n!}A^n$, find an explicit expression for the continuous transformations (or Lie group elements) $U(\alpha_a) = \mathrm{e}^{\mathrm{i}\alpha_a\sigma_a}, a = x, y, z$. Do not assume α_a to be small.
c. Explicitly write out the multiplication of $U(\alpha_x)$ and $U(\alpha_y)$.

1.7.3 Representation Theory

Groups and algebras are abstract mathematical concepts. They consist of abstract elements that are defined only by the way they can be multiplied or added together. To tie these elements to operators carrying out symmetry transformations, we need **representations** of the abstract group structures.

In practice, we almost always consider a representation to be a set of matrices which have the same structure, in terms of rules for multiplication and addition, as the group or algebra they represent. As an example, consider the cyclic group C_n. On an abstract level, this group is defined as a set of n elements written as r^k with k between zero and n, together with the multiplication rule $r^k r^l = r^{k+l \bmod n}$. If we consider a regular n-sided polygon centred at the origin of two-dimensional space, however, we might want to identify the abstract operation r with a rotation of the polygon over an angle of $2\pi/n$ around its centre. The representation of r is then a 2×2 matrix acting on vectors (x, y), which describe coordinates in the two-dimensional space. Explicitly, it would be given as

$$r^k \mapsto \begin{pmatrix} \cos(2\pi k/n) & \sin(2\pi k/n) \\ -\sin(2\pi k/n) & \cos(2\pi k/n) \end{pmatrix}. \tag{1.55}$$

This is called a real and two-dimensional representation because r^k is given as a 2×2 matrix with real elements.

We can also describe the polygon in a different way. Instead of using Cartesian coordinates of the form (x, y), we can draw the polygon at the centre of the complex plane, in which points are denoted by complex numbers $z = x + iy$. The same rotations of the same polygon would in that case be represented by $r^k \mapsto \mathrm{e}^{\mathrm{i}2\pi k/n}$. This is called a complex, one-dimensional representation because r^k is written as a 1×1 matrix with complex elements. We thus see that the abstract notion of rotating a polygon can be represented in different ways, depending on how we choose to describe the polygon. The important point to notice, is that regardless of which representation we choose, its elements satisfy the same properties, or rules of multiplication and addition, as the abstract group elements.

In most cases, which representation of a group is being used, will be obvious from the context in which it appears. Consider, for example, a theory with two complex fields ψ_1 and ψ_2, and the Lagrangian,

$$\mathcal{L} = \frac{1}{2}|\partial_\nu \psi_1|^2 + \frac{1}{2}|\partial_\nu \psi_2|^2 - V(|\psi_1|^2 + |\psi_2|^2). \tag{1.56}$$

Here, V is some function that depends only on the combination of fields $|\psi_1|^2 + |\psi_2|^2$. You will probably notice that this Lagrangian in invariant under transformations of the fields ψ_1 and ψ_2 that keep the value of $|\psi_1|^2 + |\psi_2|^2$ fixed. This immediately suggest a natural representation for the combination of the two fields as a complex-valued vector $\psi = (\psi_1, \psi_2)$, whose length is held fixed by the symmetry transformations. The group of operations that act on complex two-component vectors and keeps their length fixed, is $SU(2)$, and its natural representation in this case is in terms of two-dimensional matrices:

$$\psi \to \mathrm{e}^{-\mathrm{i}\alpha_a \sigma_a} \psi, \qquad \psi^* \to \mathrm{e}^{\mathrm{i}\alpha_a \sigma_a} \psi^*. \tag{1.57}$$

Here, the σ_a are Pauli matrices, which generate the Lie algebra of $SU(2)$, see Exercise 1.7. Writing out the components of the vectors and matrices explicitly, the symmetry transformations are as follows:

$$\psi_m(x) \to \psi'_m(x) = \sum_{a,n} \mathrm{e}^{-\mathrm{i}\alpha_a (\sigma_a)_{mn}} \psi_n(x). \tag{1.58}$$

Notice that here, $m, n = 1, 2$ are indices of the two-vectors ψ, while $a = x, y, z$ denotes the index of the Pauli matrices.

Within the Lagrangian formalism, it is important to keep in mind the distinction between the representations of a symmetry group and operations on Hilbert space. In the example of the $SU(2)$-invariant two-vector field, for instance, the matrices σ_a act on the fields ψ, which themselves are eigenvalues of operators acting on Hilbert space. In contrast to what we saw in the Hamiltonian formalism in Section 1.3.2, the representations of the Lie algebra generators, σ_a, in this case do not correspond directly to the symmetry generators, Q_a, describing conserved Noether charges. Like any other observable in the Lagrangian formulation, the Noether charges can always be expressed in terms of the fields ψ_n themselves. In this specific case, they are

$$\begin{aligned} Q_a &= \int \mathrm{d}^3x \left(\pi(-\mathrm{i}\sigma_a \psi) + (\mathrm{i}\psi^\dagger \sigma_a)\pi^\dagger\right) \\ &= \int \mathrm{d}^3x \left(- \mathrm{i}(\partial_t \psi^\dagger)\sigma_a \psi + \mathrm{i}\psi^\dagger \sigma_a (\partial_t \psi)\right) \\ &= \int \mathrm{d}^3x \left(- \mathrm{i}(\partial_t \psi^*_m)(\sigma_a)_{mn} \psi_n + \mathrm{i}\psi^*_m (\sigma_a)_{mn} (\partial_t \psi_n)\right). \end{aligned} \tag{1.59}$$

These conserved charges Q_a are clearly related to the representations σ_a of the Lie group generators, but they are not the same thing.

Exercise 1.8 (Representations of $U(1)$ and $\mathbb{Z}$).
The one-dimensional complex representations of the group $U(1)$ of phase rotations over an angle α are given by $\mathrm{e}^{\mathrm{i}n\alpha}$, for $n \in \mathbb{Z}$.
a. Show that this representation obeys the group properties of $U(1)$.
The one-dimensional complex representations of the group $\mathbb{Z}$ of lattice translations over a multiple n of the elementary lattice vector are given by $\mathrm{e}^{\mathrm{i}\alpha n}$ with $\alpha \in [0, 2\pi)$.
b. Show that this representation obeys the group properties of $\mathbb{Z}$.
The groups $U(1)$ and $\mathbb{Z}$ also have zero-dimensional representations.
c. Show that the **trivial representation** with only the identity element obeys the group properties of both groups (and indeed of any group).
Representations where distinct group elements are represented by distinct matrices are called **faithful**. The trivial representation is an example of a non-faithful representation.

2 Symmetry Breaking by Example

A symmetric system, described by a Hamiltonian, Lagrangian, or action that is left invariant under a unitary transformation, typically has a symmetric equilibrium configuration. In quantum physics, one can even prove that any symmetric system either has a unique and symmetric ground state or a degenerate set of ground states related by the symmetry transformation. Looking around in our everyday world, however, we rarely see any truly symmetric objects. How things manage to be in stable configurations that seemingly evade the symmetries of the laws of nature is explained by the theory of spontaneous symmetry breaking.

Spontaneous symmetry breaking (SSB) is the phenomenon in which a stable state of a system, such as its ground state or thermal equilibrium state, is not symmetric under a symmetry of its Hamiltonian, Lagrangian, or action.

In this chapter, we give a brief description of some basic notions of spontaneous symmetry breaking, emphasising the importance of singular limits in Section 2.2. We illustrate these concepts in three concrete examples of SSB, the harmonic crystal (Section 2.3), the Heisenberg antiferromagnet (Section 2.4), and the Bose–Einstein condensate (Section 2.5).

2.1 Basic Notions of Spontaneous Symmetry Breaking

If the state $|\psi\rangle$ of a system is not left invariant by a symmetry transformation $\hat{U}$ of the Hamiltonian that describes the system, the state is said to have spontaneously broken the symmetry. This single observation immediately implies that for every symmetry-broken state there exist a multitude of related states, which all share the same energy. After all, for a given transformation $\hat{U}$, the fact that the state $|\psi\rangle$ breaks the symmetry implies that it is different from $\hat{U}|\psi\rangle$. At the same time, the Hamiltonian being symmetric implies that it commutes with the symmetry transformation. The two inequivalent states $|\psi\rangle$ and $\hat{U}|\psi\rangle$ must therefore have the same energy expectation values. Continuing this way, we can define a whole set of distinct symmetry-broken states, which all have the same energy, by performing all possible symmetry transformations $\hat{U}$ on a given initial symmetry-broken state $|\psi\rangle$. For example, a rock or other solid piece of material is typically localised in a single (centre of mass) position in space, while the Hamiltonian for objects in homogeneous space is translationally invariant. Moving the rock to another position yields a distinct state, but with the same energy. The set of all these inequivalent but degenerate states consists of the rock being localised in all possible positions.

The set of symmetry-broken states related by the action of the symmetry transformation allows us to define the **order parameter** $\hat{\mathcal{O}}$ as an operator whose eigenstates are all inequivalent symmetry-broken states in that set. We will return to this definition in detail in Section 3.2. We choose the order parameter operator such that it has distinct and non-zero eigenvalues for each of the states in the set of symmetry-broken states. Finally, the order parameter operator should have zero expectation value for symmetric states. This order parameter operator, in general, does not commute with the Hamiltonian (barring a few important exceptions that will be discussed in Section 4.2). In the example of a solid object, the symmetry-broken states are all eigenstates of the position operator $\hat{X}$, which serves as an order parameter operator. The Hamiltonian is translationally invariant, implying that it commutes with the momentum operator $\hat{P}$, and therefore not with the order parameter operator.

The fact that symmetry-broken states are eigenstates of an operator that does not commute with the Hamiltonian raises a conundrum: these states cannot be eigenstates of the Hamiltonian. Somehow, all the symmetry-broken states that you see around you every day, including the table in front of you right now, are not energy eigenstates. In fact they cannot even be (thermal) mixtures of energy eigenstates, and they are therefore not in thermal equilibrium!

That these states can nonetheless exist and be stable is owing to a large degree to the singularity of the **thermodynamic limit**. The thermodynamic limit for a system of N particles and volume V is defined by taking both the limit $N \to \infty$ and $V \to \infty$, while keeping the ratio N/V fixed. This means that intensive quantities like density and temperature do not change as the limit is taken, while extensive quantities like N and entropy grow to infinity. It turns out that, even though the order parameter operator and the Hamiltonian in general do not commute, the commutator expectation value vanishes in the thermodynamic limit. Furthermore, all distinct symmetry-broken states become orthogonal to one another in that limit, and they become degenerate with symmetric, exact eigenstates of the Hamiltonian. Precisely in that limit then, the symmetry-broken states actually are eigenstates of $\hat{H}$ and may occur in thermal equilibrium. The fact that the situation for truly infinite N and V is qualitatively different from that of any finite volume, no matter how large, makes the thermodynamic limit a **singular limit**.

A main theme that will be emphasised throughout this book is that the thermodynamic limit is a mathematical idealisation, which serves as a guide to what happens in the real world, but does not actually describe it. A real system may have N and V very large, but not infinite. So the question of how real, finite-sized objects may be observed in symmetry-broken configurations remains.

In addressing this question, the first observation to be made is that the spectra of symmetric Hamiltonians have some common properties. Using a Fourier transformation, the Hamiltonian can always be separated into a centre-of-mass, or collective, part at zero wave number, and a part for finite wave numbers containing all possible information about the internal degrees of freedom. Moreover, these two parts commute. The essential observation is now that, in order to describe the breaking of a global symmetry, we only need to consider the collective part of the Hamiltonian. For example, a solid object localised in space has a collective Hamiltonian that describes its centre-of-mass position and the collective motion of all of its N atoms of mass m being displaced in unison. In free space, the collective and symmetric Hamiltonian is just that of a free particle of mass mN, and its lowest energy levels are spaced by energy differences of the order of $\Delta E \sim 1/N$. These low-energy eigenstates of the collective Hamiltonian make up what is called the **tower of states**. The states in this tower are highly collective and non-local. That is, they cannot be written as product states of the form $|\psi\rangle = \otimes_j |\psi_j\rangle$, with $|\psi_j\rangle$ a single-particle state. For the example of solid objects, the tower of states consists of eigenstates of total momentum, each of which is collectively delocalised over all of space.

Because they are non-local, the energy eigenstates in the tower are increasingly unstable to the effect of local perturbations as the system size is increased. This instability of exact energy eigenstates prevents them from being realised in our everyday world because even the weakest asymmetric interaction suffices to destabilise them completely. Stable states, on the other hand, are local, may be written as tensor products of single-particle states, and are not very sensitive to local perturbations. For the example of the solid rock, stable states are the localised eigenstates of the position operator. They are superpositions of the collective configurations in the tower of states and are not generally energy eigenstates. However, since the energy spacing within the tower of states scales as $1/N$, the energy uncertainty of the stable states is very small for large systems. Similarly, the stable states are not orthogonal, but their overlap drops as e^{-N}, so that once the system ends up in a stable state, the probability of tunnelling to different stable state is exponentially suppressed. Most importantly, these stable states are not symmetric, and they are degenerate in the sense that they have the same energy expectation value, which is very close to the energy of the exact ground state of the Hamiltonian.

Spontaneous symmetry breaking for finite-sized objects is the phenomenon that a system may exist in a stable state that is not an exact eigenstate of its symmetric collective Hamiltonian. But how does a system single out just one of the many distinct, stable, symmetry-broken states? Clearly, the symmetric Hamiltonian cannot account for this, and one is forced to consider an external perturbation, which explicitly breaks the symmetry and favours one of the stable states over all others. A large symmetric system is exceedingly sensitive to such symmetry-breaking disturbances, and a perturbation with an energy scale as small as $\sim 1/N$ suffices to single out a particular stable state. This is the spontaneous part of spontaneous symmetry breaking: no matter how small the perturbation, it will entirely determine the fate of sufficiently large systems.[1]

Considering the symmetric collective Hamiltonian together with an arbitrarily small symmetry-breaking perturbation, the combined ground state of the full system is a stable symmetry-broken state. In the thermodynamic limit, it extrapolates to an eigenstate of the order parameter operator. On the other hand, in the strict absence of any perturbations, the symmetric ground state is the true ground state for systems of any size. The fact that the state encountered in the thermodynamic limit changes qualitatively if even an infinitesimally weak external perturbation is added or removed is a clear manifestation of its singular nature.

Because the overlap between distinct symmetry-broken states is exponentially suppressed for large system sizes, we can treat a system with a spontaneously broken symmetry as if, for all practical purposes, it has a single, symmetry-broken ground state. All other symmetry-broken states are inaccessible to the system, and its entire dynamics takes place in a restricted part of Hilbert space that contains the symmetry-broken state and its excitations. In other words, configuration space (or phase space) is effectively restricted to a small subspace. As long as we are interested in the physics of the symmetry-broken phase, and not for

[1] The term 'spontaneous symmetry breaking', while accurate in this sense, is not ideal when trying to explain symmetry breaking to non-specialists. It was introduced by Baker and Glashow [14], and so far nobody has come up with a workable alternative.

instance in phase transitions, we only need to consider an effective Hamiltonian describing physics within the small symmetry-broken subspace, and we may safely disregard the rest. This an important instance of **ergodicity breaking**, where part of the phase space is not accessible on any reasonable timescales, and global thermal equilibrium can never be reached [58]. It also occurs in disordered systems such as glasses, which we discuss briefly in Section A.1.

2.2 Singular Limits

The possibility of spontaneously breaking a symmetry is closely related to the singular nature of the thermodynamic limit. Although singular limits occur throughout all parts of physics, and indeed daily life, they are not commonly encountered in standard physics curricula. To develop some feeling for them, consider the example of a classical perfect cylinder. If we sharpen such a cylinder so that it ends in a tip on one end, we get a pencil-shaped object, which we could try and balance on a table, as shown schematically in Figure 2.1. In this picture, y is the distance between the centre of mass of the pencil and the surface of the table, θ is the angle between the central axis of the pencil and a line normal to the table surface, and b is the diameter of the base area of the pencil's tip, which we assume to be flat and circular.

If we manage to perfectly balance the pencil, so that θ is strictly zero, it will be symmetric under rotations around the axis normal to the table surface. The equations of motion describing the pencil, which include the effects of inertia as well as gravity acting perpendicular to the table surface, are invariant under the same rotations. If you ever tried to balance a sharp pencil on its tip in real life, however, you probably discovered this is a very hard thing to do. In fact, if the pencil is sharp enough, it becomes practically impossible to balance it, and the

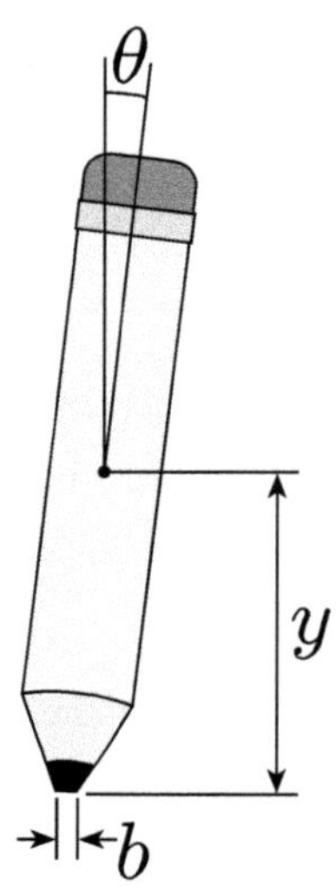

Figure 2.1 Balancing a blunted pencil becomes increasingly difficult if we sharpen it. The limits $b \to 0$ and $\theta \to 0$ do not commute and provide an example of singular limits.

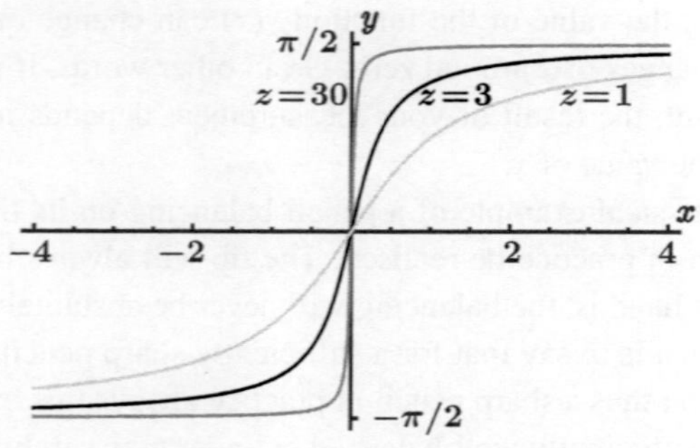

Figure 2.2

The function $y = \arctan(zx)$ has non-commuting limits $x \to 0$ and $z \to \infty$.

pencil will always fall over when released. As the pencil drops flat onto the table, it does so in a single direction and will thus no longer be symmetric under rotations around the axis normal to the table. That is, by tipping over in a specific but uncontrollable direction, the pencil has spontaneously broken its symmetry.

More precisely, the fact that it seems impossible to balance a sharp pencil in a symmetric state can be described mathematically in terms of two limits. Trying to balance the pencil perfectly corresponds to taking the limit $\theta \to 0$, while sharpening the tip of the pencil corresponds to taking $b \to 0$. The spontaneous breakdown of rotational symmetry can then be understood to be a consequence of the fact that these two limits do not commute:

$$\lim_{b\to 0}\lim_{\theta\to 0} y > 0,$$
$$\lim_{\theta\to 0}\lim_{b\to 0} y = 0. \tag{2.1}$$

That is, if we manage to really, perfectly balance a pencil (taking θ to zero first), it will stay upright in a symmetric state no matter how sharp the pencil happens to be (even if b approaches zero). On the other hand, if we really take an infinitely sharp pencil (taking b to zero first), even an infinitesimal deviation from being perfectly upright (θ approaching zero) suffices to tip it over and break the symmetry. The fact that in the limit $b \to 0$ the pencil becomes infinitely sensitive to the perturbation θ makes the limit of b going to zero a singular limit. In general, the failure of two limits to commute is both a necessary and a distinctive signal of the presence of a singular limit.[2]

Mathematically, the fact that limits do not commute implies the development of a non-analytic feature in some function. This can be illustrated by plotting the function $y = \arctan(zx)$. For any given, finite value of z, the function $y(x)$ is a smooth curve, going through the origin at $x = 0$, as shown in Figure 2.2. As long as $y(x)$ remains a smooth function, its value at x approaching zero does not depend on whether x goes to zero from above or below. As z increases, however, the function $y(x)$ becomes steeper and steeper around the origin, until in the limit $z \to \infty$ it becomes a step function. At that point, the value of $y(x)$ with x approaching zero is no longer zero and does depend on how you approach it:

$$\lim_{z\to\infty}\lim_{x\to 0} y = 0, \qquad \lim_{x\downarrow 0}\lim_{z\to\infty} y = 1, \qquad \lim_{x\uparrow 0}\lim_{z\to\infty} y = -1. \tag{2.2}$$

[2] A nice introduction of the role of singular limits in different fields of physics is given in Ref. [27].

In the limit of $z \to \infty$, the value of the function $y(x)$ can change qualitatively under even infinitesimally small changes of x around zero. Or, in other words, if the value $y(0)$ is something you could measure, the result of your measurement depends infinitely sensitively on how well you control the value of x.

Notice that in the physical example of a pencil balancing on its tip, neither of the limits $b \to 0$ and $\theta \to 0$ can in practice be realised. The tip will always be a little blunt, and no matter how steady your hand is, the balancing will never be absolutely perfect. The meaning of the singular limits then is to say that for a sufficiently sharp pencil, it becomes arbitrarily difficult to balance it, and thus a sharp pencil in practice always tips over. The other ordering of limits implies that a sufficiently well-balanced cylinder may safely be sharpened (keeping the angle θ constant) without tipping it over. Of course, the practical value of the latter order or limits is limited in real life.

There are a few particularities to be noticed about the pencil spontaneously breaking rotational symmetry. The first concerns the question of precisely which symmetry is broken. As we said before, the balanced pencil has a global rotational symmetry, meaning that we rotate the pencil in its entirety. This transformation does not involve the table. The symmetry is therefore global in the sense of affecting every single piece of the pencil, but it occurs relative to the table, which is held fixed. The rotation is thus not the global gauge freedom of rotating the universe as a whole. In this simple classical setting pointing out the difference between global transformations within a fixed reference frame and a truly global gauge freedom may seem a bit esoteric. But when you consider that both the table and the pencil are built up out of quantum mechanical atoms and molecules, which in turn consist of indistinguishable elementary particles, it is not at all obvious that a global transformation which acts only on the particles inside the pencil but not those inside the table is a natural thing to consider. This is of course the same issue as the one we discussed in Section 1.6.4 and again the conclusion is that one has to be careful in identifying the symmetry that is spontaneously broken.

Second, it should be noted that the symmetric state is unstable while the broken-symmetry states are stable, in accordance with the discussion of Section 2.1. In classical physics, the stable, symmetry-broken states are also ground states of the symmetric Hamiltonian, even for finite-sized systems, while the symmetric, balanced state has higher energy.

Finally, the symmetric Hamiltonian describing the pencil cannot account for the direction in which the cylinder will fall. For this, an external perturbation favouring a certain θ is necessary. The spontaneous aspect of the symmetry breaking lies in the fact that this perturbation may be arbitrarily small for a sufficiently sharp pencil, and hence practically undetectable and uncontrollable.

Exercise 2.1 (Classical magnet).
The fact that all symmetry-broken states of, for example, a pencil lying flat on the table are degenerate may seem like an innocent statement, but taken at face value it actually represents a serious conundrum, even within the confines of classical physics. To see this clearly, consider the example of a classical magnet, consisting of many microscopic bar magnets coupled together by nearest-neighbour interactions, so that its internal energy is

$$E = \sum_{\mathbf{x},\delta} -|J|\, \mathbf{S}_{\mathbf{x}} \cdot \mathbf{S}_{\mathbf{x}+\delta}. \tag{2.3}$$

Here $\mathbf{x}$ labels the position of the bar magnet with magnetisation $\mathbf{S_x}$, and $\boldsymbol{\delta}$ connects nearest neighbours. All bar magnets have the same size $|\mathbf{S_x}| \equiv s$. Clearly, the energy of the full magnet can be minimised by having all microscopic bar magnets point in the same direction. It is invariant however, under the simultaneous rotation of all bar magnets around their individual centres. That is, there is a global symmetry which dictates that all states of maximum magnetisation are degenerate, regardless of the direction of total magnetisation. Remember that thermal expectation values may be computed as

$$\langle M \rangle_T = \frac{\sum_{\text{states}} e^{-E(\text{state})/k_B T} M(\text{state})}{\sum_{\text{states}} e^{-E(\text{state})/k_B T}}. \tag{2.4}$$

a. Show that the equilibrium expectation value of the total magnetisation is zero for any temperature, even $T = 0$.

Notice what this implies: classical magnets in thermal equilibrium cannot have a well-defined north or south pole, at any temperature. Clearly this rigorous result is at odds with everyday experience, in which magnets do have a non-zero and permanent magnetisation. The resolution of the paradox of course lies in the spontaneous breakdown of symmetry, which we can describe mathematically by adding a small symmetry-breaking magnetic field to the system:

$$E' = \sum_{\mathbf{x},\boldsymbol{\delta}} -|J| \, \mathbf{S_x} \cdot \mathbf{S_{x+\delta}} - B\hat{n} \cdot \mathbf{S_x}. \tag{2.5}$$

Here, B is the strength of the applied magnetic field, which we will send to zero at the end of the calculation, and the unit vector $\hat{n}$ is its direction.

b. Argue that in this case the expectation value for the magnetisation at zero temperature will be $\langle M \rangle_T = Ns\hat{n}$, where N is the number of microscopic bar magnets.

We now again find a set of non-commuting limits:

$$\begin{aligned} \lim_{N\to\infty} \lim_{B\to 0} \langle M \rangle_T / N &= 0, \\ \lim_{B\to 0} \lim_{N\to\infty} \langle M \rangle_T / N &= s\hat{n}. \end{aligned} \tag{2.6}$$

c. Explain in words what these limits mean for magnets in our everyday world.

Even after breaking their symmetry, real magnets do not thermalise in the way described by Eq. (2.4). The reason for this is that states with different magnetisation, while formally connected via thermal fluctuations, are actually not accessible on ordinary time scales. All microscopic bar magnets being simultaneously rotated over the same angle in a single thermal fluctuation is exceedingly unlikely to occur for large magnets. This is a clear example of a large part of configuration space being effectively inaccessible in a system with a spontaneously broken symmetry.

2.3 The Harmonic Crystal

To see how spontaneous symmetry breaking is realised in the quantum realm, we consider the example of a particularly simple model for a quantum crystal. The Hamiltonian for a collection of atoms with mass m in which neighbouring atoms are held together by harmonic forces with characteristic frequency ω_0, is given by

$$\hat{H} = \sum_{\mathbf{x},\boldsymbol{\delta}} \frac{\hat{\mathbf{P}}^2(\mathbf{x})}{2m} + \frac{1}{2} m\omega_0^2 \left(\hat{\mathbf{X}}(\mathbf{x}) - \hat{\mathbf{X}}(\mathbf{x}+\boldsymbol{\delta})\right)^2. \tag{2.7}$$

Here, $\hat{\mathbf{P}}(\mathbf{x})$ and $\hat{\mathbf{X}}(\mathbf{x})$ are the momentum and position operator of the atom at equilibrium position $\mathbf{x}$, with commutation relations $[\hat{X}_a(\mathbf{x}), \hat{P}_b(\mathbf{x}')] = i\hbar\delta_{ab}\delta(\mathbf{x}-\mathbf{x}')$. The atomic position $\mathbf{x}$ may be in one-, two-, or three-dimensional space with components labelled by a, b, while the connections $\boldsymbol{\delta}$ between interacting atoms may cover nearest neighbours, next-nearest neighbours, or any other interatomic distance. In fact, for the following arguments, even the quadratic form of the potential is not essential, and we could straightforwardly include anharmonic interactions as well. We are thus not just considering an oversimplified model for a hypothetical piece of material, but also the family of Hamiltonians which in principle describes all solids, including the very chair on which you sit.

The crucial point to notice about this Hamiltonian is that it commutes with the operator for total (or centre-of-mass) momentum:

$$\hat{\mathbf{P}}_{\text{tot}} \equiv \sum_{\mathbf{x}} \hat{\mathbf{P}}(\mathbf{x}), \qquad \left[\hat{H}, \hat{\mathbf{P}}_{\text{tot}}\right] = 0. \tag{2.8}$$

The fact that the Hamiltonian for a crystal commutes with the total-momentum operator implies that all eigenstates of the crystal are total-momentum eigenstates. Because of Heisenberg's uncertainty principle, states in which the value of the total momentum can be known with certainty must be states in which the centre-of-mass position is entirely unpredictable. In other words, the total momentum eigenstates, and hence the eigenstates of the crystal Hamiltonian, are wave functions that are spread out over all of space. Clearly this is not the ground state you would expect for a piece of matter that you can hold in your hand. Much less that of an object on which you can safely sit.

To see why objects in our everyday world, with its translationally symmetric energy eigenstates, can in fact be localised, we first write the Hamiltonian in momentum space with wave numbers $\mathbf{k}$, and separate it into two independent parts:

$$\hat{H} = \hat{H}_{\text{CoM}} + \sum_{\mathbf{k}\neq 0} \hat{H}_{\text{int}}(\mathbf{k}). \tag{2.9}$$

The part of the Hamiltonian for non-zero values of $\mathbf{k}$ describes the internal dynamics of the crystal, in terms of all of its phonon (quantised sound) excitations, and their interactions. The centre-of-mass part at $\mathbf{k} = 0$, on the other hand, describes the collective dynamics of the crystal as a whole. It can be straightforwardly shown that in the case of the crystal, the collective part of the Hamiltonian is given by

$$\hat{H}_{\text{CoM}} = \frac{\hat{\mathbf{P}}_{\text{tot}}^2}{2mN}. \tag{2.10}$$

Here, as always, N is the number of particles. For now, we will only consider the properties of the collective part of the Hamiltonian and completely ignore the internal, phonon-related part. This is possible because the two parts of the Hamiltonian commute, which implies that good quantum numbers of one part are also good quantum numbers for the other. Moreover, we will discover shortly that the energies in the spectrum of the collective part of the Hamiltonian are far smaller than even the lowest possible excitation energy of a single phonon. At

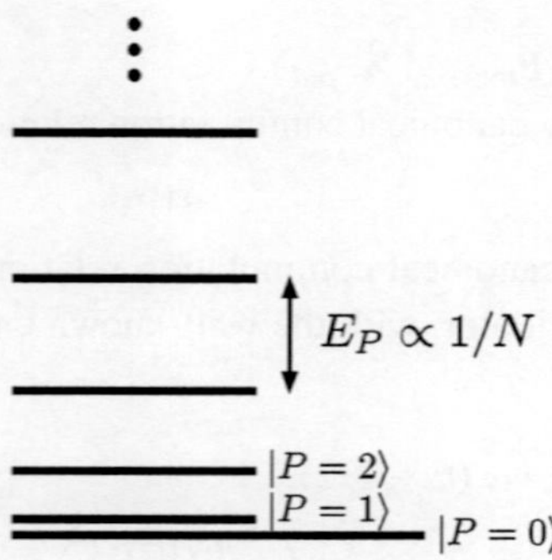

Figure 2.3

Schematic depiction of the tower of states for a harmonic crystal, assuming space has periodic boundary conditions for simplicity.

extremely low temperatures, the collective part of the Hamiltonian is therefore the only part that matters. From here on, we will focus on a one-dimensional system for the sake of simplicity, so we do not need to keep track of any spherical harmonics and other complications. This will not affect the generality of any of the conclusions.

The collective Hamiltonian in Eq. (2.10) is that of a free particle of mass mN, and its eigenstates are total-momentum states, with energies $E_P = P^2/2mN$. The resulting spectrum of collective excitations, the tower of states, is sketched in Figure 2.3. The ground state of the crystal is the state with total momentum $P = 0$. Its wave function is completely and evenly spread out over all of space, with equal amplitude and even equal phase at every possible position. The energy separating the ground state from the collective excitations with non-zero total momentum is inversely proportional to the total mass mN of the whole crystal. This means that as we consider a larger and larger crystal, it becomes easier and easier to make collective excitations. In fact, in the limit $N \to \infty$ of an infinitely large crystal, it would cost no energy at all to make collective excitations, and all states in the collective part of the spectrum become degenerate with the ground state. In that limit then, a wave packet of total momentum states with a well-defined centre-of-mass position would have the same energy expectation value as the zero-momentum state.

Of course real crystals are not infinitely large, and superpositions of momentum states do cost energy to create. But the pieces of matter we are interested in contain a very large, albeit finite, number of atoms. It is therefore reasonable to wonder just how difficult it would be to make a superposition of crystal eigenstates, which has a well-defined centre-of-mass position. In terms of the collective Hamiltonian, this amounts to adding a perturbation that tends to localise the crystal:

$$\hat{H}'_{\text{CoM}} = \frac{\hat{\mathbf{P}}^2_{\text{tot}}}{2mN} + \mu \hat{\mathbf{X}}^2_{\text{CoM}}. \tag{2.11}$$

Here, $\hat{\mathbf{X}}_{\text{CoM}} = 1/N \sum_{\mathbf{x}} \hat{\mathbf{X}}(\mathbf{x})$ is the centre of mass position of the crystal, while μ is the strength of a potential that tends to localise the crystal at the origin of our coordinate system. We will consider μ to be a very small perturbation and take the limit $\mu \to 0$ at the end of our calculation.

Exercise 2.2 (Commutator of $\hat{\mathbf{P}}_{\text{tot}}$ and $\hat{\mathbf{X}}_{\text{CoM}}$).
Show that $\hat{\mathbf{P}}_{\text{tot}}$ and $\hat{\mathbf{X}}_{\text{CoM}}$ obey canonical commutation relations.

Since $\hat{\mathbf{P}}_{\text{tot}}$ and $\hat{\mathbf{X}}_{\text{CoM}}$ obey canonical commutation relations, the perturbed Hamiltonian $\hat{H}'_{\text{CoM}}$ is that of a harmonic oscillator, with the well-known energies and eigenstates (in one dimension):

$$E_n = \hbar\omega\,(n + 1/2),$$
$$\psi_0(x) = \langle x|n = 0\rangle = \left(\frac{2mN\mu}{\pi^2\hbar^2}\right)^{1/8} \mathrm{e}^{-\sqrt{\frac{mN\mu}{2\hbar^2}}x^2}. \tag{2.12}$$

Here, we defined $\omega = \sqrt{\frac{2\mu}{mN}}$. The ground state of the harmonic crystal in the presence of the perturbation of Eq. (2.11) is thus a Gaussian wave packet. Its width in position space, $\sigma^2 = \hbar/\sqrt{2mN\mu}$, decreases as the number of particles in the crystal grows. The occurrence of spontaneous symmetry breaking is again signalled by the appearance of non-commuting limits:

$$\lim_{N\to\infty}\lim_{\mu\to 0} |\psi_0(x)|^2 = \text{constant},$$
$$\lim_{\mu\to 0}\lim_{N\to\infty} |\psi_0(x)|^2 = \delta(x). \tag{2.13}$$

As before, the presence of non-commuting limits is an indication that the thermodynamic limit $N \to \infty$ is singular. For a system of any finite size, no matter how large, the perturbation can be made small enough for the ground state wave function to be essentially spread out over the entire universe. But for an infinitely large system, any perturbation, no matter how weak, is enough to completely localise the wave function in a single position. In the thermodynamic limit, the localisation happens even in the presence of an infinitesimal potential, which in effect means that the wave function localises spontaneously. Notice that the energy of the state does not depend on the order of limits. In both cases, $E_0 = \hbar\omega/2 = 0$. This implies that indeed in the thermodynamic limit the symmetry-broken, localised state of the crystal has the same energy as the exact, plane-wave ground state.

Real materials are not infinitely large, and thus neither of the limits in Eq. (2.13) strictly speaking applies to real pieces of matter. This is not just a practical consideration, but a point of view that is increasingly advocated even in formal mathematical approaches to spontaneous symmetry breaking [104, 105]. The importance of the non-commuting limits is to signal that even if N is not yet truly infinite, the approach towards the thermodynamic limit is singular. This implies in particular that as you consider larger and larger pieces of matter, a weaker and weaker perturbation suffices to make its ground state a localised wave packet. Imagine, for example, a typical 'finite size' iron crystal, with a volume of 1 cm^3. Iron has an atomic mass of $55.8\,u = 9.27 \times 10^{-26}$ kg, and a lattice constant of $a = 2.856 \times 10^{-10}$ m. This means the iron crystal contains approximately $N = 4 \times 10^{22}$ atoms. So how strong does the symmetry-breaking field need to be in order to reasonably localise such a crystal? The units of μ are kg s^{-2} or N m^{-1} (Newton per metre). The weakest possible forces that can currently be measured are of the order of zeptonewtons (10^{-21} N). The weakest possible symmetry-breaking field for our piece of iron that would still be measurable is thus something like a zeptonewton per centimeter, or $\mu \sim 10^{-19}$ N m^{-1}. Even with such an unmeasurably weak perturbation, the width of the crystal's ground state wave function is constrained to be about 2×10^{-12} m. That is, two orders of magnitude smaller than the unit cell of the iron crystal itself. Clearly then, it is practically impossible to exclude perturbations to such an extent that it would allow any sort of everyday-sized object to be found in anything approaching a delocalised momentum eigenstate.

Exercise 2.3 (Elitzur's theorem for an ideal gas).
We stated in Section 1.5.3 that, in contrast to global symmetries, local symmetries cannot be spontaneously broken. To understand why this is the case, consider an ideal gas of N non-interacting particles with positions $\hat{X}_j$ and momenta $\hat{P}_j$, described by

$$\hat{H}_0 = \sum_j \frac{\hat{P}_j^2}{2m}. \tag{2.14}$$

For simplicity, we consider the ideal gas in one dimension only.
a. The ideal gas has a local translation symmetry. Write down the operator $\hat{U}$ which describes local translations and show that the Hamiltonian is invariant under it.
We can try and break the local symmetry by adding a perturbation that acts on one particle only:

$$\hat{H}' = \hat{H}_0 + \frac{1}{2}\kappa \hat{X}_{j=2}^2. \tag{2.15}$$

b. What is the ground state $|\psi\rangle$ of the perturbed Hamiltonian?
c. In the symmetry-broken ground state of the perturbed Hamiltonian, calculate the expectation values of both the unperturbed energy and the uncertainty in position of the perturbed particle:

$$\begin{aligned} \bar{E} &= \langle\psi|\hat{H}_0|\psi\rangle, \\ \Delta X_{j=2}^2 &= \langle\psi|\hat{X}_{j=2}^2|\psi\rangle - \left(\langle\psi|\hat{X}_{j=2}|\psi\rangle\right)^2. \end{aligned} \tag{2.16}$$

d. What are the values of $\bar{E}$ and $\Delta X_{j=2}^2$ in the limit $\kappa \to 0$?
e. Is there any limit that does not commute with the limit $\kappa \to 0$? In particular, does $\kappa \to 0$ commute with the thermodynamic limit $N \to \infty$? What do your answers imply for the local symmetry of $\hat{H}_0$?

Exercise 2.3 is suggestive of the general fact that a continuous local symmetry cannot be spontaneously broken. This result is known as **Elitzur's theorem** [50], and it has the status of a mathematical theorem. Its only assumptions are that the physics of the system is described by a minimum action principle, and that physically relevant perturbations are described by (sums of) operators with local support. These assumptions are satisfied for all presently known theories of physics. The central ingredient needed to prove the theorem is the realisation that for locally symmetric systems, there cannot exist a singular limit of the kind we encountered for global symmetries. Systems with a local symmetry therefore lack the instability that comes with global symmetry, and they are always robust against small perturbations. This is not a result that can be negotiated with by 'smart engineering' of a model or system. Spontaneously breaking a continuous local symmetry is just impossible.

2.4 The Heisenberg Antiferromagnet

A second instructive example of spontaneous symmetry breaking in quantum physics, is that of the Heisenberg antiferromagnet. Consider a cubic lattice with a spin-half degree of freedom on every site and isotropic interactions between neighbouring spins:

$$\hat{H} = J \sum_{\langle i,j \rangle} \hat{\mathbf{S}}_i \cdot \hat{\mathbf{S}}_j. \tag{2.17}$$

Here, $\hat{\mathbf{S}}_i = \hbar \sum_{s,s'} |s_i = s\rangle \, \boldsymbol{\sigma}_{s,s'} \, \langle s_i = s'|$ is the spin operator on site i, represented in the basis of spin-z states $|s_i \in \pm\rangle$ by the Pauli matrices $\boldsymbol{\sigma}_{s,s'}$. They obey the usual $SU(2)$ commutation relations $[\hat{S}_i^a, \hat{S}_j^b] = \mathrm{i}\hbar\delta_{ij}\epsilon_{abc}\hat{S}_i^c$. The indices i and j label all sites of a cubic lattice, and $\langle i,j \rangle$ indicates that we sum over nearest-neighbour pairs of spins only. Expanding the product of spin vectors, the Hamiltonian becomes

$$\begin{aligned} \hat{H} &= J \sum_{\langle i,j \rangle} \hat{S}_i^z \hat{S}_j^z + \hat{S}_i^y \hat{S}_j^y + \hat{S}_i^x \hat{S}_j^x \\ &= J \sum_{\langle i,j \rangle} \hat{S}_i^z \hat{S}_j^z + \frac{1}{2}\left(\hat{S}_i^+ \hat{S}_j^- + \hat{S}_i^- \hat{S}_j^+\right). \end{aligned} \tag{2.18}$$

The operators $\hat{S}_i^{\pm} = \hat{S}_i^x \pm \mathrm{i}\hat{S}_i^y$ are the spin raising and lowering operators. The strength of the interaction between neighbouring spins is denoted by the coupling constant J. If it is negative, the ground state energy is minimised whenever neighbouring spins point in the same direction. The ground state is then a ferromagnet, with all spins in the entire material pointing in the same direction, which is spontaneously chosen. We will come back to the very special case of spontaneous symmetry breaking in ferromagnets in Exercise 3.3.

For positive J, the energy is lowered by neighbouring spins being anti-parallel. If we divide the cubic lattice into two sublattices, as shown in Figure 2.4, then all nearest-neighbour pairs of spins can be made anti-parallel by choosing all spins on the A-sublattice to point up and all spins on the B-sublattice to point down. Such a perfect **antiferromagnetic** arrangement of spins is known as a **Néel state**. Again, the axis along which the spins point either up or down is spontaneously chosen. Global rotations of the spins around this axis remain symmetry transformations in the Néel state (they are 'unbroken'), while global spin rotations around any other axes are spontaneously broken.

If the spins in Eq. (2.17) were classical bar magnets, the Néel state would have been an exact ground state of the antiferromagnet. The quantum mechanical Hamiltonian, however, contains terms like $\hat{S}_i^+ \hat{S}_j^-$, of which the Néel state is not an eigenstate. The Néel state can therefore not be the ground state of the Heisenberg Hamiltonian. In fact, for a three dimensional lattice, it is still an open question what the precise ground state of the antiferromagnetic Heisenberg Hamiltonian is.

The Hamiltonian of Eq. (2.17) is invariant under global, simultaneous rotations of all spins around any axis. You can check that the Hamiltonian indeed commutes with the generators $\hat{S}^a = \hat{S}^a_{\text{tot}} = \sum_i \hat{S}_i^a$ of spin-rotations around all axes $a \in x, y, z$. It follows that the Heisenberg Hamiltonian also commutes with the operator for total spin $\hat{S}^2 = \hat{\mathbf{S}} \cdot \hat{\mathbf{S}}$.

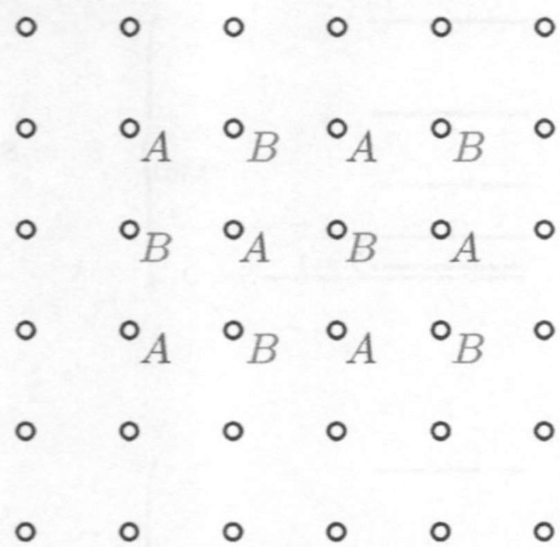

Figure 2.4
A square (or hypercubic) lattice can be divided into two inequivalent sublattices, *A* and *B*. In a perfect Néel antiferromagnet, the spins on the two sublattices point in opposite directions.

Exercise 2.4 (Noether current of the Heisenberg magnet).
The Heisenberg Hamiltonian of Eq. (2.17) is not in canonical form. That is, it is not expressed in term of operators and their conjugate momenta. We then cannot perform a Legendre transformation to obtain a Lagrangian and action, and we cannot use the method of Section 1.3.3 to obtain the Noether current. However, we already have the global symmetry generators $\hat{Q}^a = \sum_i \hat{S}^a_i$, expressed as volume 'integrals' over what must be the Noether charge densities $j^t_a(i) = \langle \hat{S}^a_i \rangle$. We can then obtain the corresponding conservation law by calculating their time derivatives explicitly.
a. Calculate the Heisenberg equation of motion $\partial_t \langle \hat{S}^a_i \rangle = \frac{i}{\hbar} \langle [\hat{H}, \hat{S}^a_i] \rangle$, using the spin commutation relations.
The vector connecting a point i on a lattice with a neighbouring point $i+\delta$ may be written as $v_{i,i+\delta}$. In this notation, the (lattice) divergence of a vector at i is given by $\sum_\delta v_{i,i+\delta}$.
b. Show that the right-hand side of the equation of motion has the form of the lattice divergence of a vector.
The vector in part b corresponds to the spatial Noether current j^m_a, and the equation of motion in part a is the lattice version of Noether's theorem $\partial_t j^t_a + \partial_m j^m_a = 0$.

2.4.1 The Symmetric Ground State

Since the Hamiltonian is invariant under global rotations of all spins, you might expect that, as in the case of a crystal, it will be sufficient to consider the collective part of the Hamiltonian in order to see if its symmetry can be spontaneously broken. This is true, but in the case of the antiferromagnet, we do need to be careful about what this collective part is precisely. Because the A and B sublattice are different, there is a collective mode which rotates all spins on the A sublattice, while the ones on the B sublattice are left invariant, and the other way around. These two modes correspond to two components in the Fourier transform of the spin rotation operators, rather than only the one component that defined the collective motion of the crystal. Together, the two Fourier components make up the collective part of the Heisenberg Hamiltonian in Eq. (2.17):

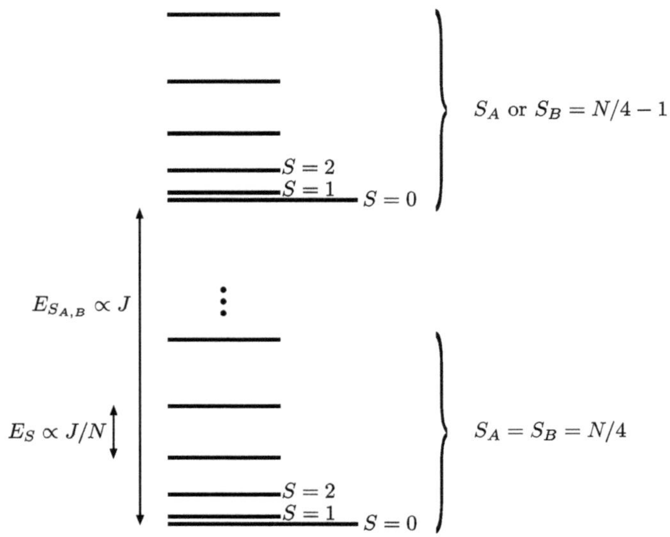

Figure 2.5 The states with lowest energies in the Lieb–Mattis Hamiltonian.

$$\hat{H} = \hat{H}_{\text{coll}} + \sum_{\mathbf{k} \neq 0, \pi/a} \hat{H}(\mathbf{k}), \qquad \hat{H}_{\text{coll}} = \frac{J}{N} \hat{\mathbf{S}}_A \cdot \hat{\mathbf{S}}_B. \tag{2.19}$$

This collective Hamiltonian is known as the Lieb–Mattis model. Here, the collective operator for the total spin on the A sublattice is defined as $\hat{\mathbf{S}}_A = \sum_{i \in A} \hat{\mathbf{S}}_i$, and similarly for $\hat{\mathbf{S}}_B$. The non-collective, internal modes that make up $\hat{H}(\mathbf{k})$ are the spin-wave analogue to the phonons in a crystal and are known as **magnons**.

The eigenstates of the collective Hamiltonian can be easily found by re-writing it in terms of the total spin operator $\hat{\mathbf{S}} = \hat{\mathbf{S}}_A + \hat{\mathbf{S}}_B$:

$$\hat{H}_{\text{coll}} = \frac{J}{2N} \left(\hat{S}^2 - \hat{S}_A^2 - \hat{S}_B^2 \right). \tag{2.20}$$

Because all operators in this expression commute, we can immediately write the eigenstates and the corresponding energies in terms of the quantum numbers for total spin and total spin on the sublattices:

$$\begin{aligned} \hat{H}_{\text{coll}} \left| S_A, S_B, S, S^z \right\rangle &= E(S, S_A, S_B) \left| S_A, S_B, S, S^z \right\rangle, \\ E(S, S_A, S_B) &= \frac{J\hbar^2}{2N} \left(S(S+1) - S_A(S_A+1) - S_B(S_B+1) \right). \end{aligned} \tag{2.21}$$

The spectrum of low-energy eigenstates is shown schematically in Figure 2.5. The ground state is the state with S_A and S_B equal to their maximal values of $N/4$, and the total spin S equal to zero. Because $S = 0$ implies that $\langle S^x \rangle = \langle S^y \rangle = \langle S^z \rangle = 0$, the ground state does not break spin-rotation symmetry. It is a unique, non-degenerate state in which the spins on each sublattice are all exactly aligned with each other and anti-aligned with neighbours on a different sublattice, but in which no direction in space is different from any other. You can think of the total spin singlet ground state as a superposition of infinitely many Néel states, pointing in all possible directions. Clearly, this is a highly non-local state.

2.4.2 The Symmetry-Broken State

Flipping one of the individual spins on, for example, the A sublattice would decrease S_A by one. Such an excitation costs an energy of order J to create, and if the temperature is low enough, these excitations will not be present. It is also possible, however, to make collective excitations, which increase the value of the total spin S by one while keeping the values of S_A and S_B fixed. Such excitations cost an energy that is only proportional to J/N, and for a large enough system, these excitations can never be neglected. In fact, in the limit of an infinitely large system, with $N \to \infty$, there is a again a tower of states labelled by different values of S, which all become degenerate with the ground state.[3] This suggests that again, in the thermodynamic limit even an infinitesimally small perturbation might have a qualitative effect and spontaneously break the spin-rotational symmetry.

The perturbation or symmetry-breaking field to consider in this case is a field that prefers the spins to be arranged into a Néel state. As before, the reasoning is that in principle we should consider all possible perturbations and find that none of them have any effect in the limit where they are infinitesimally small, except for the one that happens to stabilise the classically expected ground state. Since we already know what a classical antiferromagnet looks like, we will straight away consider the only relevant perturbation:

$$\hat{H}'_{\text{coll}} = \frac{J}{N}\hat{\mathbf{S}}_A \cdot \hat{\mathbf{S}}_B - \mu\left(\hat{S}^z_A - \hat{S}^z_B\right). \tag{2.22}$$

Since the perturbation breaks spin-rotational symmetry, it does not commute with the original Hamiltonian. The states $|S_A, S_B, S, S^z\rangle$ are therefore no longer eigenstates of the perturbed Hamiltonian. In order to find the new eigenstates, we need to calculate the matrix elements of the perturbed Hamiltonian in the basis of the old eigenstates. Because the excitations that change S_A and S_B cost far more energy than an infinitesimal perturbation is expected to provide, we can safely assume that $S_A = S_B = N/4$ throughout. And because the quantum number $S^z = S^z_A + S^z_B$ is a good quantum number even for the perturbed Hamiltonian, we can additionally fix $S^z = 0$. The only remaining relevant quantum number is then the total spin S, and we can abbreviate the eigenstates of the original Hamiltonian by writing $|S_A = S_B = N/4, S, S^z = 0\rangle \equiv |S\rangle$. Calculating the matrix elements of the symmetry-breaking field in this basis involves some rather tedious exercises in the addition of angular momentum, but the result is known to be [201]

$$\begin{aligned}\langle S'|\hat{S}^z_A - \hat{S}^z_B|S\rangle &= \delta_{S',S-1}\, S\sqrt{\frac{(N/2+1)^2 - S^2}{4S^2-1}} + \delta_{S',S+1}\, S'\sqrt{\frac{(N/2+1)^2 - S'^2}{4S'^2-1}} \\ &\approx \left(\delta_{S',S-1} + \delta_{S',S+1}\right)\frac{N}{4}.\end{aligned} \tag{2.23}$$

In the final line, we simplified the matrix element by using the fact (to be confirmed shortly) that only states with $1 \ll S \ll N$ contribute to the perturbed ground state.

The Hamiltonian in the $|S\rangle$ basis can now be explicitly written as

$$\hat{H}'_{\text{coll}} \approx \sum_S \frac{J\hbar^2}{2N}S^2\,|S\rangle\,\langle S| - \mu\frac{N}{4}\,|S\rangle\,\langle S+1| - \mu\frac{N}{4}\,|S+1\rangle\,\langle S|, \tag{2.24}$$

[3] In this context, it is also referred to as the **Anderson tower of states** after P. W. Anderson, who argued its existence already in 1952 [6].

where again we used the assumption $1 \ll S \ll N$ to simplify the diagonal term, and we left out the constant contribution to the energy from the S_A and S_B quantum numbers. We can simplify the expression further by writing the eigenstates of the perturbed Hamiltonian as $|n\rangle = \sum_S \psi_n(S) |S\rangle$ and by approximating S to be a continuous variable and taking the continuum limit in the Schrödinger equation:

$$
\begin{aligned}
\langle S|\hat{H}|n\rangle &= \langle S| E_n |n\rangle, \\
\frac{J\hbar^2}{2N} S^2 \psi_n(S) - \frac{\mu N}{4} \left(\psi_n(S+1) + \psi_n(S-1)\right) &= E_n \psi_n(S), \\
\Rightarrow \quad \left(\frac{J\hbar^2}{2N} S^2 - \frac{\mu N}{2}\right) \psi_n(S) - \frac{\mu N}{4} \frac{\partial^2 \psi_n(S)}{\partial S^2} &= E_n \psi_n(S), \\
-\frac{1}{2} \frac{\partial^2 \psi_n(S)}{\partial S^2} + \frac{1}{2} \omega^2 S^2 \psi_n(S) &= \epsilon_n \psi_n(S).
\end{aligned} \tag{2.25}
$$

In the third line, we used the discrete version of the second derivative to take the continuum limit, while in the fourth line we defined the variables $\omega^2 = 2J\hbar^2/(\mu N^2)$ and $\epsilon_n = (2E_n + \mu N)/(\mu N)$.

The final expression is just the Schrödinger equation for a harmonic oscillator with $m = \hbar = 1$, whose eigenstates and energies are well known. We should keep in mind the subtlety that in the present problem, the quantum number S can only be positive. We should therefore only accept solutions of Eq. (2.25), which have a node at the origin. That is, the eigenstates of the perturbed antiferromagnet are the (positive S part of the) harmonic oscillator eigenstates with odd values for the quantum number n, and the eigenvalue corresponding to the lowest allowed state is $\epsilon_1 = (1 + 1/2)\omega$. The ground state of the perturbed antiferromagnet is therefore the state $|n = 1\rangle$, with ground state energy:

$$
E_1 = \frac{\mu N}{2}\left(\frac{3}{2}\omega - 1\right) = \frac{3}{4}\sqrt{2\mu J\hbar^2} - \frac{\mu N}{2}. \tag{2.26}
$$

From the ground state energy, we can understand what happens in the thermodynamic limit. If we first take the limit $\mu \to 0$, then the energy of the perturbed system is just equal to the energy of the unperturbed system, and the symmetric ground state is unaffected. However, if we first take the limit $N \to \infty$ while keeping μ non-zero (but infinitesimally small), then the first term in the energy can be neglected, and we find $E_1 = -\mu N/2$, with $N \to \infty$. But this is precisely the maximum amount of energy that can only possibly by gained by having all spins on the A sublattice point upwards, and all spins on the B sublattice point downwards. The ground state in this limit must thus approach the Néel state, and the spin-rotational symmetry is spontaneously broken.

An alternative way of seeing the broken symmetry is to calculate the expectation value of the difference in magnetisation between the two sublattices in the thermodynamic limit:

$$
\begin{aligned}
\lim_{N\to\infty} \lim_{\mu\to 0} \langle n = 1| \hat{S}^z_A - \hat{S}^z_B |n = 1\rangle / N &= 0, \\
\lim_{\mu\to 0} \lim_{N\to\infty} \langle n = 1| \hat{S}^z_A - \hat{S}^z_B |n = 1\rangle / N &= \frac{1}{2}.
\end{aligned} \tag{2.27}
$$

Again, the final line indicates that for a large enough piece of antiferromagnetic material, even an infinitesimally small symmetry-breaking field suffices to completely align all of the spins. The spin-rotational symmetry of the Heisenberg Hamiltonian is thus spontaneously broken in the thermodynamic limit.

2.5 Bose–Einstein Condensation and Superfluidity

The third pedagogical example of how spontaneous symmetry breaking works is mathematically the simplest, but conceptually the most counterintuitive: the breaking of $U(1)$ phase symmetry. In Section 1.4.3, we showed that in a many-body theory, global phase rotation symmetry corresponds to the conservation of particle number. Surely, the conservation of particles cannot be spontaneously broken?

In fact, it can, and we must emphasise that this is not in any way stranger than breaking spin-rotational symmetry in a magnet (Section 2.4), or translational symmetry in a crystal (Section 2.3). Just as in these two cases, something special happens with the $\mathbf{k} = 0$ part of the Hamiltonian, which allows even infinitesimal external influences to qualitatively alter the ground state in the thermodynamic limit.

2.5.1 Thermodynamics of Free Bosons

That something unusual should happen with the particle number can be seen already in the simplest model of non-interacting bosons in thermal equilibrium, described by

$$\hat{H} = \sum_{\mathbf{k}} (\epsilon_{\mathbf{k}} - \mu)\hat{b}^{\dagger}_{\mathbf{k}}\hat{b}_{\mathbf{k}}. \tag{2.28}$$

Here $\hat{b}^{\dagger}_{\mathbf{k}}$ and $\hat{b}_{\mathbf{k}}$ are the boson creation and annihilation operators, and $\epsilon_{\mathbf{k}} = \hbar^2 k^2/(2m)$ is the dispersion relation. The presence of the chemical potential μ indicates that we consider the grand canonical ensemble, in which particles can be exchanged with an external reservoir. The value of the chemical potential μ should be chosen at any given temperature T such that the ensemble-averaged total number of bosons equals a fixed value N. In thermal equilibrium, the occupation of each individual state with momentum $\mathbf{k}$ is then given by the Bose–Einstein distribution function, $N_{\mathbf{k}} = 1/[\mathrm{e}^{(\epsilon_{\mathbf{k}}-\mu)/T} - 1]$.

Since $N_{\mathbf{k}}$ should be positive and finite for all $\mathbf{k}$ in order to have a finite total number of particles N, the chemical potential always has to be negative, $\mu < \epsilon_{\mathbf{k}=0} = 0$. Taking this into account, the highest possible density of particles that can be obtained in thermal equilibrium is given by

$$n_{\max}(T) = \lim_{\mu \to 0^-} \int \frac{d^3k}{(2\pi)^3} N_{\mathbf{k}} = \int \frac{d^3k}{(2\pi)^3} \frac{1}{e^{\hbar^2 k^2/2mT} - 1} = \left(\frac{mT}{2\pi\hbar^2}\right)^{3/2} \zeta(\tfrac{3}{2}). \tag{2.29}$$

Here, ζ is the Riemann zeta function. In the low-temperature limit, this maximum density vanishes. It thus becomes impossible to describe the equilibrium configuration as an ensemble of thermally occupied momentum states for any non-zero total density of bosons. What happens instead is that a system with fixed boson density n undergoes a phase transition as soon as $n = n_{\max}(T)$, or

$$T = T_{\mathrm{BEC}}(n) = \frac{2\pi\hbar^2}{m}\left(\frac{n}{\zeta(\frac{3}{2})}\right)^{2/3}. \tag{2.30}$$

Below T_{BEC}, the bosonic occupation numbers deviate from the standard Bose–Einstein statistics, and a **Bose–Einstein condensate** (BEC) is formed, in which the $\mathbf{k} = 0$ state is occupied by a macroscopic density $n_0 = N_0/V > 0$ of bosons.

2.5.2 Phase Coherence

To understand how the Bose–Einstein condensate is formed, it is instructive to first consider an isolated system of free bosons, with a fixed total number of particles, and no external bath to exchange any energy or particles with

$$\hat{H} = \sum_{\mathbf{k}} \epsilon_{\mathbf{k}} \hat{b}^{\dagger}_{\mathbf{k}} \hat{b}_{\mathbf{k}} . \tag{2.31}$$

The exact ground state of this Hamiltonian has all N bosons in the lowest energy single-particle state: $(\hat{b}^{\dagger}_0)^N |\text{vac}\rangle$. This state has a precise number of particles and is symmetric under global phase rotations, $\hat{b}^{\dagger}_{\mathbf{k}} \to e^{i\alpha} \hat{b}^{\dagger}_{\mathbf{k}}$.

Since the energy $\epsilon_{\mathbf{k}}$ at $\mathbf{k} = 0$ vanishes, it does not cost any energy to add or remove a particle in the ground state. However, for the system to explore states with different numbers of particles, it needs to be connected to a reservoir. We can consider the effect of interacting with such a reservoir explicitly by focussing on the collective $\mathbf{k} = 0$ part of the Hamiltonian only, and writing

$$\hat{H}_0 = -\mu \hat{b}^{\dagger}_0 \hat{b}_0 - \lambda \sqrt{V} \left(\hat{b}^{\dagger}_0 e^{i\phi} + \hat{b}_0 e^{-i\phi} \right) . \tag{2.32}$$

Here, λ is a small parameter describing the tunnelling amplitude of a particle between the system and the reservoir, and μ is the chemical potential induced by the reservoir, which determines the average density of particles at any given reservoir temperature. The volume V of the system is explicitly included here to ensure that the energy is extensive.

The constant phase ϕ in Eq. (2.32) indicates that the reservoir we consider is a coherent source of bosons. This could describe, for example, the coherent simulated emissions of photons in a laser source. Just as we needed an external magnetic field to break the spin rotation symmetry of the antiferromagnet, or a localised interaction potential to break the translational symmetry of the crystal, here we need a phase-coherent source of particles to break the phase rotation symmetry in the Bose–Einstein condensate. The Hamiltonian of Eq. (2.32) is thus entirely equivalent in structure to Eq. (2.22) (for the antiferromagnet) or Eq. (2.11) (for the crystal).

The ground state for the bosonic system in the presence of a coherent reservoir can be found by a transformation to the new operators $\hat{\beta}_0$, defined as

$$\hat{b}_0 = \hat{\beta}_0 - \frac{\lambda}{\mu} \sqrt{V} e^{i\phi} . \tag{2.33}$$

In terms of these new bosons, the Hamiltonian of Eq. (2.32) becomes

$$\hat{H}_0 = -\mu \hat{\beta}^{\dagger}_0 \hat{\beta}_0 + \frac{\lambda^2}{\mu} V . \tag{2.34}$$

While the equilibrium state expressed in terms of $\hat{\beta}_0$ particles is occupied according to the standard Bose–Einstein distribution, the original $\hat{b}_0$ operators obtain a non-zero expectation value:

$$\langle \hat{b}_0 \rangle = -\frac{\lambda}{\mu} \sqrt{V} e^{i\phi} . \tag{2.35}$$

It can be clearly seen that this expectation value in the ground state of Eq. (2.32) no longer has the phase rotation symmetry $\hat{b}^{\dagger}_{\mathrm{k}} \to e^{i\alpha} \hat{b}^{\dagger}_{\mathrm{k}}$. It is also clear that the thermodynamic limit

$V \to \infty$ in this expression does not commute with the limit of vanishing symmetry-breaking field λ.

As in the case of antiferromagnets and crystals, it is the presence of a small but non-vanishing perturbation that breaks the symmetry. In this case, the perturbation is provided by the coupling to a reservoir of coherent particles. Since the Noether current associated with $U(1)$ symmetry is the conservation of particle number, the Bose–Einstein condensate represents an equilibrium configuration in which particle number is not conserved.

In all occasions in nature where bosons without mutual interactions form a Bose–Einstein condensate that does not conserve the number of particles, such as in a laser, the phase rotation symmetry is broken explicitly by coupling to a reservoir that allows particle exchange and drives the condensation. Reaching a steady state with an average number of particles N_0 in the $\mathbf{k} = 0$ state then requires the ratio of coupling constants $-\lambda/\mu = \sqrt{N_0/V}$. The resulting accumulation of particles in the $\mathbf{k} = 0$ state is often referred to as a 'macroscopic occupation of the ground state'.

2.5.3 Interactions

Having seen explicit breaking of the phase rotation symmetry, one may wonder if the conservation of particle number can also be broken spontaneously, with vanishing coupling to a reservoir? For this, an additional ingredient is needed, namely (at least) weak interactions between the bosons. This makes intuitive sense, since if the particles do not interact at all, they could not possibly form a collective, rigid state that acts as a single whole to break a global symmetry. Also, in all situations in nature where a non-zero density of bosons occurs in equilibrium, such as in a container of ^{4}He atoms, these particles interact. The simplest interaction term we can consider is of the form

$$\hat{H}_{\text{int}} = \nu \sum_{\mathbf{x}} \hat{b}^\dagger_{\mathbf{x}} \hat{b}^\dagger_{\mathbf{x}} \hat{b}_{\mathbf{x}} \hat{b}_{\mathbf{x}} = \nu \sum_{\mathbf{x}} (\hat{b}^\dagger_{\mathbf{x}} \hat{b}_{\mathbf{x}})^2 - \hat{b}^\dagger_{\mathbf{x}} \hat{b}_{\mathbf{x}} . \tag{2.36}$$

This describes a local repulsion between bosons, or an energy cost whenever more than one boson occupies a single site. The part of the interaction that only involves $\mathbf{k} = 0$ operators is $(\nu/V)((\hat{b}^\dagger_0 \hat{b}_0)^2 - \hat{b}^\dagger_0 \hat{b}_0)$. The presence of interactions thus lifts the degeneracy of states with different numbers of particles in the zero-momentum state. Instead, we find a tower of collective low-energy states, just as in the crystal of Section 2.3 and the antiferromagnet of Section 2.4. Each of the energy eigenstates in the tower contains a precise number of particles and is invariant under the U(1) phase rotation symmetry.

To describe the spontaneous breakdown of symmetry, we again consider an explicit symmetry-breaking field and consider the limit of it having vanishing strength. As before, the symmetry-breaking field can be provided by allowing particle exchange with an external reservoir of coherent particles:

$$\hat{H}_0 = \frac{\nu}{V}((\hat{b}^\dagger_0 \hat{b}_0)^2 - \hat{b}^\dagger_0 \hat{b}_0) - \mu \hat{b}^\dagger_0 \hat{b}_0 - \lambda \sqrt{V} \left(\hat{b}^\dagger_0 \, e^{i\phi} + \hat{b}_0 \, e^{-i\phi} \right) . \tag{2.37}$$

The coupling to the reservoir introduces both a tunnelling amplitude λ and a chemical potential μ, both of which tend to zero as the coupling to the reservoir vanishes.

Unlike the collective parts of the crystal and antiferromagnetic Hamiltonians, finding a closed form expression for the eigenstates of Eq. (2.37) is not straightforward. It is already clear, however, that the final terms proportional to λ cause transitions between the number states that are eigenstates of the term proportional to ν. By taking the thermodynamic limit $V \to \infty$ before letting λ and μ vanish, the coupling to the reservoir will dominate the Hamiltonian. Its ground states will then be well approximated by the symmetry-breaking state of Eq. (2.35), so that we may expect the symmetry to be broken spontaneously.

A good variational ansatz for the full Hamiltonian turns out to be the so-called **coherent state**:

$$|\mathrm{coh}\rangle = \mathrm{e}^{-\frac{|z|^2}{2}} \sum_n \frac{z^n}{\sqrt{n!}} |n\rangle . \tag{2.38}$$

Here, z is a complex number with phase ϕ determined by the coupling to the coherent reservoir. It can be straightforwardly confirmed that this state minimises the energy if its amplitude is proportional to $\sqrt{N_0}$, so that $\langle \hat{b}_0^\dagger \hat{b}_0 \rangle \propto N_0$. At the same time, the phase of the BEC order parameter $\langle \hat{b}_0 \rangle$ will be ϕ, equal to that of the reservoir. This state thus exhibits broken phase rotation symmetry.

The difference between the model in the presence of interactions $\propto \nu$ and without does not lie in the ground state, which breaks particle conservation and singles out a particular phase value both with and without interactions. Rather, the difference lies in the excitations that one can make on top of the ground state. For the non-interacting boson gas, every single particle can be excited on its own, without affecting any other boson. Maintaining a BEC in that case requires constant application of the symmetry-breaking field, because without it, single-particle excitations will destroy the phase coherence. In the interacting BEC on the other hand, the lowest energy excitations are collective and retain the overall phase coherence. The spontaneously symmetry-broken state will then persist in the thermodynamic limit, even for vanishing coupling strength to the reservoir.

The fact that particle number conservation may be spontaneously violated might seem counterintuitive, but it is not different in principle from the breaking of translational or rotational symmetry that we encountered in Sections 2.3 and 2.4. In particular, in order to obtain a BEC with broken phase rotation symmetry and thus broken number conservation, we need a spectrum of symmetric low-energy states, a singular thermodynamic limit, and coupling to a system that already has a coherent global phase. The outside system could be provided by a continuously driven source, as in a laser, or by weak coupling to a second BEC. Either way, the outside system needs to provide a coherent source of particles that already have broken phase rotation symmetry. This is directly equivalent to the need for a symmetry-broken perturbation to obtain the spontaneously broken symmetries of crystals and antiferromagnets that we saw before. For the antiferromagnet, coupling to either an applied magnetic field or a second antiferromagnet provides a source of coherent quanta of magnetisation, which allows breaking of the spin rotation symmetry. The crystal, meanwhile, requires coupling to a potential energy or second crystal to exchange momentum with. We are so used to seeing sources of momentum all around us in daily life (e.g., you and I obviously break translational and rotational symmetry), that we hardly find it surprising to see things localised in space. We should not let the exotic sound of a 'coherent source of particles' turn us off from phase rotation symmetry breaking, however. Conceptually, it is all the same. Quantitatively, the combination of the size of a BEC and the strength of its coupling to a boson source will determine the extent to which the phase can be fixed. In practice, this may be less pronounced (with more fluctuations) for many BECs than it is for the fixing of positions in everyday objects, but nonetheless, the mechanics of symmetry breaking is the same in both cases.

2.5.4 Superfluidity

Having found a spontaneously broken phase rotation symmetry for the BEC, we can consider situations in which its phase ϕ varies with position. Following Eq. (1.21), the Noether current only depends on the gradient of the symmetry-breaking phase of the BEC,

$$\vec{j} = \frac{\hbar^2}{m}\vec{\nabla}\phi. \tag{2.39}$$

Non-uniform configurations of the phase of the condensate thus carry currents of bosons. As we will explain in detail in Chapter 4, phase variations associated with the spontaneously broken symmetry give rise a so-called Nambu–Goldstone mode, which is protected from scattering in the limit of small momenta. Consequently, the current carried by the Nambu–Goldstone mode becomes dissipationless. The prime hallmark of Bose–Einstein condensation is the existence of such **superfluidity**, meaning currents can flow without resistance. The current of Eq. (2.39) is therefore called a **supercurrent**.

Superfluidity was first discovered in 1937, by noticing that Helium-4 below the temperature $T_{\mathrm{BEC}} = 2.17$ K started creeping out of its container without apparent friction. More recently, in 1995, BECs were realised in extremely cold dilute atom gases, leading to a Nobel Prize in 2001. At the same time, we now recognise that superconductors and superfluid Helium-3 are akin to Bose–Einstein condensates of fermion pairs, as will be extensively discussed in Chapter 12.

3 The Zen of Symmetry Breaking

In Chapter 2, we became acquainted with three paradigmatic examples of spontaneous symmetry breaking (SSB): the harmonic crystal, the Heisenberg antiferromagnet, and the Bose–Einstein condensate. They all had in common a singular limit encountered upon including a weak symmetry-breaking perturbation. The theoretical limits exhibiting singular behaviour, however, are nearly impossible to realise in practice. The thermodynamic limit requires an infinitely large system and tuning the external perturbation to zero requires perfectly isolation. The true 'Zen' of spontaneous symmetry breaking lies in the fact that its concepts already apply to small, dirty, non-isolated systems. To see SSB in practice, we never need to actually realise a singular limit, and the concepts of SSB allow us to classify, quantify and realise symmetry-broken states in real-life systems.

In this chapter, we therefore introduce aspects of SSB that apply in realistic settings. We start by classifying broken-symmetry states (Section 3.1), which requires the concept of an order parameter (Section 3.2) and the corresponding classical states (Section 3.3). Correlations in the symmetry-broken state are defined by long-range order (Section 3.4), and any symmetry breaking can be detected using the Josephson effect (Section 3.5). Finally, we show that SSB occurs in systems of finite size because of the presence of a tower of states (Section 3.6) and a corresponding instability of symmetric configurations (Section 3.7).

3.1 Classification of Broken-Symmetry States

For mathematical convenience, we will from here on consider only Hamiltonians and broken-symmetry states that have some degree of translational invariance. It is sufficient for the invariance to be discrete. That is, we consider states with a periodic arrangement of unit cells on a regular lattice. The examples of the crystal and the antiferromagnetic Néel state clearly fall in this class. On a course-grained level, it also applies more generally to almost any symmetry-breaking system of interest, which is sufficient for our present purposes. Translational invariance allows for a Fourier transformation with well-defined wave numbers, which is a prerequisite for the introduction of Nambu–Goldstone modes in Chapter 4. It also ascertains that we can write the broken-symmetry state as a product state of the form $|\psi\rangle = \otimes_i |\psi_i\rangle$, with i labelling unit cells and $|\psi_i\rangle$ the same local wave function in each of them. If the translational symmetry happens to be continuous, rather than discrete, the index i may be replaced by a continuous parameter x. In the following, we will use i and x interchangeably.

Just the assumption of translational invariance is sufficient to guarantee that distinct symmetry-breaking states are orthogonal in the thermodynamic limit. To see this, consider the overlap between two normalised broken-symmetry states $|\psi\rangle$ and $|\psi'\rangle$:

$$\begin{aligned}\langle\psi'|\psi\rangle &= \left(\otimes_{x'}\langle\psi'(x')|\right)\left(\otimes_x|\psi(x)\rangle\right)\\ &= \prod_x \langle\psi'(x)|\psi(x)\rangle = \langle\psi'(0)|\psi(0)\rangle^N.\end{aligned} \tag{3.1}$$

Here, we used the vanishing overlap between local states $|\psi(x)\rangle$ and $|\psi(x')\rangle$ for $x \neq x'$ to write the overlap of product states as a product of local overlaps. In the final step, we also used translational invariance by taking the local states to be independent of position so that $|\psi(x)\rangle = |\psi(0)\rangle \; \forall x$. The power N denotes the number of unit cells. If the two symmetry-breaking states $|\psi\rangle$ and $|\psi'\rangle$ are distinct, then the local overlaps must be smaller than one, and the full inner product $\langle\psi'|\psi\rangle$ vanishes as N is taken to infinity. On the other hand, if the states are not distinct and differ only by a phase $|\psi'(x)\rangle = e^{i\varphi}|\psi(x)\rangle$, the inner product equals one, up to a phase factor. We thus find that any two symmetry-breaking states in the thermodynamic limit must be either equivalent or orthogonal.

At this point, we can apply the group theory of Section 1.7.1 to begin the classification of all possible distinct symmetry-breaking states. Recall that symmetry transformations make up a group G, with elements g. As we saw before, a specific symmetry transformation g is represented in quantum physics by a unitary operator acting on Hilbert space. With a slight abuse of notation and to avoid introducing too many symbols, we will write g both for the element in the group of symmetry transformations and for the operator it represents. We can then consider a state $|\psi\rangle$ that breaks some of the symmetries in the group G, but is invariant under others. For symmetry transformations g that are broken, $|\psi\rangle$ and $g|\psi\rangle$ are distinct, inequivalent states. Assuming some degree of translational invariance, they must then be orthogonal in the thermodynamic limit. For unbroken symmetry transformations g, on the other hand, we know from the definition of a symmetric state in Eq. (1.1) that $g|\psi\rangle = e^{i\varphi}|\psi\rangle$. The set of all such unbroken transformations together forms a subgroup $H \subset G$ called the **residual symmetry group**.

Exercise 3.1 (Subgroup of unbroken transformations).
Recall the definition of a subgroup from Section 1.7.1 and show that the set of transformations that leave $|\psi\rangle$ invariant indeed form a subgroup.

Now consider two transformations, g_1 and g_2, which happen to satisfy the relation $g_1 = g_2 h$ for some element h of the residual symmetry group. We then see that $g_1|\psi\rangle = e^{i\varphi}g_2|\psi\rangle$, which implies that $g_1|\psi\rangle$, is equivalent to $g_2|\psi\rangle$. Conversely, if g_1 and g_2 do not satisfy $g_1 = g_2 h$ for any $h \in H$, it follows that $g_1|\psi\rangle$ and $g_2|\psi\rangle$ are distinct and orthogonal broken-symmetry states.

Since the operators corresponding to a broken symmetry transform a given symmetry-breaking state into an inequivalent one, they are often said to correspond to abstract 'rotations' within the space of broken-symmetry states. Starting from the initial state $|\psi\rangle$, we can label each symmetry-breaking state $|\psi'\rangle$ that is distinct from it by some g for which $|\psi'\rangle = g|\psi\rangle$. The resulting list of labels for inequivalent broken-symmetry states consists of a subset of the elements g of the symmetry group G, in which no two elements g and g' satisfy the relation $g = g'h$ for any $h \in H$. Looking back at Section 1.7.1, this coincides precisely with the definition of the coset space G/H.

Inequivalent symmetry-broken states are classified by the cosets gH as elements of the coset space G/H, where G is the group of all symmetry transformations and $H \subset G$ is the subgroup of unbroken transformations.

Notice that this classification of symmetry-breaking states in terms of cosets only applies to the thermodynamic limit. In a system of finite size, distinct broken-symmetry states generally are not precisely orthogonal. This is easy to see, for example, in the case of a finite object breaking continuous rotational symmetry. The Hilbert space is then finite-dimensional, but there are infinitely many symmetry-broken states labelled by all possible directions in space. These correspond to the infinitely many cosets of the rotational symmetry group.

To make the classification more concrete, consider, for example, a Hamiltonian with a global phase rotational symmetry, described by the continuous group of symmetry transformations $G = U(1)$. For a stable state that breaks the phase rotations, so that there is no symmetry left, the residual symmetry group is just the trivial group $H = e$, containing only the identity operator. The coset space is then $G/H = U(1)$ and distinct or inequivalent broken-symmetry states may be labelled by their phase factors $e^{i\varphi}$. This global phase rotation actually is the broken symmetry characterising a superfluid, and superfluids with different phase values may be distinguished by observing a Josephson current between them, as we will see in Section 3.5. More generally, whenever all symmetry transformations in a group are simultaneously broken, the symmetry-breaking states are simply labelled by the elements of the full symmetry group.

As a second example, suppose that the spin-rotation symmetry in the Heisenberg antiferromagnet of Section 2.4 is broken down to only rotations around a single axis. This is the case, for example, in the classical Néel state. Then, the group describing the spin-rotational symmetry of the Hamiltonian is $G = SU(2)$, while the residual symmetry group describing the leftover rotations around a single axis is $H = U(1)$. It can be shown that the coset space G/H equals $SU(2)/U(1) \simeq S^2$, which corresponds to the set of points on the surface of a sphere. These points indicate the possible directions of the residual rotation axis, or equivalently, the possible directions the sublattice magnetisation can take. Note that while the set S^2 classifies all possible symmetry-broken states, it does not have a group structure (because there is no group multiplication that allows its elements to be combined in a way that respects the properties defined in Section 1.7.1).

For continuous groups, the classification of broken-symmetry states can also be expressed in terms of the generators of the continuous symmetry transformations, defined in Eq. (1.5). In this context, generators $\hat{Q}$ of which the broken-symmetry state under consideration is an eigenstate are called **unbroken generators** or unbroken Noether charges, and any finite transformation generated by $\hat{Q}$ is also unbroken. Conversely, generators that do not leave the state invariant are called broken. The continuous symmetry group may be broken down to either a continuous or a discrete subgroup, and even to the trivial group. The dimension of the coset space G/H for continuous groups G is said to equal the number of broken generators. The algebraic relations between broken and unbroken generators will play an important role in the classification of Nambu–Goldstone modes in Chapter 4.

3.2 The Order Parameter

Having classified all possible inequivalent symmetry-breaking states, it would be useful to have an operator whose expectation value can be used to distinguish between them. Ideally, such an operator would have expectation value zero in any symmetric state and a unique non-zero expectation value for each of the sets of equivalent broken-symmetry states. As it turns out, such an operator can be defined and is called the **order parameter operator**. Actually the term order parameter is often used to denote any quantity whose expectation value is non-zero in the broken-symmetry phase and zero in the symmetric state, without having the additional benefit of distinguishing between inequivalent symmetry-breaking configurations. A well-known example can be found in the theory of superconductivity, where the amplitude of the gap function (see Section 8.2) is often called the superconducting order parameter. In this book, we will stick to the more narrow definition, which has the added advantage that it will be instrumental in deriving the Goldstone theorem in Chapter 4.

To define a fool-proof recipe for identifying an order parameter operator, we first need to slightly update our definition of what constitutes a symmetric state.

Definition 3.1 Let $\hat{U} = \mathrm{e}^{\mathrm{i}\alpha\hat{Q}}$ be a symmetry transformation that commutes with the Hamiltonian, parametrised by a discrete or continuous variable α. A state $|\psi\rangle$ breaks this symmetry if there exists any operator $\hat{\Phi}$ such that

$$\langle\psi|[\hat{Q}, \hat{\Phi}]\,|\psi\rangle \neq 0. \tag{3.2}$$

If no such operator $\hat{\Phi}$ exists, the state is symmetric under the transformation $\hat{U}$.

This definition is consistent with our earlier intuitive definition of symmetric states being eigenstates of the symmetry transformation, because if $\hat{U}\,|\psi\rangle = \mathrm{e}^{\mathrm{i}\varphi}\,|\psi\rangle$, the left-hand side of Eq. (3.2) vanishes for any operator $\hat{\Phi}$, and the state is said to be symmetric.[1]

The operator $\hat{\Phi}$ appearing in Eq. (3.2) will actually be an operator $\hat{\Phi}(x)$ acting locally in space in all cases that we shall encounter in this book. It is called the **interpolating field**. With it, we can define the order parameter operator $\hat{\mathcal{O}}(x)$ related to a broken symmetry $\hat{Q}$, and its expectation value $\mathcal{O}(x)$, which is known as the (local) order parameter:

$$\hat{\mathcal{O}}(x) = [\hat{Q}, \hat{\Phi}(x)]\,, \qquad \mathcal{O}(x) = \langle\psi|\,\hat{\mathcal{O}}(x)\,|\psi\rangle\,. \tag{3.3}$$

Notice that if $|\psi\rangle$ is translationally invariant, then so is $\mathcal{O}(x)$. Also, $\hat{\Phi}(x)$ and therefore $\hat{\mathcal{O}}(x)$ are not necessarily Hermitian, but they can always be used to construct observable quantities, such as $\hat{\mathcal{O}} + \hat{\mathcal{O}}^\dagger$ or $\hat{\mathcal{O}}^\dagger\hat{\mathcal{O}}$. The order parameter $\mathcal{O}(x)$ is just the left-hand side of Eq. (3.2).

[1] The slightly subtle definition of symmetric states in Eq. (3.2) includes more than just the intuitive eigenstates of $\hat{U}$. If two symmetry transformations commute, it is always possible to find a set of states which are simultaneous eigenstates of both. This does not mean, however, that if multiple transformations commute with the Hamiltonian, it is always possible to find states that are simultaneous eigenstates for all of them. For that to be possible, all pairs of symmetry transformations have to also mutually commute. A (ferro)magnet with three spin-rotational symmetries, for example, cannot in general have a ground state that is a simultaneous eigenstate of all symmetry transformations. It turns out, however, that in the thermodynamic limit, it is still possible to define a ground state $|\psi\rangle$ that looks the same from all sides. For that state, no operators $\hat{\Phi}$ exist that yield a finite expectation value of $\langle\psi|[\hat{Q}, \hat{\Phi}]\,|\psi\rangle$ for any of the symmetry generators, so that this state indeed is symmetric under the new definition.

Therefore, the order parameter is automatically zero if $|\psi\rangle$ is symmetric and non-zero if $|\psi\rangle$ breaks a symmetry under the new definition of the symmetry of states. It thus clearly satisfies the requirement of distinguishing symmetric from symmetry-breaking states.

To make sure that the order parameter also distinguishes between inequivalent broken-symmetry states, we can require that any good order parameter operator $\hat{\mathcal{O}}$ has eigenvalues that map in a one-to-one fashion onto the coset space G/H introduced in Section 3.1. This way, the order parameter $O(x)$ will not only be different for distinct broken-symmetry states and equal for states related by residual symmetry transformations, but it will also inherit the structure of the coset space. In particular, states that are close to each other according to the topological structure of the Lie group G will have only a small difference in their corresponding order parameter values.

It is always possible to find an order parameter operator that satisfies these constraints, by using the fact that Eq. (3.2) does not uniquely determine the order parameter and interpolating field. For example, multiplying Φ by a constant, or taking its Hermitian conjugate, yield alternative definitions of an interpolating field that still obey Eq. (3.2). In almost all cases, a convenient choice for the order parameter operator, which maps onto the coset space G/H, is suggested by the physics of the symmetry-breaking system itself.

To make the formal definitions of the order parameter and interpolating field more concrete, consider the example of the Heisenberg antiferromagnet, which we discussed in Section 3.1. The Hamiltonian has $SU(2)$ spin-rotational symmetry, which is broken down in the antiferromagnetic state to just $U(1)$ rotations around the axis of sublattice magnetisation. Inequivalent broken-symmetry states correspond to antiferromagnetic configurations with the sublattice magnetisation pointing in different directions. All possible directions constitute the set of points on the surface of a sphere, S^2, which indeed coincides with the quotient $SU(2)/U(1) \simeq S^2$. For a specific broken-symmetry state with the sublattice magnetisation along the z-axis, the symmetry generators $\hat{S}^x$ and $\hat{S}^y$ are spontaneously broken, while $\hat{S}^z$, which generates rotations around the z-axis, is unbroken. We can then define the **staggered magnetisation** as $\hat{N}^a = \sum_i \hat{N}_i^a = (\pm 1)^i \hat{S}_i^a$, where i is a position index that is even on the A-sublattice and odd on the B-sublattice, and a denotes a direction in spin space. To describe the breaking of rotations generated by $\hat{S}^x$, we can choose $-\mathrm{i}\hat{N}^y$ to be the interpolating field, which gives $-\mathrm{i}\sum_{ij}[\hat{S}_i^x, \hat{N}_j^y] = \sum_{ij}\delta_{ij}\hat{N}_i^z = \hat{N}^z$ for the order parameter operator. Similarly, for the breakdown of rotations around the y-axis, we can choose $\mathrm{i}\hat{N}^x$ as the interpolating field, which yields the same order parameter operator. The staggered magnetisation is also the natural choice for the order parameter of an antiferromagnet because the Néel state is precisely an eigenstate of the staggered magnetisation operator.

As a second example, recall the Schrödinger field theory of Section 1.4.1. This model has a global $U(1)$ symmetry describing uniform rotations of the phase of a complex scalar field $\psi(x)$. To find out what it means for this symmetry to be broken, we can, for example, consider the field operator $\hat{\psi}(x)$ to be an interpolating field. The associated order parameter is then

$$[\hat{Q}, \hat{\psi}(x)] = \int \mathrm{d}^D x' \, \hbar [\hat{\psi}^\dagger(x')\hat{\psi}(x'), \hat{\psi}(x)] = -\hbar\hat{\psi}(x). \tag{3.4}$$

Here, we used the commutation relation $[\hat{\psi}^\dagger(x'), \hat{\psi}(x)] = -\delta^D(x - x')$. The order parameter operator thus turns out to be given by the field operator $\hat{\psi}(x)$ itself. Because an eigenstate of the order parameter is a symmetry-breaking state, the states that break the global phase rotation symmetry must be eigenstates of the operator $\hat{\psi}(x)$. But that operator is the annihilation operator for quanta of the ψ-field. Within Fock space, it is possible to construct eigenstates of the annihilation operator, and these are called **coherent states** (see also Eq. (2.38)). You can check that the state $e^{\int dx\, \phi(x)\hat{\psi}^\dagger(x)}\,|\text{vac}\rangle$ is an eigenstate of the annihilation operator $\hat{\psi}(x)$, with complex eigenvalue $\phi(x)$. Because this coherent state is an eigenstate of the field operator, $|\phi(x)|^2$ is the expectation value of the number of field quanta, or particles, at position x. Notice that if we assume translational invariance as before, ϕ is independent of x. Expanding the exponential in the definition of the coherent state, it is seen to create a superposition of infinitely many states with different numbers of excited field quanta. Conversely, an eigenstate of the number or density operator $\hat{\psi}^\dagger(x)\hat{\psi}(x)$ can be written as a superposition of infinitely many coherent states, which all have the same absolute value of $\phi(x)$, but which differ in phase. The phase and modulus of ϕ can in fact be shown to be conjugate variables, corresponding to operators with canonical commutation relations.

The symmetry-breaking state of the Schrödinger field theory is a coherent state, which has an indefinite number of particles or quanta. This is in direct accordance with the fact that the symmetry generated by $\hat{Q} \propto \int \hat{\psi}^\dagger\hat{\psi}$ is associated with the conservation of the number of field quanta. Distinct symmetry-breaking states are characterised by the phase of ϕ. Performing rotations with $\exp(i\alpha\hat{Q})$ will lead to the phase of the order parameter changing by α, such that distinct broken-symmetry states, with different phase values, can be labelled by elements of the coset space $U(1)/1 \simeq U(1)$. The formation of a state with indeterminate particle number and precise phase-value occurs, for example, in Bose–Einstein condensates such as superfluids and superconductors, as discussed in Section 2.5. There, the expectation value $\langle\hat{b}_i\rangle$ can be interpreted as the order parameter.

3.3 The Classical State

Given the definition of the order parameter operator $\hat{O}$ in Eq. (3.3), you might be tempted to believe that broken-symmetry states are simply the eigenstates of $\hat{O}$, with eigenvalues O. This would certainly justify the definition of the order parameter as the expectation value of the order parameter operator, and in most cases, it also agrees very well with our expectation for a perfectly ordered state. This is because in translationally invariant systems, the eigenstates of $\hat{O}$ are tensor products $|\psi\rangle = \otimes_x |\psi(x)\rangle$, with $|\psi(x)\rangle$ eigenstates of the local operator $\hat{O}(x)$. Such product states correspond directly to the states of a classical Hamiltonian, in which all operators are replaced by their expectation values. We will call these eigenstates of the order parameter operator **classical states**.

The symmetry-breaking states encountered in real quantum systems are typically not classical states. After all, even though a symmetry-breaking perturbation may dominate the shape of the ground state for sufficiently large systems, it is not the only contribution to the Hamiltonian. The remaining, symmetric part of the Hamiltonian also takes part in determining the ground state, and takes it away from the classical ideal. We have already seen this for the antiferromagnet in Section 2.4. The true, quantum, broken-symmetry states typically have order parameter expectation values close to those of the classical state. The quantum states can therefore be thought of as arising from a perturbation theory around the classical states. The differences between the classical state and the true quantum broken-symmetry state are then known as **quantum corrections** to the classical state. The corrections at zero momentum are related to the tower of states that will be discussed in more detail in Section 3.6. The corrections at non-zero momentum are the topic of Chapter 6.

3.4 Long-Range Order

For any given uniform ground state, the order parameter identifies whether it has broken or unbroken symmetries and distinguishes between inequivalent symmetry-breaking states. In practical calculations, however, one often does not know the exact ground state of a given Hamiltonian, and many systems exist in conditions that are not perfectly uniform. In those cases, a particularly useful alternative way of quantifying the occurrence of spontaneous symmetry breaking is through the behaviour of the so-called **two-point correlation function**:

$$C(x,x') = \langle\psi|\,\hat{O}^{\dagger}(x)\hat{O}(x')\,|\psi\rangle\,. \tag{3.5}$$

Here, $\hat{O}(x)$ is the local order parameter operator. Clearly, if $|\psi\rangle$ is uniform in space and an eigenstate of the order parameter, the correlation function is the same for any choice of the coordinates x and x' and equal to the square of the uniform order parameter. The advantage of the two-point correlation function, however, is that it can be used also in less clear-cut cases. The behaviour of this function can be divided roughly into two distinct classes, depending on its functional form as the distance $|x-x'|$ is taken to infinity:

$$C(x,x') \propto \begin{cases} \text{constant} & \text{'long-range ordered'} \\ \mathrm{e}^{-|x-x'|/l} & \text{'disordered'} \end{cases} \quad \text{as } |x-x'| \to \infty. \tag{3.6}$$

Here, l is a length scale called the **correlation length**. For long-range ordered systems, the spatial average of the local order parameter will be non-zero, and the correlation length l diverges. The presence of long-range order is therefore associated with the breaking of a symmetry. In fact, the two-point correlation function signals the propensity to break symmetry already for finite-size systems, which really have a symmetric ground state. Even though the order parameter expectation value in the exact, symmetric ground state is exactly zero, the two-point correlation function has a constant value for large separations. This can be easily understood by considering the extreme example of the two-spin singlet state $|\uparrow\downarrow\rangle - |\downarrow\uparrow\rangle$, which is the ground state of the two-site Heisenberg antiferromagnet. Clearly, the singlet state has no preferred direction of staggered magnetisation, but the two spins are definitely anti-parallel. Especially in numerical investigations, the two-point correlation function is typically easy to 'measure' even with only limited information about the spectrum and is widely used in establishing the presence of order and spontaneous symmetry breaking. We will come back to this in more detail when we discuss stability in Section 3.7.

Notice that the separation of states into long-range ordered and disordered is not exhaustive. A special case may occur in low dimensions, if the two-point correlation function is proportional to $|x-x'|^c$, for some exponent c. This is called **algebraic long-range order** or quasi long-range order. We will discuss this type of order in Section 7.4.

You may have also encountered the term 'off-diagonal long-range order' or 'ODLRO' in the literature. This term was introduced by Oliver Penrose and Lars Onsager in the 1950s [155, 156], in the context of superfluidity in helium, as a way of contrasting the superfluid order with ordering in solids. The concept is largely historical, and the distinction between ODLRO and other types of order has become obsolete with the modern definition of the order parameter in Eq. (3.3), and the corresponding classification of two-point correlation functions in Eq. (3.6). To see this, we will consider the usual definition of ODLRO in terms of the N-particle wave function $\Psi(x_1,\ldots,x_N)$, which we can use to define

$$\rho(x_1,\ldots,x_N,y_1,\ldots,y_N) = \Psi^*(x_1,\ldots,x_N)\Psi(y_1,\ldots,y_N). \tag{3.7}$$

Notice that ρ is a function of two sets of coordinates x_i and y_i. For coinciding sets of coordinates, $y_i = x_i$, the value of ρ is a diagonal element of the usual **density matrix**, giving the probability for finding the N particles at positions x_i. The diagonal elements of the two-particle reduced density matrix can then be found by integrating over all but two of the coordinates:

$$\rho_{\mathrm{D},2}(x_1,x_2) = \int \mathrm{d}x_3\cdots\mathrm{d}x_N\ \Psi^*(x_1,x_2,x_3,\ldots,x_N)\Psi(x_1,x_2,x_3,\ldots,x_N). \tag{3.8}$$

If space is uniform, the reduced density matrix is invariant under global translations and can only depend on the difference of the two coordinates, so that $\rho_{\mathrm{D},2}(x_1,x_2) = \rho_{\mathrm{D},2}(x_1-x_2)$. Diagonal long-range order is said to occur if $\rho_{\mathrm{D},2}(x_1-x_2)$ is periodic in x_1-x_2. This is the usual ordering we find for solids such as the harmonic crystal introduced in Section 2.3. It is called diagonal because we only need to consider diagonal elements of $\rho(x_i,y_i)$, with $y_i = x_i$.

Alternatively, we can consider another type of two-point function:

$$\rho_{\mathrm{O},2}(x,y) = \int \mathrm{d}x_2\cdots\mathrm{d}x_N\ \Psi^*(x,x_2,\ldots,x_N)\Psi(y,x_2,\ldots,x_N). \tag{3.9}$$

That is, we choose all wave function coordinates except the first to coincide, and integrate over them. Note that this entails one more integration than the definition of the reduced density matrix in Eq. (3.8). In fact, $\rho_{\mathrm{O},2}$ and $\rho_{\mathrm{D},2}$ represent two very different physical quantities that have little to do with one another. For superfluids in particular, $\rho_{\mathrm{O},2}$ is almost identical to the two-point correlation function of Eq. (3.5), if we choose the order parameter operator $\hat{\mathcal{O}}(x)$ to be the field operator $\hat{\psi}(x)$. Off-diagonal long-range order (ODLRO) is said to occur if $\rho_{\mathrm{O},2}(x,y)$ does not vanish as $x-y$ is taken to infinity. It was called off-diagonal because it involves off-diagonal elements of $\rho(x_i,y_i)$.

Although we need to be careful when applying Eq. (3.6) to crystals because they retain discrete translational symmetry and $C(x,x')$ approaches a periodic function rather than a true constant at large separations, the two-point function does capture the breakdown of symmetry in both crystals and superfluids in essentially the same way. In fact, both diagonal and off-diagonal long-range order are part of a much larger family of possible types of ordered states that are all classified by the behaviour of the two-point correlation function at large separation. There is nothing special about either the order occurring in crystals (DLRO) or that in superfluids (ODLRO). When off-diagonal long-range order was originally introduced, the concept of symmetry breaking for internal degrees of freedom and its embedding within the larger theory of symmetry breaking in general were not yet developed, and ODLRO was a way to capture the long-range ordering of the internal $U(1)$-phase degree of freedom contained in the N-body wave function.

3.5 The Josephson Effect

When we first discussed symmetric states in Section 1.1, we pointed out that symmetry must always be defined with respect to some reference. In the example of a crystal breaking translational symmetry, the broken translations are really defined with respect to an outside observer, who can, for example, measure the distance between the crystal and herself. More generally, in order to observe the breakdown of a symmetry in a given state, an observer needs some reference frame with respect to which the broken symmetry can be measured. For the reference frame to be able to distinguish between inequivalent symmetry-breaking states, it must itself be in a broken-symmetry state. That is, you cannot measure the position of a crystal with respect to a uniform fluid permeating all of space, and it is not possible to measure a direction of magnetisation with a piece of plastic that cannot itself be magnetised. Although this might seem obvious for crystals and magnets, one could wonder what it implies for materials hosting less intuitive forms of broken symmetry, such as the $U(1)$ phase-rotation symmetry associated with conserved particle number, which we argued in Section 3.2 to be broken in superfluids. This question found a literal manifestation in the spontaneous tunnelling current that was predicted by Josephson in 1962 to occur between two separated pieces of superconducting material [83].

Since the origin of the Josephson effect lies in the broken $U(1)$ symmetry, we will discuss it here for neutral superfluids rather than superconductors. As we saw before, a superfluid is a state with broken $U(1)$ phase-rotation symmetry and a complex order parameter $\psi = |\psi|e^{i\varphi}$. In terms of observable properties, the superfluid is characterised by its ability to host supercurrents that flow without viscosity. One way to understand this is by noticing that the supercurrent is identical to the conserved Noether current associated with the broken $U(1)$ symmetry:

$$j^n \propto i\left((\partial_n\psi^*)\psi - \psi^*(\partial_n\psi)\right). \tag{3.10}$$

The local order parameter $\psi(x)$ for the superfluid is the expectation value of the field operator. It is often referred to in the more popular literature as a 'macroscopic wave function'. Although somewhat misleading, this terminology does emphasise the fact that like a quantum wave function, the field $\psi(x)$ satisfies equations of motion with spatial derivatives, which force it to be continuous. As a consequence, the field does not abruptly vanish at the boundary of a sample, but rather falls off exponentially into the vacuum. For two samples of superfluid separated by a small gap, the order parameter fields extending into the gap from both sides can overlap. Just like for quantum mechanical wave functions, this implies the possibility for field quanta to tunnel from one sample to the other, which is the essence of the **Josephson effect**.

For a junction of width w between superfluids with constant order parameters ψ_1 and ψ_2, the order parameter field inside the junction can be written as

$$\psi(x) = Ae^{-x/\xi} + Be^{x/\xi}. \tag{3.11}$$

Here, A and B are complex constants, and the decay length of the order parameter field in the vacuum is ξ. The two superfluids are assumed to extend indefinitely in the y and z directions, while being semi-infinite in the x direction, with one having an edge at $x = -w/2$ and the other at $x = w/2$. You can think of the field inside the junction as a superposition of the exponentially decaying order parameter fields from either side. The boundary conditions are given by the field values in the two superfluids, $\psi(-w/2) = \psi_1$ and $\psi(w/2) = \psi_2$, so that we find

$$A = \frac{e^{w/2\xi}\psi_1 - e^{-w/2\xi}\psi_2}{2\sinh(w/\xi)}, \qquad B = \frac{e^{w/2\xi}\psi_2 - e^{-w/2\xi}\psi_1}{2\sinh(w/\xi)}. \tag{3.12}$$

The current density in the junction is given by Eq. (3.10), and equals $j^x = 2i(B^*A - A^*B)/\xi$, independent of the position x. Substituting the values of A and B yields the expression

$$j^x \propto i\frac{\psi_2^*\psi_1 - \psi_1^*\psi_2}{\xi\sinh(w/\xi)} = \frac{2|\psi_1||\psi_2|}{\xi\sinh(w/\xi)}\sin(\varphi_2 - \varphi_1). \tag{3.13}$$

We thus find a current per unit area j^x flowing through the junction, which is proportional to the sine of the phase difference between the two superfluid order parameters.

The flow of supercurrent without a chemical potential difference between the superfluid samples is an interesting physical observation in and of itself. In the context of symmetry breaking, however, it also gains a more fundamental interpretation. The phase of the order parameter for one superconducting sample can be determined with respect to the phase of a second sample by measuring the Josephson current flowing between them. This is precisely analogous to the way that the position of a crystal can be determined only with respect to the position of some other object with broken translational symmetry. Historically, the discovery of the Josephson effect was therefore the deciding factor in settling the debate of whether or not a symmetry was spontaneously broken in superconductors (see, for instance, Ref. [8]). More generally, a calculation analogous to that of the Josephson effect in superfluids should be possible for two pieces of material with any type of spontaneously broken symmetry. There are very few known examples of generalised Josephson effects outside of superconductivity and superfluidity, but at least conceptually, the Josephson effect offers an unambiguous and general way of measuring the order parameter of any sample with respect to a reference broken-symmetry state [19].

The Josephson effect provides a way of explicitly breaking the phase rotation symmetry of one superfluid by bringing it close to another, and thus realises the symmetry-breaking field introduced in Eq. (2.32). This does not mean, however, that superfluids cannot exist on their own. As we argued in Section 2.5, the symmetry in a superfluid can be broken spontaneously, requiring only a vanishingly weak symmetry-breaking perturbation. Again, the situation is precisely analogous to that in crystals. We are used to being surrounded by objects with broken translational symmetry in daily life. But we do not expect a comet travelling in outer space to be in a delocalised quantum state just because it is far away from any possible source of translational symmetry-breaking perturbations. We automatically assume that space rocks will still be localised and that intuition is supported by the fact that even the tiniest perturbations provided by far-off constellations suffice to localise massive objects in space. In the same way, superfluids require only a vanishingly small coupling to a coherent particle source in order to break their phase rotation symmetry. In practice, isolating a single superfluid from sources of particle fluctuations to the extent needed to completely prevent the onset of symmetry breaking is impossible for superfluids of any realistic size. Although less intuitive due to our lack of daily interaction with objects that do not conserve particle number, phase rotation symmetry can be broken by a Josephson-type coupling that is of comparable weakness to the interaction between a comet and the stars.

Exercise 3.2 (Josephson effect).
The equations describing the generalised Josephson current between any two symmetry-breaking objects can be derived from a very simple model [52]. In the Heisenberg picture, we can write the global order parameter operators of the systems at the left and right as $\hat{\Psi}_\mathrm{L}(t)$ and $\hat{\Psi}_\mathrm{R}(t)$ (without any x-dependence). The Hamiltonian is taken to be of the form

$$\hat{H} = \hat{H}_\mathrm{L} + \hat{H}_\mathrm{R} + \hat{H}_\mathrm{int}. \tag{3.14}$$

Here, the left and right systems are described by the local Hamiltonians $\hat{H}_\mathrm{L}$ and $\hat{H}_\mathrm{R}$, while the coupling between them is described by $\hat{H}_\mathrm{int}$.

For superconductors, the order parameter operators correspond to field operators $\hat{\psi}_i$, with commutation relation $[\hat{\psi}_i, \hat{\psi}^\dagger_{i'}] = \delta_{i,i'}$. The coupling is then given by

$$\hat{H}_\mathrm{int} = K(\hat{\psi}^\dagger_\mathrm{R}\hat{\psi}_\mathrm{L} + \hat{\psi}^\dagger_\mathrm{L}\hat{\psi}_\mathrm{R}). \tag{3.15}$$

Here, K is a coupling constant with units of energy.
a. Using the Heisenberg equations of motions $-\mathrm{i}\hbar\partial_t\hat{A} = [\hat{H},\hat{A}]$, derive an expression for the Josephson current $\hat{I}_\mathrm{J} = \partial_t(\hat{\psi}^\dagger_\mathrm{L}\hat{\psi}_\mathrm{L})$. Compare your result with Eq. (3.13).

Now consider two ferromagnets, whose magnetisation vector is given by the expectation value $\mathbf{M} = \langle \sum_i \hat{\mathbf{S}}_i \rangle$, with spin operators obeying the commutation relations $[\hat{S}^a_i, \hat{S}^b_{i'}] = \mathrm{i}\epsilon_{abc}\delta_{i,i'}\hat{S}^c_i$. The magnets are coupled through the interaction described by

$$\hat{H}_\mathrm{int} = K\,\hat{\mathbf{S}}_\mathrm{L} \cdot \hat{\mathbf{S}}_\mathrm{R}. \tag{3.16}$$

b. Derive an expression for the 'spin Josephson current' $\partial_t \mathbf{M}_\mathrm{L}$. Your result will agree with the much more sophisticated calculation based on a microscopic description of the tunnelling of electrons between two ferromagnets [110].

3.6 The Tower of States

As emphasised several times before, the true ground state of a finite-sized quantum system without any symmetry-breaking field present is typically symmetric. Except for symmetry-breaking states associated with conserved order parameters that will be introduced in Exercise 3.3 and discussed in more detail in Section 4.4, the true ground state of a symmetric Hamiltonian is unique and is an eigenstate of the symmetry generators. Since the symmetry transformations are global, the symmetric ground state also has a global structure. In particular, it typically contains long-range entanglement between distant parts of the system, which therefore all strongly depend on each other (as discussed in more detail in Section A.2). Furthermore, as we will see in Section 3.7, the symmetric state is unstable. In terms of the spectrum of eigenstates of the symmetric Hamiltonian, however, the long-range entangled nature of the ground state is in no way exceptional. There are a whole set of eigenstates that can be seen as low-energy, global excitations on top of the ground state, which all share the same feature.

Crudely speaking, if the order parameter can be defined in terms of some canonical observable, the symmetric Hamiltonian must contain a kinetic energy proportional to the total canonical momentum squared. This follows from the fact that the eigenstates of momentum are symmetric combinations of all possible canonical positions. Within a finite volume V, the Fourier transform of the canonical momentum operator is given by $\hat{p}(x) = \frac{1}{\sqrt{V}} \sum_k \mathrm{e}^{\mathrm{i}kx} \hat{p}_k$. The total momentum is proportional to the $k = 0$ component of this decomposition:

$$\begin{aligned} \hat{p}_{\text{tot}} &= \int \mathrm{d}x \, \hat{p}(x) = \frac{1}{\sqrt{V}} \sum_k \int \mathrm{d}x \, \mathrm{e}^{\mathrm{i}kx} \hat{p}(k) \\ &= \sqrt{V} \sum_k \delta(k) \hat{p}(k) = \sqrt{V} \hat{p}_{k=0}. \end{aligned} \tag{3.17}$$

Here, we used the representation of the delta function given by $\frac{1}{V} \int \mathrm{d}x \, \mathrm{e}^{\mathrm{i}kx} = \delta(k)$. The term in the Hamiltonian proportional to the total momentum comes from the $k = 0$ part of the usual kinetic energy operator:

$$\begin{aligned} \hat{H}_{\text{kin}} \propto \int \mathrm{d}x \, \hat{p}^2(x) &= \sum_k \hat{p}(k) \hat{p}(-k) \\ &= \frac{1}{V} \hat{p}_{\text{tot}}^2 + \sum_{k>0} \hat{p}(k) \hat{p}(-k). \end{aligned} \tag{3.18}$$

The second term in the final line combines with the potential energy to describe internal excitations such as phonons, magnons, supercurrents, and so on. The first term, on the other hand, is just the (canonical) kinetic energy of the object as a whole.

Since $V \propto N$, the modes corresponding to the total canonical momentum will be quantised in any confining potential, yielding energy eigenstates with spacing $\propto 1/N$, just as in Figures 2.3 and 2.5. This set of global eigenstates of canonical total momentum is referred to as the **tower of states**, or the Anderson tower of states, or occasionally as the thin spectrum. In the thermodynamic limit, all states in the tower become exactly degenerate.

As we will see in Chapter 4, systems with a spontaneously broken symmetry have gapless, propagating excitations, called Nambu–Goldstone modes. These modes exist at non-zero wave number and for a finite system of linear size L, the lowest possible energy they can take is proportional to $1/L$. In spatial dimension $d > 1$, the states in the $k = 0$ tower of states have energies proportional to $1/V \propto 1/L^d$ and these are therefore much lower in energy than the $k > 0$ states with even the lowest possible energy. Since all states in the tower are eigenstates of the total canonical momentum at zero wave number, they all have a global structure, and they are all symmetric. The classical symmetry-broken states are superpositions of the states in the tower, with the special property that they can be written as local product states of the form $\otimes_x |\psi(x)\rangle$. They are not energy eigenstates, but because the energy spacing between the states in the tower is so small, the classical states are very narrow wave packets in energy space, which take on a single, well-defined energy expectation value in the thermodynamic limit.

Because the existence of a tower of global excitations is so intrinsically linked to spontaneous symmetry breaking, it is reasonable to ask whether these states influence any other measurable properties of a symmetry-broken object. To see this, consider the free energy of a quantum system with symmetric Hamiltonian $\hat{H}$, which at temperature T can be calculated using

$$F = -k_B T \ln Z,$$
$$Z = \sum_{|\psi\rangle} \langle\psi|\, e^{-\frac{\hat{H}}{k_B T}}\, |\psi\rangle . \tag{3.19}$$

The sum in the partition function Z runs over all energy eigenstates $|\psi\rangle$ of the Hamiltonian. The collective part, or tower of states, associated with the Hamiltonian for an object consisting of N interacting particles generically contains $O(N)$ states. Using Eq. (3.19), the free energy associated with this part of the spectrum is then proportional to $\ln N$. Again crudely, this can be seen by defining the collective part of the Hamiltonian to be the kinetic energy associated with some total canonical momentum, so that

$$Z_{\text{coll}} = \sum_{p_{\text{tot}}} \langle p_{\text{tot}}|\, e^{-\frac{\hat{H}_{\text{coll}}}{k_B T}}\, |p_{\text{tot}}\rangle \sim \int dp_{\text{tot}}\, e^{-\frac{p_{\text{tot}}^2}{k_B T V}}$$
$$\propto \sqrt{k_B T N}. \tag{3.20}$$

Since Z_{coll} is proportional to $\sqrt{N}$, the contribution of the tower of collective states to the free energy F_{coll} is proportional to $\ln N$. The free energy F associated with the full Hamiltonian must always be proportional to N because it is an extensive quantity. The relative contribution of the collective states to the total free energy, F_{coll}/F, is then proportional to $\ln N/N$, and disappears in the limit of large system size. In other words, even though they are the only states with energies as low as $1/N$, there are so few states in the tower that they do not contribute to the free energy at any non-zero temperature, no matter how low. This part of the spectrum in fact is so 'thin' that it cannot be observed in any thermodynamic properties of the material, such as specific heat or conductivity, which are all determined by the free energy.

This observation is again general for collective states governing the spontaneous breakdown of any continuous symmetry. They always form an exceedingly thin part of the spectrum that is practically undetectable for any realistically sized system in our everyday world. Paradoxically, one of the few ways in which the presence of this part of the spectrum does have an influence on measurable quantities, is due precisely to its undetectable nature. If a material with a broken continuous symmetry is used to store quantum information, for example, using the presence or absence of a magnon in an antiferromagnet as the zero and one states of a hypothetical qubit, then the presence of $O(N)$ states beyond any experimental control acts as a sort of environment to the qubit. Even if one could entirely isolate such a system from any external influences, the qubit will decohere because the information about the magnon state becomes entangled with the unmeasurable, thin part of the spectrum [201].

3.7 Stability of States

In spontaneous symmetry breaking, the exact ground state of a system is infinitely sensitive to perturbations, which therefore always yield a broken-symmetry state in the thermodynamic limit. However, such symmetric states are also generically unstable all by themselves. To illustrate this, consider a magnetic system with some local magnetisation $\langle\hat{\sigma}^z(x)\rangle$ defined at each position. If a measurement of the magnetisation at x can influence a subsequent measurement at a far-away position y, the system is said to be **unstable against local measurements**.

In other words, stability requires the expectation value $\langle \hat{\sigma}^z(y) \rangle$ at position y to be independent of the measurement of magnetisation $\langle \hat{\sigma}^z(x) \rangle$ at a position x far away from y.

A simple example of an **unstable state** is the ground state of the transverse field Ising model:

$$\hat{H} = -J \sum_{\langle ij \rangle} \hat{\sigma}_i^z \hat{\sigma}_j^z - \mu \sum_i \hat{\sigma}_i^x. \tag{3.21}$$

This Hamiltonian is defined on any lattice, with $\langle ij \rangle$ denoting nearest neighbours, and $\hat{\sigma}_i^a$ Pauli matrices on site i. The coupling J is positive and the transverse field is represented by μ. This model has a discrete global $\mathbb{Z}_2$ symmetry of simultaneously flipping all spins in the z-direction, $\hat{\sigma}_i^z \to -\hat{\sigma}_i^z$. If the transverse field is small, $0 < \mu \ll J$, the ground state of this model is well approximated by a symmetric superposition of all spins up and all spins down:

$$|\psi_0\rangle \approx |\uparrow\uparrow\uparrow \cdots\rangle - |\downarrow\downarrow\downarrow \cdots\rangle. \tag{3.22}$$

Notice that in this state, the expectation value $\langle \hat{\sigma}_i^z \rangle$ at any position i equals zero, as expected for a system with spin flip symmetry. Measuring the z-component of the spin at position i will collapse the superposed ground state onto the component corresponding to the observed value. For example, if we happen to measure an up spin at i, the entire state after the measurement has collapsed to $|\uparrow\uparrow\uparrow \cdots\rangle$, and subsequently measuring the z-component of spin at any position j will always yield up. The expectation value $\langle \hat{\sigma}_j^z \rangle$ has thus qualitatively changed because σ_i^z was measured regardless of the distance between positions i and j, and the state of Eq. (3.22) is concluded to be unstable against local measurements.

Conversely, the broken-symmetry state $|\uparrow\uparrow\uparrow \cdots\rangle$ itself is stable against local measurements, because no measurement of the local magnetisation at any position j will influence the result of subsequent measurements at any other position. In fact, this pattern is general, and the stability of the symmetry-breaking state is a direct consequence of its long-range order, introduced in Section 3.4. Local measurements will generically rapidly collapse an unstable symmetric state onto one of the possible broken-symmetry states. The definition of stability with respect to local measurements is especially relevant when considering the embedding of any given system in its local environment. Even the weakest interactions with an environment can easily amount to an effective measurement of local observables like the magnetisation $\langle \hat{\sigma}_i^z \rangle$, and thus prevent symmetric states from being observed in any realistic setting.

The central ingredient in the definition of stability against local measurements is that in an unstable state, a single measurement influences the outcome of many subsequent measurements. To quantify the meaning of 'many', the concept of **cluster decomposition** can be used. A state is said to satisfy the cluster decomposition property if and only if for all local observables $\hat{a}(x)$ and $\hat{b}(y)$, we have

$$C_{ab}(x, y) = \langle \hat{a}(x)\hat{b}(y) \rangle - \langle \hat{a}(x) \rangle \langle \hat{b}(y) \rangle \to 0 \quad \text{for} \quad |x - y| \to \infty. \tag{3.23}$$

This means that measurements of any $\hat{a}(x)$ and $\hat{b}(y)$, provided x and y are far apart, will be independent. Cluster decomposition can therefore be considered a requirement for macroscopic stability against local measurements.

One can check that the ground state of the transverse field Ising model in Eq. (3.22) does not satisfy the cluster decomposition property. States like these are sometimes called **cat states**, in reference to Schrödinger's cat. The exact ground states of Hamiltonians susceptible to spontaneous symmetry breaking are almost always cat states. Conversely, the broken-symmetry states that may be stabilised in the thermodynamic limit by even an infinitesimal perturbation can always be written as product states, which do satisfy the cluster decomposition property.

The concept of cluster decomposition is closely related to a thermodynamic restriction on fluctuations of extensive observables. In thermodynamics, the extensive observables of two subsystems can be added together to find the extensive observable associated with the combined system. In other words, extensive observables can be written as a sum of local observables, $\hat{A} = \sum_x \hat{a}(x)$. The expectation value of $\hat{A}$ must therefore scale with the volume of the system, $\langle \hat{A} \rangle \sim \mathcal{O}(V)$. On the other hand, the variance of an observable generally scales as $\text{Var}(\hat{A}) = \langle \hat{A}^2 \rangle - \langle \hat{A} \rangle^2 \propto V^\alpha$. If the exponent α is two or greater, $\text{Var}(\hat{A})/\langle \hat{A} \rangle^2$ does not vanish in the thermodynamic limit, and the fluctuations in measured values of the observable are as large as its expectation value. The state is then said to be **thermodynamically unstable**. On the other hand, for states with $\alpha = 1$, the fluctuations vanish in comparison to the expectation value, and these states are thermodynamically stable. Product states are always of the latter type. Some special states may have $1 < \alpha < 2$. These are thermodynamically stable, but they may be fragile in other senses. The critical systems that we will discuss in Section 5.5 are an example of this class.

Unsurprisingly, a system that violates the cluster decomposition property is a superposition of macroscopically distinct states, and thus possesses large variance in the expected measurement outcomes of extensive observables. This can be easily seen by writing the variance in terms of the two-point correlation function:

$$\text{Var}(\hat{A}) = \langle \hat{A}^2 \rangle - \langle \hat{A} \rangle^2 = \sum_{x,y} \langle \hat{a}(x)\hat{a}(y) \rangle - \langle \hat{a}(x) \rangle \langle \hat{a}(y) \rangle = \sum_{x,y} C_{aa}(x,y). \tag{3.24}$$

In a product state, such as the ordered, symmetry-breaking states, the expectation value $\langle \hat{a}(x)\hat{a}(y) \rangle$ becomes equal to the product of local expectation values if x and y are far apart. The variance is then dominated by contributions with $x \sim y$, and therefore scales as $\text{Var}(A) \sim \mathcal{O}(V)$. On the other hand, for systems with long-range entanglement, such as symmetric states or cat states, $C(x,y)$ does not vanish for large $x - y$, and the variance contains significant contributions from all terms in the double sum. We then find $\text{Var}(\hat{A}) \sim \mathcal{O}(V^2)$, indicating that fluctuations in measurements of $\hat{A}$ are as large as the observed average $\langle \hat{A} \rangle$ itself.

We can thus define stability in three equivalent ways: using the stability against local measurements, examining the cluster decomposition property, and considering the variance of extensive variables. The symmetric ground state of models that exhibit SSB are generically unstable, while classical broken-symmetry states are always stable under any of these three definitions of stability. The applicability of the rule that symmetric ground states are inherently unstable reaches further than many other results presented so far. For example, Noether's theorem only applies to continuous symmetries, and the tower of states is only relevant to systems in which the order parameter does not commute with the Hamiltonian (see Exercise 3.3). However, the instability of symmetric states is a property of all systems that exhibit spontaneous symmetry breaking.

Exercise 3.3 (Heisenberg Ferromagnet).
In daily parlance, the ferromagnet is often used as the simplest example of spontaneous symmetry breaking. Unfortunately, as you will see, the properties of a ferromagnet make

it exceptional, and unlike almost any other forms of symmetry-broken states, such as anti-ferromagnets, superfluids, or even the symmetry-breaking phases in the Standard Model of particle physics.

Consider the Heisenberg Hamiltonian of Eq. (2.17), but with negative coupling $J < 0$, so that pairs of spins prefer to be aligned, rather than anti-aligned. This can easily be accommodated by, for example, having all spins point in the z-direction.
a. Show that $\hat{S}^x$ and $\hat{S}^y$ are spontaneously broken according to the definition of Eq. (3.2), by finding appropriate interpolating fields.

The order parameter operator for the state with all spins aligned in the z-direction is $\hat{S}^z$, which is itself one of the symmetry generators of the symmetric Heisenberg Hamiltonian. The order parameter operator thus commutes with the Hamiltonian. The ferromagnet is truly exceptional, however, due to the following property:
b. Show that the state with all spins aligned (in the z-direction) is an eigenstate of the Hamiltonian.
Hint: Use the second line of Eq. (2.18).

This result implies that the fully magnetised, classical state in the sense of Section 3.3 is an exact ground state of the symmetric quantum mechanical Hamiltonian, for any system size. In fact, any fully magnetised state, with all spins simultaneously pointing in any direction, is a ground state. The ground state is thus far from unique, even for finite-sized ferromagnets. There is no tower of states and there are no quantum corrections to the classical state. The system merely chooses a classical ground state that is stable against local perturbations from the full set of degenerate ground states. We will examine the ferromagnet in more detail in Section 4.4. For now, the moral of this exercise is that you should not use the ferromagnet as an archetype of spontaneous symmetry breaking. It truly is an exceptional case.

4 Nambu–Goldstone Modes

Every symmetry of the Hamiltonian or Lagrangian corresponds to a conserved quantity, regardless of whether or not the state of the system respects the symmetry. In homogeneous space, for example, both a classical ball with spontaneously broken translational symmetry and an electron in a symmetric, plane-wave state will have a conserved total momentum. The intimate relation between the conserved global quantity and the possibility of spontaneously breaking a symmetry was elucidated in Section 3.6, where we discussed the tower of states. This collective, $k = 0$, part of the spectrum consists of eigenstates of the conserved global quantity, which in the thermodynamic limit can be combined into a coherent-state superposition. Both the individual eigenstates and the symmetry-breaking superposition conserve the (expectation value of the) global quantity.

As we saw in Section 1.3.2, however, Noether's theorem has implications far beyond the global aspects of the system. For every continuous global symmetry, it guarantees the existence of a locally conserved current, obeying a local continuity equation. Again, this form of Noether's theorem holds regardless of whether or not the state of the system respects the symmetry. Moreover, the local conservation law is intimately tied to a generic property of the spectrum of systems with a spontaneously broken continuous symmetry. Rather than affecting the collective states, however, the local continuity equation impacts the excitations at non-zero wave number and guarantees the appearance of gapless modes known as **Nambu–Goldstone (NG) modes**. In particle physics and relativistic quantum field theory, these modes are referred to as Goldstone bosons, and they are said to be massless instead of gapless. The difference is purely a matter of nomenclature.

To understand the nature of the NG modes, consider the temporal component of the Noether current operator $\hat{\jmath}^t(\mathbf{x})$ related to the symmetry generator $\hat{Q} = \int d\mathbf{x}\, \hat{\jmath}^t$ that is spontaneously broken. The NG mode $|\pi(\mathbf{k}, t)\rangle$ can then be viewed as a plane-wave superposition of local excitations created by acting with the Noether current operator on the broken-symmetry state $|\psi\rangle$:

$$|\pi(\mathbf{k}, t)\rangle \propto \int \mathrm{d}^d x\, \mathrm{e}^{\mathrm{i}\mathbf{k}\cdot\mathbf{x}} \hat{\jmath}^t(\mathbf{x}, t)\, |\psi\rangle\,. \tag{4.1}$$

Goldstone's theorem, which we will introduce in Section 4.1, shows the states $|\pi(\mathbf{k}, t)\rangle$ to be gapless. That is, their energy goes to zero as $\mathbf{k} \to 0$. These low-energy excitations correspond to creating local Noether charge density, and Noether's continuity equation guarantees that such excitations will be dispersed over time. In other words, low-energy local disturbances in the order parameter will be carried away over time, much like the local excitation made by a raindrop in a puddle is carried away by surface waves. Over time, the perturbed region necessarily returns to its original symmetry-broken state, and systems with a spontaneously broken symmetry are thus endowed with a form of **rigidity** [9].

4.1 Goldstone's Theorem

Before delving into the proof for Goldstone's theorem and discussing some of its implications and more modern aspects, let us state the theorem and define its realm of applicability:

Theorem 4.1 (Goldstone's theorem). If a global, continuous symmetry is spontaneously broken in the absence of long-ranged interactions and leaving some (discrete, course grained) translational symmetry intact, then there exists a mode in the spectrum whose energy vanishes as its wave number approaches zero.

The theorem includes many assumptions, and in cases where these do not hold, the NG mode either ceases to exist or ceases to be gapless. If a symmetry is explicitly broken by an external field μ, for example, the NG mode will exist, but with a gap of size μ at $k \to 0$. If the broken symmetry is discrete, rather than continuous, there is no NG mode at all. And if the symmetry appears in conjunction with a gauge freedom encoding a long-ranged interaction, the NG mode may couple to the gauge field and develop a gap (this is called the Anderson–Higgs mechanism and will be addressed in Section 8.3). In its historic original formulation, the theorem also required Lorentz invariance, but non-relativistic versions have been derived later, which we shall address in Section 4.2.

The requirement that some translational invariance remains in the symmetry-broken state is the same as that used to prove Noether's theorem in Section 1.3.3. We again need translational invariance only on a coarse-grained level, so that momentum is a good quantum number, and modes will have a definite value of momentum. We can then define a complete set of eigenstates of the Hamiltonian, $|n, \mathbf{k}\rangle$, labelled by their momentum $\mathbf{k}$ and energy $E_n(\mathbf{k})$, with n encoding all relevant quantum numbers other than momentum. These states are orthogonal, $\langle n', \mathbf{k}'|n, \mathbf{k}\rangle = (2\pi)^d \delta_{nn'} \delta(\mathbf{k} - \mathbf{k}')$, and can be used to write a resolution of the identity:

$$\hat{\mathbb{I}} = \sum_n \int \frac{\mathrm{d}^d k}{(2\pi)^d} \; |n, \mathbf{k}\rangle \langle n, \mathbf{k}| \,. \tag{4.2}$$

We can insert this into the definition of a broken-symmetry state in terms of the interpolating field given by Eq. (3.2):

$$\begin{aligned} \langle \psi | [\hat{Q}, \hat{\Phi}] | \psi \rangle &= \sum_n \int \frac{\mathrm{d}^d k}{(2\pi)^d} \Big(\langle \psi | \hat{Q}(t) | n, \mathbf{k}\rangle \langle n, \mathbf{k}| \hat{\Phi} | \psi \rangle - \text{c.c.} \Big) \\ &= \int_\Omega \mathrm{d}^d x \sum_n \int \frac{\mathrm{d}^d k}{(2\pi)^d} \Big(\langle \psi | \hat{\jmath}^t(\mathbf{x}, t) | n, \mathbf{k}\rangle \langle n, \mathbf{k}| \hat{\Phi} | \psi \rangle - \text{c.c.} \Big) \neq 0. \end{aligned} \tag{4.3}$$

Here, c.c. indicates the complex conjugate, and in the second line the global conserved charge $\hat{Q} = \int d^d x \, \hat{\jmath}^t$ is written as an integral over the Noether charge density. Goldstone's theorem addresses the modes as k approaches zero, but it is not concerned with the tower of states at precisely $k = 0$. It is thus related to the behaviour of the Noether charge density integrated over a large, but finite part of space.[1] The primary assumption in the derivation of Goldstone's theorem is that we can take the integration volume Ω in Eq. (4.3) to be large but finite. Because the interpolating field $\hat{\Phi}$ is local, any contributions to the expectation value from outside the volume Ω are guaranteed to vanish in relativistic theories by causality. In non-relativistic, or effective, theories, it vanishes as long as the theory does not contain any long-ranged interactions. That is, all interactions should decay sufficiently quickly with distance.

With this caveat in mind, we can again use translational invariance, and write

$$\begin{aligned}
&\langle\psi|[\hat{Q},\hat{\Phi}]|\psi\rangle \\
&=\int_\Omega \mathrm{d}^d x \sum_n \int \frac{\mathrm{d}^d k}{(2\pi)^d}\left(\langle\psi|\,\mathrm{e}^{-\frac{\mathrm{i}}{\hbar}(\hat{H}t-\hat{\mathbf{P}}\cdot\mathbf{x})}\hat{\jmath}^t(0,0)\mathrm{e}^{\frac{\mathrm{i}}{\hbar}(\hat{H}t-\hat{\mathbf{P}}\cdot\mathbf{x})}\,|n,\mathbf{k}\rangle\langle n,\mathbf{k}|\,\hat{\Phi}\,|\psi\rangle - \text{c.c.}\right) \\
&=\int_\Omega \mathrm{d}^d x \sum_n \int \frac{\mathrm{d}^d k}{(2\pi)^d}\left(\mathrm{e}^{\frac{\mathrm{i}}{\hbar}(E_n t-\hbar\mathbf{k}\cdot\mathbf{x})}\langle\psi|\,\hat{\jmath}^t(0,0)\,|n,\mathbf{k}\rangle\langle n,\mathbf{k}|\,\hat{\Phi}\,|\psi\rangle - \text{c.c}\right) \\
&=\sum_n \int \mathrm{d}^d k\,\delta_\Omega(\mathbf{k})\left(\mathrm{e}^{\frac{\mathrm{i}}{\hbar}E_n t}\langle\psi|\,\hat{\jmath}^t(0,0)\,|n,\mathbf{k}\rangle\langle n,\mathbf{k}|\,\hat{\Phi}\,|\psi\rangle - \text{c.c.}\right)\neq 0.
\end{aligned} \tag{4.4}$$

In the first line, the local Noether charge $\hat{\jmath}^t(\mathbf{x},t)$ was translated in time and space using the shift operators $\mathrm{e}^{-\mathrm{i}\hat{H}t/\hbar}$ and $\mathrm{e}^{-\mathrm{i}\hat{\mathbf{P}}\cdot\mathbf{x}/\hbar}$. In the second line we set E_n to be the energy of the state $|n,\mathbf{k}\rangle$ relative to that of the state $|\psi\rangle$, and we invoked translational invariance to see that $|\psi\rangle$ is a zero-momentum state. In the final line, we defined $(2\pi)^d\delta_\Omega(\mathbf{k})\equiv\int_\Omega \mathrm{d}^d x\exp(\mathrm{i}\mathbf{k}\cdot\mathbf{x})$ to be a strongly peaked function tending towards a Dirac delta function in the limit of large integration volume. Because $|\psi\rangle$ is assumed to be a broken-symmetry state, the order parameter cannot be zero. This implies at least one state $|n,\mathbf{k}\rangle$ must exist such that the integrand in the final line also does not vanish, even for large Ω, when only contributions with momentum $\mathbf{k}$ tending to zero can contribute. This is the first part of the theorem: there must exist a state near zero momentum that is excited from the broken-symmetry state by both the local Noether charge operator $\hat{\jmath}^t(0,0)$ and the interpolating field $\hat{\Phi}$.

Noether's theorem guarantees the global charge $\langle\hat{Q}\rangle$ to be time-independent. If $\hat{\Phi}$ does not depend on time, then in the thermodynamic limit where the broken state $|\psi\rangle$ is an energy eigenstate, the order parameter $\langle[\hat{Q},\hat{\Phi}]\rangle$ is also time-independent. For the right-hand side of Eq. (4.4), we then find

$$\begin{aligned}
\partial_t\langle[\hat{Q},\hat{\Phi}]\rangle &= \partial_t\sum_n\int\frac{\mathrm{d}^d k}{(2\pi)^d}\,\delta_\Omega(\mathbf{k})\left(\mathrm{e}^{\mathrm{i}E_n t}\langle\psi|\,\hat{\jmath}^t(0,0)\,|n,\mathbf{k}\rangle\langle n,\mathbf{k}|\,\hat{\Phi}\,|\psi\rangle - \text{c.c.}\right) \\
&=\sum_n\int\frac{\mathrm{d}^d k}{(2\pi)^d}\,\delta_\Omega(\mathbf{k})\mathrm{i}E_n\left(\mathrm{e}^{\mathrm{i}E_n t}\langle\psi|\,\hat{\jmath}^t(0,0)\,|n,\mathbf{k}\rangle\langle n,\mathbf{k}|\,\hat{\Phi}\,|\psi\rangle - \text{c.c.}\right)=0.
\end{aligned} \tag{4.5}$$

We already found that there must be at least one state, the NG mode, for which the term between brackets does not vanish. The final line then implies that the NG mode must have vanishing energy, $E_n(\mathbf{k})\to 0$ as $\mathbf{k}\to 0$. This completes the proof of Goldstone's theorem: a system with a spontaneously broken symmetry has at least one excitation whose energy vanishes as its wave number approaches zero.

[1] Formally, we should consider both the limit of the volume V of our system tending to infinity and that of the integration volume Ω tending to V. Taking $V\to\infty$ before taking $\Omega\to V$, then guarantees that the point $k=0$ is excluded from any momentum integrals appearing in this section. This singular limit is discussed in detail in Ref. [107].

Notice that Goldstone's theorem is constructive, in the sense that it not only guarantees that gapless modes exist whenever a symmetry is spontaneously broken but also indicates how to find these modes. They can be excited from the symmetry-broken state by acting on it with either the local Noether charge operator or the interpolating field. In Section 10.1.2, we will discuss an alternative but equivalent derivation of Goldstone's theorem using the quantum action encountered in effective field theory.

4.2 Counting of NG Modes

The derivation of Goldstone's theorem may at first sight seem to suggest that there is always one NG mode for each broken symmetry generator. This is known not to be the case, however. For example, the Heisenberg ferromagnet has only a single NG mode, while two spin-rotation symmetries are broken. Similarly, one may be tempted to assume that the energy of NG modes always vanishes linearly in momentum, $E_n \propto k$. For relativistic systems, this certainly is the case, since Lorentz symmetry dictates that time and space derivatives appear on equal footing in the action. However, in non-relativistic systems, the Heisenberg ferromagnet again provides a counterexample to the general rule. Its single NG mode is quadratic in momentum, rather than linear.

Goldstone's theorem as we derived it earlier, just states there is at least one NG mode whenever any symmetry is broken, and it does not specify its dispersion relation other than that it is gapless. Consequently, there is no real contradiction with the observed properties of ferromagnets. How many NG modes should really be expected in any given system and what replaces the seemingly intuitive rule of one mode per broken symmetry was clarified only relatively recently [31, 32, 73, 139, 146, 176, 205, 206, 207, 209].

In the derivation of Goldstone's theorem, we found that NG modes can be excited from the broken-symmetry state by either the generator of a broken symmetry or the interpolating field. A special case arises if the interpolating field $\hat{\Phi}$ is itself also a generator of a broken symmetry. A clear example is again the Heisenberg ferromagnet, in which one of the spin-rotation operators, say $\hat{S}^z$, obtains a non-zero expectation value. The commutator of the broken generators $\hat{S}^x$ and $\hat{S}^y$ is proportional to $\hat{S}^z$ and can thus be used as an order parameter. The broken generators in this case act as interpolating fields for each other, and Eq. (4.4) shows that they must excite the same NG mode [139].

More generally, take any two symmetry generators $\hat{Q}_{a,b} = \int_x \hat{\jmath}^t_{a,b}(x)$ and consider the commutator expectation value:

$$\langle[\hat{Q}_a, \hat{\jmath}^t_b(x)]\rangle = \int \mathrm{d}^d y\, \langle[\hat{\jmath}^t_a(y), \hat{\jmath}^t_b(x)]\rangle = \int \mathrm{d}^d y\, \delta(x-y) \sum_c \mathrm{i} f_{abc} \langle \hat{\jmath}^t_c(y)\rangle$$
$$= \sum_c \mathrm{i} f_{abc} \langle \hat{\jmath}^t_c(x)\rangle = \langle[\hat{\jmath}^t_a(x), \hat{Q}_b]\rangle. \tag{4.6}$$

If the commutators in this equation have non-zero expectation value in the broken-symmetry state, the operators $\hat{\jmath}^t_a$ and $\hat{\jmath}^t_b$ are again seen to excite the same NG mode. Following Watanabe and Murayama, we call such excitations **type-B NG modes**, while 'ordinary' NG modes excited by only a single Noether charge are said to be **type-A** [207]. From Eq. (4.6), it is clear that type-B modes cannot arise for Abelian symmetry groups, in which all generators commute with one another.

To systematically count the number of NG modes of either type, it is convenient to construct the so-called Watanabe–Brauner matrix [206]:

$$\beta_{ab} = -\mathrm{i}\langle\psi|[\hat{Q}_a, \hat{j}^t_b(x)]|\psi\rangle. \tag{4.7}$$

Here, a and b label all broken symmetry generators, and $|\psi\rangle$ is a symmetry-breaking state. As explained in Section 3.2, the total number of broken generators is equal to the dimension of the coset space G/H. Notice that since the broken-symmetry state is assumed to be translationally invariant, the matrix elements β_{ab} do not depend on the position x. The numbers n_A and n_B of type-A and type-B Nambu–Goldstone modes are now given by

$$n_\mathrm{A} = \dim G/H - \operatorname{rank}\beta, \qquad n_\mathrm{B} = \frac{1}{2}\operatorname{rank}\beta. \tag{4.8}$$

Two independent proofs [73, 207] moreover show that (in almost all cases) type-A NG modes have linear dispersion, while type-B modes have quadratic dispersion. Both Eq. (4.8) and the typical dispersions of NG modes can be understood using the quantum action in a low-energy effective Lagrangian approach [31, 113, 207, 209], as we will discuss in Section 10.3.

Without going into detail here, the core of the construction consists of writing down the most general Lagrangian allowed by the symmetry of the problem, in terms of fields $\pi_a(x)$ which take values in the space G/H of broken symmetry generators. The number of these fields equals the number of broken symmetry generators, but the fields are not necessarily independent. The gapless modes in the spectrum of this low-energy effective Lagrangian will correspond to the NG modes. The lowest order terms can generically be written as [207]

$$\mathcal{L}_\mathrm{eff} = m_{ab}(\pi_a\partial_t\pi_b - \pi_b\partial_t\pi_a) + \bar{g}_{ab}\partial_t\pi_a\partial_t\pi_b - g_{ab}\nabla\pi_a \cdot \nabla\pi_b. \tag{4.9}$$

Here, m_{ab}, $\bar{g}_{ab}$, and g_{ab} are coefficients, some of which are constrained by symmetry. For example, there are no terms linear in gradients, since we assume space to be isotropic. For the same reason, it is clear that the first term in $\mathcal{L}_\mathrm{eff}$ breaks Lorentz invariance and can only be non-zero in non-relativistic systems. The coefficients m_{ab} turn out to be given precisely by the corresponding elements of the Watanabe–Brauner matrix.

Exercise 4.1 (Number of type-B NG modes).
Part of the proof of Eq. (4.8) in Ref. [207] is shown in Eq. (4.10). The matrix β_{ab} is real and antisymmetric. Then there exists an orthogonal transformation O such that $\tilde{\beta} = O\beta O^\mathrm{T}$ takes the form

$$\tilde{\beta} = \begin{pmatrix} M_1 & & & & & \\ & \ddots & & & & \\ & & M_m & & & \\ & & & 0 & & \\ & & & & \ddots & \\ & & & & & 0 \end{pmatrix}, \qquad M_i = \begin{pmatrix} 0 & \lambda_i \\ -\lambda_i & 0 \end{pmatrix}, \tag{4.10}$$

where $m = \frac{1}{2}\operatorname{rank}\beta$ and the $\lambda_1, \dots, \lambda_m$ are real and non-zero. To prove this:
a. Show that the eigenvalues of β are purely imaginary. This means that the matrix E with the imaginary eigenvalues on the diagonal can be obtained by some unitary transformation $E = U\beta U^\dagger$.

b. Show that the non-zero eigenvalues come in conjugate pairs $(\mathrm{i}\lambda_i, -\mathrm{i}\lambda_i)$. Since there are rank β non-zero eigenvalues, this implies rank β is even, so n_B in Eq. (4.8) is integer.
c. For each 2×2-submatrix e_i with a conjugate pair on the diagonal elements, find a unitary matrix w_i such that $w_i e_i w_i^\dagger = M_i$.

You now established that β is unitarily equivalent to $\tilde{\beta}$ by the unitary transformation WU, where W has the submatrices w_i on the top-left diagonal and other entries zero. Since β and $\tilde{\beta}$ are both real, they are then also orthogonally equivalent, and the orthogonal matrix O can be constructed from WU. This is done, for example, in problem 160 of Ref. [66].

From the effective Lagrangian, we can find the equations of motion for the fields $\pi_a(x)$ and the corresponding dispersion relations. In systems where m_{ab} is zero, including all relativistic systems, the effective Lagrangian describes modes with linear dispersions. More precisely, Fourier transforming the Lagrangian will introduce a frequency ω for every time derivative and a momentum k for every gradient, so the dispersion will obey $\omega^2 \propto k^2$. If the coefficients m_{ab} are not zero, their contribution to the Lagrangian will always dominate the second-order derivatives at sufficiently low energies, and the dispersion will be quadratic, $\omega \propto k^2$. Notice that terms like $\pi_a \partial_t \pi_a$ (no summation over a) are total derivatives in the Lagrangian and vanish in the action. Therefore, m_{ab} must be antisymmetric, and the first term in Eq. (4.9) can only be non-zero in systems where two fields are coupled. This corresponds to the previously found reduction in the number of gapless NG modes when two generators excite the same mode. The effective action shows that such a reduction goes hand in hand with the fact that type-B modes have quadratic dispersion.

Finally, notice that in the effective Lagrangian, it is possible for the coefficients g_{ab} to be zero. In that case, higher-order terms must be taken into account, and it is therefore possible, for instance, that type-A modes obtain a quadratic dispersion, or rather $\omega^2 \propto k^4$. An example of this situation occurs for so-called Tkachenko modes of vortex lattices in rotating superfluids [209].

4.3 Examples of NG Modes

To see how the formal considerations of Sections 4.1 and 4.2 impact the observable properties of real materials, we consider a short selection of practical examples. This list is far from exhaustive, but should give you a feeling for the extent to which Goldstone's theorem shapes the physics of all systems subject to spontaneous symmetry breaking.

Superfluid: The superfluid was described in Section 3.2 in terms of a complex scalar field theory, in which the field operator itself acts as the order parameter. The action is invariant under rotations of the phase of the field, and Noether's theorem shows this symmetry to be associated with the conservation of particle number. In the superfluid phase, the $U(1)$ phase-symmetry is spontaneously broken and the number of particles in the superfluid condensate is indeterminate. There is one broken symmetry generator and one NG mode, which may be excited by finite-wave number rotations of the phase variable in the complex scalar field. The NG mode is type-A and its low-energy dispersion is linear in momentum.

The supercurrent (a particle current that flows without viscosity) is a direct manifestation of this NG mode.

Crystal: Crystals in d spatial dimensions break the symmetries of space, which consist of d translations and $\frac{1}{2}d(d-1)$ rotations. The translation group is Abelian, so the associated NG modes are all type-A, with linear dispersions at low energies. They are called phonons, or sound waves, and there is one for each direction of space. The rigidity due to breaking translational symmetry is shear rigidity, whose non-zero value is the traditionally used quantity for distinguishing solids from liquids.

The broken rotational symmetries in the crystal do not lead to any additional NG modes. As was shown only recently [20, 115, 208], rotations and translations are not independent symmetry operations. Consequently, the NG fields excited by broken translations and broken rotations are also not independent and contain redundant degrees of freedom. Because of this, the broken rotations do not lead to independent NG modes. Intuitively, this means that if you try to excite a rotational NG mode by applying local torque stress to a crystal, you instead end up exciting transverse sound modes. In fact, Lorentz boosts are also spontaneously broken in the crystal (or any other medium), but like rotations, they do not lead to independent NG modes [144].

Antiferromagnet: The Heisenberg antiferromagnet described in Section 2.4 breaks two out of three spin-rotational symmetries, generated by say $\hat{S}^x$ and $\hat{S}^y$. The commutator in the off-diagonal elements of the Watanabe–Brauner matrix is the magnetisation $\hat{S}^z$, which vanishes in the antiferromagnetically ordered state. The NG modes excited by the broken symmetry generator are thus independent, and there are two type-A NG modes with linear dispersions at low energies, called magnons. These can be viewed as plane waves of precession for the spins on each sublattice.

Ferromagnet: The Heisenberg ferromagnet of Exercise 3.3 breaks the same spin-rotation symmetries as the antiferromagnet. This time, however, the magnetisation $\hat{S}^z$ is an order parameter, whose expectation value does not vanish in the broken-symmetry state. The Watanabe–Brauner matrix therefore has non-zero off-diagonal elements, and the modes excited by the two broken symmetry generators are not independent. There is then one type-B NG mode, with a quadratic dispersion at low energy.

Canted antiferromagnet: Adding a term to the antiferromagnetic Heisenberg Hamiltonian that favours orthogonal alignment of neighbouring spins will lead to a symmetry-broken state in which all spins uniformly cant away from their preferred directions in the Néel state. The result is a configuration with non-zero total uniform magnetisation, as well as non-zero staggered or sublattice magnetisation in a perpendicular direction. This state breaks all three spin-rotation symmetries. In this case, the broken generator of rotations around the axis of uniform magnetisation will excite one type-A NG mode, while the remaining two broken generators both excite the same type-B NG mode.

4.3.1 NG-Like Excitations

There are several systems with excitations that are clearly related to the physics of spontaneous symmetry breaking and NG modes, but that do not satisfy all of the assumptions underlying Goldstone's theorem. Again, we give a short selection of examples to give you a feeling for how NG-like excitations occur in systems that strictly speaking fall just outside of the realm of spontaneous symmetry breaking.

Gapped and pseudo NG modes: If a system with a spontaneously broken symmetry is exposed to an external field that explicitly breaks the same symmetry, there is an NG-like excitation with an energy gap proportional to the external field. This has been dubbed a massive or gapped NG mode [142, 212]. The typical example is that of spin waves in a ferromagnet exposed to a magnetic field parallel to the magnetisation.

Similarly, if a symmetry is broken explicitly by a weak coupling to other fields (or other degrees of freedom) rather than to an external field, there is a scalar particle with an energy gap, which otherwise has all the characteristics of an NG mode. This is often called a pseudo NG boson in particle physics. The most famous example is the pion in the Standard Model, a scalar particle with small mass due to chiral symmetry breaking, which will be discussed in detail in Section 11.2.

Quasi-NG mode: In some cases, the exact ground state of a system may have more symmetry than the Hamiltonian itself. If moreover that additional symmetry is broken spontaneously, an NG-like excitation emerges, which is called a quasi-NG mode. Examples occur both in particle physics (such as in technicolour and supersymmetry models) and in condensed matter physics (Helium-3 superfluids and spinor Bose–Einstein condensates).

Goldstino: In general, broken symmetry generators act as creation operators for NG modes when acting on a symmetry-breaking state. Since they generally obey some set of commutation relations, the NG modes are usually bosons. Typically, they correspond to scalar particles, although sometimes it makes sense to assign a vector or tensor structure to groups of several NG modes. In the special case that the symmetry generators happen to satisfy anti-commutation relations, the NG modes are fermions. This situation occurs for broken **supersymmetry**, which is a possible extension of the Poincaré algebra of spacetime symmetries (see Section 11.1) with fermionic supersymmetry generators. If supersymmetry is spontaneously broken, the associated fermionic NG modes are called Goldstinos.

4.4 Gapped Partner Modes

Type-B Nambu–Goldstone modes may be excited from a symmetry-breaking state by two distinct broken symmetry generators, whose commutator has a non-zero expectation value. In the effective Lagrangian description of Eq. (4.9), this was signalled by two fields π_1 and π_2 not being independent. Even though the modes are not independent, they do originate from two distinct symmetry transformations. One might wonder whether there should not be two modes associated with the two fields in the Lagrangian. In fact, there generally are two modes, but the second mode is gapped [18, 70, 87, 143]. To see this, consider the effective Lagrangian for a system with two coupled broken symmetry generators, and nothing else [94]:

$$\mathcal{L}_{\text{eff}} = 2M(\pi_1 \partial_t \pi_2 - \pi_2 \partial_t \pi_1) + \frac{1}{c^2}(\partial_t \pi_1)^2 + \frac{1}{c^2}(\partial_t \pi_2)^2 - (\nabla \pi_1)^2 - (\nabla \pi_2)^2. \tag{4.11}$$

The dispersion relations for the two modes described by this Lagrangian are

$$\omega_\pm = \sqrt{c^2k^2 + M^2c^4} \pm Mc^2. \tag{4.12}$$

That is, there is one gapless NG mode with dispersion $\omega_- = \frac{k^2}{2M} + \mathcal{O}(k^4)$, and one **gapped partner mode** with dispersion $\omega_+ = 2Mc^2 + \frac{k^2}{2M} + \mathcal{O}(k^4)$. In the limit $M \to 0$, the two modes decouple and we obtain a degenerate, linear dispersion $\omega_\pm = ck$. This corresponds to having two type-A NG modes, which is consistent with M being the expectation value of the off-diagonal element in the Watanabe–Brauner matrix, which then becomes diagonal if M vanishes. In the opposite limit of $c \to \infty$, the gap goes to infinity, and we are effectively left with only a single, gapless mode. This corresponds to having only terms with single time derivatives in Eq. (4.11). An example for such a situation is provided by the Heisenberg ferromagnet, which has no gapped mode in its spectrum. Physically, the fact that the second mode disappears altogether for the ferromagnet can be understood by realising that its NG modes are always excited by a lowering of the maximally polarised spins. That is, in both $\hat{S}^x = (\hat{S}^+ + \hat{S}^-)/2$ and $\hat{S}^y = (\hat{S}^+ - \hat{S}^-)/2\mathrm{i}$ only the $\hat{S}^-$ part actually excites a mode. Acting with $\hat{S}^+$ on a maximally polarised ferromagnet annihilates the state and does not correspond to a physical excitation, as shown in Exercise 3.3. The actions of the two broken symmetry generators on the symmetry-breaking state are thus entirely equivalent, and there really is only a single excitation associated with them. More generally, non-zero values for both M and c always indicate the presence of both a type-B NG mode and an accompanying gapped partner mode.

At first sight, the existence of a gapped mode excited by a broken symmetry generator may seem to invalidate the proof of Goldstone's theorem in Section 4.1, in which we argued that all modes excited by broken generators must be massless to ensure that the order parameter is time-independent. Although there is no general proof of which we are aware, it has been checked in several cases that this paradox is resolved by observing that the terms in Eq. (4.5) always contain a product of two matrix elements, of the form

$$E_n \langle\psi|\, \hat{\jmath}^t(0,0)\, |n,\mathbf{k}\rangle \langle n,\mathbf{k}|\, \hat{\Phi}\, |\psi\rangle . \tag{4.13}$$

In all verified cases where $\hat{\jmath}^t(0,0)$ excites a gapped partner mode with non-zero energy E_n, it turns out that $\langle\psi|\, \hat{\Phi}\, |n,\mathbf{k}\rangle$ is proportional to the energy of the accompanying gapless mode, which vanishes for $\mathbf{k} \to 0$. The existence of the gapped partner mode therefore does not contradict the time-independence of the order parameter.[2]

Exercise 4.2 (Heisenberg ferrimagnet).
A **ferrimagnet** can be described by the Heisenberg Hamiltonian $\hat{H} = J\sum_{\langle i,j\rangle} \hat{\mathbf{S}}_i \cdot \hat{\mathbf{S}}_j$ on a square lattice with positive coupling J, but for a system in which the spins on the two sublattices have different sizes. We consider spin-S_A particles on the A sites, and spin-S_B particles on the B sites.
a. Calculate the average magnetisation per unit cell $\langle \hat{S}^z_{\mathrm{tot}}\rangle/(N/2)$ for the Néel-like state in which all spins are in eigenstates of the $\hat{S}^z_j$ operators, with eigenvalues $m_A = S_A$ and $m_B = -S_B$. This state breaks the spin-rotation symmetries generated by S^x and S^y.
b. Calculate the matrix elements of the Watanabe–Brauner matrix defined in Eq. (4.7) (take the average per unit cell).
c. The local magnetisation $\hat{S}^z_i$ is an order parameter operator for this state, but it turns out

[2] We thank Tomas Brauner for discussions concerning this point.

that there is also a second order parameter: calculate the expectation value per unit cell for the staggered magnetisation operator $\hat{N}_i^z = (-1)^i \hat{S}_i^z$.

We thus find that the breaking of spin-rotational symmetry in the ferrimagnet may be described by (at least) two distinct order parameters, only one of which commutes with the Hamiltonian. Clearly then, it is not sufficient to find an order parameter operator that does not commute with the Hamiltonian to claim that a symmetry-breaking system is of type-A. One should calculate all the elements of the Watanabe–Brauner matrix, as you did in this exercise.
With a bit more effort, you can see that the ferrimagnet has one quadratically dispersing NG mode and one gapped partner mode, whose gap scales with the difference in spin sizes, $\Delta \propto |S_A - S_B|$. If the spins on the two sublattices have equal sizes, the system is an ordinary antiferromagnet with two gapless, linearly dispersing modes (as described by Eq. (4.12) in the limit $M \to 0$). In the opposite limit, as Δ grows, exciting the gapped partner mode costs ever more energy, and the spectrum looks more and more like that of a ferromagnet with only a single gapless, quadratically dispersing, NG mode.

4.5 The Tower of States and Type-B NG Modes

NG modes may be seen as the $k > 0$ cousins of the collective excitations with zero wave number that make up the tower of states in systems with a spontaneously broken symmetry. The close relation between the internal and collective modes persists in the distinction between systems with type-A and those with type-B modes. We have seen that the Heisenberg antiferromagnet and the harmonic crystal, for example, have unique ground states and a tower of low-energy states, all of which are unstable. This is general for systems with type-A NG modes. We have also seen in Exercise 3.3 that the ferromagnet has a macroscopically degenerate ground state and no tower of states. The number of exact ground states is infinite in the thermodynamic limit, but of order N for finite systems, with N the number of particles.

As it turns out, in systems with a type-B NG mode and a gapped partner mode (which have both M and c non-zero in Eq. (4.11)) there is a ground state degeneracy, and still no tower of states. While we are not aware of a general proof, the relation of the ground state degeneracy to the collective modes can be seen in the explicit example of an antiferromagnet on the Lieb lattice.[3] The Hamiltonian is the same as that of the usual Heisenberg antiferromagnet, Eq. (2.17), with spin-S particles on both sublattices. On the Lieb lattice, however, there are twice as many A sites as there are B sites, as shown in Figure 4.1.

The classical ground state is a Néel-type state with spins pointing antiparallel on the two sublattices. It has non-zero magnetisation $\langle \hat{S}^z \rangle = S(N_A - N_B) = \frac{1}{3}SN$, with $N_{A,B}$ denoting the number of sites on the A and B sublattices, and N the combined total number of sites. Because the magnetisation is non-zero, the NG modes of the Lieb-lattice antiferromagnet will be type-B. Just like the usual square-lattice antiferromagnet discussed in Section 2.4,

[3] We thank Hal Tasaki for pointing out this example, which is discussed also in his textbook [191].

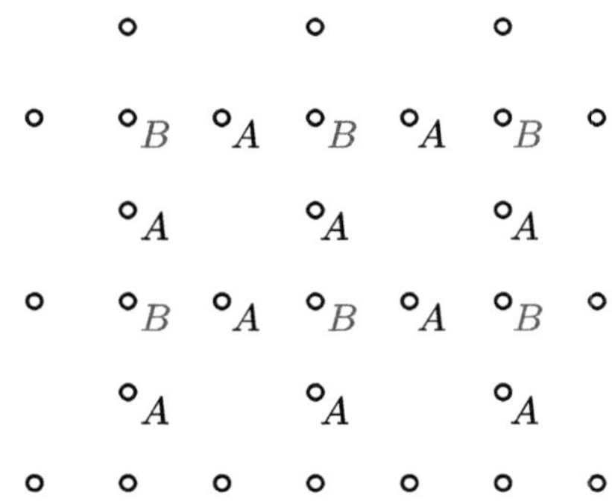

Figure 4.1 The Lieb lattice. Each A site has only B neighbours and vice versa, but there are twice as many A sites as there are B sites.

	NG mode dispersion (to leading order)	Ground state degeneracy	Tower of states
Type-A	$\propto k$	No	Yes
Type-B ferro	$\propto k^2$	Yes, $\mathcal{O}(N)$	No
Type-B ferri	$\propto k^2, M + k^2$	Yes, $\mathcal{O}(N)$	No

Table 4.1 Tower of states and NG modes.

any exact eigenstates of the model have non-vanishing overlap with those of a corresponding Lieb–Mattis model (the $k = 0$ part of the Hamiltonian, with infinte-range interactions). Using that fact, the ground state can be shown to have total spin $S_{\text{tot}} = S(N_A - N_B)$, and degeneracy $2S_{\text{tot}} + 1$, which is of order N. The energy required to excite any of the remaining $k = 0$ states is of order $\mathcal{O}(J)$, in contrast to the usual antiferromagnet with type-A NG modes, where $k = 0$ excitations have energy vanishing as $\mathcal{O}(J/N)$. In other words, there is no tower of states. The relations between the types of NG modes and the spectrum of collective excitations are summarised in Table 4.1.

4.6 Rigidity

As mentioned in the introduction to Chapter 4, Noether's theorem guarantees that any local accumulation of Noether charge will be dispersed over time. In systems with a spontaneously broken symmetry, this implies that over time, any local perturbations of the order parameter are carried away and spread out, resulting in a restoration of the local order. This resistance to deformations of the order parameter endows symmetry-breaking objects with a type of generalised rigidity.

To see how this relates to the familiar notion of rigidity used for solid objects in everyday life, recall the derivation of Goldstone's theorem in Section 4.1. It shows that in the presence of a field proportional to the Noether charge, acting with the time evolution operator on a symmetric state will cause the creation of NG modes. The energy or wavelength at which modes are excited, depends on the strength of the applied field, but Goldstone's theorem guarantees that even at arbitrarily low energies, NG modes are always available. In the so-called linear response limit, where we consider the effect of an applied field with vanishing strength, the only perturbations to the order parameter that can generically be created, are the NG modes with diverging wave length. In a crystal, which breaks translational symmetry, the NG modes are phonons or quantised sound waves. Exciting a phonon with infinite wave length corresponds to a collective translation of the entire crystal, and we thus recover the intuitive definition of rigidity: the linear response of a crystal to external forces is not to deform, but rather to move as a whole. In other words, pushing a crystal softly enough doesn't cause any dents or bending, but moves it along as a single, rigid object. Notice that even if the system is perturbed locally, for example, kicking a rock with the tip of your shoe, this will yield a global response, causing the rock as a whole to fly away.

The argument for the rigidity of crystals depends only on Goldstone's theorem and is therefore general to any symmetry-breaking state. Whenever a system with a spontaneously broken symmetry is subjected to an applied field perturbing its order parameter with vanishing energy, the response will be a global, collective change in the value of the order parameter. For crystals with broken translational symmetry, an externally applied force causes the translation of the crystal as a whole. Similarly, for magnetic systems, with broken spin rotational symmetry, the order can be perturbed by applying magnetic fields. The linear response will be a global rotation of all spin magnetic moments, without rotating the crystal structure they happen to be in.

In each case, the field causing a global response in the low-energy limit is proportional to the generator of a broken symmetry and is said to be **conjugate** to the order parameter. Continuing this reasoning, a superfluid or superconductor with broken phase rotational symmetry should display rigid phase rotations when subjected to a gradient in chemical potential. This is indeed the case, and the global phase rotations correspond to global supercurrents or flows. The fact that these do not encounter any resistance or viscosity can now be understood as a manifestation of the diverging lifetime of low-energy NG modes, which will be discussed in more detail in Section 10.3.2.

A second way of exciting a global, rigid response in an ordered system is given by the generalisation of the Josephson effect, which we encountered in Section 3.5. Any two systems that break the same type of symmetry will interact through a coupling of their order parameters [19]. The interaction energy is minimised whenever the two systems have the same order parameter value. For two superconductors, the Josephson coupling tends to cause a rigid alignment of their global phases. In this case, the interaction is mediated by a tunnelling supercurrent that runs between the superconductors as their phases turn, and thus transports Noether charge from one superconductor to the other. Generalising the Josephson effect to other types of broken symmetry, we can understand the interaction between two ferromagnets as the rigid alignment of their global magnetic moments, caused by transporting magnetisation between them. Similar reasoning can be applied to any type of spontaneously broken symmetry, yielding a tendency to align the order parameters of any pair of symmetry-breaking objects interacting through a Josephson-type coupling.

Regardless of whether a perturbation in the order is caused by a conjugate field or a Josephson coupling, Goldstone's theorem guarantees that for any object with a spontaneously broken symmetry, low-energy excitations of the order occur at diverging wave lengths, and cause a global change, or rigid response, in the value of the order parameter.

5 Phase Transitions

So far, we discussed properties of equilibrium phases of matter in which a symmetry of the action or Hamiltonian is spontaneously broken. For any symmetric Hamiltonian, however, there must also be equilibrium configurations that respect the symmetry. After all, at infinite temperature, the thermal density matrix of any system is proportional to the identity matrix, which is invariant under all possible symmetry transformations. To find a long-range ordered state at low temperatures, some symmetries must then be broken at a specific **critical temperature**, T_c, as the system cools down. That is, there must be a **phase transition** from the symmetric high-temperature state to the symmetry-breaking low-temperature state. The study of phase transitions is a major field in and of itself, and is introduced in detail in several textbooks [39, 58, 225]. Here, we give a limited overview of some of the central concepts most relevant to the study of spontaneous symmetry breaking.

5.1 Classification of Phase Transitions

Every ordered, symmetry-broken phase of matter has a non-zero expectation value of its order parameter operator. In contrast, the order parameter will be zero in the corresponding symmetric, or disordered, phase. We can therefore distinguish between order and disorder by considering the value of the order parameter. Although the distinction is sharp, in the sense that the order parameter is either zero or not, we can still distinguish between different types of phase transitions by considering the way in which the order parameter goes from zero to non-zero as a system is cooled through the phase transition.

- In **discontinuous** or **first-order** phase transitions, the order parameter jumps discontinuously from zero to a non-zero value at the critical temperature. Similarly, there is a sudden change in entropy at the transition point, and going through the transition requires the release of latent heat. First-order transitions often show hysteresis in the thermal evolution of the order parameter as the system is cycled across the phase transition. Phase transitions between different states characterised by the same broken symmetries, such as the gas to liquid transition, are almost always first order. The converse is not true, however, and transitions that involve the breakdown of a symmetry can readily be discontinuous, with the liquid-to-solid transition a famous example.
- In **continuous** or **second-order** phase transitions the order parameter increases continuously from zero as the critical temperature is traversed. The entropy also changes continuously, while in contrast, the correlation length and related energy scales diverge at the critical temperature. In fact, at the critical temperature of a second-order phase transition, systems become scale-invariant, in the sense that their physical properties no longer

depend on the length (or energy) scale at which they are probed. Many symmetry-breaking phase transitions are second order, with the onsets of superfluidity, (anti)ferromagnetism, and several phases of liquid crystals as famous examples.

The terminology of first- and second-order phase transitions stems from the now largely obsolete Ehrenfest classification of phase transitions, in which a transition is said to be of n-th order if the nth derivative of the free energy with respect to temperature is discontinuous at the critical temperature. While not yet experimentally accessible at Ehrenfest's time, it has now become clear that for instance the heat capacity, which is proportional to the second derivative of the free energy, is not just discontinuous at the critical temperature of many systems, but in fact diverges upon approaching the critical temperature [58]. This complicates the direct application of Ehrenfest's rules, and we therefore prefer a classification based only on the behaviour of the order parameter. This has the additional advantage that the continuous or discontinuous evolution of the order parameter can be directly generalised to **quantum phase transitions**, which are phase transitions occurring at zero temperature as function of some other parameter such as pressure, density, or magnetic field (see Appendix A.3).

To summarise the behaviour of the order parameter as a function of multiple parameters, including temperature, it is often useful to draw a **phase diagram**. This is a plot with the externally controllable parameters such as temperature and pressure on each of the axes. Within this parameter space, the different symmetric and symmetry-breaking phases are indicated, while phase transitions are denoted by lines or surfaces separating the different phases. A line of second-order phase transitions is sometimes called a critical line. A line of first-order transitions can end in a point. If the transition exactly at this point is continuous, then that point is called a critical endpoint. As an example, the phase diagram of Helium-4 is shown in Figure 5.1, which contains all these features.

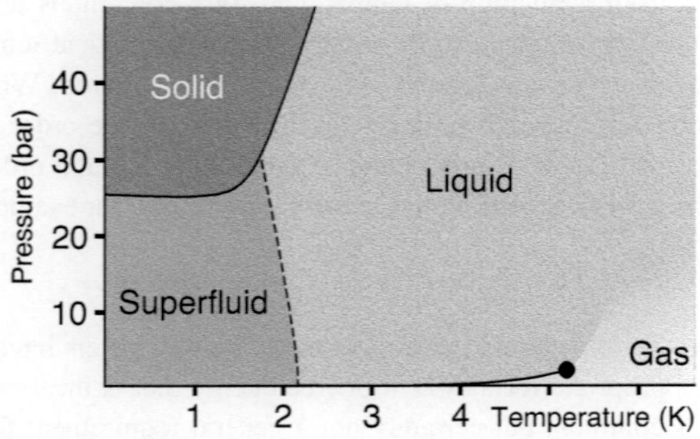

Figure 5.1

Phase diagram of Helium-4. The different equilibrium phases as a function of pressure and temperature are indicated. Solid lines represent the first-order transitions between liquid and solid phases (involving the breaking of a symmetry), and between liquid and gas (without breaking any symmetry). The liquid–gas transition line ends in a critical endpoint (large dot). The scale invariance at this point is signalled for instance by its critical opalescence. At temperatures beyond the critical endpoint, it is no longer possible to clearly distinguish between a liquid and a gas, and there is no abrupt transition, but a smooth crossover between the two phases. The dashed line denotes the second-order phase transition from a liquid to a superfluid.

5.2 Landau Theory

The modern view of phase transitions, also known as **Landau theory**, can be said to have originated with the **equation of state** introduced by Van der Waals to explain the transition between gas and liquid phases of matter [78, 103, 196]. It posits that below a critical temperature, T_c, the pressure of a collection of interacting particles will decrease when volume is decreased, due to attractive forces between the particles. This is in contrast to what we expect for any gas, and the critical temperature thus separates two distinct phases of matter. At the phase transition, no symmetry is broken, as both the ideal gas and the ideal liquid are symmetric under spatial transformations. Nevertheless, the liquid and gas have preferred densities that do not depend on the volume of the container they are in, and we can use this to define an 'order parameter' o, which has different values on either side of the transition. It is not an order parameter in the sense of Section 3.2, because no symmetries are broken and this parameter does not and cannot take values in any coset space G/H of inequivalent symmetry-broken states. Nevertheless, the role of the preferred density in delineating between gas and liquid phases, is the same as that of symmetry-based order parameters distinguishing order from disorder in other transitions.

5.2.1 Continuous Phase Transitions

In general, equations of state may be expressed in terms of thermodynamic quantities such as pressure, entropy, and so on. These thermodynamic quantities can always be written as derivatives of the free energy. The central idea of Landau theory is that the free energy of any system undergoing a phase transition can formally be written as a function of the order parameter, which is itself a function of temperature. In a continuous phase transition, the order parameter $o(T)$ is guaranteed to be small close to the critical temperature, and one can carry out a Taylor expansion around $o(T) = 0$ in that region. Which terms in such an expansion are non-zero depends strongly on the nature of the order parameter and the symmetries of the system. In the simplest case of a real-valued, scalar order parameter o, the expansion of the free energy according to Landau theory could, for example, read

$$\mathcal{F}_L[o,T] = \mathcal{F}_L[0,T] + \frac{1}{2}r(T)o^2 + \frac{1}{4}u(T)o^4 + \ldots \tag{5.1}$$

This form of the expansion, with odd powers of the order parameters having vanishing coefficients, would be appropriate for a system or free energy that is invariant under $o \to -o$. Such a symmetry is common, but certainly not a general requirement for the free energy. Notice that writing an expansion of the form of Eq. (5.1) is not an obvious thing to do. We know from thermodynamics that the (Gibbs) free energy is really only a function of temperature and pressure. The variables o and T on which the free energy $\mathcal{F}_L[o,T]$ depends are thus not independent. In fact the order parameter is a function of T, and its value is fixed for any given temperature. Nevertheless, it is formally possible to treat o as an independent parameter and write an expansion in terms of the form of Eq. (5.1). We can then heuristically think of $\mathcal{F}_L[o,T]$ with values of o not equal to the equilibrium value $o(T)$ as describing the energy of a system slightly out of equilibrium due to a perturbation of the order parameter.

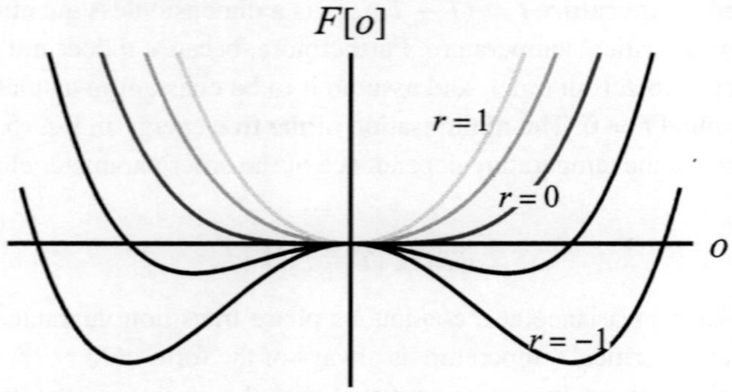

Figure 5.2

The Landau free energy of Eq. (5.1) as a function of the order parameter o during a continuous phase transition described by Eq. (5.2), with $u > 0$ for several values of r. The minimum of $\mathcal{F}_L$ moves continuously from zero to non-zero values as r decreases from positive to negative values.

Because in equilibrium the free energy is always minimised, any perturbations away from its equilibrium value must increase the free energy. This implies that given an expansion of the type of Eq. (5.1), the equilibrium value of $o(T)$ at any particular T can be found by minimising the free energy with respect to the order parameter while keeping the values of the expansion coefficients $r(T)$ and $u(T)$ fixed (see also Figure 5.2). Because the expectation value of a local order parameter should be finite in any physical situation, the free energy functional $\mathcal{F}_L[o,T]$ should always be bounded from below. This is guaranteed if the highest-order term appearing in the Landau expansion has an even power and a positive coefficient. In the case of Eq. (5.1), that means we should have $u(T) > 0$. If we somehow determine or measure the value of $u(T)$ and find it to be negative, a proper description of the system requires us to carry out the expansion of the free energy to higher order, continuing until we arrive at a term with positive prefactor.

Assuming for now that $u(T)$ is positive, then if $r(T)$ happens to also be positive, the lowest free energy can be found in the symmetric state, with $o(T) = 0$. However, if $r(T)$ happens to be negative, minimisation of the free energy yields a non-zero order parameter:

$$\frac{\partial \mathcal{F}_L[o,T]}{\partial o} = -|r(T)|o + u(T)o^3 = 0 \quad \Rightarrow \quad o = \pm\sqrt{\frac{|r(T)|}{u(T)}}. \tag{5.2}$$

A phase transition can now be described as a process in which a variation of temperature leaves $u(T)$ positive, but causes $r(T)$ to change sign. The order parameter then obtains a non-zero value at the temperature for which $r(T)$ goes through zero, which defines the critical temperature. In Figure 5.2 the free energy is plotted for different values of $r(T)$. The equilibrium order parameter value for which the free energy is minimal is seen to smoothly change from zero to non-zero values as $r(T)$ changes from positive to negative values, indicating a continuous phase transition.

Close to the critical temperature T_c, the function $r(T)$ can be Taylor expanded to first order in $T - T_c$, and will be of the form

$$r(T) \approx r_0 \frac{T - T_c}{T_c} \equiv r_0 t. \tag{5.3}$$

Here, the **reduced temperature** $t = (T - T_c)/T_c$ is a dimensionless quantity that measures the distance from the critical temperature. Furthermore, because u does not change sign, we can Taylor expand it to zeroth order, and assume it to be constant in a small enough region of temperature around $t = 0$. The minimisation of the free energy in Eq. (5.2) then gives an explicit prediction for the temperature dependence of the order parameter close to the critical temperature:

$$o(T) \propto (T - T_c)^{\frac{1}{2}}. \tag{5.4}$$

As it turns out, scale invariance at a continuous phase transition guarantees that the order parameter close to the critical temperature is always of the form $o(T) \sim (T - T_c)^\beta$, where β is called a **critical exponent** (not to be confused with the inverse temperature $\beta = 1/k_B T$). Similar critical exponents are associated with the behaviour of the specific heat at a continuous phase transition, $C_V \sim |T - T_c|^\alpha$, with the susceptibility, $\chi \sim |T - T_c|^\gamma$, and so on. We will come back to these exponents in Section 5.5, where the values of the critical exponents are found to depend solely on the form of the Landau free energy, highlighting the importance of symmetry in the analysis of phase transitions.

5.2.2 Discontinuous Phase Transitions

Contrary to continuous phase transitions, there is no guarantee that an expansion of the free energy in powers of the order parameter makes sense close to a discontinuous phase transition. Nevertheless, Landau theory turns out to describe these types of transitions as well. Consider a situation in which the prefactor u of the quartic term is negative, and the expansion of the free energy is carried out to sixth order:

$$\mathcal{F}_L[o, T] = \mathcal{F}[0, T] + \frac{1}{2} r(T) o^2 + \frac{1}{4} u(T) o^4 + \frac{1}{6} w(T) o^6 + \ldots \tag{5.5}$$

Here, we assume $w(T)$ is positive for all temperatures of interest, to ensure that the free energy is bounded from below. For simplicity, consider $r(T) = w(T) = 1$. Then there is a jump in the location of the minimum of the free energy as u decreases continuously from $u > -2$ to $u < -2$, indicating a discontinuous phase transition. This is shown pictorially in Figure 5.3. At discontinuous phase transitions, there is no scale invariance, so that thermodynamic quantities are not expected to have an algebraic form, and there are no critical exponents.

5.3 Symmetry Breaking in Landau Theory

Landau theory can describe both continuous and discontinuous phase transitions in terms of an order parameter, whose value changes from zero to non-zero at the critical temperature. To see how this description of phase transitions is related to spontaneous symmetry breaking, consider the shape of the free energies in Figures 5.2 and 5.3. At the lowest temperatures, there are two minima with equal energies, related by the transformation $o \to -o$. That is, the free energy in these figures has a discrete $\mathbb{Z}_2$ symmetry, and when the system realises a specific ground state, with either a positive or a negative value of the order parameter, the symmetry is spontaneously broken.

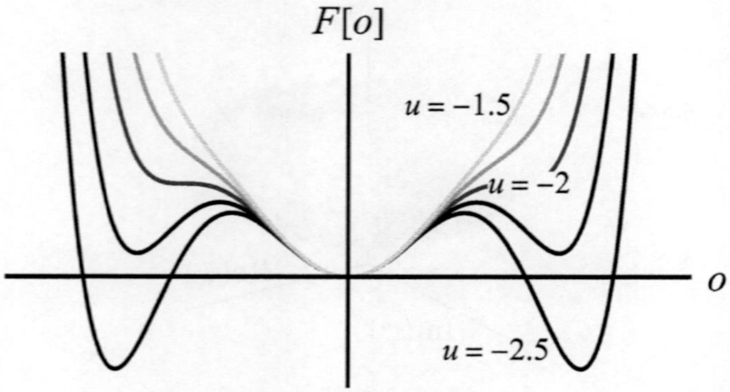

Figure 5.3

The Landau free energy of Eq. (5.5) as a function of the order parameter o during a discontinuous phase transition, with $r = w = 1$, for several values of u. The location of the minimum of $\mathcal{F}_L$ jumps from zero to non-zero values as u crosses the value -2.

5.3.1 The Mexican Hat Potential

As an example of how the spontaneous breakdown of a continuous symmetry is manifested within the framework of Landau theory, consider a complex order parameter ψ and a Hamiltonian or action that is invariant under continuous $U(1)$ rotations of the phase of the order parameter. One possible way for a symmetric configuration to transition into one with broken phase rotation symmetry is described using a Landau free energy expansion of the form

$$\mathcal{F}_L[\psi, T] = \mathcal{F}[0, T] + \frac{1}{2}r(T)\psi^*\psi + \frac{1}{4}u(T)(\psi^*\psi)^2. \tag{5.6}$$

This expression is invariant under phase rotations of the field ψ, since it only depends on powers of its squared amplitude. In analogy to the theory for a real scalar field in Eq. (5.1), we can assume u to be constant close to the critical temperature, and $r(T)$ to be linear function changing sign at $t = 0$. At high temperatures, for $r > 0$, the free energy has a single minimum at $\psi = 0$, while at low temperatures, for $r < 0$ the minimum will be obtained for configurations with non-zero amplitude $|\psi|$ and arbitrary phase. The amplitude of ψ goes from zero to non-zero at the phase transition and may be called a Landau order parameter, as it corresponds to the small parameter in the Landau free energy expansion. The order parameter of Section 3.1, however, which takes values in the coset space $G/H = U(1)/1 \simeq U(1)$, is the full complex number ψ, with different phase values denoting distinct symmetry-breaking states. An example of such an order parameter is the one we encountered in Section 3.2, where the field operator ψ in a complex scalar field theory obtains an expectation value $\langle\psi\rangle \neq 0$ in the ordered state.

The form of the free energy is depicted in Figure 5.4. It is colloquially referred to as the **Mexican hat potential** because of its shape, and it is the prototypical example in discussions of spontaneous symmetry breaking in the context of Landau theory. There is a continuum of states with amplitude $|\psi| = \sqrt{|r|/u}$ and arbitrary value for the phase that all minimise the free energy. The free energy is thus invariant under $U(1)$ phase-rotations, but only single points on the circle of minima represent stable configurations. Upon traversing the phase transition, one of the states in the circle, and hence a definite value for the phase, will be spontaneously chosen.

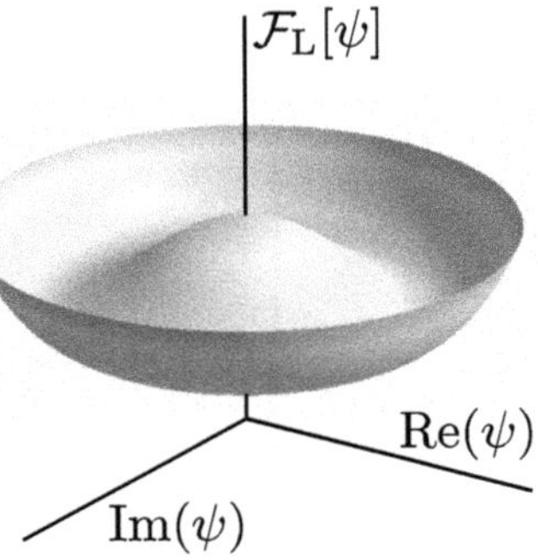

Figure 5.4 Plot of the so-called Mexican hat free energy in Eq. (5.6) for $r < 0$.

Having realised a minimum-energy state with a spontaneously chosen order parameter, the free energy in Figure 5.4 shows that excitations that change the phase of the order parameter but not its amplitude, do not cost any energy. The potential, or free energy, along the circle is flat, and this is therefore known as a **flat direction** in field space [186]. The Nambu–Goldstone mode associated with the broken $U(1)$ symmetry is precisely a long-wavelength modulation of the phase of the field ψ. That is, it is an excitation for which the order parameter oscillates along the flat direction in order parameter space. This result readily generalises to more involved instances of symmetry breaking, so that in any low-temperature free energy with a symmetry, Nambu–Goldstone modes correspond to excitations oscillating in flat directions.

5.3.2 From Hamiltonian to Landau Free Energy

The Landau free energy $\mathcal{F}_L$ is constructed purely as an expansion in terms of the order parameter, with symmetry dictating both the form of the order parameter and whether or not any given terms are allowed to be non-zero. The values of the coefficients in the expansion are typically determined by fitting predictions for thermodynamic quantities to experimental data, and it may therefore seem like the macroscopic Landau theory is entirely phenomenological, and disconnected from the detailed microscopic theory defined by a Lagrangian or Hamiltonian. In fact, this is not the case. It is often possible to find both the terms which appear in the Landau expansion, and the values of the corresponding coefficients, starting from a microscopic description.

Rather than giving a general and abstract procedure, we will illustrate the connection by considering the specific example of the Heisenberg antiferromagnet of Section 2.4, defined by the Hamiltonian:

$$\hat{H} = J \sum_{i,\delta} \hat{\mathbf{S}}_i \cdot \hat{\mathbf{S}}_{i+\delta}. \tag{5.7}$$

Here, i runs over all N sites in a d-dimensional hypercubic lattice, and δ runs over the $z = 2d$ connections to nearest neighbours. A good choice for the antiferromagnetic order parameter operator is the staggered magnetisation, $\hat{\mathcal{O}}_i = (-1)^i \hat{\mathbf{S}}_i$. We consider a vector-valued order parameter here because the $SU(2)$-symmetric Hamiltonian does not prefer any particular axis for the spins to align with. The expectation value of the order parameter operator,

$\mathbf{m} = \langle(-1)^i\hat{\mathbf{S}}_i\rangle$, will be zero in the disordered phase and non-zero in the translationally invariant antiferromagnetic state.

The first step towards formulating a macroscopic description of the antiferromagnet is to rewrite all operators in terms of an average, called the **mean-field value**, plus deviations from that average. The spin operators $\hat{\mathbf{S}}_i$, for example, can be written as a sum of the order parameter $\mathbf{m}$ plus deviations $\delta\hat{\mathbf{S}}_i$:

$$\hat{\mathbf{S}}_i = (-1)^i\mathbf{m} + \delta\hat{\mathbf{S}}_i. \tag{5.8}$$

Using this notation, the Heisenberg Hamiltonian becomes

$$\begin{aligned}\hat{H} &= J\sum_{i\delta}\left((-1)^i\mathbf{m} + \delta\hat{\mathbf{S}}_i\right)\cdot\left((-1)^{i+1}\mathbf{m} + \delta\hat{\mathbf{S}}_{i+\delta}\right)\\ &= -JNz|\mathbf{m}|^2 - 2Jz\sum_i(-1)^i\mathbf{m}\cdot\delta\hat{\mathbf{S}}_i + J\sum_{i\delta}\delta\hat{\mathbf{S}}_i\cdot\delta\hat{\mathbf{S}}_{i+\delta}.\end{aligned} \tag{5.9}$$

This form of the Hamiltonian allows us to make a rigorous approximation. If the system is ordered, the expectation values of the deviations $\langle\delta\hat{\mathbf{S}}_i\rangle$ will be small, which suggests that we may neglect terms in the Hamiltonian that are quadratic or higher order in the deviations. The resulting Hamiltonian is called the mean-field Hamiltonian:

$$\begin{aligned}\hat{H}[m] &= -JNz|\mathbf{m}|^2 - 2Jz\sum_i(-1)^i\mathbf{m}\cdot\delta\hat{\mathbf{S}}_i\\ &= +JNz|\mathbf{m}|^2 - 2Jz\sum_i(-1)^i\mathbf{m}\cdot\hat{\mathbf{S}}_i.\end{aligned} \tag{5.10}$$

In the second line, we reintroduced the original spin operators, using $\delta\hat{\mathbf{S}}_i = \hat{\mathbf{S}}_i - (-1)^i\mathbf{m}$. Note that this results in a sign change of the first term.

The mean-field Hamiltonian depends on the order parameter $|\mathbf{m}|$, whose value we do not know. Nevertheless, it is only linear in spin operators and can therefore be solved exactly. In the present context, this means that we can compute the partition function $Z = \mathrm{Tr}e^{-\beta\hat{H}}$, and from there the thermodynamic free energy $\mathcal{F} = -\frac{1}{\beta}\log Z$. We can find both of these as functions of the order parameter $\mathbf{m}$ without specifying its value.

Because the Hamiltonian is proportional to $\mathbf{m}\cdot\hat{\mathbf{S}}_j$, the energy eigenstates can be chosen to coincide with those of $\hat{S}^z_j$ by defining the z-axis to be parallel to $\mathbf{m}$. The partition function can then be computed by explicitly summing over all eigenvalues of the operators $\hat{S}^z_j$. For the case of $\hat{S}^z_j$ being spin-$\frac{1}{2}$ operators, this yields

$$\begin{aligned}Z &= \sum_{S^z_1=\pm1/2}\ \sum_{S^z_2=\pm1/2}\cdots\sum_{S^z_N=\pm1/2} \mathrm{e}^{-\beta JNz|\mathbf{m}|^2}\mathrm{e}^{2Jz\beta|\mathbf{m}|S^z_1}\mathrm{e}^{2Jz\beta|\mathbf{m}|S^z_2}\cdots\mathrm{e}^{2Jz\beta|\mathbf{m}|S^z_N}\\ &= \mathrm{e}^{-\beta JNz|\mathbf{m}|^2}\left(2\cosh Jz\beta|\mathbf{m}|\right)^N.\end{aligned} \tag{5.11}$$

Taking the logarithm, the free energy associated with this partition function is

$$\mathcal{F} = JNz|\mathbf{m}|^2 - \frac{N}{\beta}\log\left(2\cosh Jz\beta|\mathbf{m}|\right). \tag{5.12}$$

As expected, we find an expression for the free energy in terms of the mean-field order parameter. To compare this expression to the form of a Landau free energy, we perform a Taylor expansion of $\mathcal{F}$ around $\mathbf{m} = 0$:

$$\mathcal{F}/N = -\frac{\log 2}{\beta} + Jz\,(1 - Jz\beta/2)\,|\mathbf{m}|^2 + \frac{1}{12}(Jz)^4\beta^3|\mathbf{m}|^4 + \ldots \tag{5.13}$$

This has exactly the shape of the Landau free energy of Eq. (5.1).

At high temperatures, or low β, the coefficients of both the quadratic and quartic terms are positive, and the free energy is minimised for $|\mathbf{m}| = 0$. Therefore, we do not expect to find an antiferromagnetically ordered state, and the spin-rotational symmetry is unbroken. As temperature is lowered, the coefficient of the quadratic term decreases, until it goes through zero at $Jz\beta = 2$. From that point onwards, the free energy is minimised by a non-zero value for the order parameter, indicating that the symmetry is broken, and long-range antiferromagnetic order is established.

Although the expansion of the free energy yields precisely the macroscopic description of the antiferromagnetic phase transition, we expected, you may object that the analysis is not internally consistent. We started out by defining $\mathbf{m}$ as the ground state expectation value of the order parameter operator, but we ended up claiming to find the value of $\mathbf{m}$ from the minimisation of the free energy, which does not explicitly involve taking any expectation value. As it turns out, however, minimising the mean-field free energy is precisely equivalent to computing the mean-field expectation value of the order parameter operator.

Exercise 5.1 (Mean-field order parameter).
The expectation value of the order parameter is found self-consistently in any mean field theory. The consistency condition allows us to derive the mean field equations in two different ways.
a. Show that in the ground state of the mean-field Hamiltonian of Eq. (5.10), the thermal expectation value of the magnetisation $|\mathbf{m}| = |\langle(-1)^i\hat{\mathbf{S}}_i\rangle|$ is given by $\frac{1}{2}\tanh Jz\beta|\mathbf{m}|$.
b. Use the exact expression for the Landau free energy in Eq. (5.12) to compute the magnetisation $\mathbf{m}$ by minimising $\mathcal{F}$. (Hint: Compute $\partial(\mathcal{F}/N)/\partial|\mathbf{m}| = 0$.)

The expectation value of the order parameter operator with respect to the mean-field ground state is thus the same as the equilibrium value with respect to the Landau free energy.

One of the great triumphs of theoretical physics is the connection made by Gor'kov, in a similar fashion to what we showed above for the Heisenberg antiferromagnet, between the microsopic Bardeen–Cooper–Schrieffer theory of superconductivity and the phenomenological Ginzburg–Landau theory. An accessible derivation can be found in the book by De Gennes [43].

5.4 Spatial Fluctuations

In all expansions of the free energy that we considered so far, we assumed that the local order parameter had the same value everywhere in space. This is certainly appropriate when looking for the equilibrium state of a translationally invariant system, be it symmetric or symmetry breaking. More generally, however, there is no reason not to consider configurations of a system that are described by a spatially varying order parameter. As you might expect, the free energy of such inhomogeneous states may be used to study the role of fluctuations near a phase transition.

As before, we start from an expansion of the free energy in powers of the order parameter, which is assumed to be small. This time, however, we simultaneously expand in powers of the spatial derivatives of the order parameter, which are also assumed to be small. That is, we consider a system near a symmetry-breaking phase transition, with smooth or long-wavelength fluctuations. The study of free energy expansions including spatial derivatives is often called **Ginzburg–Landau theory**, although this term is also used by some to refer exclusively to a theory of superconductivity, including gauge fields, that we will encounter in Chapter 8.

The minimal extension of the Landau theory for second-order phase transitions in Eq. (5.1), including the lowest possible power of spatial derivatives, is given by

$$\mathcal{F}_{\mathrm{GL}}[o(\mathbf{x}),T] = \mathcal{F}[0,T] + \int d^d x \left\{ \frac{c^2}{2}|\nabla o(\mathbf{x})|^2 + \frac{r}{2}[o(\mathbf{x})]^2 + \frac{u}{4}[o(\mathbf{x})]^4 + \ldots \right\}. \tag{5.14}$$

This expression for the Ginzburg–Landau free energy is no longer a function of a single order parameter value, but rather a functional, which depends on the values of the order parameter field $o(\mathbf{x})$ at all positions. Notice that if $o(\mathbf{x})$ is dimensionless, the coefficient c^2 must have units of energy density times length squared, and all coefficients may depend on temperature.

5.4.1 Coherence Length

One thing Ginzburg–Landau theory can tell us, is what the typical size of fluctuations in the order parameter field will be. To see this, we use the trick of adding a small local perturbation to the potential [62] and seeing what configuration of the order parameter field is established in response. In general, an externally applied potential $\mu(\mathbf{x})$ coupling to the order parameter can be introduced as

$$\mathcal{F}_{\mu}[o,T] = \mathcal{F}_{\mathrm{GL}}[o,T] - \int d^d x \mu(\mathbf{x}) o(\mathbf{x}). \tag{5.15}$$

For now, consider a perturbation that only affects the system at a single location $\mathbf{x}_0$, so that the potential is a delta function, $\mu(\mathbf{x}) = \mu_0 \delta(\mathbf{x}-\mathbf{x}_0)$. As usual, the equilibrium configuration of the order parameter field is the one that minimises the free energy. Since the order parameter is now itself a position-dependent function, the minimum is found by demanding that the functional derivative $\frac{\delta \mathcal{F}}{\delta o(\mathbf{x})} = 0$ equals zero. For Eq. (5.15), this results in

$$-c^2 \nabla^2 o(\mathbf{x}) + r o(\mathbf{x}) + u[o(\mathbf{x})]^3 = \mu_0 \delta(\mathbf{x}-\mathbf{x}_0). \tag{5.16}$$

Because we consider a situation close to homogeneous order, we assume that the deviations of the order parameter field from its uniform average value $\bar{o} = \langle o \rangle$ are small. Writing $o(\mathbf{x}) = \bar{o} + \delta o(\mathbf{x})$ then yields

$$-c^2 \nabla^2 \delta o(\mathbf{x}) + r\bar{o} + r\delta o(\mathbf{x}) + u\bar{o}^3 + 3u\bar{o}^2 \delta o(\mathbf{x}) + \mathcal{O}\left((\delta o)^2\right) = \mu_0 \delta(\mathbf{x}-\mathbf{x}_0). \tag{5.17}$$

The weak and local potential $\mu_0 \delta(\mathbf{x}-\mathbf{x}_0)$ will not affect the average value of the order parameter, $\bar{o}$, which is therefore equal to the value we found in the uniform Landau theory of Eq. (5.2). That is, the average order parameter should be zero for temperatures above the critical temperature, while at low temperatures we expect to find $\bar{o} = \sqrt{|r|/u}$. Substituting these averages, the equation for the equilibrium configuration in the presence of the local potential becomes

$$\begin{aligned} -c^2\nabla^2\delta o(\mathbf{x}) + r\delta o(\mathbf{x}) &= \mu_0\delta(\mathbf{x}-\mathbf{x}_0) \qquad &\text{for } T > T_c, \\ -c^2\nabla^2\delta o(\mathbf{x}) - 2r\delta o(\mathbf{x}) &= \mu_0\delta(\mathbf{x}-\mathbf{x}_0) \qquad &\text{for } T < T_c. \end{aligned} \tag{5.18}$$

These are ordinary differential equations for the response $\delta o(\mathbf{x})$ to a local perturbation $\mu_0\delta(\mathbf{x}-\mathbf{x}_0)$, and can be straightforwardly solved. In three spatial dimensions, the solution is given by

$$\delta o(\mathbf{x}) = \frac{\mu_0}{4\pi c^2}\frac{\mathrm{e}^{-|\mathbf{x}-\mathbf{x}_0|/\xi}}{|\mathbf{x}-\mathbf{x}_0|}, \qquad \begin{cases} \xi = \sqrt{\frac{c^2}{r}} & T > T_c, \\ \xi = \sqrt{\frac{c^2}{-2r}} & T < T_c. \end{cases} \tag{5.19}$$

The deviation $\delta o(\mathbf{x})$ of the order parameter field from its average value thus falls off exponentially in all directions from the point of the perturbation. It does so with a characteristic length scale, ξ, which is called the **coherence length**.[1] This is the length scale over which fluctuations of the order parameter, or in other words, deviations of the magnitude of $o(\mathbf{x})$ from $\bar{o}$, persist. It is sometimes referred to as the **healing length**, because the order parameter field returns to its average value $\bar{o}$ within this length scale from the position of a local external perturbation. It is also the typical size of spontaneously generated thermal fluctuations. At length scales larger than the coherence length, perturbations and fluctuations have little effect, and the order parameter field is well approximated by it average value. The order and broken symmetry completely determine the way the system looks at those scales. On the other hand, at scales shorter than the coherence length, the local configuration of the order parameter field is dominated by perturbations, and the average order parameter will be hard to distinguish among the microscopic fluctuations. In a way, the long-range order **emerges** from the underlying local physics on length scales larger than the coherence length.

Notice that the coherence length is not the same as the correlation length associated with the two-point correlation function of Eq. (3.5). The former indicates the size of a single fluctuation, while the latter corresponds to the likelihood of two distant regions behaving the same way. In a long-ranged ordered system, the correlation length may be infinitely long, while the coherence length remains finite.

The parameter r in the definition of the coherence length in Eq. (5.19) depends on temperature. In fact, it is the parameter that goes from positive to negative values in the Landau description of a second-order phase transition. As the transition temperature is approached, r must therefore go to zero, and the coherence length will diverge. Such divergences turn out to be a general feature of second-order phase transition, in which not only the coherence length, but all relevant length (and energy) scales diverge as the system advances towards the phase transition. This will be discussed in more detail in Section 5.5.

5.4.2 Ginzburg Temperature

The divergence of fluctuations as the phase transition is approached poses a problem for the expansion of the free energy in Ginzburg–Landau theory, because it was based on the

[1] Eq. (5.19) differs by a factor of $1/\sqrt{2}$ from a definition commonly used in superconductivity [192].

assumption that variations in the order parameter field are small. As temperature is tuned towards its critical point, there must therefore be a value $T_G \neq 0$ at which the Ginzburg–Landau theory is no longer applicable. Remarkably, this so-called **Ginzburg temperature** can be determined from within Ginzburg–Landau theory itself. To do this, consider the two-point correlation function $\langle o(\mathbf{x})o(0)\rangle$. If the system is long-range ordered, the order parameter should not vary significantly as a function of position. In other words, variations of the order parameter should be small compared to the value of the order parameter itself:

$$\int \mathrm{d}^3x \, \langle o(\mathbf{x})o(0)\rangle_T - \langle o(\mathbf{x})\rangle\langle o(0)\rangle_T \ll \int \mathrm{d}^3x \, \bar{o}^2. \tag{5.20}$$

Here, $\langle A\rangle_T$ denotes the thermal average of A, defined as $\langle A\rangle_T = \mathrm{Tr}\; A\mathrm{e}^{-\beta\mathcal{F}}/Z$, where the trace signifies a sum over all possible configurations of the order parameter field. The **Ginzburg criterion** now states that whenever this inequality is violated, Ginzburg–Landau theory breaks down. Because we know the size of a single fluctuation is just the coherence length of Eq. (5.19), the integrals in the inequality will be zero beyond the coherence length. The Ginzburg criterion thus really determines whether a local fluctuation is sufficient to destroy the local order, rather than finding out whether many fluctuations together destroy the global long-ranged order. The latter criterion would instead give the critical temperature associated with a phase transition at which $\bar{o}$ vanishes.

The right-hand side of the inequality is just an integral over the constant average value of the order parameter $\bar{o}^2 = |r|/u$. To evaluate the left-hand side, we can use an incarnation of the fluctuation–dissipation theorem (see Section 9.1.3), which relates the thermal average of fluctuations to derivatives of the free energy in the presence of perturbations [58, 62]. Within Ginzburg–Landau theory, this relation can be seen simply as a property of the free energy of Eq. (5.15), combined with the definition of the thermal average:

$$\begin{aligned}
\frac{1}{\beta}\frac{\delta}{\delta\mu(\mathbf{x}')}\langle \delta o(\mathbf{x})\rangle_T &= \frac{1}{\beta}\frac{\delta}{\delta\mu(\mathbf{x}')}\langle o(\mathbf{x}) - \bar{o}\rangle_T \\
&= \frac{1}{\beta}\frac{\delta}{\delta\mu(\mathbf{x}')}\mathrm{Tr}\left[(o(\mathbf{x}) - \bar{o})\,\mathrm{e}^{-\beta\mathcal{F}_\mu}/Z\right] \\
&= \mathrm{Tr}\left[(o(\mathbf{x}) - \bar{o})\,o(\mathbf{x}')\,\mathrm{e}^{-\beta\mathcal{F}_\mu}/Z\right] \\
&= \langle o(\mathbf{x})o(\mathbf{x}')\rangle_T - \langle o(\mathbf{x})\rangle_T\langle o(\mathbf{x}')\rangle_T.
\end{aligned} \tag{5.21}$$

Here, we used $\bar{o} = \langle o(\mathbf{x})\rangle_T$ in going to the final line. The thermal average $\langle \delta o(\mathbf{x})\rangle_T$ appearing on the left-hand side of the equation can be identified with the thermal equilibrium configuration of Eq. (5.19). Writing $\mu_0 f(\mathbf{x} - \mathbf{x}_0) = \int d^dx\, \mu(\mathbf{x}')f(\mathbf{x} - \mathbf{x}')$ in that expression, we can evaluate the functional derivative and substitute $1/\beta = k_\mathrm{B}T$ to find

$$\langle o(\mathbf{x})o(0)\rangle_T - \langle o(\mathbf{x})\rangle_T\langle o(0)\rangle_T = \frac{k_\mathrm{B}T}{4\pi c^2}\frac{\mathrm{e}^{-|\mathbf{x}|/\xi}}{|\mathbf{x}|}. \tag{5.22}$$

Finally taking the limit $\mu_0 \to 0$ so that $\mathcal{F}_\mu \to \mathcal{F}_\mathrm{GL}$, this expression relates the correlation function due to thermal fluctuations in an unperturbed system to the length scale over which local perturbations are dissipated.

This form of the fluctuation–dissipation relation can be inserted into the Ginzburg criterion of Eq. (5.20). In three dimensions we can then use $\int \mathrm{d}^3x\; \mathrm{e}^{-x/\xi}/x = \int \mathrm{d}\Omega \int \mathrm{d}x\; x\mathrm{e}^{-x/\xi}$ and $\partial_\xi \mathrm{e}^{-a/\xi} = (a/\xi^2)\mathrm{e}^{-a/\xi}$ to evaluate the integral, which we approximate to run from $x = 0$ to $x = \xi$:

$$\frac{k_B T}{c^2}(1 - 2/\mathrm{e})\xi^2 \ll \frac{4}{3}\pi\xi^3\bar{\sigma}^2,$$
$$\frac{3 - 6/\mathrm{e}}{4\pi}\frac{k_B T}{c^2\xi\bar{\sigma}^2} \ll 1. \tag{5.23}$$

Close to the phase transition, the temperature in the numerator is approximately T_c. In the denominator, we can use our results from the uniform Landau theory to substitute $\xi = c/\sqrt{2|r|}$ and $\bar{\sigma}^2 = |r|/u$. We also know that at the phase transition, r changes sign, so that close to the transition, we can also write $r \approx r_0 t$ with t the reduced temperature. Putting everything together, the entire fraction on the left-hand side is then seen to diverge as $1/\sqrt{t}$ as the critical temperature is approached. There must therefore be a region of temperatures around the critical temperature for which the Ginzburg criterion is violated, and Ginzburg–Landau theory breaks down. The Ginzburg temperature indicating the approximate size of this region can be defined as the reduced temperature t_G at which the fraction in Eq. (5.23) equals one. This results in

$$\sqrt{t_G} = \sqrt{2}\frac{3 - 6/\mathrm{e}}{4\pi}\frac{k_B T_c u}{\sqrt{r_0}c^3}. \tag{5.24}$$

The Ginzburg temperature is not an exact, quantitative bound up to which Ginzburg–Landau theory can be trusted. Rather, it gives an indication of the order of magnitude for the reduced temperature at which thermal fluctuations become important. For reduced temperatures of the order of t_G and below, the approximations on which Ginzburg–Landau theory is based are not justified, and more sophisticated methods should be used to describe the system. The applicability of Ginzburg–Landau theory to any realistic situation may seem precarious, since the theory is based on the assumption that the average order parameter is not too large, but also breaks down when it becomes too small, close to the transition. In practice, it turns out that the regime over which the theory is reliable is actually very large for most symmetry-breaking systems. To give you a feeling, the Ginzburg temperature in superconductors may vary from $t_G \sim 10^{-16}$ for strongly type-I superconductors to $t_G \sim 10^{-4}$ for strongly type-II superconductors.

During the analysis of the coherence length and Ginzburg temperature we discarded in Eq. (5.17) all terms with powers of the fluctuation higher than one. This approximation is similar to the one we made in our discussion of the Heisenberg antiferromagnet in Section 5.3.2, and as in that case, it implies that the Ginzburg–Landau theory is an example of a mean-field theory. The expression for the Ginzburg temperature in Eq. (5.23) is the result for three spatial dimensions. In general, the left-hand side is proportional to $t^{\frac{d-4}{2}}$ [58]. In dimensions $d \geq 4$, the suppression of local order due to fluctuations thus no longer diverges, and the results of mean-field theory are robust all the way up to the critical temperature. The spatial dimension below which thermal fluctuations qualitatively affect the phase transition and invalidate a mean-field description, is called the **upper critical dimension**. For the Ginzburg–Landau theory of this section, the upper critical dimension is four. This is to be contrasted with the lower critical dimension introduced in Section 6.2, at and below which fluctuations are so violent they prevent the establishment of long-range order altogether. For the Ginzburg–Landau theory of Eq. 5.14 the lower critical dimension turns out to be two. At non-zero temperatures, we thus find that only in three spatial dimensions there can exist a phase transition, in which local fluctuations destroy a long-range ordered phase.

5.5 Universality

The Landau theory of phase transitions is based on an expansion of the free energy in powers of the local order parameter. Both the nature of the order parameter and the allowed terms in the expansion are entirely determined by the symmetries of the phases on either side of the transition, which therefore also determine many observable properties of the phase transition. The temperature dependence of the order parameter near the phase transition, for example, was shown in the mean-field theory of Eq. (5.4) to be a power law with exponent $\beta = \frac{1}{2}$. This value of this critical exponent depends only on the fact that we considered a second-order phase transition involving a real and scalar order parameter, both of which follow directly from the symmetry being broken at the phase transition.

Notice the profound implication of this observation: knowing only the symmetries on either side of the phase transition, and nothing whatsoever about the microscopic Hamiltonian, we can already deduce observable properties of the system near its phase transition. This is an example of **universality** in physics because it implies that observable properties near a phase transition characterised by a certain symmetry are shared among many, potentially very different physical implementations of that symmetry. Groups of models leading to the same universal behaviour near phase transitions can then be collected into **universality classes**. For instance, the Ising model is in the same universality class as the liquid–gas transition, and the superfluid transition in Helium-4 is in the same universality class as the *XY*-model of Eq. (6.17). On a more practical level, the same universality also guarantees that the experimental measurement of, for example, the temperature dependence of specific heat near a phase transition does not depend on any microscopic details like impurities in the sample, stray magnetic fields, or the fact that the material being measured is not described in full detail by a simplified theoretical model. Universal quantities can be measured and compared with theoretical predictions despite such practical difficulties.

Precisely at a second-order phase transition or critical point, the universality becomes even stronger. The correlation function of Eq. (5.22) depends on the coherence length ξ. At the phase transition, ξ diverges, and the correlation function becomes proportional to $1/|\mathbf{x}|$. This function is an example of a **scale invariant** function, which does not define a typical length, and looks the same at every scale. It should be contrasted with functions like $\cos(x/x_0)$, or $x^2(x^2 - x_0^2)$, which depend on a parameter x_0 that defines a characteristic length scale. The direct physical consequence of observables being described by scale invariant functions, is that they will look the same regardless of the scale at which they are measured. A famous example is the **critical opalescence** at the critical endpoint of the liquid–gas transition (see Figure 5.1). The normally transparent water suddenly becomes opaque there, and it does so for light at all possible wavelengths. The reason is the presence of scale-invariant fluctuations, which cover all length scales and therefore scatter light at all wavelengths, including scales far beyond any related to atomic or molecular properties.

The correlation function at such a phase transition is in fact a so-called homogeneous function, satisfying

$$C(|\mathbf{x}|) = b^{\kappa} C(b|\mathbf{x}|). \tag{5.25}$$

Here, b may be any real number, and κ is a characteristic exponent, which in the case of a correlation function proportional to $1/|\mathbf{x}|$ equals one. The fact that the correlation function is homogeneous is a key ingredient in the theory of renormalisability and the **renormalisation group** description of phase transitions. These approaches are a way of describing the temperature region immediately around the critical temperature, where the coherence length diverges and Ginzburg–Landau theory breaks down. Crudely, the idea is to first identify some small length a in the microscopic model, which could, for example, be the lattice constant. We can then coarse-grain or average over any physical excitations or fluctuations that occur at length scales between a and ba, where $b > 1$. Like the original Hamiltonian, the coarse grained description can be used to formulate an effective description in terms of a Landau free energy, but this time using the rescaled coordinate $x' = bx$. Because none of the symmetries of the model are affected by the coarse-graining, the Landau free energy will look the same as for the original model, but with different values for its parameters. Using Eq. (5.25), the coherence length in the coarse-grained model can then be expressed as a function of the original one, $\xi' = f(\xi)$. If the coarse-grained coherence length happens to be smaller in the course-grained model than in the original model, the effect of fluctuations is lower on the course-grained level and the critical temperature may be approached more closely before the Ginzburg–Landau theory becomes invalid. Repeating the coarse-graining many times, we can even hope to approach the critical temperature arbitrarily closely.

This procedure is known as the real-space renormalisation group. Several related procedures exist, including renormalisation in momentum space, or even in terms of the order parameter fields themselves. All these approaches capture the effects of thermal fluctuations near phase transitions in many realistic settings. The study of the renormalisation group is consequently a major field of study on its own, as described in several textbooks [58, 72, 225, 230].

PART II

CONCEPTS

6 Corrections to the Order Parameter

The eigenstates of the local order parameter operator, which are the classically expected symmetry-broken states, are not in general eigenstates of their corresponding symmetric Hamiltonian. Except for systems with conserved order parameters and type-B NG modes, such as ferromagnets and ferrimagnets, symmetric Hamiltonians have unique, symmetric ground states. That the states actually realised in nature nevertheless typically bear a large resemblance to the classically expected states is owing to their stability, in the sense of Section 3.7. Even the classical states, however, are not exactly the states found in actual quantum materials. This may be attributed to the effect of excitations with finite wave number reducing the perfect local order of the classical states that are preferred by the $k = 0$, collective part of the Hamiltonian in the thermodynamic limit. The differences between the classically expected and actually encountered state at zero temperature are known as **quantum corrections**. At non-zero temperature, these may be supplemented by thermal fluctuations, which further suppress the value of local order parameter.

Notice the distinction made between thermal fluctuations and quantum corrections. The latter are often referred to as 'quantum fluctuations' in the literature. The reasoning being that the effect of quantum corrections in reducing the order parameter amplitude is quite similar to that of the thermal fluctuations. Indeed, they may even lead to a quantum phase transition at zero temperature, which is in some ways analogous to the usual, thermally induced, phase transition (see also the discussion in Section A.3). We believe this terminology is misleading. Thermal fluctuations describe actual small, random fluctuations within a thermal state due to, for example, Brownian motion. At zero temperature, on the other hand, nothing ever fluctuates. The ground state may be unique all the way from a maximally ordered phase to just before a quantum phase transition, despite the expectation value of the local order parameter strongly decreasing. For any given value of system parameters, nothing about the ground state evolves or fluctuates in time. We therefore prefer the term quantum corrections to denote the difference between the actual quantum ground state and the perfectly ordered, classical state.

Rather than describing the generic properties of quantum corrections to a general ordered state, we will first study the specific example of the Heisenberg antiferromagnet in Section 6.1. This allows us both to introduce some of the commonly encountered mathematical techniques in studying the actually realised ground states of ordered systems and to show in detail a realistic description of a system in which quantum corrections are of significant magnitude. In contrast, in most ordered states that we encounter in daily life, quantum corrections to the classical state are 'utterly negligible' [9]. For example, a chair or table, even at low temperatures, is extremely well described by its classical state with infinitely well-defined position.

Unlike these everyday objects, quantum corrections and thermal fluctuations in low-dimensional systems are generically so large that they altogether prevent the formation of any ordered state and the spontaneous breaking of continuous symmetries. Section 6.2 introduces the Mermin–Wagner–Hohenberg–Coleman theorem, which explains both the effect of dimensionality, and the link between quantum corrections and thermal fluctuations.

6.1 Linear Spin-Wave Theory

There is no known exact expression for the broken-symmetry state realised by the Heisenberg antiferromagnet in the thermodynamic limit, in dimensions higher than one. One thing we can do to describe it, however, is to start from the classical state (the eigenstate of the Néel order parameter operator) and use the variational principle to look for deviations that lower the energy expectation value. This is done in a systematic wave in **spin-wave theory**, which is based on a reformulation of the Heisenberg Hamiltonian in terms of boson operators acting on the classical state, so that each bosonic excitation lowers the order parameter expectation value. In so-called linear spin-wave theory, the bosonic Hamiltonian is additionally linearised, allowing for a ground state to be identified by direct diagonalisation. We can thus find the exact ground state of an approximate Hamiltonian and trust that it serves as an approximate ground state to the exact Hamiltonian. As the name suggests, linear spin-wave theory considers only the first non-trivial order in a systematic expansion of the order parameter. For Heisenberg spin-S antiferromagnets on bipartite lattices with z nearest neighbours, it turns out that the results obtained this way are good beyond expectation, even for $S = 1/2$ and $z = 4$, where neither $1/S$ nor $1/z$ are expected to be very good expansion parameters.

The Hamiltonian for the spin-S Heisenberg antiferromagnet on a bipartite lattice in d dimensions is given by Eq. 2.17:

$$\hat{H} = J \sum_{\langle \mathbf{i},\mathbf{j} \rangle} \left(\hat{S}^x_{\mathbf{i}} \hat{S}^x_{\mathbf{j}} + \hat{S}^y_{\mathbf{i}} \hat{S}^y_{\mathbf{j}} + \hat{S}^z_{\mathbf{i}} \hat{S}^z_{\mathbf{j}} \right) = \frac{J}{2} \sum_{\mathbf{j},\boldsymbol{\delta}} \left(\hat{S}^x_{\mathbf{j}} \hat{S}^x_{\mathbf{j}+\boldsymbol{\delta}} + \hat{S}^y_{\mathbf{j}} \hat{S}^y_{\mathbf{j}+\boldsymbol{\delta}} + \hat{S}^z_{\mathbf{j}} \hat{S}^z_{\mathbf{j}+\boldsymbol{\delta}} \right). \tag{6.1}$$

Here, the coupling constant J is positive, so that neighbouring spins prefer to anti-align. The lattice vectors $\boldsymbol{\delta}$ run over the z connections of any site to all of its nearest neighbours, and the factor $1/2$ is included to avoid double counting. Because the lattice is bipartite, it can be divided into sublattices labelled A and B. For the square lattice, we saw this before in Figure 2.4. Anticipating antiferromagnetic Néel order, we expect a broken-symmetry state may be realised in which the spins on the A-sublattice have positive magnetisation, while spins on the B-sublattice have negative magnetisation. To avoid the inconvenience of having to keep track of the local magnetisation direction on each sublattice, we introduce rotated spin operators $\hat{\mathbf{N}}_{\mathbf{j}}$, defined by the components:

$$\begin{aligned} \hat{N}^x_{\mathbf{j}\in A} &= \hat{S}^x_{\mathbf{j}}, & \hat{N}^y_{\mathbf{j}\in A} &= \hat{S}^y_{\mathbf{j}}, & \hat{N}^z_{\mathbf{j}\in A} &= \hat{S}^z_{\mathbf{j}}; \\ \hat{N}^x_{\mathbf{j}\in B} &= \hat{S}^x_{\mathbf{j}}, & \hat{N}^y_{\mathbf{j}\in B} &= -\hat{S}^y_{\mathbf{j}}, & \hat{N}^z_{\mathbf{j}\in B} &= -\hat{S}^z_{\mathbf{j}}. \end{aligned} \tag{6.2}$$

That is, the coordinate system for spins on the B sublattice is rotated over an angle of π around the x-axis with respect to the coordinate system used for spins on the A-sublattice. The new spin operators $\hat{\mathbf{N}}_{\mathbf{j}}$ are still proper spin-S operators, obeying the commutation relations associated with the $SU(2)$ algebra, $[\hat{N}^a_{\mathbf{i}},\hat{N}^b_{\mathbf{j}}] = \mathrm{i}\epsilon_{abc}\hat{N}^c_{\mathbf{j}}\delta_{\mathbf{i},\mathbf{j}}$.[1]

The Heisenberg Hamiltonian expressed in terms of rotated spins is

$$H = \frac{J}{2}\sum_{\mathbf{j},\delta}\left(\tfrac{1}{2}\hat{N}^+_{\mathbf{j}}\hat{N}^+_{\mathbf{j}+\delta} + \tfrac{1}{2}\hat{N}^-_{\mathbf{j}}\hat{N}^-_{\mathbf{j}+\delta} - \hat{N}^z_{i}\hat{N}^z_{\mathbf{j}+\delta}\right). \tag{6.3}$$

In writing this, we made use of the fact that on bipartite lattices, sites on the A-sublattice always have nearest neighbours on the B-sublattice, and the other way around. The operators $\hat{N}^{\pm}_{\mathbf{j}} = \hat{N}^x_{\mathbf{j}} \pm \mathrm{i}\hat{N}^y_{\mathbf{j}}$ are the raising and lowering operators for rotated spins. The eigenstate of $\hat{N}^z_{\mathrm{tot}} = \sum_{\mathbf{j}}\hat{N}^z_{\mathbf{j}}$ with maximal eigenvalue is a classical Néel state, but the first two terms in Eq. (6.3) clearly show that it is not an eigenstate of the Hamiltonian.

6.1.1 Spin Waves and NG Bosons

To characterise the difference between the exact ground state and the Néel state, we consider local excitations of the fully developed Néel order, which are created by applying the spin-lowering operator $\hat{N}^-_{\mathbf{j}}$. These lower the value of the local staggered magnetisation by the same quantised amount each time they act and are similar in their effect to the ladder operators of a harmonic oscillator, or more generally, to boson creation operators. This similarity can be made exact by formally expressing the transformed spin operators in terms of boson operators $\hat{a}_{\mathbf{j}}$ and $\hat{a}^\dagger_{\mathbf{j}}$:

$$\begin{aligned}
\hat{N}^+_{\mathbf{j}} &= \sqrt{2S}\sqrt{1-\frac{1}{2S}\hat{n}_{\mathbf{j}}}\;\hat{a}_{\mathbf{j}},\\
\hat{N}^-_{\mathbf{j}} &= \sqrt{2S}\hat{a}^\dagger_{\mathbf{j}}\sqrt{1-\frac{1}{2S}\hat{n}_{\mathbf{j}}},\\
\hat{N}^z_{\mathbf{j}} &= S-\hat{n}_{\mathbf{j}}.
\end{aligned} \tag{6.4}$$

The boson operators obey canonical commutation relations $[\hat{a}_{\mathbf{i}},\hat{a}^\dagger_{\mathbf{j}}] = \delta_{\mathbf{i},\mathbf{j}}$ and $\hat{n}_{\mathbf{j}} = \hat{a}^\dagger_{\mathbf{j}}\hat{a}_{\mathbf{j}}$, while the square root is defined by its power series expansion. It can be checked that the expression of $\hat{\mathbf{N}}_{\mathbf{j}}$ in terms of bosons still respects the $SU(2)$ algebra of spin operators. The bosons introduced in this way of writing spin operators are known as **Holstein–Primakoff bosons**. The Néel state can be defined as the bosonic vacuum, $\hat{a}_{\mathbf{j}}\,|\text{Néel}\rangle = 0$, so that it has the correct eigenvalue of S for all of the $\hat{N}^z_{\mathbf{j}}$ operators. States with non-zero numbers of bosons correspond to states with non-maximal values of the Néel order parameter.

The Holstein–Primakoff transformation is exact, but the square roots in its definition prevent a simple diagonalisation of the Heisenberg Hamiltonian written in terms of Holstein–Primakoff bosons. We therefore make a linear approximation of the square roots, keeping only the first terms in their power series expansion, so that none of the approximate expressions are more than bilinear in $\hat{a}$ and $\hat{a}^\dagger$. That is, we approximate the spin raising and lowering

[1] Note that in Section 3.2 and Exercise 4.2, we introduced the order parameter operators $\hat{N}^a_{\mathbf{j}} = (-1)^{j^x+j^y+j^z}\hat{S}^a_{\mathbf{j}}$. While that definition is simpler to write than the current one, it is important for spin wave theory that the $\hat{N}^a_{\mathbf{j}}$ operators satisfy the standard $SU(2)$ commutation relations. Here, we therefore use Eq. (6.2) to define rotated spin operators.

operators as $\hat{N}_{\mathbf{j}}^{+} \approx \sqrt{2S}\hat{a}_{\mathbf{j}}$ and $\hat{N}_{\mathbf{j}}^{-} \approx \sqrt{2S}\hat{a}_{\mathbf{j}}^{\dagger}$, while the expression for $\hat{N}_{\mathbf{j}}^{z}$ remains unaltered. This can really only be expected to be a good approximation for large S and low numbers of boson excitation $\langle \hat{a}_{\mathbf{j}}^{\dagger}\hat{a}_{\mathbf{j}} \rangle$, but the resulting approximate antiferromagnetic ground state and its excitations will turn out to give quite accurate results even for spin-$\frac{1}{2}$ systems.

The linear approximation for the rotated spin operators yields the approximate form of the Heisenberg Hamiltonian:

$$\hat{H} \approx \frac{1}{2}JS\sum_{\mathbf{j},\boldsymbol{\delta}}\left(\hat{a}_{\mathbf{j}}\hat{a}_{\mathbf{j}+\boldsymbol{\delta}} + \hat{a}_{\mathbf{j}}^{\dagger}\hat{a}_{\mathbf{j}+\boldsymbol{\delta}}^{\dagger}\right) - \frac{1}{2}zNJS^2 + zJS\sum_{\mathbf{j}}\hat{a}_{\mathbf{j}}^{\dagger}\hat{a}_{\mathbf{j}}. \tag{6.5}$$

Here, N is the total number of sites, and z is again the number of nearest neighbours, also known as the coordination number. To diagonalise the approximate Hamiltonian, we first use the fact that is invariant under lattice translations, by writing it in terms of Fourier transformed operators $\hat{a}_{\mathbf{j}} = \frac{1}{\sqrt{N}}\sum_{\mathbf{k}} \mathrm{e}^{\mathrm{i}\mathbf{k}\cdot\mathbf{j}}\hat{a}_{\mathbf{k}}$. The momentum-space bosons still obey the canonical commutation relations $[\hat{a}_{\mathbf{k}},\hat{a}_{\mathbf{k}'}^{\dagger}] = \delta_{\mathbf{k},\mathbf{k}'}$, and the terms appearing in the Hamiltonian become

$$\sum_{\mathbf{j}}\hat{a}_{\mathbf{j}}^{\dagger}\hat{a}_{\mathbf{j}} = \frac{1}{N}\sum_{\mathbf{j},\mathbf{k},\mathbf{k}'} \mathrm{e}^{\mathrm{i}(-\mathbf{k}\cdot\mathbf{j}+\mathbf{k}'\cdot\mathbf{j})}\hat{a}_{\mathbf{k}}^{\dagger}\hat{a}_{\mathbf{k}'} = \sum_{\mathbf{k}}\hat{a}_{\mathbf{k}}^{\dagger}\hat{a}_{\mathbf{k}}, \tag{6.6}$$

$$\sum_{\mathbf{j},\boldsymbol{\delta}}\hat{a}_{\mathbf{j}}\hat{a}_{\mathbf{j}+\boldsymbol{\delta}} = \frac{1}{N}\sum_{\mathbf{j},\boldsymbol{\delta},\mathbf{k},\mathbf{k}'} \mathrm{e}^{\mathrm{i}(\mathbf{k}\cdot\mathbf{j}+\mathbf{k}'\cdot\mathbf{j}+\mathbf{k}'\cdot\boldsymbol{\delta})}\hat{a}_{\mathbf{k}}\hat{a}_{\mathbf{k}'} = \sum_{\mathbf{k},\boldsymbol{\delta}} \mathrm{e}^{-\mathrm{i}\mathbf{k}\cdot\boldsymbol{\delta}}\hat{a}_{\mathbf{k}}\hat{a}_{-\mathbf{k}}, \tag{6.7}$$

$$\sum_{\mathbf{j},\boldsymbol{\delta}}\hat{a}_{\mathbf{j}}^{\dagger}\hat{a}_{\mathbf{j}+\boldsymbol{\delta}}^{\dagger} = \frac{1}{N}\sum_{\mathbf{j},\boldsymbol{\delta},\mathbf{k},\mathbf{k}'} \mathrm{e}^{\mathrm{i}(-\mathbf{k}\cdot\mathbf{j}-\mathbf{k}'\cdot\mathbf{j}-\mathbf{k}'\cdot\boldsymbol{\delta})}\hat{a}_{\mathbf{k}}^{\dagger}\hat{a}_{\mathbf{k}'}^{\dagger} = \sum_{\mathbf{k},\boldsymbol{\delta}} \mathrm{e}^{\mathrm{i}\mathbf{k}\cdot\boldsymbol{\delta}}\hat{a}_{\mathbf{k}}^{\dagger}\hat{a}_{-\mathbf{k}}^{\dagger}. \tag{6.8}$$

Here, we used the definition of the delta function $\frac{1}{N}\sum_{\mathbf{j}} \mathrm{e}^{\mathrm{i}(\mathbf{k}-\mathbf{k}')\cdot\mathbf{j}} = \delta_{\mathbf{k},\mathbf{k}'}$. To simplify notation, we introduce $\gamma_{\mathbf{k}} \equiv \frac{1}{z}\sum_{\boldsymbol{\delta}} \mathrm{e}^{\mathrm{i}\mathbf{k}\cdot\boldsymbol{\delta}}$. Notice that for the square lattice, $\gamma_{\mathbf{k}}$ is real and reduces to a sum over cosines. The Hamiltonian in momentum space now reads

$$\hat{H} = -\frac{1}{2}NJzS^2 + JzS\sum_{\mathbf{k}}\left[\hat{a}_{\mathbf{k}}^{\dagger}\hat{a}_{\mathbf{k}} + \frac{1}{2}\gamma_{\mathbf{k}}\left(\hat{a}_{\mathbf{k}}\hat{a}_{-\mathbf{k}} + \hat{a}_{\mathbf{k}}^{\dagger}\hat{a}_{-\mathbf{k}}^{\dagger}\right)\right]. \tag{6.9}$$

The products of two creation operators and two annihilation operators in the expression of Eq. (6.9) still hinder a simple identification of the ground state. The appearance of these terms once again indicates that the Néel state, or bosonic vacuum, is not an eigenstate of the Heisenberg Hamiltonian. To find the exact ground state of the linearised Hamiltonian, we perform a so-called **Bogoliubov transformation** and introduce a second set of boson creation and annihilation operators $\hat{b}_{\mathbf{k}}^{\dagger}$ and $\hat{b}_{\mathbf{k}}$:

$$\hat{a}_{\mathbf{k}} = \cosh(u_{\mathbf{k}})\,\hat{b}_{\mathbf{k}} + \sinh(u_{\mathbf{k}})\,\hat{b}_{-\mathbf{k}}^{\dagger}, \qquad \hat{a}_{\mathbf{k}}^{\dagger} = \cosh(u_{\mathbf{k}})\,\hat{b}_{\mathbf{k}}^{\dagger} + \sinh(u_{\mathbf{k}})\,\hat{b}_{-\mathbf{k}}. \tag{6.10}$$

Here, $u_{\mathbf{k}}$ is an unknown but real function of $\mathbf{k}$, obeying $u_{\mathbf{k}} = u_{-\mathbf{k}}$. The new operators again satisfy the canonical boson commutation relations $[\hat{b}_{\mathbf{k}},\hat{b}_{\mathbf{k}'}^{\dagger}] = \delta_{\mathbf{k},\mathbf{k}'}$. The approximate Hamiltonian can be written in terms of the new boson operators:

$$\begin{aligned}\hat{H} = -\frac{1}{2}JNzS^2 + JzS\sum_{\mathbf{k}}\Big[&\sinh^2(u_{\mathbf{k}}) + \tfrac{1}{2}\gamma_{\mathbf{k}}\sinh(2u_{\mathbf{k}})\\ &+ (\cosh(2u_{\mathbf{k}}) + \gamma_{\mathbf{k}}\sinh(2u_{\mathbf{k}}))\hat{b}_{\mathbf{k}}^{\dagger}\hat{b}_{\mathbf{k}}\\ &+ \tfrac{1}{2}(\gamma_{\mathbf{k}}\cosh(2u_{\mathbf{k}}) + \sinh(2u_{\mathbf{k}}))(\hat{b}_{\mathbf{k}}^{\dagger}\hat{b}_{-\mathbf{k}}^{\dagger} + \hat{b}_{\mathbf{k}}\hat{b}_{-\mathbf{k}})\Big].\end{aligned} \tag{6.11}$$

This expression for the Hamiltonian would be diagonal if the terms in the final line vanish. Since the Bogoliubov transformation was introduced in Eq. (6.10) in terms of an arbitrary function $u_{\mathbf{k}}$, we are free to now consider the particular choice for $u_{\mathbf{k}}$ that renders the off-diagonal terms in the Hamiltonian zero. That is, we choose $u_{\mathbf{k}}$ such that $\gamma_{\mathbf{k}} \cosh(2u_{\mathbf{k}}) + \sinh(2u_{\mathbf{k}})$ equals zero. Using the general relation $\cosh^2(x) - \sinh^2(x) = 1$, this amounts to

$$\sinh(2u_{\mathbf{k}}) = \frac{-\gamma_{\mathbf{k}}}{\sqrt{1-\gamma_{\mathbf{k}}^2}}, \qquad \cosh(2u_{\mathbf{k}}) = \frac{1}{\sqrt{1-\gamma_{\mathbf{k}}^2}}. \tag{6.12}$$

Notice that the expression for $u_{\mathbf{k}}$ is ill defined at zero wave number, where $\gamma_{\mathbf{k}=0} = 1$. This collective, centre-of-mass part of the Hamiltonian at zero wave number corresponds to the tower of states, just as in Eqs. (2.9), (2.19), and (3.18). The diagonalisation of the Hamiltonian using a Bogoliubov transformation here forces us to treat these collective excitations separately from the internal, finite-wave number excitations corresponding to quantum corrections and NG modes.

Writing the approximate Hamiltonian in terms of the Bogoliubov transformed excitations, diagonalising it by our choice of $u_{\mathbf{k}}$, and omitting the $\mathbf{k} = 0$ part, we finally find the linear spin wave Hamiltonian:

$$\begin{aligned} \hat{H} &\approx -\tfrac{1}{2}JNzS^2 - \tfrac{1}{2}JNzS + JzS\sum_{\mathbf{k}}\sqrt{1-\gamma_{\mathbf{k}}^2}\left(\hat{b}_{\mathbf{k}}^\dagger \hat{b}_{\mathbf{k}} + \tfrac{1}{2}\right) \\ &= \underbrace{-\tfrac{1}{2}JNzS^2}_{\text{Classical}} \underbrace{- \tfrac{1}{2}JzS\sum_{\mathbf{k}}\left(1-\sqrt{1-\gamma_{\mathbf{k}}^2}\right)}_{\text{Quantum corrections}} \underbrace{+ JzS\sum_{\mathbf{k}}\sqrt{1-\gamma_{\mathbf{k}}^2}\,\hat{b}_{\mathbf{k}}^\dagger \hat{b}_{\mathbf{k}}}_{\text{NG modes}}. \end{aligned} \tag{6.13}$$

This Hamiltonian consists of three parts. The first two parts describe the energy expectation value in the ground state, containing both the energy expected for the classical state, and its quantum corrections. The final term contains all excitations that can propagate with non-zero wave number and describes the NG modes. Their energy is positive, so they will be absent in the ground state, which corresponds to the vacuum defined by $\hat{b}_{\mathbf{k}} |0\rangle = 0$. If excited, the NG modes can reduce the expectation value of the local order parameter beyond the reduction coming from the quantum corrections. This happens, for example, at non-zero temperatures, where the occupation number of the bosonic NG modes follows the Bose–Einstein distribution function.

On a square lattice with lattice spacing a, the dispersion relation for the NG modes can be approximated at low wave numbers to be

$$E_{\mathbf{k}} = zJS\sqrt{1-\gamma_{\mathbf{k}}^2} \approx 2\sqrt{d}JS|\mathbf{k}|a + \dots, \tag{6.14}$$

where we used $z = 2d$. As expected for type-A NG modes, the dispersion is linear in wave number. Because the antiferromagnet breaks two spin rotations, we should in fact expect to find two type-A NG modes, and it may seem like our linear spin wave description is missing an entire branch of excitations. In fact, this is a consequence of introducing rotated spin operators in Eq. (6.2). In terms of the rotated spin operators, all spins in the lattice look equivalent, and the unit cell is thus half as large as it was for the original spins. In reciprocal space, this implies a doubling of the Brillouin zone. In the dispersion of Eq. (6.14), half the excitations are thus folded out to higher wave numbers. The dispersion going linearly to zero at $k_x = k_y = \cdots = \pi$, corresponds to the second branch of NG modes, which would be

	Absolute energy E/JN	Relative energy gain	$S=\frac{1}{2}$		$S=1$	
$d=2$	$-2S(S+0.1579)$,	$-15.8/S$ %	-0.658,	-31.6%	-2.32,	-15.8%
$d=3$	$-3S(S+0.0972)$,	$-9.72/S$ %	-0.896,	-19.4%	-3.29,	-9.72%

Table 6.1 Quantum correction to the ground state energy for the Heisenberg antiferromagnet on cubic lattices.

folded back to $\mathbf{k}=0$ if we return to using the same, non-rotated, coordinate system at every lattice site.

6.1.2 Quantum Corrections

Coming back to the first two terms in the Hamiltonian of Eq. (6.13), we see that the first part is just the energy expectation value of the classical Néel state. The second part is negative and lowers the energy below that of the classical state. It represents the difference in ground state energy between the exact quantum ground state (of the approximate Hamiltonian) and the classical state, or in other words, it shows the quantum correction to the ground state energy. The ground state energy can be evaluated numerically in the continuum limit, by replacing sums with integrals. The results for square and cubic lattices are listed in Table 6.1. Particularly for spin-$\frac{1}{2}$ antiferromagnets, the quantum corrections to the ground state energy are seen to be substantial.

Like the energy, the expectation value of the order parameter is affected by quantum corrections. To see this, we can start from the expression in Eq. (6.4), for the staggered magnetisation in terms of the original Holstein–Primakoff bosons:

$$\langle 0|\frac{1}{N}\sum_{\mathbf{j}}\hat{N}^z_{\mathbf{j}}|0\rangle = S-\frac{1}{N}\sum_{\mathbf{j}}\langle 0|\hat{a}^\dagger_{\mathbf{j}}\hat{a}_{\mathbf{j}}|0\rangle = S-\frac{1}{N}\sum_{\mathbf{k}}\langle 0|\hat{a}^\dagger_{\mathbf{k}}\hat{a}_{\mathbf{k}}|0\rangle\,. \tag{6.15}$$

The ground state $|0\rangle$ appearing in this equation is the vacuum of the Bogoliubov transformed particles $\hat{b}_{\mathbf{k}}$, rather than the original Holstein–Primakoff bosons $\hat{a}_{\mathbf{k}}$. Using the Bogoliubov transformation of Eq. (6.10), and the fact that the ground state does not contain any $\hat{b}$-excitations, the staggered magnetisation may be written as

$$\begin{aligned}\langle 0|\frac{1}{N}\sum_{\mathbf{j}}\hat{N}^z_{\mathbf{j}}|0\rangle &= S-\frac{1}{N}\sum_{\mathbf{k}}\sinh^2(u_{\mathbf{k}})\\ &\approx S+\frac{1}{2}-\frac{1}{(2\pi)^d}\int_{-\pi}^{\pi}\mathrm{d}^d k\,\frac{1}{2}\frac{1}{\sqrt{1-\gamma_{\mathbf{k}}^2}}.\end{aligned} \tag{6.16}$$

Here, we again took the continuum limit in the second line. The integral diverges in one dimension, indicating that the quantum corrections to the order parameter in that case are strong enough to suppress the order altogether. This in fact turns out to be a general phenomenon, to which we return in Section 6.2.

The integral in the final line of Eq. (6.16) may be evaluated numerically. The results are shown in Table 6.2, and they show that the quantum corrections to the order parameter take it substantially away from its value in the classical Néel state, especially for low-spin

	Order parameter value $\langle \hat{N}^z_{\text{tot}} \rangle / N$	Relative correction	$S = \frac{1}{2}$		$S = 1$	
$d = 2$	$S - 0.1966$,	$-19.7/S\ \%$	0.303,	−39.3%	0.803,	−19.7%
$d = 3$	$S - 0.0784$,	$-7.8/S\ \%$	0.422,	−15.6%	0.922,	−7.8%

Table 6.2 Quantum corrections to the order parameter for the Heisenberg antiferromagnet on cubic lattices.

antiferromagnets. The sizes of these quantum corrections in the Heisenberg antiferromagnet are exceptional. In most ordered systems in three dimensions, quantum corrections are tiny.

The linear spin wave approximation we used here to estimate the effect of quantum corrections turns out to give unexpectedly good results compared to more precise, variational methods. The best estimates for the ground state energy of the Heisenberg antiferromagnet, for example, are within a few percent of the results found here [120].

Exercise 6.1 (Quantum corrections for the *XY*-model).
The ***XY*-model** describes interactions between rotors in a plane, which have a global $U(1)$ rotational symmetry, and which can be written in terms of spin operators as

$$\hat{H}_{XY} = J \sum_{\langle \mathbf{i},\mathbf{j} \rangle} \left(\hat{S}^x_{\mathbf{i}} \hat{S}^x_{\mathbf{j}} + \hat{S}^y_{\mathbf{i}} \hat{S}^y_{\mathbf{j}} \right). \tag{6.17}$$

For simplicity, we assume $J < 0$.

You will calculate the quantum corrections for the *XY*-model in d dimensions using linear spin-wave theory [11]. We start with a trick: by rotating the coordinate frame on all sites and calling the rotated spin operators $\hat{S}'$, the Hamiltonian can be expressed as $\hat{H}_{XY} = J \sum_{\langle \mathbf{i},\mathbf{j} \rangle} \left(\hat{S}'^x_{\mathbf{i}} \hat{S}'^x_{\mathbf{j}} + \hat{S}'^z_{\mathbf{i}} \hat{S}'^z_{\mathbf{j}} \right)$.

a. In the rotated frame, write the operators $\hat{S}'^x_{\mathbf{j}}$ in terms of the raising and lowering operators $\hat{S}'^{\pm}_{\mathbf{j}}$.

Notice that when they act on a state, the raising and lowering operators may cause spins to rotate outside of the two-dimensional plane in which the rotors are defined. This is clearly an unphysical aspect of the description in terms of spin operators. For now, however, we ignore this uneasy aspect of the model and continue the calculation, hoping the end result will still capture some of the essential physics.

b. Write the Hamiltonian in terms of Holstein–Primakoff bosons, using

$$\begin{aligned} \hat{S}'^{+}_{\mathbf{j}} &= \sqrt{2S}\sqrt{1 - \frac{1}{2S}\hat{n}_{\mathbf{j}}}\, \hat{a}_{\mathbf{j}} \approx \sqrt{2S}\hat{a}_{\mathbf{j}}, \\ \hat{S}'^{-}_{\mathbf{j}} &= \sqrt{2S}\, \hat{a}^{\dagger}_{\mathbf{j}} \sqrt{1 - \frac{1}{2S}\hat{n}_{\mathbf{j}}} \approx \sqrt{2S}\hat{a}^{\dagger}_{\mathbf{j}}, \\ \hat{S}'^{z}_{\mathbf{j}} &= S - \hat{n}_{\mathbf{j}}. \end{aligned} \tag{6.18}$$

c. First, perform a Fourier transformation on the Hamiltonian, and then the Bogoliubov transformation of Eq. (6.10).
d. Choose the function $u_{\mathbf{k}}$ such that the Hamiltonian becomes diagonal in the Bogoliubov-transformed operators.
e. Numerically evaluate the ground state energy density E/N and the order parameter density $\langle \hat{S}'^{z}_{\text{tot}} \rangle / N$ in two and three dimensions for general S.

The results for the order parameter should be $S - 0.0609$ in two dimensions and $S - 0.0225$ in three dimensions. The quantum corrections are significant, but not as large as for the antiferromagnet.

6.2 Mermin–Wagner–Hohenberg–Coleman Theorem

The lowering of the order-parameter expectation value in the actually realised broken-symmetry state, as compared to the classically expected state, is a general property of systems undergoing spontaneous symmetry breaking. In low dimensions ($d = 1$ for the antiferromagnet), quantum corrections may even preclude the existence of a non-zero order parameter altogether. A similar thing happens at elevated temperatures, where **thermal fluctuations** may prevent the onset of spontaneous symmetry breaking. This thermal limit to ordering is known as the **Mermin–Wagner–Hohenberg theorem**, while the zero-temperature absence of spontaneous symmetry breaking in low dimensions is known within quantum field theory as the **Coleman theorem**.

The calculation showing the divergence of quantum corrections in the Heisenberg antiferromagnet cannot be directly generalised to apply to all systems with spontaneous symmetry breaking. We will therefore consider the effective Lagrangian of Eq. (4.9), which may be considered a course-grained description of a symmetry-breaking system with NG modes, but no other low-energy excitations. We will find that possessing either type-A or type-B NG modes has significant implications for the way in which quantum corrections affect the broken symmetry. Either way, the highest dimension in which quantum corrections or thermal excitations prevent the onset of long-range order, is called **lower critical dimension** d_{L}. Spontaneous symmetry breaking of continuous symmetries can therefore only occur in dimensions d above the lower critical dimension, $d > d_{\mathrm{L}}$.

In the remainder of this section, we will work out the lower critical dimensions for systems with various types of NG modes, both at zero and non-zero temperature. The analysis will require the use of imaginary-time contour integrals and may be skipped by readers not interested in the technical analysis. They can directly go to the results summarised in Table 6.3 and Section 6.2.5.

6.2.1 The Variance of the Order Parameter

The spontaneous breakdown of symmetry, and the associated emergence of long-range order, is described in general by some local order parameter operator $\hat{\mathcal{O}}$ obtaining a non-zero expectation value. For the order to survive the effect of quantum corrections and that of thermal fluctuations, the variance of the order parameter should be smaller than its expectation value in the symmetry-breaking ground state or in thermal equilibrium. We thus consider the variance of the local order parameter, defined as

$$\begin{aligned}\mathrm{Var}\langle\mathcal{O}\rangle = \langle\hat{\mathcal{O}}(\mathbf{x},t)^2\rangle - \langle\hat{\mathcal{O}}(\mathbf{x},t)\rangle^2 &= \lim_{\mathbf{x}'\to\mathbf{x},t\to t'} \langle\hat{\mathcal{O}}(\mathbf{x},t)\hat{\mathcal{O}}(\mathbf{x}',t')\rangle - \langle\hat{\mathcal{O}}(\mathbf{x},t)\rangle\langle\hat{\mathcal{O}}(\mathbf{x}',t')\rangle \\ &= \lim_{\mathbf{x}'\to\mathbf{x},t'\to t} \langle\delta\hat{\mathcal{O}}(\mathbf{x},t)\delta\hat{\mathcal{O}}(\mathbf{x}',t')\rangle. \end{aligned} \tag{6.19}$$

In the second line, we expanded the order parameter around its expectation value as $\hat{\mathcal{O}}(\mathbf{x},t) = \langle\hat{\mathcal{O}}(\mathbf{x},t)\rangle + \delta\hat{\mathcal{O}}(\mathbf{x},t)$, and we assumed that the average of the fluctuations vanishes, so that $\langle\delta\hat{\mathcal{O}}(\mathbf{x},t)\rangle = 0$.

At very low temperatures or energies, the gapless NG modes $\hat{\pi}_a(\mathbf{x},t)$ dominate the fluctuations of the order parameter, and the expression reduces to

$$\lim_{\mathbf{x}'\to\mathbf{x},t'\to t} \langle\delta\hat{\mathcal{O}}(\mathbf{x},t)\delta\hat{\mathcal{O}}(\mathbf{x}',t')\rangle = \lim_{\mathbf{x}'\to\mathbf{x},t'\to t} \sum_a \langle\hat{\pi}_a(\mathbf{x},t)\hat{\pi}_a(\mathbf{x}',t')\rangle + \dots \tag{6.20}$$

A more precise expression of the variance in terms of NG modes can be found in [209], but this approximate form suffices to understand their role in establishing the lower critical dimensions. Notice that $\langle\hat{\pi}_a(\mathbf{x},t)\hat{\pi}_a(\mathbf{x}',t')\rangle$ precisely coincides with the definition for the real-space propagator of the NG mode.

Because we are interested in long-ranged ordered symmetry-breaking systems, with some form of translational invariance, it will be most convenient to calculate the propagator in momentum space. Taking a Fourier transformation, it becomes

$$\begin{aligned} G(0) &\equiv \lim_{\mathbf{x}'\to\mathbf{x},t'\to t} \sum_a \langle\hat{\pi}_a(\mathbf{x},t)\hat{\pi}_a(\mathbf{x}',t')\rangle \\ &= \lim_{\mathbf{x}'\to\mathbf{x},t'\to t} \int \frac{\mathrm{d}\omega}{2\pi} \int \frac{\mathrm{d}^d k}{(2\pi)^d}\, \mathrm{e}^{\mathrm{i}\mathbf{k}(\mathbf{x}-\mathbf{x}')-\mathrm{i}\omega(t-t')} \sum_a \langle\hat{\pi}_a(-\mathbf{k},-\omega)\hat{\pi}_a(\mathbf{k},\omega)\rangle \\ &= \int \frac{\mathrm{d}\omega}{2\pi} \int \frac{\mathrm{d}^d k}{(2\pi)^d} \sum_a \langle\hat{\pi}_a(-\mathbf{k},-\omega)\hat{\pi}_a(\mathbf{k},\omega)\rangle. \end{aligned} \tag{6.21}$$

An expression for the thermal average at non-zero temperatures can be found from this ground state expectation value, by applying the trick of analytic continuation to imaginary time ($t \to -\mathrm{i}\tau$), and introducing bosonic Matsubara frequencies $\omega \to \mathrm{i}\omega_n$ [93, 119, 137]. Here, ω_n may take the discrete values $2n\pi/(\hbar\beta)$, with n integer and $\beta = 1/k_\mathrm{B}T$ and the integral becomes $\int \mathrm{d}\omega/(2\pi) \to 1/(\hbar\beta)\sum_n$. This yields an expression for the NG mode propagator of the form

$$G(0) = \frac{1}{\hbar\beta} \sum_n \int \frac{\mathrm{d}^d k}{(2\pi)^d} \sum_a \langle\hat{\pi}_a(-\mathbf{k},-\omega)\hat{\pi}_a(\mathbf{k},\omega)\rangle\big|_{\omega=\mathrm{i}\omega_n}. \tag{6.22}$$

This expression now applies to non-zero temperatures, while the zero-temperature result is recovered in the limit $\beta \to \infty$. We will explicitly carry out the sum over Matsubara frequencies [137]. The dependence of the remaining integral over momentum $\mathbf{k}$ on the number of spatial dimensions d will then determine the lower critical dimension.

The propagator of the NG modes can be found using the effective Lagrangian of Eq. (4.9). Upon Fourier transforming the NG fields $\pi(\mathbf{x},t)$ to $\pi(\mathbf{k},\omega)$ this effective Lagrangian takes the form:

$$\mathcal{L}_\mathrm{eff} = \pi_a(-\mathbf{k},-\omega)\left[-\mathrm{i}\omega m_{ab} + \overline{g}_{ab}\omega^2 - g_{ab}k^2\right]\pi_b(\mathbf{k},\omega), \tag{6.23}$$

where the part in the square brackets is the inverse propagator $(G_{ab})^{-1}$. Inverting this to the propagator G_{ab} allows us to calculate the quantum and thermal corrections to the order parameter $\langle \delta \hat{\mathcal{O}}^2 \rangle$.

6.2.2 Matsubara Summation

Before calculating the variance of a local order parameter in the case of either type-A or type-B NG modes, we will give a brief and incomplete introduction to the general technique of Matsubara summation. The objective will be to analytically evaluate the summation in Eq. 6.22, which is of the general form $G(0) = \frac{1}{\hbar\beta} \sum_n g(\mathrm{i}\omega_n)$, independent of the type of NG modes.

The common trick for evaluating the sum over Matsubara frequencies is to change it into a sum over the poles of a function in the complex plane [72, 137]. To do so, we first replace the function $g(\mathrm{i}\omega_n)$ by the function $g(z)$, which has the exact same functional form, but in which the purely imaginary Matsubara frequencies are replaced by a general complex variable z. Having done this so-called analytic continuation, we can use the Cauchy residue theorem to replace the sum by a contour integral. In general, the residue theorem relates a contour integral in the complex plane to a sum over the poles enclosed within the contour:

$$\oint_C \mathrm{d}z f(z) = 2\pi\mathrm{i} \sum_n \mathrm{Res}\, f(z_n). \tag{6.24}$$

Here, z_n are the complex coordinates of poles of the function $f(z)$, and Res indicates the residue at those poles. For simple poles, of order one, the residue is $\mathrm{Res}\, f(z_n) = \lim_{z\to z_n} (z - z_n) f(z)$.

To write the sum in the expression for $G(0)$ as a contour integral, we need to find a function whose poles coincide with the bosonic Matsubara frequencies $\mathrm{i}\omega_n$. One such function happens to be the function $F(z) = (\mathrm{e}^{\hbar\beta z} - 1)^{-1}$. You can check that indeed it has poles of order one at the coordinates of the Matsubara frequencies $z_n = \mathrm{i}2\pi n/\hbar\beta = \mathrm{i}\omega_n$. Assuming $g(z)$ itself has no poles at z_n, we can then write all terms $g(\mathrm{i}\omega_n)$ that are summed in the expression for $G(0)$ as the residues of a function in the complex plane:

$$\frac{1}{\hbar\beta} g(z_n) = \lim_{z\to z_n} \left((z - z_n) \frac{g(z)}{\mathrm{e}^{\hbar\beta z} - 1} \right) = \mathrm{Res}_{z=z_n} \left(g(z) F(z) \right). \tag{6.25}$$

Putting together Eqs. (6.24) and (6.25), we can now write the Matsubara summation as a contour integral:

$$\begin{aligned} \frac{1}{\hbar\beta} \sum_n g(\mathrm{i}\omega_n) &= \frac{1}{\hbar\beta} \sum_n g(z_n) && \text{analytic continuation} \\ &= \sum_n \mathrm{Res}_{z=z_n} \frac{g(z)}{\mathrm{e}^{\hbar\beta z} - 1} && \text{the trick} \\ &= \frac{1}{2\pi\mathrm{i}} \oint_C \mathrm{d}z \frac{g(z)}{\mathrm{e}^{\hbar\beta z} - 1} && \text{residue theorem.} \end{aligned} \tag{6.26}$$

The contour C in the final line tightly encloses the imaginary axis and therefore all poles of the function $F(z)$ (see the left of Figure 6.1). The function $g(z)$ however, may also have poles

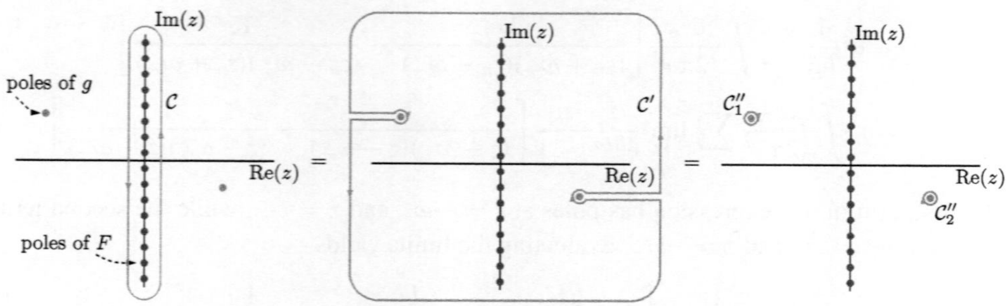

Figure 6.1

The contour integrals used in evaluating the sum over Matsubara frequencies, for a propagator with two poles in the complex plane.

of itself. For sake of clarity, consider an example in which $g(z)$ has two poles in the complex plane, as indicated in Figure 6.1. We can then use the fact that a contour integral over an analytic function without any enclosed poles yields zero, to deform the contour C into the contour C' without changing the value of the integral (see the middle of Figure 6.1). Next, we use the fact that any circular contour integral with infinite radius must vanish because normalisability requires all physical fields to decay sufficiently fast as they approach infinity. This allows us to equate the original contour integral enclosing poles on the imaginary axis to the sum of integrals along the contours C_1'' and C_2'', encircling the poles of the function $g(z)$ in the complex plane (see the right of Figure 6.1). As a result, the Matsubara summation may be written as:

$$
\begin{aligned}
\frac{1}{\hbar\beta}\sum_n g(\mathrm{i}\omega_n) &= \frac{1}{2\pi\mathrm{i}}\oint_C \mathrm{d}z \frac{g(z)}{\mathrm{e}^{\hbar\beta z}-1} \\
&= -\sum_j \frac{1}{2\pi\mathrm{i}}\oint_{C_j} \frac{g(z)}{\mathrm{e}^{\hbar\beta z}-1} && \text{shift contour} \\
&= -\sum_j \mathrm{Res}_{z=z_j}\frac{g(z)}{\mathrm{e}^{\hbar\beta z}-1} && \text{residue theorem} \\
&= -\sum_j \lim_{z\to z_j}\frac{(z-z_j)g(z)}{\mathrm{e}^{\hbar\beta z_j}-1}. && (6.27)
\end{aligned}
$$

The minus sign in the second line appears because the contours C'' run clockwise around their enclosed poles, rather than the counterclockwise direction with which the contour C encloses the imaginary axis. The expression in the final line contains a sum over just a small number of poles z_j of the function $g(z)$ and is much easier to evaluate than the original expression involving a sum over infinitely many Matsubara frequencies.

As an example of how to apply this method for evaluating Matsubara summations, consider the effective Lagrangian of Eq. (4.11), describing a system with two coupled broken symmetry generators. As indicated by Eq. (4.12), it yields NG modes with dispersions $\omega_\pm = \sqrt{c^2k^2+M^2c^4} \pm Mc^2$. The definition of the propagator in Eq. (6.21) can then be shown to yield

$$
G(0) = \int\frac{\mathrm{d}\omega}{2\pi}\int\frac{\mathrm{d}^d k}{(2\pi)^d}\left[\frac{c^2}{(\omega+\omega_+)(\omega-\omega_-)}+\frac{c^2}{(\omega-\omega_+)(\omega+\omega_-)}\right] \tag{6.28}
$$

$$\rightarrow c^2 \frac{1}{\hbar\beta} \sum_n \int \frac{\mathrm{d}^d k}{(2\pi)^d} \left[\frac{1}{(z_n + \omega_+)(z_n - \omega_-)} + \frac{1}{(z_n - \omega_+)(z_n + \omega_-)} \right]$$
$$= c^2 \int \frac{\mathrm{d}^d k}{(2\pi)^d} \sum_j \lim_{z \to z_j} \frac{z - z_j}{\mathrm{e}^{\hbar\beta z_j} - 1} \left[\frac{1}{(z + \omega_+)(z - \omega_-)} + \frac{1}{(z - \omega_+)(z + \omega_-)} \right].$$

The first term in this expression has poles at $z = -\omega_+$ and $z = \omega_-$, while the second term has poles at $z = \omega_+$ and $z = -\omega_-$. Evaluating the limits yields

$$G(0) = c^2 \int \frac{\mathrm{d}^d k}{(2\pi)^d} \frac{1}{\omega_+ + \omega_-} \left[\frac{1}{\mathrm{e}^{\hbar\beta\omega_+} - 1} - \frac{1}{\mathrm{e}^{-\hbar\beta\omega_+} - 1} + \frac{1}{\mathrm{e}^{\hbar\beta\omega_-} - 1} - \frac{1}{\mathrm{e}^{-\hbar\beta\omega_-} - 1} \right]$$
$$= \int \frac{\mathrm{d}^d k}{(2\pi)^d} \frac{c}{\sqrt{k^2 + M^2 c^2}} \left[n_\mathrm{B}(\hbar\omega_+) + n_\mathrm{B}(\hbar\omega_-) + 1 \right]. \tag{6.29}$$

In the final line, we introduced the Bose–Einstein distribution function $n_\mathrm{B}(\varepsilon) = (\mathrm{e}^{\beta\varepsilon} - 1)^{-1}$, and substituted the expressions for the NG mode dispersions. This general expression can be further analysed by evaluating the momentum space integral, but we will leave it here as an example of Matsubara summations, and instead focus on the particular cases of systems with only type-A or only type-B NG modes.

Exercise 6.2 (Propagator for NG modes).
a. Write the low-energy effective Lagrangian of Eq. (4.11) in terms of the Fourier-transformed field $\pi(\mathbf{k}, \omega)$. Your result will have the form $\mathcal{L}_\mathrm{eff} = \pi_a(-\mathbf{k}, -\omega)$ $(G^{-1})_{ab}\pi_b(\mathbf{k}, \omega)$.
b. Invert the matrix G^{-1} and compute the trace of the propagator, $\mathrm{Tr}\, G_{ab}(\mathbf{k}, \omega)$.
c. By finding the poles of the propagator $G(\mathbf{k}, \omega)$ deduce the energies of the NG modes $\omega_\pm(\mathbf{k})$.
d. Use the energies of the NG modes to rewrite the propagator $G(\mathbf{k}, \omega)$ in the form of Eq. (6.28).

6.2.3 Type-A NG Modes

Applying the general technique of Matsubara summation to the particular problem of evaluating the propagator of NG modes, we start from the effective Lagrangian of Eq. 4.9, which for the case of a single type-A mode can be simplified to

$$\mathcal{L}_\mathrm{eff} = \frac{1}{c^2} \partial_t \pi\, \partial_t \pi - \nabla\pi \cdot \nabla\pi. \tag{6.30}$$

Using Fourier transforms to write the Lagrangian density in terms of momentum and frequencies, it becomes

$$\mathcal{L}_\mathrm{eff} = \pi(-\mathbf{k}, -\omega) \left(\frac{1}{c^2}\omega^2 - k^2 \right) \pi(\mathbf{k}, \omega). \tag{6.31}$$

As usual, the propagator of the NG modes, which according to Eq. (6.21) corresponds to the variance of the order parameter, is found by inverting the quadratic part of the Lagrangian:

$$G(0) = \frac{1}{\hbar\beta}\sum_n \int \frac{\mathrm{d}^d k}{(2\pi)^d} \left\langle \hat{\pi}(-\mathbf{k}, -\omega)\hat{\pi}(\mathbf{k}, \omega)\right\rangle\Big|_{\omega=\mathrm{i}\omega_n}$$
$$= \frac{1}{\hbar\beta}\sum_n \int \frac{\mathrm{d}^d k}{(2\pi)^d} \frac{c^2}{(\mathrm{i}\omega_n)^2 - c^2k^2}. \tag{6.32}$$

This defines the function $g(\mathrm{i}\omega_n)$ that needs to be summed over the Matsubara frequencies. We then directly apply the procedure of Eqs. (6.26) and (6.27) and write the Matsubara summation as a sum over the poles of $g(z)$, which in the present case lie at $z_\pm(k) = \pm ck$:

$$G(0) = \int \mathrm{d}^d k \sum_{j=\pm} \mathrm{Res}_{z=z_j} \frac{1}{\mathrm{e}^{\hbar\beta z} - 1} \frac{c^2}{(z - z_+(k))(z - z_-(k))}$$
$$= \int \mathrm{d}^d k \left(\frac{1}{\mathrm{e}^{\hbar\beta z_+(k)} - 1} \frac{c^2}{z_+(k) - z_-(k)} + \frac{1}{\mathrm{e}^{\hbar\beta z_-(k)} - 1} \frac{c^2}{z_-(k) - z_+(k)} \right)$$
$$= \int \mathrm{d}^d k \frac{c}{2k} \left(\frac{1}{\mathrm{e}^{\hbar\beta ck} - 1} - \frac{1}{\mathrm{e}^{-\hbar\beta ck} - 1} \right)$$
$$= \int \mathrm{d}\Omega \int \mathrm{d}k\, k^{d-2} c \left(n_{\mathrm{B}}(\hbar\beta ck) + \frac{1}{2} \right). \tag{6.33}$$

In the final line, $n_{\mathrm{B}}(\varepsilon)$ is again the Bose–Einstein distribution function, and we introduced spherical coordinates for the momentum integral. The angular part, $\int \mathrm{d}\Omega$, evaluates to the surface area of a d-dimensional hypersphere, since the integrand only depends on $k = |\mathbf{k}|$.

At zero temperature, equivalent to β approaching infinity, the contribution from the Bose factor n_{B} vanishes. The remaining integral technically diverges, but we can introduce a high-momentum cut-off by taking the upper limit of the integral to be finite, corresponding to the inverse lattice spacing or some other inverse length scale set by the microscopic lattice that was ignored in the effective, coarse grained Lagrangian that we started with.

For small momentum, however, the momentum integral still diverges in dimensions lower than two, due to the factor k^{d-2}. Such a low-momentum divergence is called an **infrared divergence**. The fact that $G(0)$ diverges indicates that the variance of the order parameter diverges as well. The variance is then necessarily larger than the expectation value of the order parameter itself. This shows that in dimensions lower than two, the classical symmetry-broken state is unstable once we include the quantum corrections. In other words, the quantum corrections due to the presence of type-A NG modes preclude spontaneous symmetry breaking and the formation of long-range order in dimensions lower than two, even at zero temperature. This is known in the quantum field theory literature as the Coleman theorem.

For non-zero temperatures, we can expand the Bose–Einstein distribution function in Eq. (6.33) for small values of $\hbar\beta ck$, since the dominant contribution to the integral will come from low k values, for any non-zero value of β. We thus use the expansion:

$$\frac{1}{\mathrm{e}^x - 1} = \frac{1}{x} + \frac{1}{2} + \mathcal{O}(x). \tag{6.34}$$

Substituting this into the expression for $G(0)$, we find that the momentum integral is now over a function proportional to Tk^{d-3}. The presence of thermal fluctuations thus has an even larger effect on the formation of order than the quantum corrections at zero temperature, and the variance of the order parameter diverges in a system with only type-A NG modes in fewer than three spatial dimensions for any non-zero temperature. This is known as the Mermin–Wagner–Hohenberg theorem.

Exercise 6.3 (Thermal fluctuations of the Heisenberg antiferromagnet).
In Section 6.1, we calculated the quantum corrections at zero temperature for the Heisenberg antiferromagnet, which has two type-A NG modes. Here, we will continue with linear spin-wave theory to calculate the thermal fluctuations at non-zero temperature.
a. Starting from Eq. (6.15) and using the Bogoliubov transformation of Eq. (6.10), write the expectation value for the antiferromagnetic order parameter in terms of the linear spin-wave operators $\hat{b}_{\mathbf{k}}^{\dagger}$ and $\hat{b}_{\mathbf{k}}$.
b. Replace expectation values of the spin-wave operators with Bose–Einstein distribution functions to write the expectation value of the order parameter at finite temperature in terms of an integral over $n_{\mathrm{B}}(E_{\mathbf{k}})$. Compare this with Eq. (6.33).
c. Show that in $d = 2$ there is an infrared divergence, precluding antiferromagnetic order at non-zero temperature, and that in $d = 3$ there is no infrared divergence. The linear spin-wave theory thus gives results in accordance with the Mermin–Wagner–Hohenberg–Coleman theorem.

6.2.4 Type-B NG Modes

Interestingly, the lower critical dimension for systems with only type-B NG modes is different from that for systems with only type-A modes. Recall that type-B modes arise whenever there are terms with a single time derivative in the effective Lagrangian. For definiteness, we consider here the simplest type-B system, without gapped partner modes, which is described by

$$\mathcal{L}_{\text{eff}} = 2m\left(\pi_1 \partial_t \pi_2 - \pi_2 \partial_t \pi_1\right) - \nabla\pi_a \cdot \nabla\pi_a. \tag{6.35}$$

Type-B systems with gapped partner modes behave in the same way as those without as far as their lower critical dimension is concerned. Notice that there are necessarily two NG fields in the effective Lagrangian, π_1 and π_2, coupled by the terms with time derivatives. Using Fourier transforms, we can again write the Lagrangian density in terms of momentum and Matsubara frequencies and express it in matrix form:

$$\mathcal{L}_{\text{eff}} = \begin{pmatrix}\pi_1(-\mathbf{k},-\omega) & \pi_2(-\mathbf{k},-\omega)\end{pmatrix} \begin{pmatrix} -k^2 & -2im\omega \\ 2im\omega & -k^2 \end{pmatrix} \begin{pmatrix}\pi_1(\mathbf{k},\omega) \\ \pi_2(\mathbf{k},\omega)\end{pmatrix}. \tag{6.36}$$

As before, the propagators for the NG fields are found by inverting the quadratic part of the Lagrangian and taking the trace over the resulting matrix:

$$G(\mathbf{k}, \mathrm{i}\omega_n) = \frac{2k^2}{(2m\mathrm{i}\omega_n)^2 - k^4} = \frac{1}{2m}\left(\frac{1}{\mathrm{i}\omega_n - \frac{k^2}{2m}} - \frac{1}{\mathrm{i}\omega_n + \frac{k^2}{2m}}\right). \tag{6.37}$$

Because both fields contribute to the variance of the order parameter, we can directly use this combined expression of $g(\mathrm{i}\omega_n)$ in the summation over Matsubara frequencies in Eqs. (6.26) and (6.27). In this case, the poles of the analytically continued function $g(z)$ lie at $z_\pm = \pm k^2/2m$:

$$G(0) = \int \mathrm{d}^d k \sum_{i=\pm} \mathrm{Res}_{z=z_i} \frac{1}{\mathrm{e}^{\hbar\beta z} - 1} \frac{1}{2m}\left(\frac{1}{z - \frac{k^2}{2m}} - \frac{1}{z + \frac{k^2}{2m}}\right)$$

$$
\begin{aligned}
&= \int \mathrm{d}^d k \frac{1}{2m} \left(\frac{1}{\mathrm{e}^{\hbar\beta \frac{k^2}{2m}} - 1} - \frac{1}{\mathrm{e}^{-\hbar\beta \frac{k^2}{2m}} - 1} \right) \\
&= \int \mathrm{d}^d k \frac{1}{2m} \coth\left(\tfrac{1}{2}\hbar\beta \frac{k^2}{2m} \right).
\end{aligned}
\tag{6.38}
$$

To evaluate the variance at zero temperature, we need to take the limit $\beta \to \infty$ in the hyperbolic cotangent in the final line. Notice that the momentum k of the NG modes may be arbitrarily small, but not zero, because the exact $k = 0$ modes form the tower of collective states. Since for type-B NG modes the parameter m is non-zero, the hyperbolic cotangent then necessarily evaluates to one in the zero-temperature limit. The momentum integral is then over a constant and has no infrared divergence. In fact, if we again introduce an ultraviolet (high-momentum) cutoff on the integral to represent the discreteness of the atomic lattice, the variance is finite in any spatial dimension. Spontaneous symmetry breaking and long-range order may thus occur at zero temperature in any dimension for systems with only type-B NG modes.

Due to the absence of an infrared divergence, the zero-temperature quantum corrections for type-B NG modes are finite and possibly small. In some special cases, such as the Heisenberg ferromagnet, quantum corrections can even be completely absent. This happens if the classical state is an eigenstate of the full quantum theory. This is a rare occurrence however, and in general systems with type-B NG modes will have small but non-zero quantum corrections, as, for example, in a ferrimagnet.

At non-zero temperatures, the dominant contribution to the momentum integral will again come from low momenta. Expanding the integrand for small k in this case leads to $\coth(x) = 1/x + \ldots$. The thermal population of type-B NG modes then induces a variance of the order parameter according to

$$
\begin{aligned}
G(0) &= \int \mathrm{d}^d k \frac{2}{\hbar\beta k^2} + \ldots \\
&= \int \mathrm{d}\Omega \int \mathrm{d}k\, k^{d-3} \frac{2T}{\hbar k_\mathrm{B}}.
\end{aligned}
\tag{6.39}
$$

This integral diverges due to the low-k contributions in dimensions lower than three. That is, even though there are no quantum corrections to the order at zero temperature, the thermal population of type-B NG modes at non-zero temperatures yields the same lower critical dimension as thermal population of type-A NG modes. Starting from Eq. (6.29) the same result can be found for type-B systems with gapped partner modes.

6.2.5 Summary

In summary, the Mermin–Wagner–Hohenberg–Coleman theorem indicates that, compared to the classical state, the inclusion of both quantum corrections (at any temperature) and thermal fluctuations (at non-zero temperatures) in general reduces the value of the order parameter $\langle \hat{\mathcal{O}} \rangle$. The magnitude of the reduction is determined by the local propagator of NG modes:

$$
\langle \delta\hat{\mathcal{O}}^2 \rangle = G(0) = \frac{1}{\hbar\beta} \sum_n \int \frac{\mathrm{d}^d k}{(2\pi)^d} \sum_a \langle \hat{\pi}_a(-\mathbf{k}, -\omega)\hat{\pi}_a(\mathbf{k}, \omega) \rangle \big|_{\omega = \mathrm{i}\omega_n}.
\tag{6.40}
$$

	NG mode dispersion at low k	Ground state degeneracy	Tower of states	Quantum corrections to ground state	Lower critical dimension d_{L}	
					$T=0$	$T>0$
Type-A	$\propto k$	No	Yes	Yes	1	2
Type-B ferro	$\propto k^2$	Yes, $\mathcal{O}(N)$	No	No	0	2
Type-B ferri	$\propto k^2, M+k^2$	Yes, $\mathcal{O}(N)$	No	Yes	0	2

Table 6.3 The effect of NG modes on the emergence of long-range order.

Following Goldstone's theorem of Section 4.1, the NG modes are gapless, which implies that in low-dimensional systems the local propagator $G(0)$ might diverge. An infrared divergence exists if the propagator for small momenta behaves like

$$G(0) \sim \int dk\, k^{d+1-d_{\mathrm{L}}} + \dots. \tag{6.41}$$

An integral like this diverges if the dimension d is equal to or smaller than the lower critical dimension d_{L}. The value of the lower critical dimension depends on whether the NG modes are of type-A or type-B and whether or not temperature is non-zero.

The lower critical dimensions for systems with only type-A or only type-B NG modes are summarised in Table 6.3. As an interesting aside, notice that the same methodology can also be applied to calculate the lower critical dimension when the dispersion relation of type-A NG modes is not linear. For instance, for the Tkachenko modes mentioned in Section 4.2, which are type-A modes with $\omega \propto k^2$, the lower critical dimension will be two at zero temperature and four at any non-zero temperature. Systems with such modes therefore cannot order at any non-zero temperature even in three dimensions [181].

Topological Defects 7

Fluctuations of the local order parameter from its average value play an important role in establishing the stability of the broken-symmetry state. Their proliferation at sufficiently high temperatures can cause the long-range order to melt and induce a phase transition, while the Mermin–Wagner–Hohenberg theorem shows that in sufficiently low dimensions, fluctuations can even prevent the occurrence of long-range order altogether. So far, all fluctuations we considered consisted of Nambu–Goldstone modes. That is, they represented low-amplitude, wave-like modulations of the order parameter around its average value. One may wonder if there exist any other types of fluctuation that can influence the order of a broken-symmetry state. In fact, a whole other class of such alternative excitations exist, known as topological excitations or **topological defects**. Contrary to the NG modes, topological excitations take the order parameter far from its average value over large areas, and cannot be smoothly removed.

7.1 The Meaning of Topology and Defect

The most intuitive example of a topological defect is that in a state of broken $U(1)$ symmetry. Examples of systems with $U(1)$ order include superconductors and superfluids, as well as the *XY* magnet of Eq. (6.17). Their degrees of freedom can be visualised by vectors in a two-dimensional plane. In the ordered state, all vectors point in the same, spontaneously chosen, direction. Nambu–Goldstone modes then correspond to plane wave excitations, in which the directions of the vectors oscillate around the average as the system is traversed (see Figure 7.1a). For long-wavelength excitations, neighbouring vectors are never far from parallel, and the energy cost of creating NG-modes may be arbitrarily low.

It is also possible, however, to modulate the direction of vectors while traversing a circle, rather than propagating along a straight line. As shown in Figure 7.1b, it is even possible to consider modulations in which the vectors are taken far from their original direction, while still ensuring that almost all neighbouring vectors remain close to parallel. The local energy cost for creating this configuration is thus low almost everywhere, despite it containing substantial rotations of individual vectors, which complete a full rotation as the circle is traversed. This **topological defect** configuration, which is called a **vortex**, is fundamentally different from the plane waves representing NG-modes.

The distinction can be quantified by considering a closed contour such as the solid line in Figure 7.1b. Because the contour is closed and the vectors are almost parallel everywhere along the contour, the vectors must smoothly rotate over an integer multiple of 2π as the contour is traversed. This integer is called the **winding number**. Its value does not depend on where precisely the contour is drawn, as long as it encircles the centre of the vortex configuration. It is therefore a property of the vortex itself, called its **topological charge**

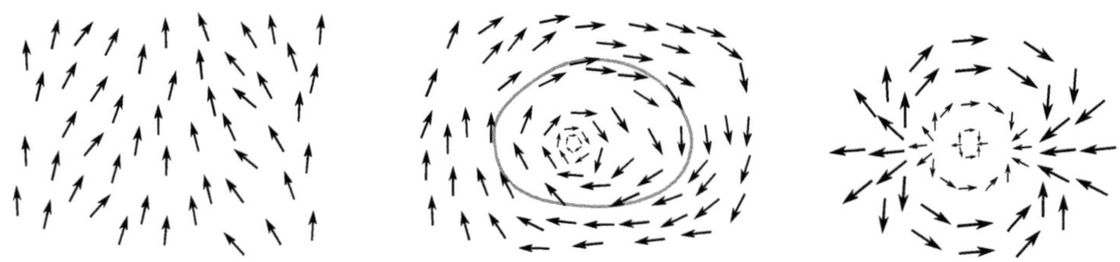

(a) Nambu–Goldstone mode (b) $N = 1$ vortex (c) $N = 2$ vortex

Figure 7.1 (a) The symmetry-broken state of an *XY*-model, perturbed by NG-modes. (b) A single vortex with winding number $N = 1$. Following the contour, the unit vectors wind by 2π, independent of the position and shape of the contour, as long as it encloses the vortex core. At the vortex core, the phase in not defined, indicating the presence of a singularity. If the amplitude of vectors is allowed to vary, it will typically shrink to zero at the vortex core. (c) A single vortex with winding number $N = 2$.

or topological invariant. The charge is topological in the sense that it its value is invariant under smooth deformations of the contour used to determine it. Similarly, any smooth variations of the vector-field configuration do not alter the winding number along a fixed contour.

Somewhere within the contour, there must be a singularity in the order parameter field because the winding of the vectors is independent of the contour size. Shrinking the contour as much as possible, it must then end up in a single point at which the direction of the order parameter vector is undefined. This point signifies a **defect** in the order parameter field. In actual physical systems, the existence of a singularity or undefined point is avoided, either because the order occurs in a discrete crystal lattice or because the amplitude of the order parameter can go to zero at the singular point, as in a superfluid. The **core** of the vortex, defined as the area around the singularity over which the amplitude of the order parameter is substantially suppressed, has a radial size of the order of the coherence length ξ, as defined in Eq. (5.19).

Exercise 7.1 ($U(1)$ vortices).
The order parameter of system with broken $U(1)$ symmetry can be described by a complex number $\psi(\mathbf{x}) = A(\mathbf{x})e^{i\phi(\mathbf{x})}$. The phase $\phi(\mathbf{x})$ determines the direction of the $U(1)$ vector depicting the order parameter. The winding number of any configuration is then given by

$$n = \frac{1}{2\pi} \oint_C \vec{\nabla}\phi(\mathbf{x}) \cdot \vec{dl}. \tag{7.1}$$

Here, C is a curve surrounding the area whose winding number it measures. By convention, we consider the integral to traverse the curve C in clockwise direction. For convenience, we consider $d = 2$ dimensions, where a vortex is a point object.
a. Verify that $n = 0$ if the curve C does not enclose a vortex.
b. The configuration $\phi(\mathbf{x}) = \arctan \frac{y}{x}$ has a vortex at its origin $(x, y) = (0, 0)$. Calculate its winding number.

c. Find an expression for the field $\phi(\mathbf{x})$ that corresponds to the configuration displayed in Figure 7.1b. Then, using Eq. (7.1), calculate its winding number.

Exercise 7.2 (Vortex core).
The Ginzburg–Landau theory for $U(1)$ symmetry breaking (introduced in Eq. (5.14)), up to fourth order and in $d = 2$, is given by

$$\mathcal{F}_{\mathrm{GL}} = \int d^2x \left\{ \frac{\hbar^2}{2m^*} |\nabla\psi(\mathbf{x})|^2 + \alpha\, |\psi(\mathbf{x})|^2 + \frac{\beta}{2} |\psi(\mathbf{x})|^4 \right\}. \tag{7.2}$$

Here, m^* is an effective mass. In the ordered phase, $\alpha < 0$ and $\beta > 0$. Consequently, a stable, ordered configuration satisfies

$$-\frac{\hbar^2}{2m^*}\nabla^2\psi(\mathbf{x}) + \alpha\psi(\mathbf{x}) + \beta|\psi(\mathbf{x})|^2\psi(\mathbf{x}) = 0. \tag{7.3}$$

In a homogenous system $|\psi|^2 = |\alpha|/\beta \equiv \overline{\psi}^2$.
a. Introduce the normalised order parameter field $\tilde{\psi} = \psi/\overline{\psi}$ and express Eq. (7.3) in terms of the normalised field and the coherence length $\xi = \hbar/\sqrt{2m^*|\alpha|}$.

Now consider a particular solution of the Ginzburg–Landau equations that includes a vortex configuration, of the form

$$\psi(r,\theta) = \overline{\psi} f(r) e^{i\theta}. \tag{7.4}$$

Here, (r,θ) are polar coordinates, and $f(r)$ is a positive, real function.
b. Write out the differential equation for the function $f(r)$.

Near the centre of the vortex, we can use the fact that the order parameter amplitude will be small and consider a linear approximation to the differential equation by neglecting the term of order $(f(r))^3$.
c. The equation is now exactly solvable. Find the solution and notice that the order parameter indeed vanishes at the centre of the vortex core.

7.1.1 Topological Defects in Different Dimensions

Topological defects or excitations, like the $U(1)$ vortex, exist in ordered states in any dimension. A one-dimensional chain of vectors that prefer a parallel alignment, for example, may have all vectors to the left of the origin pointing up and all vectors to the right pointing down. There is then a zero-dimensional topological defect, called a domain wall, at the origin. An example of this type will be discussed in Section 7.3 in the context of the Ising model. Likewise, a two-dimensional system with $U(1)$ order may host a vortex. This gives rise to a new type of phase transition, which we will discuss in Section 7.4. Three-dimensional vectors in a three-dimensional volume may be arranged to all point outwards from the centre, like

Defect	Defect dimension	Homotopy	Examples
Soliton	d	π_d	Skyrmion, soliton, Shankar monopole
Domain wall	$d-1$	π_0	Ising domain wall
Vortex	$d-2$	π_1	$U(1)$ vortex, dislocation, cosmic string
Monopole	$d-3$	π_2	Hedgehog, 't Hooft-Polyakov monopole
Instanton	$d-4$	π_3	Non-Abelian instanton

Table 7.1 Classification of topological defects in d dimensions.

the needles on a hedgehog, in what is known as a monopole configuration. More generally, for any d-dimensional system, defects of dimension $d-1$ are often referred to as domain walls, of $d-2$ as vortices, and of $d-3$ as monopoles, although the nomenclature may vary across subfields in physics. Vortices, such as those in Figure 7.1b have cores that are pointlike objects in two dimensions and linelike in three dimensions. In Table 7.1 we list the main types of topological defects, along with their associated homotopy groups, which will be discussed in Section 7.2.

Starting from a perfectly ordered configuration, creating a single topological defect involves significantly changing the orientation of the order parameter almost everywhere in the system. This makes it extremely unlikely that such defects are introduced spontaneously by thermal or other fluctuations. Systems with isolated topological defects can be created, however, when defects are forced in from the outside. Consider, for example, a superfluid in a container. The superfluid order makes its flow dissipationless and proportional to $j(\mathbf{x}) = \nabla\phi(\mathbf{x})$, where $\phi(\mathbf{x})$ is a $U(1)$ vector denoting the complex phase of the order parameter. As a consequence, the flow within the superfluid must be irrotational, since $\nabla \times j(\mathbf{x}) = \nabla \times \nabla\phi(\mathbf{x}) = 0$. If we now start to spin the container, trying to apply an external torque on the superfluid, at first nothing will happen. The superfluid will remain perfectly still inside container, seemingly ignoring its rotation. Once the externally applied torque exceeds the energy cost of forming a single vortex, however, a topological defect will move into the system from the side and cause the phase of the order parameter $\phi(\mathbf{x})$ to wind throughout the entire superfluid. This way, a quantised amount of angular momentum, proportional to the winding number of the total vortex configuration, may be imposed on the superfluid.

While a single topological defect is not easily created, it is very stable once formed. Removing the defect typically requires changing the order parameter over an extensive part of the system. Furthermore, local disturbances to the order parameter do not alter the topological charge. This stability of topological defects has applications for instance in quantum computation and information storage [130].

In stark contrast to the effort required to create a single, isolated topological defect, it is straightforward and common for them to arise in topologically neutral combinations. For vortices, the total topological charge of multiple defects can be found by considering a contour that simultaneously encloses all vortex cores. If the phase of the order parameter does not wind along this contour, the defects together form a neutral configuration, such as the one depicted in Figure 7.2. Such neutral combinations only affect the orientation of the order parameter within an isolated part of the system, allowing them to be created as thermal excitations.

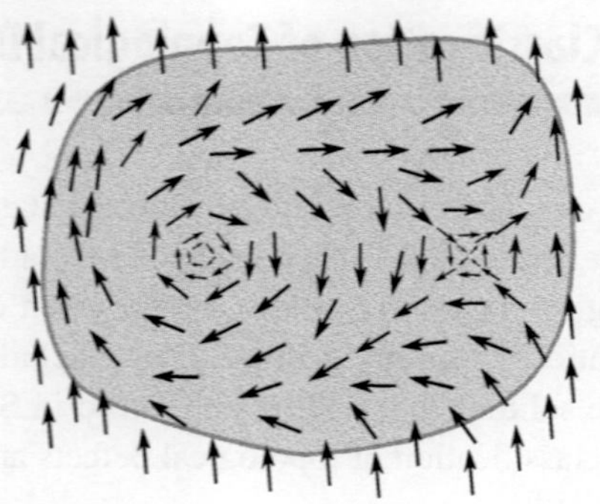

Figure 7.2

Two vortices with opposite topological charge. The total configuration is topologically neutral, because along the contour enclosing both vortex cores, the vector field does not wind at all. Furthermore, far away from the cores, all vectors point approximately in the same, spontaneously chosen, direction.

Exercise 7.3 (Winding number).
Recall the complex order parameter and the winding number introduced in Exercise 7.1. The configuration of the complex order parameter field $\psi(\mathbf{x}) = A(\mathbf{x})e^{i\phi(\mathbf{x})}$ depicted in Figure 7.2 can be described as a superpositions of a vortex at position $(-m, 0)$ and an antivortex at position $(m, 0)$. This results in the field configuration $\phi(\mathbf{x}) = \arctan\frac{x-m}{y} + \arctan\frac{-(x+m)}{y}$.

Verify that a contour around the left vortex gives $n = 1$, around the right vortex $n = -1$ and around both $n = 0$.

Exercise 7.4 (Hedgehogs).
A **hedgehog** is a topological defect in $d = 3$ dimensions, for which the order parameter is an element of S^2, namely a three-dimensional unit vector $\mathbf{n}(\mathbf{x})$. A typical hedgehog configuration is given by $\mathbf{n}(\mathbf{x}) = \frac{1}{|\mathbf{x}|}(x, y, z)$.
a. Draw the 3D field configuration of a hedgehog.

Calculating the topological charge of a hedgehog configuration is somewhat more involved than that of a vortex. Generalising Eq. (7.1) requires first the definition of a so-called field strength:

$$\Phi_i(\mathbf{x}) = \frac{1}{2}\epsilon_{ijk}\epsilon_{a_0a_1a_2}n^{a_0}\partial_j n^{a_1}\partial_k n^{a_2}. \tag{7.5}$$

This is a generalisation of the quantity $\nabla\theta$ used in the case of a $U(1)$ order parameter to the case of three-dimensional vectors. The topological charge of the hedgehog is given by a surface integral over this field Φ:

$$n = \frac{1}{4\pi}\oint_{\Sigma} \Phi \cdot d\mathbf{A}. \tag{7.6}$$

Here, Σ is any surface enclosing the hedgehog core.
b. Use Eq. (7.6) to calculate the topological charge of the hedgehog configuration.

7.2 Classification of Topological Defects

Which topological defects may arise in a particular ordered state is determined entirely by the broken symmetries that define its order parameter. The mathematics that underpins the classification of all topological excitations allowed by a given combination of broken and unbroken symmetries is that of **homotopy groups**. These include the topological defects of Section 7.1.1 as well as the solitons that will be introduced in Section 7.2.3. A much more detailed description of the classification of topological defects and homotopy groups can be found in Refs. [125, 138].

7.2.1 Homotopy Groups

In Section 3.1, we showed that the set of all possible distinct ways of breaking a symmetry, corresponds to the coset space G/H. Here, G is the group of all symmetry transformations of the symmetric system, and H is the subgroup of unbroken transformations in the ordered state. For example, a ferromagnetic Hamiltonian has full $SU(2)$ spin-rotation symmetry, while only $U(1)$ rotation symmetry around the magnetisation axis is left unbroken in its ordered state. The coset space $SU(2)/U(1) \simeq S^2$ then describes the set of all possible directions that the magnetisation can point to. More generally, distinct ordered states, or distinct ways of breaking a symmetry, correspond to different points in the coset space. As a result, the local order parameter $O(x)$ is a mathematical map from real space (denoted by X) to the coset space G/H:

$$O(x): X \rightarrow G/H. \tag{7.7}$$

That is, the order parameter assigns an element of the coset space, or a direction of the order parameter, to each point in space. It also assigns an order parameter amplitude, but since the amplitude does not affect the classification of topological excitations, we will ignore it in the this section.

To see how the map to the coset space can be used to distinguish different types of topological defects, consider again the example of the $U(1)$-valued order parameter of Section 7.1. One of its possible topological defects is given by the vortex in $d = 2$, shown in Figure 7.1b, which we argued can be described in terms of the winding of the order parameter along a contour surrounding the defect. Mathematically, this means we consider the evolution of the order parameter along a circle S^1 enclosing the vortex core, as described by the map:

$$O(x): S^1 \rightarrow U(1). \tag{7.8}$$

Here, $U(1)$ signifies the coset space G/H for this particular order parameter and S^1 indicates the fact that we consider the order parameter along a circular contour, rather than throughout all of space.

To establish which distinct types of maps of the form of Eq. (7.8) may exist, we first need to know which maps may be considered equivalent. That is, we need to define which evolutions of the order parameter along the contour are possible, given a single type of topological defect enclosed within it. Formally, two different order parameter configurations $O_i(x)$ and

G/H	π_0	π_1	π_2	π_3	Examples
S^0	$\mathbb{Z}_2$	0	0	0	Ising magnet
$U(1) \cong S^1$	0	$\mathbb{Z}$	0	0	Superfluids, XY magnets
S^2	0	0	$\mathbb{Z}$	$\mathbb{Z}$	Heisenberg magnet
S^3	0	0	0	$\mathbb{Z}$	Electroweak symmetry breaking
$SO(3)$	0	$\mathbb{Z}_2$	0	$\mathbb{Z}$	Coplanar antiferromagnet, ^{3}He A
$SU(n), n > 2$	0	0	0	$\mathbb{Z}$	Chiral symmetry breaking
$\mathbb{R}P^2$	0	$\mathbb{Z}_2$	$\mathbb{Z}$	$\mathbb{Z}$	Nematic liquid crystal
$\mathbb{T}^n$	0	$\mathbb{Z}^n$	0	0	Harmonic crystal

Homotopy groups.

Table 7.2

$O_f(x)$ of the type of Eq. (7.8) are said to be **homotopic** if there exists a continuous map $F(x,t)\colon S^1 \times [0,1] \to U(1)$ such that

$$F(x,0) = O_i(x) \text{ and } F(x,1) = O_f(x). \tag{7.9}$$

In words, the configurations are homotopic if the initial order parameter configuration $O_i(x)$ can be continuously deformed into the final $O_f(x)$ configuration. If such a continuous map exists, the two order parameter configurations are called smoothly connected, or equivalent under homotopy.

Not all possible order parameter configurations are smoothly connected. For example, the true long-range ordered state with constant order parameter, or uniformly ordered $U(1)$ vectors, cannot be smoothly connected to a configuration with a single vortex. Formally, this can be seen by describing the order parameter as a periodic function $O(\theta)$, with $\theta \in [0, 2\pi)$ indicating positions along the circular contour. Since the order parameter in this example is a $U(1)$ vector, its value is also an angle, defined modulo 2π. Uniform order is described by $O_i(\theta) = \phi$, with $\phi \in [0, 2\pi)$ a constant angle. Similarly, the configuration with a single $n = 1$ vortex shown in Figure 7.1b may, for example, correspond to $O_f(\theta) = \phi + \theta$. The two configurations O_i and O_f are inequivalent under homotopy because it is not possible to define a continuous function $F(\theta, t)$ that coincides with O_i at $t = 0$ and with O_f at $t = 1$ while also remaining periodic modulo 2π along the θ axis.

Starting from any order parameter configuration, the set of all configurations that are equivalent to it under homotopy defines a **homotopy class**. Any two configurations that cannot be continuously deformed into one another using a map of the form of Eq. (7.8) are then in distinct homotopy classes. The set of all homotopy classes is known as the **first homotopy group** or the **fundamental group**, and is denoted π_1. In the example of $U(1)$ vectors, order parameter configurations with equal winding number can always be continuously connected, while those with distinct winding numbers cannot. Each homotopy class can therefore be labelled by the number of times the map $O(x)$ winds around $U(1)$. Because the set of all possible winding numbers is just the set of all integers, the fundamental group of $U(1)$ is given by

$$\pi_1(U(1)) = \mathbb{Z}. \tag{7.10}$$

The topological distinction between vortices with different charges thus allows us to obtain a classification labelled by integer winding numbers, without having to explicitly solve any integrals like the ones in Exercise 7.1.

This idea can be extended to any type of contour, and any type of order parameter set G/H. In general, the relevant contour around a D-dimensional topological defect in d dimensions is S^{d-D-1}, a $(d-D-1)$-dimensional sphere. To classify these topological defects, we say that two order parameter configurations on the n-sphere, described by two maps $S^n \to G/H$, are equivalent under homotopy if there exists a continuous map connecting them. The set of homotopy classes containing equivalent configurations is then called the nth homotopy group $\pi_n(G/H)$. It follows that D-dimensional topological defects in d dimensions are categorised by the homotopy group $\pi_{d-D-1}(G/H)$.

In the example of XY-vectors, a one-dimensional contour in a two-dimensional system is used to characterise a zero-dimensional, point-like defect. The classes of equivalent defect configurations are thus given by $\pi_1(U(1))$. The hedgehog of three-dimensional unit vectors in $d = 3$ dimensions, on the other hand, is enclosed by the two-sphere and is thus described by $\pi_2(S^2)$. Table 7.2 shows an overview of some of the most relevant homotopy groups, applying to different order parameter spaces and dimensions. It is important to remember that the use of nomenclature like domain wall, vortex, monopole, and so on, differs from field to field. In practice, it is therefore advisable to always specify the dimensionality of both the defect and the system under consideration.

7.2.2 Examples of Defects and Their Homotopy Groups

To obtain some intuition in the use of homotopy groups to classify topological defects, it is instructive to consider some explicit examples in the context of concrete physical systems.

7.2.2.1 Ising Domain Walls

The most basic example is that of domain walls in the one-dimensional Ising model, which are classified by $\pi_0(\mathbb{Z}_2) = \mathbb{Z}_2$. The zeroth homotopy group π_0 in this example is the set of all inequivalent configurations on a line element bounded by two zero-dimensional points. Because there is no way to continuously deform Ising spins, equivalent configurations in this case are those that can be connected while keeping the spins on the boundary points fixed. The two classes of inequivalent configurations are then the ones where the boundary spins are either aligned or anti-aligned. Those with anti-aligned spins enclose an odd number of domain walls, while those with aligned spins enclose an even number. Equivalently, each link in the chain is either a domain wall, or it is not.

Notice that the configurations with all spins up and the one with all spins down are considered equivalent because they differ only by a global choice of what to call up and what to call down. That is, global coordinate transformations do not affect the homotopy groups. More generally, this also means that, for example, $\pi_0(\mathbb{Z}_3) = \mathbb{Z}_3$, with inequivalent configurations having boundary states that are either equal, one step removed, or two steps removed, regardless what the involved states are. Put formally, $\pi_0(G)$ counts the disconnected components of G. This implies that domain walls are in fact not topological defects when the order parameter set is fully connected, such as in superfluids, magnets or in electroweak symmetry breaking. In those cases, a domain wall can always be moved through a boundary point without the value of the order parameter at the boundary changing in a discontinuous way.

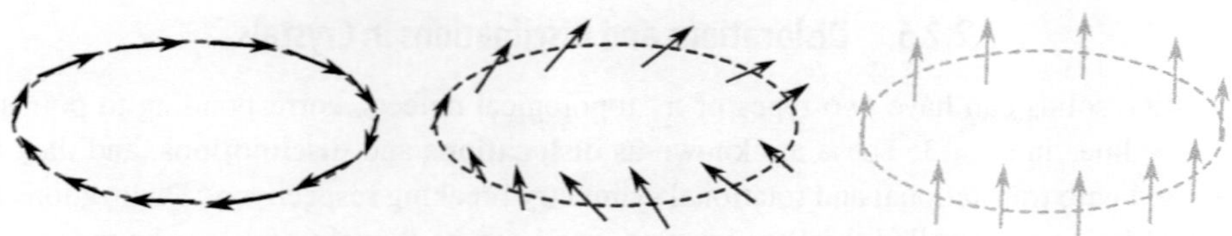

Figure 7.3

Removal of a vortex by rotating vectors out of the plane. This is sometimes called 'escape in the third dimension', and shows that the vortex is not a stable topological defect for order parameters represented by a three-dimensional vector.

7.2.2.2 *XY* Vortices

Moving on to broken $U(1)$ symmetry, such as the XY order parameters of the previous section, vortices are point-like defects in $d = 2$, while they are line-like in $d = 3$. In both cases, they are classified by $\pi_1(U(1)) = \mathbb{Z}$. The second homotopy group, $\pi_2(U(1)) = 0$, is trivial, indicating that it is impossible to obtain monopole-type defects in $U(1)$-ordered states in three dimensions.

7.2.2.3 Vortices and Monopoles in Ferromagnets

Ferromagnetic configurations of three-dimensional spins break $SU(2)$ symmetry down to $U(1)$, so that the order parameter takes values on the two-sphere $S^2 \simeq SU(2)/U(1)$, as explained in Section 3.1. Any closed loop on the surface of that sphere can be smoothly contracted to a point. This implies that any map $O(x): S^1 \to S^2$ can be continuously deformed into just a constant value, and there are no inequivalent order parameter configurations classified by a loop. In other words, the first homotopy group in this case is trivial. Physically, this means there can be no stable vortices in Heisenberg ferromagnets. If a vortex is introduced in such a magnet, all spins can be smoothly rotated to a perpendicular direction to remove the apparent singularity, as indicated in Figure 7.3.

It is possible, however, to have zero-dimensional defects in a three-dimensional Heisenberg ferromagnet, by arranging spins in a hedgehog or monopole configuration, pointing radially outward from the origin everywhere. Such monopoles are classified by the homotopy group $\pi_2(S^2) = \mathbb{Z}$, where in this case, the integer index counts the number of times the order parameter configuration on a two-dimensional surface covers the entire two-sphere of possible order parameter orientations.

7.2.2.4 Monopoles and Cosmic Strings in the Standard Model

Within the Standard Model, topological monopoles can also exist due to the electroweak symmetry breaking . One example is the **'t Hooft-Polyakov monopole**, which can be shown to behave as a source of magnetic field. It has the same structure as a hedgehog and is thus classified by $\pi_2(G/H)$ in $d = 3$ dimensions. Unlike the hedgehog in ferromagnets, however, the presence of gauge fields renders the energy of a 't Hooft–Polyakov monopole finite even in the thermodynamic limit. This also happens with $U(1)$ vortices in a charged superconductor, which we discuss in Chapter 8.

Another type of topological defect associated with the Standard Model, and one that is being actively searched for in our observable universe is a line defect called a **cosmic string**. These are discussed in more detail in Section 14.3.

7.2.2.5 Dislocations and Disclinations in Crystals

Crystalline solids can have two types of π_1 topological defects, corresponding to points in $d = 2$ or lines in $d = 3$. These are known as **dislocations** and **disclinations**, and they are associated with translational and rotational symmetry breaking respectively. Dislocations are classified by a vector called the Burgers vector and can be thought of as the boundary of a line or sheet of additional atoms inserted into an otherwise regular lattice. The disclination, on the other hand, is characterised by an angle and corresponds to the tip of a two- or three-dimensional wedge of superfluous material.

Famously, the idea that neutral pairs of dislocations in hexagonal lattices might proliferate led to the prediction of a new phase of matter called a hexatic liquid crystal. It is liquid in the sense that it is translationally symmetric, but possesses so-called hexatic order because the rotational symmetry remains broken down to C_6, or sixfold discrete rotations [67, 140, 226]. The transition between the crystalline and hexatic liquid-crystalline phases is referred to as **dislocation-mediated melting** and is similar in spirit to the BKT phase transition discussed in Section 7.4.

7.2.2.6 Instantons

Besides static topological defect configurations, one may also imagine time-dependent topological excitations. A defect that exists only at one event (i.e., at one point in spacetime) is called an **instanton**. These appear, for example, in Yang–Mills theories, as well as in theories of nucleation at first-order phase transitions. In four-dimensional spacetime, they can be enclosed by a three-dimensional contour, so they are classified by the third homotopy group $\pi_3(G/H)$. For spontaneously broken $SU(N)$ symmetry, for example, this homotopy group can be non-trivial and instantons play an important role in $SU(N)$ theories. Ref. [180] provides a good introduction to this topic.

Exercise 7.5 (Defects in a nematic liquid crystal).
In a nematic liquid crystal, the order parameter is a director (represented by a line element or two-headed arrow) rather than a vector. Consequently, the order parameter coset space is $G/H = \mathbb{R}P^2$, and the vortex classification in two dimensions is given by $\pi_1(\mathbb{R}P^2) = \mathbb{Z}_2$.
Show explicitly that there are no more than two inequivalent vortex configurations in a $d = 2$ nematic, by first drawing a single vortex, then adding a second vortex, and observing that the total final configuration is equivalent under homotopy to no vortices at all.

Exercise 7.6 (Homotopy of a crystal).
The coset space relevant for broken translation symmetry in a d-dimensional crystalline solid is a d-dimensional torus.
a. Explain why the homotopy group describing point-like defects in a two-dimensional crystal is given by $\mathbb{Z}^2$, consisting of two copies of $\pi_1(S^1)$.

b. Argue why in general, the homotopy groups of D-dimensional defects in a d-dimensional torus, consists of d copies of $\pi_{d-D-1}(S^1)$.

7.2.3 Topological Solitons and Skyrmions

In our discussion of topological excitations so far, we neglected a special category of topological objects, often called **topological solitons**. This name is sometimes reserved for systems in one spatial dimension only, but we will adopt it here for all dimensions. Solitons are topological objects in the sense that they have a topological charge, which takes quantised values that cannot be altered by smooth deformations of the order parameter configuration. They are again classified by homotopy groups. In contrast to the usual topological defects, however, they do not require any singularity in the order parameter configuration, and they always have only a finite energy, which does not scale with system size, and which is strongly localised near the centre of the soliton.

An example of a soliton in an XY magnet on a one-dimensional line is shown on the left side of Figure 7.4. Importantly, the line element in this figure actually represents an infinitely long line, which has been compressed onto a finite length. This can be done, for example, by drawing the order parameter configuration $O(x)$ not directly as a function x with $-\infty < x < \infty$, but instead as a function of $y = \tanh(x)$ with $-1 < y < 1$. The order parameter in the soliton configuration is undisturbed and constant almost everywhere along the uncompressed line, so that the XY vectors point in almost the same direction at any point sufficiently far from the origin. This makes it possible for Figure 7.4 to give a good representation of the order parameter field even on the compressed line, with the vectors pointing up representing a much larger part of the uncompressed line than the vectors pointing in other directions.

The soliton is topological because it cannot be removed by smooth rotations of the vector field. It is straightforward to check that these will only deform the soliton or move it along the line, and because the line is actually infinitely long, the soliton cannot be removed this way. To understand the origin of the topological character, connect the two endpoints of the

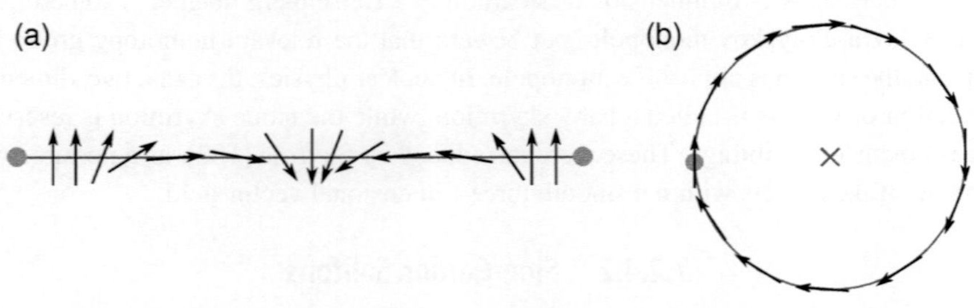

Figure 7.4

A topological soliton of charge one. (a) The configuration in real space, on an infinite line projected onto a finite length. The spins at the boundaries (large dots) point in the same direction. As the line is traversed from left to right, the spins wind smoothly over a 2π angle. (b) The dots at the boundary of the line are mapped to the same point on a circle, illustrating the topological nature of the soliton. The cross indicates the position of the associated singularity, which lies outside of the one-dimensional space on which the order parameter field is defined.

compressed line, turning it into a circle. This will not introduce any discontinuities in the order parameter, since it has the same direction at both endpoints. The resulting configuration of XY vectors along the circle, shown on the right side of Figure 7.4, is the same as that for a circle enclosing a vortex in a two-dimensional order parameter field. The vortex core lies in the centre of the circle, so that the configuration along the circle is smooth and well defined everywhere. Removing the rotation of the order parameter field along the circle would correspond to moving the vortex core outside of it. But this cannot be done smoothly, since it requires the core crossing the circle.

The possible distinct types of solitons in a $U(1)$ order parameter configuration along a one-dimensional line correspond to the distinct types of vortex configurations of a $U(1)$ field in two dimensions. Like vortices, these solitons are therefore classified by the homotopy group $\pi_1(U(1)) = \mathbb{Z}$. Similar types of topological solitons also exist in higher dimensions. Using the fact that d-dimensional space can be compactified into S^d and proceeding analogously to the $d = 1$ case, topological solitons in d dimensions are found to be classified by the homotopy group $\pi_d(G/H)$. Table 7.2 thus summarises the classification of both solitons and topological defect.

7.2.4 Examples of Solitons and Their Homotopy Groups

An intuition for the possible types of solitons may be gained by considering explicit examples in the context of concrete physical systems.

7.2.4.1 Skyrmions

The two-dimensional plane may be compactified into a two-sphere, using the stereographic projection. We will consider an order parameter field consisting of three-dimensional unit vectors on this sphere, corresponding, for example, to the magnetisation of a Heisenberg ferromagnet. it is then possible to arrange the vectors on the sphere with all spins pointing radially outward, in the same way that would have resulted from a hedgehog or monopole configuration in three-dimensional space, with the monopole core inside the sphere. Folding the sphere back out into a flat plane, the resulting vector configuration on the plane is called a **skyrmion**. This type of topological soliton appears in quantum Hall systems and some magnetic materials. A two-dimensional skyrmion in a Heisenberg magnet is sometimes also called a Belavin–Polyakov monopole, but beware that the relevant homotopy group is still $\pi_2(S^2)$, and the soliton is not itself a monopole. In nuclear physics, the same two-dimensional configuration of vectors is called a baby skyrmion, while the name skyrmion is reserved for its three-dimensional siblings. These were introduced by Skyrme [182], as a possible way of creating pointlike objects within a smooth three-dimensional vector field.

7.2.4.2 Sine-Gordon Solitons

The **sine-Gordon model** is a $1 + 1$-dimensional field theory defined by the action

$$S_{\text{sG}}[\phi] = \int d^2x \left[\frac{1}{2} \partial_\mu \phi \partial^\mu \phi + m^2 \left(\cos(\phi) - 1 \right) \right]. \tag{7.11}$$

Each constant field $\phi = 2\pi n$ with $n \in \mathbb{Z}$ is a possible ground state of this action. Solitons can then be formulated in direct analogy to the construction in Figure 7.4, and are often called kinks in this context. Their topological charge can be defined as $q = \frac{1}{2\pi}\int dx\partial_x\phi(x)$, which measures the rotation of the field ϕ along the x-axis.

What sets the particular solitons found within the sine-Gordon model apart from other systems, however, is that these configurations appear as analytic solutions to the Euler–Lagrange equations. This is possible because of the **integrability** of the sine-Gordon model, which has an infinite set of conserved quantities. As a direct consequence, solitons in the sine-Gordon model are noninteracting and pass through each other without changing their individual sizes and shapes. Moreover, the integrable structure of the sine-Gordon model allows for more general stable and strongly localised solutions, such as a bound structure of a kink and anti-kink, known as a breather. Some authors use the name soliton solely to refer to the stable localised excitations of integrable models and prefer the term solitary waves for the more general topological objects that exist also in nonintegrable systems.

7.2.4.3 Winding Numbers in Topological Insulators

Noninteracting electronic quasiparticles in a crystalline solid are described by a Bloch Hamiltonian at each point in momentum space. The complex components of the Bloch eigenvectors can be viewed as a map from the Brillouin zone, a d-dimensional torus, to the complex numbers. That map can contain windings akin to those of the XY field in Figure 7.4. An electronic structure with such a topological soliton cannot be smoothly deformed into one without. Band insulators in which the occupied electronic bands contain a form of winding are therefore designated **topological insulators**, and their winding numbers are often called topological invariants.

7.2.4.4 Shankar Monopoles

Superfluids are characterised by a $U(1)$ order parameter and, therefore, have no monopoles in three dimensions. However, superfluid ^{3}He in the A-phase, which we will discuss in more detail in Section 12.3.4, consists of spin-triplet, neutral Cooper pairs. The order parameter in this case describes the broken $U(1)$ phase-symmetry as well as the broken spin-rotation symmetry of the triplet pairs. As a result, it has an $SO(3)$ structure, which allows a soliton configuration in $d = 3$, because $\pi_3(SO(3)) = \mathbb{Z}$. These objects are called **Shankar monopoles** because they are pointlike, but bear in mind that they are not true monopoles, in the sense of containing a singularity.

Exercise 7.7 (Skyrmions vs. hedgehogs).
Using the compactification of two-dimensional space, a skyrmion in $d = 2$ can be understood as the field configuration on a two-sphere surrounding a hedgehog in $d = 3$.
a. Write an explicit expression for the hedgehog field in $d = 3$ and then use the stereographic projection to find an expression for the skyrmion field configuration in $d = 2$.

Generically, the energy of a topological excitation in a field $\phi(\mathbf{x})$ can be argued to be proportional to $E = \int d^dx(\nabla_a\phi_b(\mathbf{x}))^2$.
b. Show that the energy of a hedgehog monopole in $d = 3$ diverges, while the energy of a skyrmion in $d = 2$ is finite.

Exercise 7.8 (Sine-Gordon solitons).
a. Find the differential equation satisfied by any static field $\phi(x)$ that extremises the sine-Gordon action of Eq. (7.11).
b. Verify that the field configuration $\phi(x) = 4 \arctan e^{m(x-x_0)}$ is a solution to that differential equation.
c. Calculate the topological charge $q = \frac{1}{2\pi} \int dx \partial_x \phi(x)$ and confirm that the configuration is a topological soliton.

7.3 Topological Melting: The $d = 1$ Ising Model

To see the importance of topological defects in the study of long-range order and spontaneously broken symmetries, consider a one-dimensional chain with classical Ising spins that can point only up or down. A ferromagnet made of Ising spins has only two possible broken-symmetry ground states, either with all spins up, or all down. This is an example of discrete symmetry breaking, rather than the continuous symmetries that we mostly focus on in other sections. The Hamiltonian for the classical Ising ferromagnet is given by

$$H = -J \sum_i \sigma_i^z \sigma_{i+1}^z. \tag{7.12}$$

The ground state is twofold degenerate and has energy $E_0 = -JN$, where N is the number of links in the chain. Starting from a spontaneously chosen ground state with all spins up, the simplest excitation to create is a single spin flip, as shown in Figure 7.5b. This costs an energy $E_{\mathrm{f}} - E_0 = 4J$ because the flipped spin is now aligned antiferromagnetically with two neighbours, causing two links to change their local energy from $-J$ to $+J$. The single spin flip can be thought of as a localised version of the spin waves of the Heisenberg ferromagnet. Because Ising spins cannot be continuously rotated from up to down, a localised spin-flip is the best one can do, and what used to be a massless NG mode in the system with continuous symmetry, is a gapped excitation in the discrete case.

The spin flip reduces the total magnetisation by just 2, so that even a large but not-extensive number of spin flips does not significantly affect the total magnetisation in the thermodynamic limit. The one-dimensional Ising ferromagnet thus seems to be thermodynamically stable in sense of Section 3.7, even at non-zero temperatures. This conclusion, however, turns out the be wrong because it neglects the topological defects of the ferromagnetic state. For the one-dimensional chain, a topological defect is a **domain wall** created by splitting the chain in two segments and taking all the spins to be up in one segment and down in the other, as shown in Figure 7.5c.

Since only a single pair of neighbouring spins is now aligned antiferromagnetically, the energetic cost of the domain wall is only $E_{\mathrm{dw}} - E_0 = 2J$. That is, in this special one-dimensional case, the cost of a domain wall is lower than that of a spin flip. More importantly, a single domain wall involves the reorientation of a macroscopic number of spins, and thus strongly affects the average magnetisation. Even if we include only states with a single domain wall in the low-energy effective model, the consequences are drastic. In a chain with $N + 1$ spins and a domain wall at position j, the magnetisation is $M = 2j - N$. Since there

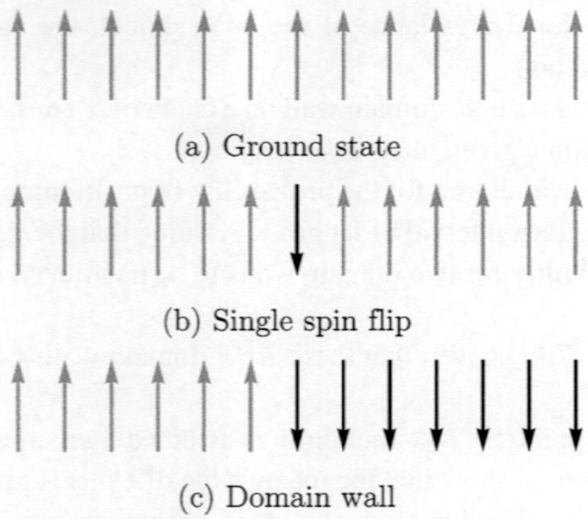

Figure 7.5

(a) The ferromagnetic ground state of the one-dimensional classical Ising model. (b) A single spin flip, which only marginally affects the total magnetisation. (c) A single domain wall, which is a topological defect with macroscopic effect on the total magnetisation. In one dimension, the domain wall has lower energy than the spin flip.

are N possible links on which a single domain wall may occur, the entropy associated with this configuration is $S = \ln N$. The free energy of a single wall is then $F_{\text{dw}} = 2J - k_{\text{B}}T \ln N$. For large systems at any non-zero temperature, the entropic gain outweighs the energetic cost of introducing a domain wall into the system. The thermal expectation value of the magnetisation therefore vanishes completely in the thermodynamic limit, and ferromagnetic order cannot occur in the one-dimensional Ising chain at any non-zero temperature. This is the simplest example of **topological melting**, in which the local order is destroyed by the proliferation of topological defects rather than NG modes.

Exercise 7.9 (Domain walls in the Ising model).
This exercise follows an elegant argument in Ref. [172] to showcase the effect of topological defects in the $d = 1$ classical Ising model defined by $H = -J\sum_j \sigma_{j+1}\sigma_j$, with σ_j the value of the classical Ising variable on site $j \in [1, N]$. First, consider states with only a few thermally occupied spin flips on top of the maximally magnetised configuration. Consider temperatures $k_B T \ll J$ so that the probability of spin flips occurring on neighbouring sites can be neglected and each spin flip raises the energy by $4J$.
a. Ignoring domain walls altogether, give an expression for the thermal expectation value of the magnetisation in the limit of large N.
b. Using similar approximations, but now ignoring single spin flips and instead considering only domain wall excitations raising the energy by $2J$, show that the thermal expectation value for the domain wall density is approximately $e^{-2J/k_{\text{B}}T}$.
c. Given that the density of domain walls is non-zero for any $T > 0$, argue that the thermal expectation value of the magnetisation vanishes for any non-zero temperature.

To make sure that the results so far are not merely a peculiarity of global operators like the total magnetisation, we also consider the classical correlation function $C_j(x) = \sigma_{j+x}\sigma_j$.

Because of the translational invariance of the Ising model, the value of the correlation function cannot depend on j.

d. Randomly positioning a single domain wall on a chain of N sites, what is the probability for it to be located within a given interval of length x?

e. Give an approximate expression for the probability of positioning n out of a total of N_{dw} domain walls within a given interval of length x. Assume that the density of domain walls is low, so that the possibility for two domain walls to be positioned at the same site can be ignored.

f. What is the value of $C_j(x)$, given that there are n domain walls between positions j and $j+x$?

g. Given that there are precisely N_{dw} randomly distributed domain walls in an Ising chain, use the binomial theorem to show that the mean value of $C_j(x)$ is given by $(1-2x/N)^{N_{\text{dw}}}$. Take into account the fact that domain walls are indistinguishable.

Using the analysis so far, the thermal expectation value of the classical correlation function can be written as $C(x) \approx 1/Z \sum_{N_{\text{dw}}} (1-2x/N)^{N_{\text{dw}}} e^{-2JN_{\text{dw}}/k_\text{B}T}$, with $Z = \sum_{N_{\text{dw}}} e^{-2JN_{\text{dw}}/k_\text{B}T}$. Here, we assumed that only terms in which the density of domain walls is low carry any appreciable thermal weight, so that each domain wall can be assumed to raise the energy by $2J$. The same approximation also allows us to extend the upper limits on the summations to infinity, and to write $(1-2x/N)^{N_{\text{dw}}} \approx e^{-2xN_{\text{dw}}/N}$.

h. Use an identity for the geometric series to evaluate the summation and show that the correlation function is approximately $C(x) \approx 1 - 2(x/N)e^{-2J/k_\text{B}T}$ for small x/N and low temperatures. True long-range order thus only occurs in the classical Ising chain at precisely zero temperature.

7.4 Berezinskii–Kosterlitz–Thouless Phase Transition

The Mermin–Wagner–Hohenberg–Coleman theorem of Section 6.2 shows that thermal fluctuations will destroy long-range order at any non-zero temperature in two-dimensional systems that have type-A NG modes at zero temperature. Rather than the exponentially decaying correlation functions that we saw in Eq. (3.6) to be characteristic of truly disordered states; however, it has been shown that some two-dimensional systems have low-temperature correlation functions that decay as a power law:

$$C(x,x') \propto |x-x'|^{-c} \qquad \text{as } |x-x'| \to \infty. \tag{7.13}$$

Here, c is a system-dependent real exponent. These types of correlation functions are not long-ranged, but they are also qualitatively different from the short-ranged, exponentially decaying correlation functions that always prevail at sufficiently high temperatures. States with power law correlation functions are therefore said to exhibit **algebraic long-range order**. Going from low to high temperatures, there must be a critical temperature at which algebraic long-range order gives way to true disorder. Because a power law cannot be analytically continued to an exponential function, the correlation function becomes non-analytic at the critical temperature, which thus corresponds to a true phase transition rather than a smooth crossover. This phase transition occurs despite the fact that no symmetry is truly broken in either the low- or the high-temperature phase.

As it turns out, the phase transition in this case is described by a form of topological melting: the **unbinding** of pairs of topological defects. Topologically neutral configurations of defect–antidefect pairs, such as that in Figure 7.2 can occur as finite-energy excitations in an otherwise ordered background. This implies that a system starting out in an ordered state at zero temperature will develop a thermal population of such pairs at non-zero temperatures. The energy cost associated with a defect–antidefect pair scales with the separation between their cores, and at low temperatures no defect will have sufficient energy to wander far from its antidefect partner. The low-temperature phase then is characterised by a thermal population of 'bound' defect pairs, in what turns out to correspond to a state of algebraic long-range order.

As temperature is raised, the pairs become more prolific, and defects within a pair become further separated from their partners. At some temperature, the average separation between partners becomes as large as the distance between pairs. At that point, an individual defect can no longer be associated with any particular antidefect, and individual, unbound excitations may freely roam the system. This picture, put forward by Berezinskii [25] as well as Kosterlitz and Thouless [96, 97], is now known as the **BKT phase transition**. The name is applied in particular to systems with $U(1)$ or XY-symmetry, but the phenomenon is much more general. The only requirement for it to occur, is that stable point-like topological defects may be formed in two-dimensional systems with a spontaneously broken symmetry at zero temperature.

The reason that defect pairs must unbind at sufficiently high temperatures can be conveyed using a heuristic argument due to Kosterlitz and Thouless [97], inspired by the 1D Ising model. They show that in a system of XY-spins, the energy of single, isolated vortex is proportional to $E_{\text{one defect}} \propto \ln(L/\xi)$, where L is the linear system size and ξ the coherence length. There are about L^2/ξ^2 ways to put an object of area ξ^2 into a system with area L^2. The entropy associated with a single defect is therefore $S_{\text{one defect}} \approx k_B \ln(L^2/\xi^2) \approx 2k_B \ln(L/\xi)$. Notice that the energy and entropy scale with system size in the same way. This means the free energy associated with a single, isolated defect is

$$F_{\text{one defect}} = E - TS \propto (J - k_B T)\ln(L/\xi). \tag{7.14}$$

Here, J is an energy scale that depends on the microscopic model. At low temperatures, the energy cost of creating a defect is higher than the entropy gain, and isolated defects will not occur. At sufficiently high temperatures, however, the entropic term outweighs the energetic one, and isolated defects proliferate throughout the system, destroying any type of order.

Although this argument nicely shows that a thermal phase transition is unavoidable, it is only part of the story. It neglects the physics of defect–antidefect pairs screening the interaction between other pairs of defects. This allows defects to drift further apart than they would without other pairs around and lowers the energy cost of creating additional defect pairs. This creates a self-enforcing cycle that culminates in the proliferation of defects at the critical temperature.

The BKT phase transition cannot be described within the usual Landau paradigm of phase transitions discussed in Section 5.2. It is sometimes said to be an 'infinite-order' phase transition because the free energy and all of its derivatives remain continuous throughout the transition. It does have distinct critical exponents, which are used to experimentally identify

BKT transitions, and which may be calculated using an appropriate version of the renormalisation group. Evidence for BKT transitions was first found in films of superfluid helium, and later in anisotropic magnets, ultracold atomic gases, colloidal discs, and even thin-film superconductors.

Exercise 7.10 (Energy of a vortex).
Consider a $U(1)$ order parameter field on a circular, two-dimensional space with radius R, whose energy is proportional to $E \propto \int d^2x(\nabla\phi)^2$. An $n = 1$ vortex configuration can be described by $\phi(\mathbf{x}) = \arctan(y/x)$. As explained in Section 7.1, the singularity at the vortex core will be avoided in any physical system either by the amplitude being suppressed, or due to the discreteness of an underlying lattice. We can take this into account, and avoid a so-called ultraviolet divergency, by approximating the spatial integral in the expression for E to start at a non-zero radius ξ, corresponding to the vortex core radius.
a. Calculate the energy of the single-vortex configuration.

A configuration with a vortex-antivortex pair can be described by $\phi(\mathbf{x}) = \arctan(y/(x - m/2)) - \arctan(y/(x + m/2))$, where m is the vortex-antivortex distance.
b. Calculate the energy of the pair configuration as a function of the pair size m. Notice that for large m, your result is the same as that for the energy of two oppositely charged particles subject to Poisson's equation in $d = 2$.

7.5 Duality

The traditional picture of a phase transition, following Landau, starts from the symmetric, disordered state and describes the emergence of a broken symmetry. As shown by the BKT transition, it may sometimes also be useful to take a complementary approach and consider how the proliferation of topological defects leads towards a disordered state starting from the ordered, symmetry-breaking phase. In some symmetry-breaking transitions, it is possible to take the complementary approach one step further, and treat the topological defects as emergent particles in their own right. The transition from the low to the high temperature phase can then be described as the Bose–Einstein condensation of defects, spontaneously breaking an associated symmetry. Since the defects condense at high temperature, while the order parameter field hosting the defects started out being ordered at low temperatures, the low- and high-temperature phases can both be considered ordered phases, albeit in very different ways. In fact, the order parameter for the defect condensate acts as a 'disorder parameter' for the original particles, and vice versa.

Creating any type of topological defect involves reorienting the order parameter field throughout all of space. A creation operator for a topological defect must therefore always be extremely non-local in terms of the creation and annihilation operators used to define the order parameter field. Nevertheless, it sometimes happens to be the case that writing the Hamiltonian for a symmetry-breaking system in terms of defect operators results in a form that is as convenient as the expression in terms of the usual creation and annihilation operators. You can then choose to describe the physics of the system either in terms of the original particles or in terms of topological defects acting as particles. Both pictures give the same predictions for any physical observables, but one is often easier to apply in the low temperature phase, and the other in the high-temperature phase.

The mathematical map between two descriptions of the same system is called a **duality** mapping. The first example of a duality in physics was established by Kramers and Wannier for the two-dimensional Ising model [98]. They wrote the Hamiltonian in terms of domain walls rather than spins, which enabled Onsager to then solve the model exactly in 1944 [150]. The existence of a duality mapping usually allows one to explore the properties of a critical point more easily or more thoroughly than without. This has met with considerable success in the description of $U(1)$-symmetry breaking in two and three dimensions, where it goes under the name of boson–vortex duality. The approach has been extended to systems involving multiple species of particles, including fermions, in a so-called web of dualities [88, 179]. In all cases, the phase transition described by a duality mapping can be viewed as the unbinding or condensation of topological defects.

Exercise 7.11 (Self-duality of the transverse field Ising model).
The one-dimensional quantum Ising model in the presence of a transverse field has the following Hamiltonian:

$$\hat{H} = -J\sum_i \hat{\sigma}^z_i \hat{\sigma}^z_{i+1} - h\sum_i \hat{\sigma}^x_i. \tag{7.15}$$

Here $J, h > 0$ and $\hat{\sigma}^{x,y,z}_i$ are Pauli matrices acting on site i, which are defined as

$$\hat{\sigma}^x = \begin{pmatrix} 0 & 1 \\ 1 & 0 \end{pmatrix}, \qquad \hat{\sigma}^y = \begin{pmatrix} 0 & -\mathrm{i} \\ \mathrm{i} & 0 \end{pmatrix}, \qquad \hat{\sigma}^z = \begin{pmatrix} 1 & 0 \\ 0 & -1 \end{pmatrix}. \tag{7.16}$$

Note that this is a factor $1/2$ different from our earlier definition of Pauli matrices (Exercise 1.7). The Pauli matrices in this form satisfy, on each site, $[\hat{\sigma}^a, \hat{\sigma}^b] = 2\mathrm{i}\epsilon_{abc}\hat{\sigma}^c$ and $\{\hat{\sigma}^a, \hat{\sigma}^b\} = 2\delta_{ab}$.

Because the model describes Ising spins, the eigenstates of $\hat{\sigma}^x$ should be interpreted as superpositions of up and down spin states, rather than spins pointing in the x direction. The model has a ferromagnetically ordered ground state for $J > h$, which becomes precisely $\prod_x |\uparrow\rangle_x$ in the $J/h \to \infty$ limit. For $J < h$, on the other hand, the ground state is a disordered paramagnet, which becomes precisely $\prod_x \frac{1}{\sqrt{2}}(|\uparrow\rangle_x + |\downarrow\rangle_x)$ in the $J/h \to 0$ limit.

The model can also be described in terms of a set of dual operators, defined as $\hat{\mu}^x_i = \hat{\sigma}^z_i \hat{\sigma}^z_{i+1}$ and $\hat{\mu}^z_i = \prod_{j\leq i} \hat{\sigma}^x_j$. In all of the following, we assume the model to be defined on an infinitely long chain, so we can ignore its boundaries.

a. Show that $\hat{\mu}^{x,z}_i$ satisfy the standard commutation relations for Pauli matrices, namely $(\hat{\mu}^{x,z}_i)^2 = 1$, $\{\hat{\mu}^x_i, \hat{\mu}^z_i\} = 0$, and $[\hat{\mu}^x_i, \hat{\mu}^z_j] = 0$ whenever $i \neq j$.

b. One of the $\hat{\mu}$-operators creates or destroys a domain wall, while the other measures the presence of a domain wall. Which one is which?

c. Construct the inverse transformation, by expressing the $\hat{\sigma}$-operators in terms of $\hat{\mu}$-operators.

d. Write the Hamiltonian in terms of the $\hat{\mu}$-operators.

Going from spin operators ($\hat{\sigma}$) to domain wall operators ($\hat{\mu}$), the quantum Ising Hamiltonian retains the same form, albeit with a swap of the model parameters J and h. This means that the paramagnetic phase in terms of spins, may equally well be described as an ordered phase in terms of domain wall creation operators, and the other way around.
An extensive review of the Ising self-duality can be found in Ref. [95].

Gauge Fields 8

Our focus so far has mainly been on systems with a global symmetry in the absence of gauge invariance. Although gauge invariance can never be broken, as explained in Section 1.6.2, its presence in models with a global symmetry does affect the physical phenomenology of their symmetry-breaking phases and phase transitions. The central ingredient in the interplay between gauge freedom and symmetry is the **Anderson–Higgs mechanism**. This effect causes excitations in the theory to become massive and will be discussed in detail in Section 8.3. It is universal in the sense that it applies to any case of spontaneous symmetry breaking in a gauge theory.

We will introduce the core concepts of symmetry breaking in gauge theories using the explicit example of $U(1)$ symmetry breaking (introduced in Section 2.5), in the presence of a $U(1)$ gauge field. The real-world realisation of this is given by superconductors, which have a global $U(1)$ phase symmetry as well as a local $U(1)$ gauge freedom. In actual superconducting materials, additional physics in the form of pair formation among fermionic degrees of freedom play an important role, which we will discuss this in Chapter 12. For now, we will ignore this aspect and focus solely on the interplay between gauge invariance and symmetry breaking, by describing superconductors simply as charged superfluids. Every aspect of symmetry breaking in the presence of gauge fields that is present in charged superfluids, also applies to more involved cases, such as the non-Abelian gauge theories encountered in the Standard Model. In particular, electroweak symmetry breaking will be discussed in Chapter 11.

8.1 Charged Superfluids

Real-world superconducting materials are metals that are cooled to very low temperatures, where they go through a phase transition and become superconductors. This instability of the metallic state can be understood in terms of the famous Bardeen–Cooper–Schrieffer (BCS) theory [17]. A key aspect of BCS theory is the pairing of electrons into **Cooper pairs**. A Cooper pair consists of two fermions, which as a whole behaves like a boson. These pairs can thus Bose-condense at sufficiently low temperatures, and the superconducting state can be viewed as a superfluid of Cooper pairs. The symmetry-breaking transition in superconductors is thus the Bose-condensation of Cooper pairs, which spontaneously breaks the global $U(1)$ symmetry associated with the conservation of the number of Cooper pairs. Because the bosons in this case are electrically charged, the dissipationless flow of the superfluid is actually a resistance-free electric supercurrent. To discuss the broken symmetry, its relation to gauge freedom and its observable consequence, however, we can largely ignore the fact that Cooper pairs really consist of two electrons. We therefore delay the discussion of how Cooper pairs are formed and how internal structure affects the final symmetry-broken state to Chapter 12, and for now start straight away from a (metallic) normal fluid of charged bosons.

We previously encountered the Landau potential of Eq. (5.6), which describes the free energy of a complex order parameter field $\psi(x)$, whose ordered state we argued corresponds to a neutral superfluid. The effect of having charged Cooper pairs rather than neutral bosons in such a theory can be seen only if we allow for fluctuations in their density, since we assume the total charge of the field $\psi(x)$ to be balanced by an equal and opposite background charge. For real superconductors, this signifies the cancellation of the total electronic (or Cooper pair) charge by the positive charge of the ionic lattice. Within the Landau potential, the squared amplitude of the field, $|\psi(x)|^2$, represents the density of bosons and thus the local charge. Adding the lowest order term in an expansion of gradients of the density, as we did for the Ginzburg–Landau theory of Section 5.4, we would expect a contribution to the energy of the form $\frac{\hbar}{2m^*}|\nabla\psi|^2$, with m^* is the mass of a single boson. Because fluctuations in the density of these bosons are charged, however, they both create and are affected by electromagnetic fields. A convenient minimal way of introducing the coupling between such fields and local charge fluctuations was suggested by Peierls, and consists of making the substitution $\nabla\psi \to (\nabla - \mathrm{i}\frac{e^*}{\hbar}\mathbf{A})\psi$, where $\mathbf{A}$ is the electromagnetic vector potential and e^* is the electric charge of an isolated boson. For Cooper pairs containing two electrons, this means $e^* = 2e$ with e the electron charge. For convenience, we assume the electromagnetic scalar potential V to be zero. The full free energy density of the Ginzburg–Landau theory for a charged superfluid is then

$$\mathcal{F}_{\mathrm{GL}} = \frac{\hbar^2}{2m^*}|(\nabla - \mathrm{i}\frac{e^*}{\hbar}\mathbf{A})\psi|^2 + \frac{1}{2}r|\psi|^2 + \frac{1}{4}u|\psi|^4 + \frac{1}{2\mu_0}(\nabla \times \mathbf{A})^2. \tag{8.1}$$

Notice that this includes the potential energy of the electromagnetic field itself, with μ_0 the magnetic constant, and that we consider only time-independent fields since we are interested in understanding equilibrium phases. This free energy density $\mathcal{F}_{\mathrm{GL}}$ can be used to explain a large part of the phenomenology of any charged superfluid, including their dissipationless current, the Meissner effect, vortex topological defects, and the Josephson effect. All of these are intimately related to symmetry breaking and will be discussed here. For other aspects or more detailed treatments, many excellent textbooks on superconductivity may be consulted [10, 43, 164, 192].

Exercise 8.1 (Peierls substitution).
The reason that $\hbar$ appears in the Peierls substitution stems from the fact that electromagnetic fields couple to charged matter via quantum electrodynamics.
Verify that $\frac{e^*}{\hbar}\mathbf{A}$ has units of inverse length, to confirm that the Peierls substitution is dimensionally correct.

The free energy density $\mathcal{F}_{\mathrm{GL}}$ is invariant under the global $U(1)$ symmetry transformation:

$$\psi(x) \to \mathrm{e}^{-\mathrm{i}\alpha}\psi(x). \tag{8.2}$$

As always, this global symmetry is associated with a conserved Noether current, which may be obtained either through the usual Noether procedure, or directly by applying the Peierls substitution to the Noether current of the neutral superfluid in Eq. (1.21). Either way, the resulting expression for the Noether current is

$$\mathbf{j} = \mathrm{i}\frac{\hbar^2}{2m^*}\left((\nabla\psi^*)\psi - \psi^*(\nabla\psi)\right) - \frac{e^*\hbar}{m^*}\psi^*\psi\mathbf{A} = \frac{\hbar^2}{m^*}|\psi|^2\left(\nabla\varphi - \frac{e^*}{\hbar}\mathbf{A}\right). \tag{8.3}$$

Here we wrote $\psi = |\psi|\mathrm{e}^{\mathrm{i}\varphi}$. The conserved Noether charge transported by this current turns out to be the number of bosons. But because the bosons are charged, a current of them also corresponds to an actual electric current. This is easily confirmed by considering the usual definition of the electric current, $\mathbf{j}_\mathrm{e} = \delta\mathcal{F}/\delta\mathbf{A}$, which shows that it is indeed related to the Noether current through $\mathbf{j}_\mathrm{e} = \frac{e^*}{\hbar}\mathbf{j}$. As always, the Noether current is manifested in the ordered phase by NG modes, whose lifetime goes to infinity in the long-wavelength limit. In the case of a charged superfluid, the infinitely long-lived current of charged bosons is called a **supercurrent**.

Besides the global symmetry, $\mathcal{F}_\mathrm{GL}$ is also invariant under a local $U(1)$ gauge transformation that simultaneously affects both the order parameter field $\psi(x)$ and the electromagnetic vector potential:

$$\psi(x) \to \mathrm{e}^{-\mathrm{i}\alpha(x)}\psi(x), \qquad \mathbf{A}(x) \to \mathbf{A}(x) - \frac{\hbar}{e^*}\nabla\alpha(x). \tag{8.4}$$

This gauge freedom is the result of having introduced a superfluous degree of freedom in the Peierls substitution, namely the longitudinal component of the electromagnetic vector potential, which does not contribute to the observable electric and magnetic fields. Because of the minimal coupling between the order parameter field and the electromagnetic vector potential in Eq. (8.1), gauge transformations now affect the longitudinal component of $\mathbf{A}$ as well as the phase of the field $\psi(x)$. Stated differently, any occurrence of the longitudinal component of $\mathbf{A}$ in $\mathcal{F}_\mathrm{GL}$ can be undone by a local rotation of the phase of the order parameter field.

As emphasised in Section 1.6, gauge transformations are not symmetries. They are simply consistency requirements, and any physical observable derived from the free energy of Eq. (8.1) must be invariant under the transformations of Eq. (8.4). In particular, this gauge invariance can never be broken, neither spontaneously nor explicitly. At this point, however, we should notice that for constant $\alpha(x) = \alpha$, the gauge transformation of Eq. (8.4) appears to coincide with the global symmetry transformation of Eq. (8.2), which we argued is spontaneously broken in the ordered phase. In fact, the situation is the same as the one we encountered in Section 1.6.4, where the spin-rotational symmetry that is broken in a ferromagnet seemed to coincide with the unbreakable global gauge freedom of choosing a coordinate system. Both for the ferromagnet and the charged superfluid, the distinction between gauge freedom and symmetry becomes clear once we use a more careful definition of the global symmetry with respect to an external reference. One way of doing this is to consider a gauge-invariant definition for the order parameter.

8.2 The Gauge-Invariant Order Parameter

Ignoring spatial variations in the Cooper pair density, the Landau potential for the charged superfluid is the same as that for the neutral one in Eq. (5.6). In the neutral case, we saw that for $r < 0$ the minimum of the potential is at $\psi \neq 0$, and a global $U(1)$ symmetry is broken by choosing a particular phase for the minimum energy configuration: $\psi = e^{i\phi}|\psi|$. Consequently, the field variable $\psi(x)$ itself could be used as the order parameter for the superfluid. Because the global symmetry being broken should not be affected by the local fluctuations that make a superconductor or charged superfluid different from the neutral version, it is tempting to also consider the field ψ as an order parameter for the charged case. However, the quantity $\psi(x)$ is not invariant under the gauge transformation of Eq. (8.4). Because any physical quantity must be gauge invariant, ψ cannot be an order parameter for the charged superfluid.

The are three ways of dealing with the fact that $\psi(x)$ is not gauge invariant, each of which is used in practice and throughout the literature.

1. Ignore the complication and simply use $\psi(x)$ as the order parameter, even though it is not gauge invariant. This is not as silly as is sounds. In many cases of practical interest, no external electromagnetic field is present, and induced fields are negligible. Then, the vector potential is approximately zero, and $\mathcal{F}_{\text{GL}}$ reduces to the free energy of a neutral superfluid. Simply ignoring any electromagnetic fields, then suffices to correctly predict many physical properties. In particular, the original publication of BCS theory used an order parameter of this form (although written differently) [17]. The articles introducing Ginzburg–Landau theory for superconductivity [57] and the Josephson effect [83] treated the order parameter in a similar way as well.
2. Choose a particular **gauge fix**. Just as we regularly do for calculations in electromagnetism, we can impose additional, arbitrary, constraints on the vector potential and phases of ψ to remove the freedom of doing gauge transformations. That is, whenever we encounter some configuration of ψ and $\mathbf{A}$, we can choose to always apply the particular gauge transformation $\psi \to \psi'$, $\mathbf{A} \to \mathbf{A}'$, which makes the transformed fields ψ' and $\mathbf{A}'$ satisfy the additional constraints. The constraints are often chosen to ensure a mathematically convenient or aesthetically pleasing form of the fields. For the theory of charged superfluids, there are two particularly useful choices of constraints. One is the so-called **unitary gauge**, which demands the phase of the field ψ to be zero everywhere, so that all degrees of freedom reside in the vector potential. The second is the **Coulomb gauge** (also called the **London gauge**), which imposes $\nabla \cdot \mathbf{A} = 0$ everywhere, so that the longitudinal degree of freedom is carried exclusively by the phase of the field ψ.

   Because gauge fixing constraints can be implemented by a gauge transformation, they are guaranteed not to affect the values of any physically observable quantities. Choosing a gauge fix is thus always an allowed thing to do at any step within a calculation. One should always remember, however, that the choice of gauge cannot affect any final physical predictions. For example, choosing to work within the unitary gauge may seem to make ψ more acceptable as an order parameter. However, simply choosing the phase of ψ to be zero in a calculation does not mean that we predict any measurement will actually find it to be zero. The physical predictions based on any calculation must be gauge invariant, even if we choose to do the calculation within a particular gauge fix.
3. Define a gauge-invariant but non-local order parameter. It is simply not possible to have a gauge-invariant local order parameter $O(x)$ for charged superfluids that depends only the values of fields within a small neighbourhood of x. It is possible,

however, to define a gauge-invariant non-local order parameter, following a proposal by Dirac [48]:

$$\psi_{\rm D}(x) = \psi(x)\, {\rm e}^{{\rm i}\int {\rm d}^3y\, \mathbf{Z}(y-x)\cdot\mathbf{A}(y)}. \tag{8.5}$$

Here, $\mathbf{Z}(x)$ is defined as a function satisfying $\nabla \cdot \mathbf{Z}(x) = \frac{e^*}{\hbar}\delta(x)$. In other words, $\mathbf{Z}$ is proportional to the electric field that would emanate from a point charge at $x = 0$. The order parameter $\psi_{\rm D}(x)$ is non-local in the sense that you need to integrate over all of space to find its value at any particular location. The Dirac order parameter $\psi_{\rm D}(x)$ reduces to simply the field $\psi(x)$ in the Coulomb gauge fix $\nabla \cdot \mathbf{A} = 0$. Knowing that a gauge-invariant formulation of the order parameter exists, it is thus possible to impose a gauge fix and simply work with the field $\psi(x)$ as a local order parameter. Doing so, however, you should always keep in mind that a gauge fix was in fact imposed and check that any physical predictions following from your calculations are gauge invariant.

Exercise 8.2 (Dirac order parameter).
Verify that the Dirac order parameter is invariant under the simultaneous transformation of ψ and $\mathbf{A}$ given by Eq. (8.4).

Using the Dirac order parameter, the difference between the global symmetry transformation of Eq. (8.2) and the uniform part of the local gauge freedom of Eq. (8.4) may be clarified. The symmetry that is broken in the charged superfluid is the global $U(1)$ phase rotation symmetry of the field $\psi_{\rm D}(x)$. Doing Exercise 8.2, you may have noticed that $\psi_{\rm D}(x)$ is invariant under any gauge transformation, except for the global transformation with constant $\alpha(x) = \alpha$. The reason for this is that in the definition of the Dirac field in Eq. (8.5), the local phase of $\psi(x)$ is effectively measured, in a gauge invariant way, with respect to the phase of the field at infinity, where it is taken to be zero. The global transformation with $\alpha(x) = \alpha$, however, changes the phase of the field everywhere, including at infinity. This situation is precisely analogous to the way that rotating all spins in the universe will have no measurable effect on a ferromagnet that breaks spin-rotation symmetry. The orientation of the spins within one magnet can only be measured with respect to the direction of the magnetic field produced by a second. And likewise, the position of a crystal breaking translational symmetry is defined only with respect to a reference frame provided, for example, by the surrounding lab.

In direct analogy to the magnet and crystal, the symmetry that is spontaneously broken in a charged superfluid must be a rotation of the phase of $\psi_{\rm D}(x)$ that is constant throughout one piece of superconducting material, but which leaves the phase of an external reference superconductor fixed. The observable describing such relative phase differences within the Dirac order parameter field is the gauge-invariant correlation function:

$$C_{\rm D}(x,x') = \langle \psi_{\rm D}(x)\psi_{\rm D}^\dagger(x')\rangle = \langle \psi(x)\, {\rm e}^{{\rm i}\int {\rm d}^3y\, (\mathbf{Z}(y-x)-\mathbf{Z}(y-x'))\cdot\mathbf{A}(y)}\, \psi^\dagger(x')\rangle. \tag{8.6}$$

In particular, when x and x' are taken to be points within two spatially separated charged superfluids, the correlation function is precisely proportional to the current measured in the **Josephson effect**, introduced in Section 3.5. The Josephson current thus provides a gauge invariant global order parameter akin to the total magnetisation of a ferromagnet or the centre-of-mass position of a crystal. The local quantity from which it is built consists of the phase of $\psi_{\rm D}(x)$, which is defined with respect to an external coordinate system, just like the local magnetisation within a magnet or the position of atoms within a crystal. The Josephson effect can even be used to measure the local value of the order parameter, by using a superconducting tip in a scanning-tunnelling experiment and registering the local value of the Josephson current.

Exercise 8.3 (Josephson junction array).
Consider a so-called Josephson junction array consisting of a one-dimensional chain of superconducting islands. For simplicity, assume the electromagnetic field to be zero everywhere (this does not affect any of the results). Each island can be described by two coarse-grained observables describing the average number of Cooper pairs on a site, $\hat{n}_j$, and the average phase of the Dirac field on each site, $\hat{\theta}_j$. These conjugate variables obey $[\hat{\theta}_j, \hat{n}_{j'}] = \mathrm{i}\delta_{j,j'}$. Notice that the gauge transformation of Eq. (8.4) leaves the Dirac field invariant, except for global phase rotations. Considering well-separated islands, we can describe the rotation of one average phase $\hat{\theta}_j$ with respect to the phases of other islands as a global phase rotation of the island at position j. A gauge transformation applied to the entire chain then amounts to rotations of the individual phases on each island, along with the associated transformation of the electromagnetic vector potential in the space between islands.
The Hamiltonian describing the array as a whole is given by

$$\hat{H} = \sum_j \frac{1}{2} C\hat{n}_j^2 - J\cos\left(\hat{\theta}_j - \hat{\theta}_{j+1} + \psi_j^{j+1}\right), \qquad \psi_j^{j+1} \equiv \frac{e^*}{\hbar}\int_j^{j+1} A^x(x')\,dx'. \tag{8.7}$$

Here, C and J are parameters known as the charging and Josephson energies, and $e^* = 2e$ is the charge of a Cooper pair. The phase ψ_j^{j+1} comes from the Peierls substitution, with A^x the x-component of the electromagnetic vector potential. Although each individual island is always superconducting, the chain as a whole has a superconducting transition temperature that depends on the ratio J/C.
a. Verify that the Hamiltonian is invariant under the simultaneous transformation of Eq. (8.4).

The Hamiltonian can be simplified by introducing new operators $\hat{\phi}_j$:

$$\hat{\phi}_j = \hat{\theta}_j - \sum_{i=2}^{j} \psi_{i-1}^{i}, \qquad [\hat{\phi}_j, \hat{n}_{j'}] = i\delta_{j,j'}. \tag{8.8}$$

Note that $\hat{\phi}_j$ is non-local in terms of ψ_j^{j+1}.
b. Show that in terms of the new operators, the Hamiltonian becomes approximately

$$\hat{H} \approx \sum_j \left[\frac{1}{2} C\hat{n}_j^2 + \frac{1}{2} J\left(\hat{\phi}_j - \hat{\phi}_{j+1}\right)^2\right]. \tag{8.9}$$

c. Use Fourier and Bogoliubov transformations to diagonalise the Hamiltonian for $k \neq 0$. Show that its spectrum is given by $\hbar\omega(k) = \sqrt{4JC}\left|\sin\left(\frac{ka}{2}\right)\right|$.

The modes in this spectrum are the Nambu–Goldstone modes associated with the chain as a whole being a superconductor. They appear here as gapless modes because we neglected the dynamics of the electromagnetic field, which we will come back to in Section 8.3. The collective, $k = 0$ part not included in the NG-spectrum, is described by $\hat{H}_{k=0} = C\hat{n}_{\mathrm{tot}}^2/2N$, with N the number of sites in the chain and $\hat{n}_{\mathrm{tot}} = \sum_j \hat{n}_j$ the total number of particles in the entire chain. The average phase across the chain is given by $\hat{\phi}_{\mathrm{ave}} = 1/N \sum_j \hat{\phi}_j$, with $[\hat{\phi}_{\mathrm{ave}}, \hat{n}_{\mathrm{tot}}] = \mathrm{i}$.

d. Show that the collective Hamiltonian $\hat{H}_{k=0}$ is invariant under the symmetry transformation $\hat{U} = \mathrm{e}^{\mathrm{i}\alpha \hat{n}_{\mathrm{tot}}}$.

The global phase rotation symmetry $\hat{U}$ of the collective Hamiltonian $\hat{H}_{k=0}$ can be broken by introducing a symmetry-breaking field:

$$\hat{H}'_{k=0} = \frac{1}{2N} C\hat{n}_{\mathrm{tot}}^2 + \frac{1}{2} J' \hat{\phi}_{\mathrm{ave}}^2. \tag{8.10}$$

This could be interpreted as a coupling of the chain to an additional, external piece of superconductor with a fixed global phase. The operator $\hat{\phi}_{\mathrm{ave}}$ then measures the relative phase difference between the chain and the external superconductor.

e. Show that a sufficiently long chain of superconducting islands will spontaneously break the global phase rotation symmetry $\hat{U}$ (at zero temperature).

8.3 The Anderson–Higgs Mechanism

The charged superfluid spontaneously breaks a continuous symmetry, and you might therefore reasonably expect it to host Nambu–Goldstone modes on top of its ordered state. The gauge fields that feature so prominently in the theory of charged superfluids, however, mediate long-ranged Coulomb interactions between the Cooper pairs. Since Goldstone's theorem does not apply in the presence of long-ranged interactions, there is then no guarantee that any gapless modes will exist in the ordered state. In fact, coupling the gapless NG mode of a neutral superfluid to the gapless photon of Maxwell electromagnetism makes both excitations massive within the superconducting state.

The quickest way to see how this happens is to rewrite the Ginzburg–Landau free energy of Eq. (8.1) as

$$\begin{aligned} \mathcal{F}_{\mathrm{GL}} &= \frac{1}{2\mu_0}(\nabla \times \mathbf{A})^2 + \frac{\hbar^2}{2m^*}|\psi|^2(\nabla\varphi - \frac{e^*}{\hbar}\mathbf{A})^2 + \frac{\hbar^2}{2m^*}(\nabla|\psi|)^2 + \frac{1}{2}r|\psi|^2 + \frac{1}{4}u|\psi|^4 \\ &= \frac{1}{2\mu_0}(\nabla \times \tilde{\mathbf{A}})^2 + \frac{e^{*2}}{2m^*}|\psi|^2\tilde{\mathbf{A}}^2 + \frac{\hbar^2}{2m^*}(\nabla|\psi|)^2 + \frac{1}{2}r|\psi|^2 + \frac{1}{4}u|\psi|^4. \end{aligned} \tag{8.11}$$

In the first line, we explicitly wrote the field in terms of its amplitude and phase, $\psi = |\psi|\mathrm{e}^{\mathrm{i}\varphi}$, while in the second line, we defined $\tilde{\mathbf{A}} \equiv \mathbf{A} - \frac{\hbar}{e^*}\nabla\varphi$. Notice that the newly defined field $\tilde{\mathbf{A}}(x)$ is invariant under the gauge transformations of Eq. (8.4), which means that, like the scalar amplitude field $|\psi|$, it is a physical and observable degree of freedom. Furthermore, this new field is proportional to the Noether current of Eq. (8.3).

In the superconducting phase, the parameter r is negative and the free energy has a minimum for non-zero values of the field amplitude, $\langle|\psi|\rangle \neq 0$. The second term in Eq. (8.11) can then be interpreted to be a mass term for the field $\tilde{\mathbf{A}}$ (you can check this by explicitly minimising $\mathcal{F}_{\mathrm{GL}}$ for a fixed value of $|\psi|$). In the ordered phase then, both of the fields appearing in the free energy, the vector field $\tilde{\mathbf{A}}$ as well as the amplitude field $|\psi|$, are massive in the sense of having a gapped dispersion.

In this manifestly gauge-invariant description, the would-be NG mode $\varphi(x)$ appears to have disappeared completely. This is described in various places in the literature by noting that φ cannot be a physical degree of freedom because it can be 'removed' from the free energy functional by imposing the unitary gauge $\varphi \equiv 0$. It is also sometimes said 'to be in an unphysical part of Hilbert space' because when using the Coulomb gauge $\nabla \cdot \mathbf{A} \equiv 0$, the field φ does not couple to any observables. In a neutral superfluid, however, $\varphi(x)$ is a real propagating mode, and physical degrees of freedom cannot simply disappear when including interactions with additional fields in a theory.

In fact, the φ excitation has not disappeared. Rather, it is included in the newly defined vector field $\tilde{\mathbf{A}}$. Before coupling to the field ψ, we were free to choose the Coulomb gauge $\nabla \cdot \mathbf{A} = 0$, which eliminates the longitudinal component as a degree of freedom. However, the new field $\tilde{\mathbf{A}}$ is invariant under gauge transformations, so none of its components can be removed or fixed by employing gauge freedom. In a way, the degree of freedom carried by φ can be said to be transferred to or represented by the longitudinal component of the $\tilde{\mathbf{A}}$ field.

A field-theory formulation of the same argument would be to observe that a massless vector field in three dimensions carries two degrees of freedom: the two transverse polarisations. A massive vector field like $\tilde{\mathbf{A}}$ in Eq. (8.11), however, carries three degrees of freedom, which include the longitudinal component. An alternative and more detailed derivation can be done within the formalism of Hamiltonian constraints mentioned in Section 1.6.4, for which an accessible treatment can be found in Ref. [104].

The transformation from the fields A_μ and φ to the new vector field $\tilde{\mathbf{A}}$ is sometimes described as 'the vector field has eaten the Goldstone boson and obtained its degree of freedom'. The vector field is also said to 'have gotten fat by becoming a massive field'. This mechanism for the emergence of a massive vector field is called the **Anderson–Higgs mechanism** [7, 74]. It also goes by several other names, depending on the subfields of physics in which it is discussed: in addition to Anderson (who proposed the mechanism first in 1962) and Higgs (1964), also Brout–Englert (1964) and Guralnik–Hagen–Kibble (1964) are considered pioneers of the mechanism.

8.3.1 The Meissner Effect

The first consequences of excitations being massive inside a charged superfluid can be seen by considering the equation of motion obtained by varying the free energy $\mathcal{F}_{\mathrm{GL}}$ with respect to the electromagnetic vector potential $\mathbf{A}$:

$$\frac{1}{\mu_0}\nabla \times (\nabla \times \mathbf{A}) = \frac{e^*\hbar}{m^*}|\psi|^2(\nabla\varphi - \frac{e^*}{\hbar}\mathbf{A}). \tag{8.12}$$

This expression is just Ampère's law $\frac{1}{\mu_0}\nabla \times \mathbf{B} = \mathbf{J}$, with the electric current given by the supercurrent $\mathbf{J} = \mathbf{j}_\mathrm{e}$ defined below Eq. (8.3). Notice that the supercurrent vanishes outside the superconducting phase, where the amplitude $|\psi|$ is zero. In deriving these equations of motion, we considered only static field configurations. Including dynamic terms, Eq. (8.12) becomes a wave equation for $\mathbf{A}$ (and hence $\mathbf{E}$ and $\mathbf{B}$), which yields the dispersion relation for photons inside the charged superfluid.

It is surprising at first sight that the supercurrent $\mathbf{j}_e$ is proportional to the vector potential $\mathbf{A}$, rather than to an electric field $\mathbf{E}$, as would be expected for normal currents. The physical implication of this unusual proportionality can be found by expressing Eq. (8.12) in terms of the magnetic field $\mathbf{B}$. For this, we take the curl of both sides, use the fact that the curl of a gradient vanishes and substitute $\nabla \cdot \mathbf{B} = 0$. We then find the expression:

$$\left(\nabla^2 - \frac{1}{\lambda_L^2}\right)\mathbf{B} = 0, \qquad \lambda_L = \sqrt{\frac{m^*}{\mu_0 e^{*2}|\psi|^2}}. \tag{8.13}$$

Here, λ_L defines the **London penetration depth**. The solution to this equation for the field $\mathbf{B}$ is an exponentially decaying function with length scale λ_L. That is, magnetic fields can only penetrate the charged superfluid up to a distance λ_L from the surface, and magnetic fields, even static ones, are expelled from the interior of the charged superfluid. This is known as the **Meissner effect**.

Because the supercurrent $\mathbf{j}_e$ is proportional to the vector potential, it obeys an equation similar to that for the magnetic field:

$$\left(\nabla^2 - \frac{1}{\lambda_L^2}\right)\mathbf{j}_e = 0. \tag{8.14}$$

This does not mean that charged superfluids do not carry supercurrents. On the contrary, the dissipationless persistent current is the primary hallmark of superconductivity. However, a stationary flow of Cooper pairs can only exist within a region of width λ_L from the edge, while wave-like excitations in the bulk of the charged superfluid are gapped and decay exponentially in time. The ability to excite 'screening' supercurrents close to the surface in fact enables the expulsion of externally applied magnetic fields: the magnetic field created by the surface supercurrent precisely cancels the external field. Because of this, superconductors are sometimes called perfect diamagnets.

Exercise 8.4 (Meissner effect).
a. Start from the Ginzburg–Landau free energy in the first line of Eq. (8.11), with $\mathbf{A}(\mathbf{x})$ independent of time. Show that minimising the free energy with respect to $\mathbf{A}(\mathbf{x})$ corresponds to Eq. (8.12).
b. Take the curl of your result to arrive at Eq. (8.13).

Imagine the edge of a three-dimensional volume of charged superfluid, such that $x > 0$ is inside the superfluid and $x < 0$ outside. On the outside, consider a uniform magnetic field $\mathbf{B} = B\hat{\mathbf{z}}$.
c. What is the magnetic field inside the charged superfluid?
d. Calculate the supercurrent inside the charged superfluid.

Notice that the current only flows close to the surface of the superconductor. This layer of screening supercurrents enables the Meissner effect in the bulk of the charged superfluid.

8.3.2 The Higgs Boson

Besides the phase variable or would-be NG-mode and the electromagnetic vector potential, which combine in a charged superfluid to form the massive vector field $\tilde{\mathbf{A}}$, the free energy of Eq. (8.11) also contains the amplitude field $|\psi(x)|$. Excitations of this propagating, massive degree of freedom are called the **amplitude mode** of the superconducting state. In the Mexican-hat potential of Figure 5.4, it correspond to oscillations perpendicular to the flat direction. Although the existence and physical properties of this mode are straightforward to understand theoretically, its experimental detection in superconducting materials is challenging and has been achieved only relatively recently [153].

In the context of elementary particle physics, the amplitude mode accompanying the Anderson–Higgs effect in the Standard Model is known as the **Higgs boson**. Amplitude modes have no a priori connection to gauge freedom. In fact neutral superfluids, charge density waves, and various other phases of matter all have amplitude modes. However, if gauge fields couple to the order parameter ψ, which is often called the Higgs field in this context, and the Higgs field has an expectation value $\langle\psi\rangle \neq 0$, then the Anderson–Higgs mechanism ensures the gauge fields to be massive. In the Standard Model of elementary particles, the W and Z-gauge bosons of the electroweak interaction become massive in this way, as we will discuss in Chapter 11.

To be clear about the terminology: the order parameter field ψ is called the Higgs field. Excitations of the amplitude $|\psi|$ on top of its equilibrium expectation value are called Higgs bosons. The gauge fields, corresponding to $\mathbf{A}$ in the charged superfluid, become massive by being coupled to the Higgs fields, and the NG mode is simultaneously converted into an additional massive component of the gauge field. This combined mass generation is called the Anderson–Higgs mechanism.

8.4 Vortices

In neutral superfluids, the broken phase-rotation symmetry allows for topological defects in the form of vortex excitations. As discussed in Chapter 7, these may enter the superfluid when external torque is applied, and they have a topological charge equal to the winding number of the phase. Since the presence of a vortex affects the orientation of the order parameter throughout the superfluid, vortices exert long-range forces on each other, and the energy of a single vortex grows with the system size.

In a charged superfluid, or superconductor, the combination of the phase degree of freedom with the electromagnetic vector potential into a single massive vector field changes the nature of the vortex excitations. Just like in neutral superfluids, the amplitude of the order parameter vanishes at the vortex core. Additionally, however, the Meissner effect, which normally expels magnetic fields from the bulk of the charged superfluid, is not operative in the core region and magnetic fields can penetrate the superfluid up to a screening length from the core. Figure 8.1 shows the resulting field profile for the vortex excitation of a charged superfluid.

To find out how much magnetic flux can penetrate the superconductor through a single vortex core, consider once more the relation between supercurrent and vector potential of

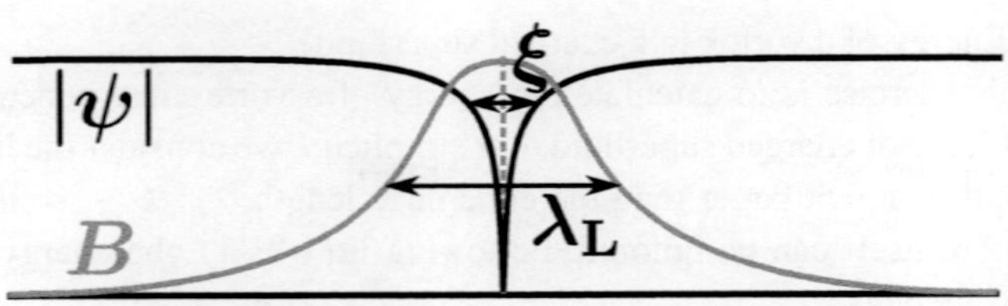

Figure 8.1

Cross section of a vortex in a charged superfluid. The amplitude of the order parameter field $|\psi|$ falls to zero at the core with a decay length set by the coherence length $\xi = \sqrt{\hbar^2/m^*|r|}$, while a magnetic field $\boldsymbol{B}$ penetrating the charged superfluid from the core decays of a length scale equal to the London penetration depth λ_L.

Eq. (8.12). We can integrate both sides of the equation along a closed contour $\mathcal{C}$ encircling the vortex core. If the contour is taken sufficiently far from the vortex core, the magnetic field will be completely screened by the Meissner effect, as shown in Figure 8.1, and the left-hand side is zero. We are then left with the equation

$$\frac{\hbar}{e^*}\oint_{\mathcal{C}} \mathrm{d}\mathbf{x}\cdot\nabla\varphi = \oint_{\mathcal{C}} \mathrm{d}\mathbf{x}\cdot\mathbf{A} = \int_{\mathcal{S}} \mathrm{d}\mathbf{S}\cdot\mathbf{B}. \tag{8.15}$$

Here, we used Stokes' theorem in the final equality, and $\mathcal{S}$ is the area enclosed by $\mathcal{C}$. The right-hand side is just the total magnetic flux through $\mathcal{S}$. On the left-hand side, we see that $\oint \nabla\varphi = 2\pi n$ is precisely the winding number of the vortex. Therefore, the magnetic flux penetrating the superconductor through vortices is quantised in units of $\Phi_0 = h/e^*$, called the **flux quantum**. Notice that both the magnetic field on the right-hand side and the phase winding on the left-hand side are gauge invariant observables, even though $\mathbf{A}$ and φ by themselves are not.

The decay of the magnetic field radially outward from the vortex core can be understood intuitively by realising that close to the vortex core, and electric supercurrents can flow that oppose the externally applied magnetic field, leading to its decay with distance from the core. The decaying fields in turn imply a cancelation of the supercurrent further from the core, which therefore also decays. Of course this is precisely the same physics that we discussed before, for the surface of a superconductor, and the length scale over which both the magnetic field and supercurrent decay is just the London penetration depth, λ_L. In stark contrast to the situation in a neutral superfluid, the order parameter field outside of an area of radius λ_L around the vortex core is virtually unaffected by the topological defect. Well-separated vortices in a charged superfluid therefore have no interaction, and the energy of a single vortex does not depend on the system size. In fact, for a three-dimensional charged superfluid with $\lambda_L \gg \xi$, you will show in Exercise 8.5 that the energy of a straight vortex line of length l_z is given by [192]

$$E_{\text{vortex}} \approx l_z\left(\frac{n\Phi_0}{\lambda_L}\right)^2 \ln\frac{\lambda_L}{\xi} = l_z\frac{\hbar^2}{4\pi m^*}n^2|\psi|^2\ln\frac{\lambda_L}{\xi}. \tag{8.16}$$

Here, n is the winding number of the vortex. As expected, the energy of the vortex scales as $\ln\lambda_L$, rather than $\ln L$, because the order parameter is affected only up to distances of the order of the London penetration depth.

Exercise 8.5 (Energy of a vortex in a charged superfluid).
The goal of this exercise is to calculate the energy of a vortex line penetrating a three-dimensional volume of charged superfluid. For simplicity, we consider the limit where the penetration depth is much larger than the coherence length, $\lambda_L \gg \xi$, so that the energy of the vortex core itself can be ignored. Following Eq. (8.11), the energy of the vortex configuration is given by

$$E = \int_{r>\xi} d^3r \left[\frac{1}{2\mu_0} (\nabla \times \mathbf{A})^2 + \frac{\hbar^2}{2m^*} |\psi|^2 (\nabla\varphi - \frac{e^*}{\hbar}\mathbf{A})^2 \right]. \tag{8.17}$$

Here, only the area outside the vortex core $r > \xi$ is included in the integral.
a. Use the definition of the supercurrent, and Ampère's law $\frac{1}{\mu_0}\nabla \times \mathbf{B} = \mathbf{J}$, to rewrite Eq. (8.17) in terms of the magnetic field $\mathbf{B}$.

You just showed that the energy of the vortex depends only on the magnetic field penetrating the charged superfluid, and not on any part of the order parameter field. From here on, consider a specific vortex configuration in which the vortex core is a line of length l_z along the z-direction at position $x = y = 0$. The magnetic field lines will point in the z-direction and will only depend on the distance $r = \sqrt{x^2 + y^2}$ from the vortex core. For $r > \xi$, the magnetic field will be screened and obey Eq. (8.13). For $r < \xi$, we can approximate the magnetic field by a Dirac delta function at the centre of the vortex core. Combining these, the total field configuration can be written as

$$\mathbf{B} = \Phi_0 \delta^2(\mathbf{r}) - \lambda_{\mathrm{L}}^2 \nabla \times (\nabla \times \mathbf{B}), \tag{8.18}$$

where $\Phi = h/e^*$ is the flux quantum.
b. Show that final term in the expression equals the $\lambda_{\mathrm{L}}^2 \nabla^2 \mathbf{B}$ term of Eq. (8.13).
c. Integrate Eq. (8.18) over an area with radius much larger than λ_{L} to show that the total flux in this vortex configuration is exactly one flux quantum.
d. Now integrate Eq. (8.18) over an area with a radius much smaller than $r \ll \lambda_{\mathrm{L}}$. Argue why the total flux in this region should be negligible and use this result to find the shape of the magnetic field $B_z(r)$ inside the vortex.
e. Use the vector identity $\nabla \cdot (\mathbf{B} \times (\nabla \times \mathbf{B})) = (\nabla \times \mathbf{B})^2 - \mathbf{B} \cdot (\nabla \times (\nabla \times \mathbf{B}))$ to show that the energy is approximately given by

$$E \approx \frac{l_z \lambda_{\mathrm{L}}^2}{2\mu_0} \left[B_z(r) \frac{\partial B_z(r)}{\partial r} 2\pi r \right]_{r=\xi}. \tag{8.19}$$

This result shows that the vortex energy depends only on the magnetic field strength at the edge of the vortex core defined by $r = \xi$.
f. Finally, insert your result for $B_z(r)$ and show that it yields Eq. (8.16).

8.4.1 Types of Superconductivity

The total free energy of a vortex configuration in a charged superfluid contains both an energy cost associated with the local suppression of the superconducting order (the vortex core energy) that does not depend on the magnetic field and the energy cost of Eq. (8.16) due to the magnetic field configuration. Assuming that $\lambda_L \gg \xi$, there is a critical value for the field applied to a charged superfluid, $H_{c1} \approx E_{\text{vortex}}/\Phi_0 l_z$, above which it becomes energetically favourable to host vortex lines with magnetic fields running through them, rather than to expel the field completely. The resulting state is called an Abrikosov vortex lattice, and it is the distinctive feature of so-called **type-II superconductivity**.

Increasing the applied field further, beyond H_{c1}, the vortices become more and more closely packed, until the point where their cores start to overlap, and superconductivity is destroyed throughout the material. Since the total flux is distributed over vortices that each carry a single flux quantum, the cores cover the entire superconductor when $\Phi_0/H \propto \xi^2$, so that superconductivity breaks down for fields higher than $H_{c2} \propto \Phi_0/\xi^2 > H_{c1}$. The two critical magnetic field strengths in type-II superconductors are known as their lower and upper critical fields.

If the London penetration depth is small compared to the correlation length, $\lambda_L < \sqrt{2}\xi$, the estimate for the vortex energy E_{vortex} in Eq. (8.16) breaks down. A more careful analysis will show that in that case, it is energetically favourable to create defects with the highest possible vorticity, rather than separating them into isolated flux quanta. An Abrikosov vortex lattice can then not be formed, and the bulk remains in a Meissner state, expelling any magnetic fields towards the edges of the sample. The externally applied magnetic field can be increased in strength until it is large enough to destroy the superconducting order altogether at a single critical field strength H_c. This type of behaviour is known as **type-I superconductivity**.

8.5 Charged BKT Phase Transition

The coupling of the phase degree of freedom to the electromagnetic field in a charged superfluid alters the characteristics of its phase transition as compared to a neutral superfluid. The situation in two dimensions is particularly interesting, bringing together several topics discussed in this book. The main physics can be described in terms of heuristic arguments, by comparing the properties of vortices in charged superfluids with their neutral counterparts.

Recall that neutral systems with $U(1)$-symmetry, such as superfluids, undergo a BKT phase transition in two dimensions, as described in Section 7.4. This comes about from the combinations of two ingredients. First, the Mermin–Wagner–Hohenberg–Coleman theorem of Section 6.2 precludes the formation of true long-range order in two dimensions, due to the divergence of thermal corrections to the order parameter by gapless NG modes. Algebraic long-range order is possible; however, because vortex–antivortex pairs are bound together at low temperatures. Second, the energy cost of these pairs is balanced by their entropy, which both scale logarithmically with system size. This leads to a critical temperature and phase transition at the temperature where vortex pairs can unbind.

In the case of the charged superfluid, both of these arguments need to be adjusted. First of all, the Anderson–Higgs mechanism of Section 8.3 renders all excitations in the superconductor massive and in particular does not allow for gapless NG modes. This means there is no divergence in the corrections to the order parameter of Eq. (6.20). Thus, at first sight, there seems to be no obstruction to having truly long-range ordered two-dimensional superconductors even at non-zero temperatures.

However, we should also take into account the fact that the vortex excitations of the neutral superfluid are altered by the electromagnetic field in the superconducting state. In Section 8.4, we saw that the winding of the order parameter around a vortex core is counteracted by the penetrating magnetic field in a charged superfluid, so that the energy of an isolated vortex is finite and does not scale with system size, being instead of the order of $\ln \lambda_{\mathrm{L}}/\xi$.

The entropy of a vortex just counts the number the possible locations for its core, so the form $S \propto \ln L^2/\xi^2$ is still valid for vortices in charged superfluids. On the other hand, the energy of superconducting vortices no longer scales with system size. This implies that for large systems with $L \gg \lambda_{\mathrm{L}}$, the entropic gain of having vortices outweighs their energetic cost even at the lowest possible temperatures. In other words, the BKT transition temperature at which vortex–antivortex pairs unbind is pushed all the way to zero, and seems to preclude superconducting order in two dimensions at any non-zero temperature.

In reality, both considerations are a bit beside the point. Since we live in a three-dimensional world, even the thinnest superconductors still couple to three-dimensional electromagnetic fields. Vortices can thus interact over large distances through the electromagnetic fields that they create in the vacuum surrounding the superconducting film. This can be described by introducing an effective penetration depth of magnetically mediated interactions within the two-dimensional charged superfluid, which turns out to be [59]

$$\lambda_{\mathrm{L,2D}} = \frac{\lambda_{\mathrm{L}}^2}{w} = \lambda_{\mathrm{L}}\frac{\lambda_{\mathrm{L}}}{w}. \tag{8.20}$$

Here, w is the thickness of the superconducting film. If it is small compared to λ_{L}, the effective penetration depth becomes very large. Recall from the definition of λ_{L} that it is inversely proportional to the charge e^* of the Cooper pairs making up the charged superfluid. This means that having a large penetration depth is equivalent to having Cooper pairs with very small charges, or in other words, to having a very weak coupling between electromagnetic fields and the particles making up the charged superfluid. A truly two-dimensional superconductor embedded in a three-dimensional vacuum therefore behaves just like a neutral superfluid. The gap of the would-be NG modes becomes very small, and thermal fluctuations can prevent the formation of true long-range order. Simultaneously the energy cost of vortices grows and, as λ_{L} surpasses L, it starts to scale with the system size, pushing the BKT transition temperature up from zero. This explains the experimental observation of BKT transitions in both thin type-II superconductors and in Josephson junction arrays [59].

8.6 Order of the Superconducting Phase Transition

Even in three dimensions, where true long-range order certainly exists, the superconducting phase transition is affected by the coupling between the phase and the electromagnetic vector potential. Both of these fields can fluctuate, and both types of fluctuations can alter the critical behaviour away from the mean-field expectation, in the style of Sections 5.4 and 5.5.

One way to investigate the consequences of the fluctuations in the vector potential is to 'integrate out' the gauge field. Recall that the partition function contains a sum over all possible configurations. It is then possible, at least formally, to split the sum into two parts and first perform only the partial sum over all possible configurations of the gauge fields while keeping the order parameter field fixed. This turns out to be a natural thing to do in the path integral formulation of quantum field theory. The result will be a new, effective theory for the order parameter field in which the gauge field does not explicitly appear, but which may include different terms and changed values of coefficients as compared to the original theory.

As a shortcut to this formal procedure, we can follow Ref. [69] and start from the free energy of the Ginzburg–Landau model in three dimensions in Eq. (8.1). We then expand the minimal coupling term $|(\nabla - ie^*\mathbf{A}/\hbar)\psi|^2$ and approximate both the vector potential and its powers by their expectation values, calculated for a constant value of the order-parameter field ψ. Since there are no static electromagnetic fields in the superconducting state, the terms linear in $\mathbf{A}$ vanish, whereas the fluctuations proportional to $\langle \mathbf{A}\cdot\mathbf{A}\rangle$ do not. In the free energy of Eq. (8.1), we thus include the term

$$\begin{aligned}\int \mathrm{d}^3x \left(\frac{e^{*2}}{2m^*}|\psi|^2\mathbf{A}^2\right) &\to \frac{e^{*2}}{2m^*}|\psi|^2 \int \mathrm{d}^3x\,\langle \mathbf{A}(\mathbf{x})^2\rangle \\ &= \frac{e^{*2}}{2m^*}|\psi|^2 \int \frac{\mathrm{d}^3k}{(2\pi)^3}\,\langle \mathbf{A}(\mathbf{k})\cdot\mathbf{A}(-\mathbf{k})\rangle. \end{aligned} \tag{8.21}$$

The expectation value for the fluctuations can be computed for fixed $|\psi|$ from the London equation in (8.13), giving the two-point correlation function for the vector potential in the Coulomb gauge, in momentum space:

$$\langle A_i(\mathbf{k})A_j(-\mathbf{k})\rangle = \mu_0 \frac{\delta_{ij} + \lambda_\mathrm{L}^2 k_i k_j}{k^2 + 1/\lambda_\mathrm{L}^2}. \tag{8.22}$$

Using this expression to carry out the integral in Eq. (8.21) results in

$$\int \mathrm{d}^3x \left(\frac{e^{*2}}{2m^*}|\psi|^2\mathbf{A}^2\right) \to \frac{3e^{*2}\mu_0}{4\pi m^*}|\psi|^2\Lambda - \frac{3e^{*3}\mu_0^{3/2}}{8m^{*3/2}}|\psi|^3. \tag{8.23}$$

Here, we introduced a cut-off momentum scale Λ, which we take to be much larger than $1/\lambda_\mathrm{L}$. As this expression replaces the term proportional to $|\psi|^2\mathbf{A}^2$ in the Ginzburg–Landau theory, we obtain a modified, effective, theory formulated entirely in terms of the order parameter field. The first term in Eq. (8.23) is proportional to $|\psi|^2$ and renormalises the value of r in the original theory. The second term, however, introduces a new term in the effective theory proportional to $|\psi|^3$. This will cause the minimum of the free energy to develop at non-zero $|\psi|$, similar to what is shown in Figure 5.3. The phase transition in the presence of fluctuations is thus first order (discontinuous), rather than second order (continuous), as would have been expected from a mean-field theory without fluctuations. This is called a **fluctuation-induced first-order** phase transition. In field theory, this way of arriving at a discontinuous phase transition is known as the **Coleman–Weinberg mechanism**.

Exercise 8.6 (Integrating out the vector potential).
In this exercise, we will derive Eq. (8.23). Starting from the quadratic term in the vector potential, $\int \mathrm{d}^3x \left(-\frac{e^{*2}}{2m^*}|\psi(x)|^2\mathbf{A}(x)^2\right)$, we can first approximate $|\psi(x)|$ to be constant and equal to $|\psi|$. In what is left, the squared vector potential can be approximated by its expectation value, $\langle \mathbf{A}(\mathbf{x})^2\rangle$, to end up in the first line of Eq. (8.21).
a. Use a Fourier transform to show that the first and second line of Eq. (8.21) are equivalent.

b. The next step is to derive the propagator of Eq. (8.22). For this, Fourier transform the London equation of Eq. (8.12) to momentum space and bring all the terms dependent on $A_i(\mathbf{k})$ to the left-hand side, which should then have the general form $P_{ij}(\mathbf{k})A_j(\mathbf{k}) = \text{r.h.s.}$ The propagator is then the inverse of $P_{ij}(\mathbf{k})$. Show that this is equal to Eq. (8.22).
c. Integrate the massive photon propagator of Eq. (8.22) over all momenta, up to a cut-off Λ. Your result, up to a constant, should be equal to Eq. (8.23).

In practice, it turns out that experiments on many superconductors show critical properties that are very close to the mean-field predictions, including for instance the value of the critical exponent β being $1/2$, as predicted by Eq. (5.4). One reason for this is that the interval where fluctuations are important, as quantified by the Ginzburg temperature t_G of Eq. (5.24), is very small, so that much of the experimentally accessible parameter regime falls outside the actual critical region. But this is not the whole story. As mentioned in Section 5.4, the Ginzburg temperature for strongly type-II superconductors (with $\lambda_L \gg \xi$) can be sizable. These same conditions also allow for the emergence of stable vortices, as discussed in Section 8.4. Using a duality mapping of the kind introduced in Section 7.5, it can be shown that presence of topological defects, which is not taken into account in the calculations above, again alters the effective theory in the critical region, and causes the superconducting phase transition in three dimensions to stay second order as long al $\lambda_L/\xi \gtrsim 1$ [92]. This prediction has been confirmed in numerical simulations.

9 Dynamics of Symmetry Breaking

So far, we discussed symmetry-broken ground states and finite-temperature equilibrium states. We saw that such states have two types of excitations: Nambu–Goldstone modes and topological excitations. In this chapter, rather than looking at equilibrium properties, we will discuss how symmetry-broken systems evolve over time after they are excited. We will consider excitation mechanisms ranging from a small pinch to a drastic change of environment.

9.1 Introduction to Dynamics

The most useful concept by far when talking about dynamics is that of a **relaxation time**. This is the property of the equilibrium state that describes how fast the system will respond to external disturbances and return to the equilibrium configuration. In this section, we will introduce the relaxation time and show how it relates to the coherence length of Section 5.4.1, the correlation length of Section 3.4, and the Nambu-Goldstone excitations of Chapter 4.

9.1.1 The Collapse of the Unstable Symmetric State

Before defining the relaxation time, it is worthwhile to consider what we already know about the formation of a symmetry-breaking state. Recall that in Section 3.7, we argued that a symmetric state is unstable against local measurement, as well as in the sense of the cluster decomposition property, and the variance of extensive variables. This leaves open the question of how spontaneous symmetry breaking unfolds **dynamically**, as a function of time. If we prepare an unstable symmetric configuration, then how long will it take to evolve into a state of spontaneously broken symmetry?

The answer is simple: it will never do so spontaneously. After all, despite being unstable, the symmetric state is an eigenstate of the symmetric Hamiltonian and will therefore not change in time unless an external perturbation is introduced. Moreover, even for initial states that are not quite symmetric, the equations of motion generated by the Hamiltonian are still symmetric and preserve the value of the order parameter. Therefore, starting, for example, from a magnet with zero magnetisation, the system is forced to remain non-magnetised as long as the Hamiltonian retains its symmetry.

We thus need an **external perturbation**, both to allow dynamical evolution away from eigenstates of the unperturbed Hamiltonian and to allow for a change of the (otherwise symmetry-protected) order parameter value. This can come in the form of a so-called **source term**, such as an externally applied field, or of a **thermal bath** that exchanges energy and/or particles with the system, or of **noise**, described as random, externally imposed fluctuations.

These three types of perturbations are not independent. Modelling the influence of a thermal bath, for example, is often done by adding noise to the time evolution. Likewise, the fluctuation–dissipation theorem links the effect of a source term to that of a bath.

Exercise 9.1 (No spontaneous symmetry breaking).
Consider the transverse-field Ising model introduced in Section 3.7, deep in the symmetry-breaking phase $J \gg \mu$.
a. What is the global symmetry of the model?
b. Use a perturbative expansion in small μ/J to show that the ground state of a finite-size system can be approximated by a superposition of all spins up and all spins down: $|\psi_0\rangle \sim |\uparrow\uparrow\cdots\rangle - |\downarrow\downarrow\cdots\rangle$. Verify that this ground state is symmetric under the global symmetry of the model.
c. Consider the system being in the symmetric state $|\psi_0\rangle$ at time $t = 0$, while a perturbation $h\hat{\sigma}_i^z$ acts on some site i for all times $t > 0$. Show explicitly that the system will not become spontaneously magnetised over time.

9.1.2 From Noise to a Thermal Bath

The concept of noise in the dynamics of physical systems was introduced originally in the context of the Brownian motion of classical particles. The Langevin equation describes how a particle suspended in a fluid experiences both a viscous force and random kicks from the surrounding fluid particles. In turns out that the order parameter of a symmetry-broken phase in contact with an external environment behaves in a very similar way. To describe this, we allow the order parameter to depend on time, $\psi(\mathbf{x}) \to \psi(\mathbf{x}, t)$. An effective drift force then pushes the system towards a minimum of the Ginzburg–Landau free energy $\mathcal{F}[\psi]$, while small random perturbations $\eta(\mathbf{x}, t)$ kick it in unpredictable directions. This can be described by an evolution of the form

$$\frac{\partial\psi(\mathbf{x},t)}{\partial t} = -\Gamma_0 \frac{\delta\mathcal{F}[\psi]}{\delta\psi(\mathbf{x},t)} + \eta(\mathbf{x},t). \tag{9.1}$$

Here, we wrote the free energy in dimensionless form for convenience. The strength of the force pushing towards equilibrium is given by the relaxation rate Γ_0. The second term models the effect of random perturbations or noise $\eta(\mathbf{x}, t)$, described as a randomly fluctuating field with correlations:

$$\begin{aligned} \langle \eta(\mathbf{x},t) \rangle &= 0, \\ \langle \eta(\mathbf{x},t)\eta(\mathbf{x}',t') \rangle &= 2\Gamma_0 \delta(\mathbf{x}-\mathbf{x}')\delta(t-t'). \end{aligned} \tag{9.2}$$

Notice the relation between the relaxation rate and the correlations in the noise, which stems from the fluctuation–dissipation theorem, and is a consequence of both the relaxation and the fluctuations originating from the influence of the same environmental processes. For the special case of **Gaussian white noise**, all expectation values of products of η can be expressed in terms of the two-point correlations defined earlier. The model of Eq. (9.1), sometimes referred to as 'Model A' in the literature [77], applies to any system with a scalar order parameter that is not conserved by the system-bath coupling. In particular, it describes the dynamics of the transverse-field Ising model coupled to a bath, which we saw in Exercise 9.1 to be a static model in isolation.

Exercise 9.2 (Ising magnet with noise).
The Landau free energy for the Ising magnet is $\mathcal{F}[m] = -\frac{r}{2}m^2 + \frac{u}{4}m^4$. In the absence of noise, any symmetric initial state $m = 0$ will remain symmetric. The goal of this exercise is to numerically simulate how noise can dynamically break the symmetry.
a. Write a code that generates a random series $\eta(t_n)$ for every $t_n = n\,\Delta t$, with $n \in \mathbb{Z}$ and Δt a fixed time interval. Make sure the values are normal distributed with variance given by Eq. (9.2) for some choice of Γ_0. Do you need to scale the variance of the noise with your choice of Δt?
b. Use a discretised version of the time derivative to calculate $m(t)$ starting with $m(t = 0) = 0$ and using the time evolution of Eq. (9.1). For convenience, set $r = 1$ and $u = 1$. Repeat this several times, to show that each noise realisation gives a different $m(t)$.
c. Compute the mean-square magnetisation over several realisations, $\overline{m}^2(t) = \langle m^2(t)\rangle$, and extract a typical timescale for the symmetry breaking to happen. An example of typical results is shown in Figure 9.1.
d. Use your numerics to explore how the average time for symmetry breaking changes as a function of Δt and Γ_0.
e. Similarly, find out how the average time for symmetry breaking changes as a function of r and u. Show that as you approach $r = 0$, the symmetry-breaking time scale increases.

Equation (9.1) is an example of a stochastic differential equation (SDE), with one of its terms depending on a random process. Because of the random term, the solutions of an SDE are stochastic too. For many physical situations, however, we are more interested in the average and variance of observables than in the properties of any one particular solution. Such quantities can be described in terms of the probability distribution function $P[\psi(\mathbf{x}), t]$, which signifies the probability (density) that the stochastic noise term in the SDE yields the particular realisation $\psi(\mathbf{x}, t)$ of the order parameter dynamics.

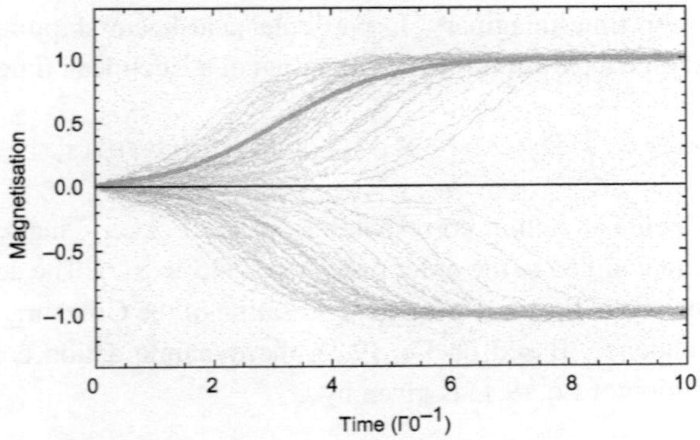

Figure 9.1 Starting from an initially symmetric state with zero magnetisation, an Ising magnet can become magnetised due to random perturbations (noise). Here, we show, following Exercise 9.2, the magnetisation as a function of time for 100 different noise realisations. The thick line indicates the average magnetisation $\sqrt{\overline{m}^2(t)}$.

The stochastic dynamics of Eq. (9.1) leads to a deterministic differential equation for the probability distribution. This is known as the **Fokker–Planck equation**, which has the general shape

$$\frac{\partial P[\psi,t]}{\partial t} = \int \mathrm{d}^d x \left(-\frac{\delta}{\delta\psi(\mathbf{x},t)} \left(D_1 P[\psi,t]\right) + \frac{\delta^2}{\delta\psi(\mathbf{x},t)^2} \left(D_2 P[\psi,t]\right) \right). \tag{9.3}$$

Here, D_1 is known as the drift coefficient, while D_2 is the diffusion coefficient. For the specific case of Eq. (9.1), the diffusion coefficient is equal to the relaxation rate, $D_2 = \Gamma_0$, while the drift term is given by the force pushing the order parameter field towards the free energy minimum:

$$D_1 = -\Gamma_0 \frac{\delta\mathcal{F}[\psi]}{\delta\psi(\mathbf{x},t)}. \tag{9.4}$$

A full derivation, from the Langevin equation to the Fokker–Planck equation, can be found, for example, in Ref. [230].

Exercise 9.3 (Probability distribution function for the order parameter).
a. Using your code of Exercise 9.2, calculate the distribution $P[m,t]$ at various times t.
b. Show that at late times, $P[m,t] \sim \exp(-\mathcal{F}[m]/k_B T)$, and specify the value of $k_B T$ in terms of model parameters.

This result shows how interactions with a bath cause the collection of all possible late-time states to approach a thermal ensemble characterised by Boltzmann weights.

It is instructive to interpret the right-hand side of the Fokker–Planck equation as a sort of 'Fokker–Planck Hamiltonian', H_{FP}, acting on the distribution $P[\psi,t]$:

$$\partial_t P[\psi,t] = -\hat{H}_{\mathrm{FP}} P[\psi,t]. \tag{9.5}$$

This equation then looks like a Schrödinger equation in imaginary or **Euclidean time**. We can therefore use all the techniques of quantum mechanics to look for solutions to this equation, provided we keep time imaginary. In particular and again skipping all details of the derivation, we can define a partition function in terms of a Euclidean time path integral:

$$Z[J] = \int \mathcal{D}\psi \exp\left[-S[\psi] + \int \mathrm{d}^d x \,\mathrm{d}t\, J(\mathbf{x},t)\psi(\mathbf{x},t) \right]. \tag{9.6}$$

Here, $S[\psi]$ is the Euclidean action corresponding to the Fokker–Planck Hamiltonian, and $J(x,t)$ is a source term coupled to the order parameter field $\psi(\mathbf{x},t)$. The action $S[\psi]$ is sometimes called the **dynamic action** and is the generalisation of the Ginzburg–Landau functional including time dependence. Based on Eq. (9.3), the dynamic action corresponding to the stochastic time evolution of Eq. (9.1) is given by

$$S[\psi] = \frac{1}{4\Gamma_0} \int \mathrm{d}^d x\, \mathrm{d}t \left\{ \partial_t \psi(\mathbf{x},t) + \Gamma_0 \frac{\delta\mathcal{F}[\psi]}{\delta\psi(\mathbf{x},t)} \right\}^2 - \Gamma_0^2 \int \mathrm{d}^d x\, \mathrm{d}t\, \frac{\delta^2\mathcal{F}[\psi]}{\delta\psi(\mathbf{x},t)^2}. \tag{9.7}$$

Substituting this into the partition function of Eq. (9.6), we can use the minimal action principle in the usual way to identify stable solutions for $\langle\psi(\mathbf{x},t)\rangle$, or to calculate correlation functions by taking derivatives with respect to the source term. For example, the two-point connected correlation function is given by

$$\begin{aligned} C(\mathbf{x},t;\mathbf{x}',t') &= \langle\psi(\mathbf{x},t)\psi(\mathbf{x}',t')\rangle - \langle\psi(\mathbf{x},t)\rangle\langle\psi(\mathbf{x}',t')\rangle \\ &= \frac{\delta^2}{\delta J(\mathbf{x},t)\delta J(\mathbf{x}',t')}\log Z[J]\bigg|_{J=0}. \end{aligned} \tag{9.8}$$

Note that in these expressions, $\langle\dots\rangle$ denotes an average over all possible realisations of the noise field $\eta(\mathbf{x},t)$, rather than a quantum mechanical expectation value.

The appearance of a Euclidean time path integral in this formulation of noise-averaged observables is reminiscent of thermal quantum dynamics. In fact, a Hamiltonian system in thermal equilibrium is described by the thermal density matrix $\hat{\rho} = \mathrm{e}^{-\beta\hat{H}}$. The corresponding partition function can also be written as a Euclidean time path integral:

$$Z = \mathrm{Tr}\left[\mathrm{e}^{-\beta\hat{H}}\right] = \int \mathcal{D}\psi\, \mathrm{e}^{-S_\beta[\psi]}. \tag{9.9}$$

This establishes a formal equivalence between systems described by a thermal density matrix, and ones described by noisy Langevin dynamics. This connection is discussed in more detail in Chapter 10, in the context of effective field theories.

Exercise 9.4 (Correlation function from path integral).
Show that the final line of Eq. (9.8) is equivalent to

$$\begin{aligned} C(\mathbf{x},t;\mathbf{x}',t') = &\frac{1}{Z[0]}\int \mathcal{D}\psi\ \psi(\mathbf{x},t)\psi(\mathbf{x}',t')e^{-S[\psi]} \\ &-\left(\frac{1}{Z[0]}\int \mathcal{D}\psi\ \psi(\mathbf{x},t)\mathrm{e}^{-S[\psi]}\right)\left(\frac{1}{Z[0]}\int \mathcal{D}\psi\ \psi(\mathbf{x}',t)e^{-S[\psi]}\right). \end{aligned} \tag{9.10}$$

9.1.3 Fluctuation–Dissipation Theorem

The formal connection between dynamics in the presence of a thermal bath and time evolution driven by white noise points to the intuitive notion that even in equilibrium the order parameter field of any particular system will not be static, but constantly exhibits small and local fluctuations. As we will see, these fluctuations are quantified by the connected correlation function $C(\mathbf{x},t;\ \mathbf{x}',t')$, which we encountered before in Section 3.7 when discussing the stability of symmetric states, in Section 6.2 when discussing the Mermin–Wagner–Hohenberg–Coleman theorem; and in Section 5.4 when discussing spatial fluctuations of the order parameter. For the connection to thermal fluctuations, we will be specifically interested in the temporal part of the correlation function.

In general, the random local fluctuations encountered in thermal equilibrium are deeply connected to the response of a system to applying an external field. To see this, we go back to the Hamiltonian formalism and consider a system described by the Hamiltonian $\hat{H}_0$, to which we add an additional external perturbation:

$$\hat{H}(t) = \hat{H}_0 + \int \mathrm{d}^d x \mu(\mathbf{x},t)\hat{\psi}(\mathbf{x}). \tag{9.11}$$

Here, the applied field $\mu(\mathbf{x}, t)$ couples to the operator $\hat{\psi}(\mathbf{x})$, which for the present discussion we will assume to be an order parameter operator. For example, μ could represent a magnetic field coupling to the magnetic moment $\hat{\psi}$ or an electric field coupling to an electric charge.

To see how the applied field μ affects the time-dependent expectation value $\langle\hat{\psi}(\mathbf{x}, t)\rangle$, we expand the time evolution operator $\hat{U}(t, t_0) = T\exp\left[-\frac{\mathrm{i}}{\hbar}\int_{t_0}^{t} \mathrm{d}t'\hat{H}(t')\right]$ in the applied field μ:

$$\hat{U}(t, t_0) = \mathrm{e}^{-\frac{\mathrm{i}}{\hbar}\hat{H}_0(t-t_0)} - \frac{\mathrm{i}}{\hbar}\int_{t_0}^{t} \mathrm{d}t'\, \mathrm{e}^{\frac{\mathrm{i}}{\hbar}\hat{H}_0(t-t')}\left[\int \mathrm{d}^d x\, \mu(\mathbf{x}, t')\hat{\psi}(\mathbf{x})\right] e^{-\frac{\mathrm{i}}{\hbar}\hat{H}_0(t'-t_0)} + \dots \tag{9.12}$$

Inserting this into the expression for the expectation value yields:

$$\begin{aligned}\langle\hat{\psi}(\mathbf{x}, t)\rangle &= \langle\hat{\psi}(\mathbf{x}, t)\rangle|_{\mu=0} - \frac{\mathrm{i}}{\hbar}\int \mathrm{d}^d x' \int_{t_0}^{t} \mathrm{d}t' \mu(\mathbf{x}', t')\langle[\hat{\psi}(\mathbf{x}, t), \hat{\psi}(\mathbf{x}', t')]\rangle|_{\mu=0} + \dots \\ &\equiv \langle\hat{\psi}(\mathbf{x}, t)\rangle|_{\mu=0} + \int d^d x' \int_{t_0}^{t} \mathrm{d}t' \mu(\mathbf{x}', t')\chi(\mathbf{x}, t;\mathbf{x}', t') + \dots \end{aligned} \tag{9.13}$$

This is the core equation of **linear response theory**, describing the response of a system to an externally applied field. To guarantee causality, the **susceptibility** χ appearing in the second line is often defined to explicitly include a Heaviside step function, so that it is non-zero only for $t' > t$:

$$\chi(\mathbf{x}, t;\mathbf{x}', t') = -\frac{\mathrm{i}}{\hbar}\Theta(t - t')\,\langle[\hat{\psi}(\mathbf{x}, t), \hat{\psi}(\mathbf{x}', t')]\rangle|_{\mu=0}. \tag{9.14}$$

The susceptibility is also often written in terms of momentum and frequency, using translation and time-translation invariance to argue that it only depends on the differences $\mathbf{x} - \mathbf{x}'$ and $t - t'$:

$$\chi(\mathbf{k}, \omega) = \int \mathrm{d}^d x\, \mathrm{d}t\; \mathrm{e}^{\mathrm{i}(\omega t - \mathbf{k}\cdot\mathbf{x})}\chi(\mathbf{x}, t;\, \mathbf{0}, 0). \tag{9.15}$$

Using these definitions, the change to the order parameter field $\langle\hat{\psi}(\mathbf{k}, \omega)\rangle$ due to the application of a perturbation $\mu(\mathbf{k}, \omega)$ is given by

$$\langle\hat{\psi}(\mathbf{k}, \omega)\rangle - \langle\hat{\psi}(\mathbf{k}, \omega)\rangle|_{\mu=0} = \chi(\mathbf{k}, \omega)\mu(\mathbf{k}, \omega). \tag{9.16}$$

In the limit of a vanishingly weak perturbation μ, this equation shows that the susceptibility equals the derivative of the order parameter with respect to the applied field. This coincides with the classical statistical mechanics notion of susceptibility in which, for example, the magnetic susceptibility is defined as the derivative of the magnetisation with respect to the applied field.

The susceptibility $\chi(\mathbf{k}, \omega)$ has both real and imaginary components. It can be straightforwardly verified that the causal structure, which imposes $\chi(t - t') = 0$ for $t < t'$, guarantees the susceptibility to be analytic in the upper half plane. In turn, this implies that its real and imaginary parts are related by a so-called Kramers–Kronig relation.

Physically, the imaginary part of a susceptibility describes the degree to which an externally applied perturbations is absorbed and **dissipated** by the system. Because of this, Im χ is also known as the **loss function**. Using causality, invariance under time translations, and the fact that $\hat{\psi}$ is an observable, the loss function can be written in the convenient form

$$\mathrm{Im}\,\chi(\omega) = \frac{1}{2\hbar}\int_{-\infty}^{\infty} \mathrm{d}t\, \mathrm{e}^{\mathrm{i}\omega t}\langle\hat{\psi}(t)\hat{\psi}(0) - \hat{\psi}(0)\hat{\psi}(t)\rangle|_{\mu=0}. \tag{9.17}$$

Here, we dropped all position and momentum dependence for notational convenience. For the special case of the response of a system in thermal equilibrium, we can use Eq. (9.9) to write the expectation value of an operator $\hat{A}$ as $\langle \hat{A} \rangle = Z^{-1}\mathrm{Tr}\left[\mathrm{e}^{-\beta \hat{H}}\hat{A}\right]$. Applying this to the loss function, we find

$$\begin{aligned} \langle \hat{\psi}(0)\hat{\psi}(t)\rangle &= Z^{-1}\mathrm{Tr}\left[\mathrm{e}^{-\beta\hat{H}}\hat{\psi}(0)\hat{\psi}(t)\right] \\ &= Z^{-1}\mathrm{Tr}\left[\mathrm{e}^{-\beta\hat{H}}\hat{\psi}(t-\mathrm{i}\hbar\beta)\hat{\psi}(0)\right] \\ &= \langle \hat{\psi}(t-\mathrm{i}\hbar\beta)\hat{\psi}(0)\rangle . \end{aligned} \tag{9.18}$$

Here, we used both the cyclic property of the trace operator and the fact that $\hat{\psi}(t) = \mathrm{e}^{\mathrm{i}\hat{H}t}\hat{\psi}\mathrm{e}^{-\mathrm{i}\hat{H}t}$ to go from the first to the second line. Substituting this result back into Eq. (9.17) yields

$$\mathrm{Im}\,\chi(\omega) = \frac{1}{2\hbar}(1-\mathrm{e}^{-\beta\hbar\omega})\int_{-\infty}^{\infty}\mathrm{d}t\,\mathrm{e}^{\mathrm{i}\omega t}\langle\hat{\psi}(t)\hat{\psi}(0)\rangle . \tag{9.19}$$

The integrand on the right-hand side is just the two-point correlation function $C(x,t;x',t') = \langle \hat{\psi}(x,t)\hat{\psi}(x',t')\rangle\rangle$ that we used in Section 3.4 to characterise the presence or absence of long-range order. Rewriting the expression then finally yields the **fluctuation–dissipation theorem**:

$$C(\mathbf{k},\omega) = 2(n_{\mathrm{BE}}(\omega)+1)\,\mathrm{Im}\,\chi(\mathbf{k},\omega). \tag{9.20}$$

Here, $n_{\mathrm{BE}}(\omega) = (e^{\beta\omega}-1)^{-1}$ is the Bose–Einstein distribution.

The fluctuation–dissipation theorem is extremely powerful. It applies to any Hamiltonian system and any stochastic system with Gaussian white noise. For all of these, it allows us to calculate the out-of-equilibrium dynamics describing the dissipation of an applied field, using only the equilibrium correlation function in the presence of thermal fluctuations. In particular, this means we can understand much of the dynamics of order parameters upon perturbing a symmetry-broken system, just by calculating correlations in thermal equilibrium. Keep in mind, however, that the fluctuation–dissipation theorem only applies to small perturbations of an initial state in thermal equilibrium. In other cases, such as the dynamics following a quench, the evolution can be much wilder.

Exercise 9.5 (Fluctuation–dissipation theorem).
a. Show that $U(t,t_0) = T\exp\left[-\frac{\mathrm{i}}{\hbar}\int_{t_0}^{t}\mathrm{d}t' H(t')\right]$ is indeed a solution of the time-dependent Schrödinger equation.
b. Show explicitly that an expansion of $U(t,t_0)$ to first order in μ yields Eq. (9.12).
c. Use Eq. (9.12) to derive Eq. (9.13).
d. Consider the case of the initial state being in thermal equilibrium, and use the spectral representation $\langle \hat{A}\rangle = Z^{-1}\mathrm{Tr}\left[\mathrm{e}^{-\beta\hat{H}}\hat{A}\right]$ to write $\chi(\mathbf{k}=0,\omega)$ in the basis of eigenstates of the Hamiltonian. Show that the imaginary part is non-zero only for 'resonances', in which the frequency ω matches an energy-level difference in the system.
e. Show that for high temperatures, the fluctuation–dissipation theorem of Eq. (9.20) becomes approximately

$$C(\mathbf{k},\omega) \approx 2\frac{k_B T}{\omega} \mathrm{Im}\chi(\mathbf{k},\omega). \tag{9.21}$$

Compared to what should the temperature be high for this approximation to hold?

f. We define the **equal-time correlation** function as $C(\mathbf{k}, t = 0) = \int \frac{d\omega}{2\pi} C(\mathbf{k},\omega)$. Use the Kramers–Kronig relations to show that at high temperature the fluctuation–dissipation theorem relates the equal-time correlation function to the zero-frequency susceptibility:

$$C(\mathbf{k}, t = 0) \approx k_B T \chi(\mathbf{k}, \omega = 0). \tag{9.22}$$

9.1.3.1 How to Break a Symmetry

The relations between dynamics driven by noise, thermal equilibrium, and the dissipation of externally applied perturbations are generic for any classical or quantum many-body system. In general, the fluctuation–dissipation theorem implies that we only need to calculate thermal equilibrium correlation functions to understand dissipative out-of-equilibrium dynamics. Those, in turn, can be calculated using path integral methods. None of this requires any symmetry to be broken, but it also implies that the dynamics of order parameter fields away from equilibrium can be calculated using 'standard' tools of quantum field theory.

In the next section, we will use these methods to introduce the relaxation time, which is a central concept in the dynamics of symmetry breaking. Using that, we can go back to the thought experiment of starting with a symmetric state and evolving it to a symmetry-broken configuration. The practical way to imagine this is by starting at a high temperature and then lowering the temperature into the symmetry-broken phase. Depending on whether the phase transition we cross is second or first order, we get different results: either the Kibble–Zurek mechanism of Section 9.3 or the nucleation dynamics of Section 9.4.

9.2 Relaxation Time

How quickly a system is able to respond to external influences determines the time scale required for a symmetry to be broken. This is characterised by the **relaxation time** τ, which is commonly defined in terms of the susceptibility χ as [77]

$$\tau(\mathbf{k}) = \left. \frac{-\mathrm{i}}{\chi(\mathbf{k},\omega)} \frac{\partial \chi(\mathbf{k},\omega)}{\partial \omega} \right|_{\omega=0}. \tag{9.23}$$

As you will show in Exercise 9.6, this definition for the relaxation time equals the the exponential decay time of the susceptibility if it evolves like $\chi(t) \sim \mathrm{e}^{-t/\tau(\mathbf{k})}$. In fact, even if the susceptibility decays exponentially only at late times, the definition in terms of the $\omega = 0$ component ensures that τ closely approximates the decay time.

Exercise 9.6 (Relaxation time).
Consider an exponentially decaying susceptibility $\chi(t) = \mathrm{e}^{-t/\tau}\Theta(t)$. Using a Fourier transform, show that the relaxation time defined by Eq. (9.23) indeed equals the decay time τ.

Exercise 9.7 (Typical susceptibility).
A typical shape for the susceptibility function is $\chi(\omega) = (-i\omega/\Gamma + r)^{-1}$. What is the relaxation time in this case?

Exercise 9.8 (Relaxation time in terms of correlation function).
Use the fluctuation–dissipation theorem to show that the relaxation time can also be defined in terms of correlation functions

$$\tau(\mathbf{k}) = \frac{C(\mathbf{k}, \omega = 0)}{2C(\mathbf{k}, t = 0)}. \tag{9.24}$$

Hint: Use the fact that the limit $\omega \to 0$ means that $\omega \ll T$ holds for any non-zero T, so that we can apply the high-temperature limit.

9.2.1 Calculating the Relaxation Time

As an example of how the relaxation time is calculated in practice, consider a scalar order parameter field $\psi(\mathbf{x}, t)$ described by the Ginzburg–Landau equation:

$$\mathcal{F}[\psi] = \int \mathrm{d}^d x \, \mathrm{d}t \left\{ \frac{\rho_s}{2}[\nabla\psi(\mathbf{x}, t)]^2 + \frac{r(T)}{2}[\psi(\mathbf{x}, t)]^2 + \frac{u(T)}{4}[\psi(\mathbf{x}, t)]^4 \right\}. \tag{9.25}$$

Substituting this into the definition of the dynamic action given by Eq. 9.7, the susceptibility can be calculated using the full arsenal of quantum field-theoretical techniques and approximations.

The most straightforward of these techniques is the **Gaussian approximation**. This means we expand the action around its minimum in terms of perturbations of the order parameter field. In the symmetric phase, the order parameter equals $\psi_0 = 0$ by definition, while in the specific case given by Eq. (9.25), the order parameter in the symmetry-broken phase is $\psi_0 = \pm\sqrt{r/u}$. We then write $\psi(\mathbf{x}, t) = \psi_0 + \delta\psi(\mathbf{x}, t)$ and neglect all terms higher than quadratic in $\delta\psi(\mathbf{x}, t)$. The resulting path integral will have a quadratic action and can thus be solved using the standard formulas for Gaussian integration.

In the symmetric phase, this approximation allows us to neglect the interaction term proportional to u and yields the effective action:

$$S[\psi] \approx \frac{1}{4\Gamma_0} \int \frac{\mathrm{d}^d k \, \mathrm{d}\omega}{(2\pi)^{d+1}} \left(\omega^2 + \Gamma_0^2 (r + \rho_s k^2)^2 \right) |\psi(\mathbf{k}, \omega)|^2. \tag{9.26}$$

Here, we used a Fourier transform to express the integral in momentum and frequency space. Writing thermal averages in terms of the weights $\mathrm{e}^{-\beta S[\psi]}$, this yields the correlation function:

$$C(\mathbf{k}, \omega) = \frac{2k_B T \Gamma_0}{\omega^2 + \Gamma_0^2 (r + \rho_s k^2)^2}. \tag{9.27}$$

Focussing on the regime of small frequencies, the susceptibility then becomes

$$\chi(\mathbf{k}, \omega) \approx \left(-i\omega/\Gamma_0 + r + \rho_s k^2 \right)^{-1}. \tag{9.28}$$

In turn, this yields a relaxation time given by

$$\tau(\mathbf{k}) = \Gamma_0^{-1}(r + \rho_s k^2)^{-1}. \tag{9.29}$$

The relaxation time is thus determined by the strength Γ_0 of the overall coupling to the bath, as well as the energy $E = r + \rho_s k^2$ of elementary excitations.

At the critical point $r = 0$, the energy of elementary excitations vanishes in the long-wavelength limit. Consequently, the relaxation time diverges when approaching a phase transition. This signifies the **critical slowing down** of all dynamics at that point.

Exercise 9.9 (Fluctuation–dissipation for the relaxation time).
Use the fluctuation–dissipation theorem to derive Eq. (9.28) from Eq. (9.27).

9.2.2 Dynamical Exponent

At a continuous phase transition, both the coherence length defined in Section 5.4 and the relaxation time of Section 9.2 diverge. This means that a small, local perturbation applied to a system close to the phase transition will not only affect the order parameter over large distances but also take a very long time to decay. The divergence of both scales as the phase transition is approached can be described in terms of **critical exponents**:

$$\begin{aligned} \xi &\sim |T - T_c|^{-\nu}, \\ \tau &\sim \xi^z \sim |T - T_c|^{-z\nu}. \end{aligned} \tag{9.30}$$

The **dynamical critical exponent** z indicates how the divergence of the relaxation time is related it to the coherence length ξ, which itself diverges with the correlation length exponent ν.

In the scalar field example with the Landau free energy functional given by Eq. (9.25), the parameter r goes to zero linearly at the transition, $r(T) \sim (T - T_c)$. The expression for the relaxation time in Eq. (9.29) then gives

$$\tau \sim |T - T_c|^{-1}. \tag{9.31}$$

This supplements the results of Section 5.4, where we showed that the coherence length diverges as $\xi \sim |T - T_c|^{-1/2}$. The critical exponents for this example are thus $\nu = 1/2$ and $z = 2$. In deriving the divergences of both the relaxation time and the coherence length, we used the Gaussian approximation and ignored all terms in the effective action that are higher than second order in the fluctuations. As explained in Section 5.3.2, this makes the derivation an example of a mean-field theory. It turns out that including higher orders in the effective action is a highly nontrivial task. They can be incorporated using the renormalisation group (RG) theory of phase transitions mentioned in Section 5.5. A brief overview of some of the values for the dynamical critical exponent z found using renormalisation group analyses for different types of dynamics is shown in Table 9.1. The labelling of models with letters ranging from A through J is a historical classification associated with the symmetry and conservation of the order parameter.

Model	Symmetry	Exponent[a]	Examples
A	$\mathbb{Z}_2$	$2+c\eta, c=6\log\left(\frac{4}{3}\right)-1$	Ising model
C	$\mathbb{Z}_2$	$2+\alpha/\nu$	Structural transitions
E	$U(1)$	$\frac{1}{2}d$	XY magnet, superfluid
F	$U(1)$	$\frac{1}{2}(d+\max(\alpha,0)/\nu)$	Superfluid ^{4}He
G	$SU(2)$	$\frac{1}{2}d$	Heisenberg antiferromagnet
J	$SU(2)$	$\frac{1}{2}(d+2-\eta)$	Heisenberg ferromagnet

[a] The exponents are obtained using RG scaling [77] and express the dynamical critical exponent z in terms of other exponents, notably the dimension d, the anomalous dimension η, the specific heat exponent α and the correlation length exponent ν.

Dynamical Critical Exponents. **Table 9.1**

Exercise 9.10 (Ising dynamical critical exponent).
In this exercise, we will calculate the dynamical critical exponent of the two-dimensional classical Ising model. Consider an $L \times L$ square lattice, with on each site a spin $\sigma = \pm 1$. The energy of a spin configuration $\mathcal{C}$ is given by

$$E[\mathcal{C}] = -\sum_{\langle ij\rangle} \sigma_i \sigma_j. \tag{9.32}$$

Here, $\langle ij\rangle$ indicates neighbouring sites on the lattice. The Boltzmann weight of each configuration is $P[\mathcal{C}] = Z^{-1}\mathrm{e}^{-\beta E[\mathcal{C}]}$, with $\beta = 1/k_\mathrm{B}T$ the inverse temperature and $Z = \sum_{\mathcal{C}} \mathrm{e}^{-\beta E[\mathcal{C}]}$ the partition function.
The thermal dynamics of the classical model can be modelled using a so-called **Monte Carlo heat bath** algorithm. This algorithm creates a sequence of spin configurations $\mathcal{C}_1 \to \mathcal{C}_2 \to \ldots$ known as a Markov chain. The algorithm is stochastic and uses a conditional probability $P_\mathrm{M}[\mathcal{C}_\mathrm{new}|\mathcal{C}_\mathrm{old}]$ of accepting a newly produced configuration $\mathcal{C}_\mathrm{new}$, depending on the previous configuration $\mathcal{C}_\mathrm{old}$. The conditional probability is defined in such a way that

$$\frac{P_\mathrm{M}[\mathcal{C}|\mathcal{C}']}{P_\mathrm{M}[\mathcal{C}'|\mathcal{C}]} = \frac{P[\mathcal{C}']}{P[\mathcal{C}]}. \tag{9.33}$$

If this condition is met, the simulation is said to obey **detailed balance**.
a. Show that if the detailed balance condition is met, then the probability of picking a particular configuration $\mathcal{C}$ from a generated sequence is given by its Boltzmann weight $P[\mathcal{C}]$.

In the Monte Carlo heat bath algorithm, a new spin configuration is generated by a two-step stochastic process. First, a random site i is chosen. Then, the spin on this site, σ_i, is flipped with probability

$$p = \frac{P[\mathcal{C}']}{P[\mathcal{C}] + P[\mathcal{C}']}. \tag{9.34}$$

Here, $\mathcal{C}'$ is the configuration that would result if spin σ_i is indeed flipped.

b. Show that this algorithm satisfies the detailed balance condition. Also argue that a numerical calculation of p only requires $O(1)$ numerical steps.

To model dynamics, each configuration generated by the heat bath algorithm (regardless of whether or not the spin flip is accepted) is considered a time step, so that the generated sequence defines the evolution of the spin configuration, $\sigma_i(t)$, as a function of time as well as space. This allows us to calculate, for example, the autocorrelation function defined as

$$C(t,t') = \frac{1}{L^2}\sum_i \langle \sigma_i(t)\sigma_i(t')\rangle. \tag{9.35}$$

Here, $\langle \cdots \rangle$ means averaging over different realisations of the simulation.
c. Write a Monte Carlo heat bath algorithm to simulate the dynamics of the classical Ising model and show that at high temperatures and late times, $t \gg t' \gg 1$, you find

$$C(t,t') \sim \exp\left(-\frac{(t-t')}{L^2\tau}\right). \tag{9.36}$$

Here, τ is the relaxation time, which should be independent of system size.

Close to the critical temperature T_c, the relaxation time diverges, but so does the coherence length. For any finite-size system, the maximum coherence length scales with the system size $\xi_{max} \sim L$. Based on Eq. (9.30), we then expect the relaxation time τ to scale with system size as $\tau \sim L^z$ at the critical temperature. The dynamical critical exponent for the Ising model can thus be calculated by a finite-size calculation at fixed $T = T_c = 2/\log(1+\sqrt{2})$.
d. Use your Monte Carlo heat bath algorithm to calculate $\tau(L)$ at the critical temperature and extract an estimate for z.

For small L, we can actually calculate $\tau(L)$ exactly. The idea is to view the conditional probability $P = P_M[\mathcal{C}|\mathcal{C}']$ as a matrix in the space of all 2^{L^2} configurations.
e. Construct the matrix P for $L = 2,3,4$. Make sure that the sum over all elements in a particular row always equals one, $\sum_j P_{ij} = 1$.

The matrix P has one eigenvalue equal to one, corresponding to the static equilibrium state. The relaxation time is given by the second largest eigenvalue, λ_2, which is shown in Ref. [147] to obey

$$\tau = -\frac{1}{L^2\log\lambda_2}. \tag{9.37}$$

f. Use your matrices P to calculate τ for $L = 2,3,4$ and extract the dynamical critical exponent z.

A two-loop RG calculation gives the value $z = 2$ for the dynamical critical exponent [68], while a more recent five-loop RG calculation gives $z = 2.14$ [4]. Direct Monte Carlo simulations have yielded $z = 2.13$ [114], while the method based on the eigenvalues of P was applied in Ref. [147], where it yielded $z = 2.1657$.

9.3 Continuous Transitions: Kibble–Zurek Mechanism

At a continuous phase transition, both the coherence length and the relaxation time necessarily diverge. In Section 5.5, we argued that because of this, there is a universality in the phenomenology of continuous phase transitions. This reasoning applies also beyond equilibrium properties, and we will see a similar universality in the dynamics of continuous phase transitions.

Returning to the questions posed at the onset of this chapter, of how a symmetry can be broken dynamically, consider cooling a system through its critical temperature. To be specific, we will consider cooling a heat bath in which the system of interest is immersed and with which it is initially in equilibrium. Starting at high temperatures and a symmetric configuration, consider cooling at a constant rate, τ_Q^{-1}. The inverse of this rate, τ_Q, is also known as the **quench time**, after the rapid cooling of hot metals in metallurgy. Close to the critical temperature T_c at which we expect the equilibrium configuration of the system to undergo a continuous phase transition, we can linearly approximate the time evolution of changing bath temperature as

$$T(t) = T_c \left(1 - \frac{t}{\tau_Q}\right). \tag{9.38}$$

Here, t is a time coordinate chosen in such a way that $T(t)$ is above the critical temperature for $t < 0$, while for positive times, $t > 0$, the equilibrium configuration of the system will be an ordered phase.

As long as the relaxation time $\tau_{rel}(T)$ of the system is small compared to the inverse rate at which the temperature of the bath changes, the system can constantly adjust to the changing temperature. The dynamics is then **adiabatic**, meaning that the system is always approximately in the equilibrium configuration corresponding to the instantaneous temperature of the bath. However, because the relaxation time diverges at the phase transition, there is unavoidably a regime in which the relaxation time is (much) larger than the inverse cooling rate τ_Q. Before the system reaches the phase transition, it is thus guaranteed to fall out of equilibrium.

The dynamics of a quench through a continuous phase transition was first studied by Kibble in the context of cosmology [89, 90, 91], and later extended by Zurek in the context of superfluid helium [232, 233]. Kibble and Zurek both argued that the system will not be able to adjust its order parameter configuration to the changing bath temperature if the relaxation time becomes too large and that it will remain 'frozen', or approximately static in that regime. Both authors then made the approximation of assuming the dynamics to be strictly adiabatic up to a point $|t| = t^*$ close to the transition, then completely static for as long as the relaxation time remains sufficiently high, and then strictly adiabatic again. Figure 9.2 gives a schematic depiction of these three regions.

The time at which the system freezes is called the **freezing time**, t^*. To estimate it, imagine starting in a disordered thermal equilibrium well above the transition temperature. If a thermal fluctuation or external perturbation introduces an ordered region into the system, it will take approximately τ_{rel} for the perturbation to be removed and the system to return to a symmetric state. If the temperature drops below the critical temperature within that time, the perturbation will not be completely removed. What remains of it will be stable at the

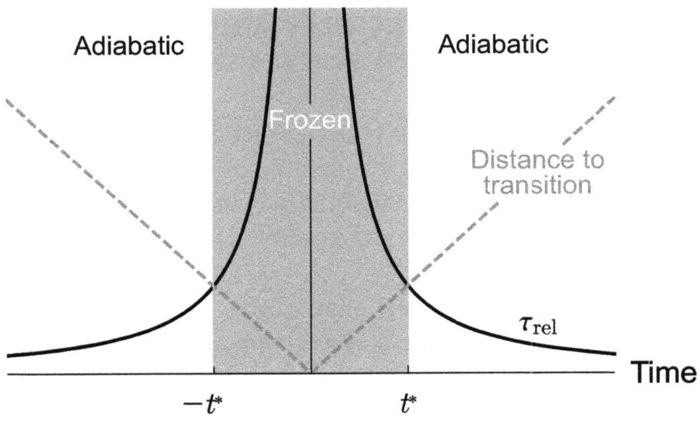

Figure 9.2 The Kibble–Zurek mechanism. The temperature during a quench changes approximately linearly close to a second-order phase transition, as indicated by the dashed line, while the relaxation time $\tau_{\rm rel}$ diverges (solid black line). When the relaxation time is small, the dynamics is adiabatic. For large relaxation times, on the other hand, the system cannot adjust to the changing bath temperature and the system dynamics is frozen. The corresponding freezing time t^* determines the size of domains and the defect density after the quench.

lower temperature, and remain. A similar argument can be made for disordering perturbations introduced below the critical temperature. At the time $|t| = t^*$, the relaxation time of the system equals the time it will take the bath to reach the transition temperature, so that any thermal fluctuations present at that point will not dissipate before being stabilised by the changing bath temperature. With the approximation of Eq. (9.38), this means

$$\tau_{\rm rel}(T(t^*)) = t^*. \tag{9.39}$$

This is a self-consistent equation, in which we explicitly wrote the dependence of the relaxation time $\tau_{\rm rel}$ on temperature, which in turn depends on the time. Rewriting Eq. (9.38) as $|t|/\tau_Q = |T - T_{\rm c}|$ and using $\tau_{\rm rel} \sim |T - T_{\rm c}|^{-z\nu}$, we find the relation between the freezing and quench times:

$$t^* \sim (\tau_Q)^{\frac{z\nu}{1+z\nu}}. \tag{9.40}$$

Because the system freezes at time $t = -t^*$, the system at $t = 0$, and even at $t = t^*$, will not be ordered over lengths longer range than the correlation length at the freezing time. Because the correlation length and the relaxation time are related by Eq. (9.30), this means the maximum size of ordered regions is given by

$$\xi^* \sim (\tau_Q)^{\frac{\nu}{1+z\nu}}. \tag{9.41}$$

At the point of exiting the frozen regime and beginning to equilibrate with a bath colder than the critical temperature, the system is thus not long-range ordered, but rather consists of domains with a typical size of order ξ^*. Continuing the dynamics, the system will be able to equilibrate with the bath again and evolve adiabatically to ever lower temperatures. The domains that were present right after the frozen period will then move, merge, and change size. Typically, the correlation length changes during this process. Any topological defects that were present at the touching points between differently ordered domains, however, will be protected. For example, if the phase transition corresponds to a $U(1)$ phase degree of freedom being broken in two dimensions, there could be a point where three domains meet in such a way that the order parameters define a non-zero winding number around the meeting point, as explained in Section 7.1. This vortex is then topologically protected and cannot be removed by any further cooling of the system. The density of topological defects of dimension D expected in a d-dimensional system after cooling it through a continuous phase transition can thus be expressed as

$$cn \sim (\xi^*)^{D-d} \sim (\tau_Q)^{(D-d)\frac{\nu}{1+z\nu}}. \tag{9.42}$$

This is the main prediction of the **Kibble–Zurek mechanism**. Using it, the density of observed cosmological defects can tell us the rate at which universe-wide symmetry breaking events occurred. Similarly, quench experiments on superfluid helium show a vortex-density consistent with Eq. (9.42). Ref. [91] gives a review of dynamical symmetry breaking across cosmological and condensed matter settings.

9.4 Discontinuous Phase Transitions: Nucleation

When quenching through a discontinuous rather than a continuous phase transition, there is no diverging length scale and hence no appearance of frozen domains. On the contrary, following the generic evolution of the Landau free energy depicted in Figure 9.3, the symmetric phase is found to be metastable even below the critical temperature, allowing the system to be supercooled and remain fully symmetric. Similarly, the symmetry-broken phase can be kept stable above its melting temperature, resulting in superheating. The dynamics of discontinuous phase transitions is therefore dominated by the evolution out of metastable configurations.

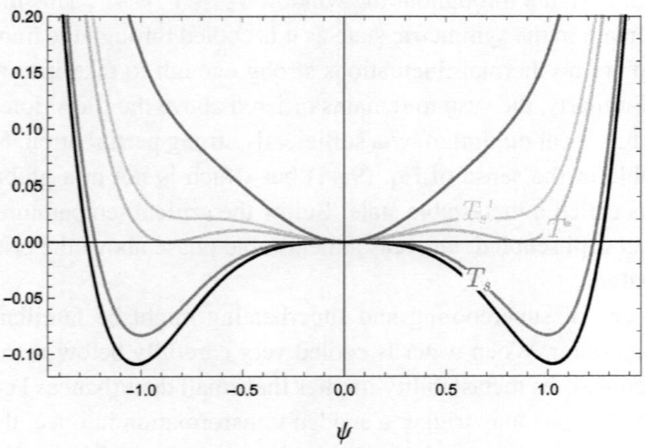

Figure 9.3

The generic form of the free energy for a discontinuous phase transition, following Eq. (9.43), at various temperatures: the critical temperature T^*, the spinodal temperatures T_S and T_S', and the high- and low-temperature regimes.

9.4.1 Metastable States

The occurrence of **metastable states** in the neighbourhood of a discontinuous phase transition can be understood from the generic form of a Landau free energy in which the coefficient of the fourth order term is negative. This necessitates expanding to sixth order, so that, for example, the free energy of a scalar order parameter ψ becomes

$$\mathcal{F}[\psi] = \frac{1}{2} r_0 (T - T_s)\psi^2 - \frac{1}{4} u\psi^4 + \frac{1}{6} w\psi^6 + \dots . \tag{9.43}$$

Here, r_0, u and w are real and positive parameters, while T_s will shortly be identified as the spinodal temperature. The form of the free energy for different temperatures is shown in Figure 9.3. A state is called **stable** if it sits at a local minimum of the free energy. Because Eq. (9.43) depends only on the order parameter, stable states are defined by

$$\frac{\partial \mathcal{F}[\psi]}{\partial \psi} = 0 \quad \text{and} \quad \frac{\partial^2 \mathcal{F}[\psi]}{\partial \psi^2} > 0. \tag{9.44}$$

At any temperature, $\psi_0 = 0$ is an extremum, but it is a minimum only for $T > T_s$. On the other hand, for all $T < T_s + u^2/(4r_0 w) \equiv T_s'$, there exists a set of local minima defined by

$$\psi_\pm^2 = \frac{u \pm \sqrt{u^2 - 4r_0 w(T - T_s)}}{2w} = \frac{u \pm 2\sqrt{r_0 w(T_s' - T)}}{2w}. \tag{9.45}$$

This means that in the temperature regime $T_s < T < T_s'$, the free energy of Eq. (9.43) has three local minima and hence three stable solutions. Not all solutions result in the same value for the free energy, however, and either the symmetric solution ψ_0 or the symmetry-broken solutions $\psi_\pm$ correspond to true ground states at a global minimum of the free energy.

At the critical temperature $T^* = T_s + 3u^2/(16 r_0 w)$, the symmetric and symmetry-broken solutions have exactly the same free energy. A system in perfect thermal equilibrium will therefore be symmetric for $T > T^*$, while it has a broken symmetry for $T < T^*$. However, the broken-symmetry minima are stable even in the regime $T^* < T < T_s'$ and similarly the symmetric minimum is stable throughout the window $T_s < T < T^*$. This makes it possible for the system to remain in the symmetric state as it is cooled through the transition temperature, as long as there are no thermal fluctuations strong enough to take it to a symmetry-broken configuration. Similarly, the system remains ordered above the transition temperature unless it is taken out of its local minimum by a sufficiently strong perturbation. More generally, any state that is stable in the sense of Eq. (9.44) but which is not at a global minimum of the free energy, it is called a metastable state. Below the critical temperature, metastable states are referred to as **supercooled**, whereas a metastable phase above the critical temperature is called **superheated**.

The phenomena of supercooling and superheating might be familiar from their occurrence in ordinary water. When water is cooled very carefully below zero degrees Celsius, it becomes supercooled. Its metastability implies that small disturbances keep the water liquid, while large perturbations may trigger a sudden transformation into ice, the stable crystalline phase of water. In fact, most of the water in the universe, in interstellar space, comets or distant cold moons, exists in the form of supercooled liquids. Superheated water behaves similarly, as can become painfully clear when heating water in a microwave. The superheated water is a liquid with a temperature higher than 100°C, in which a sufficiently strong perturbation can induce a sudden and violent boiling.

Exercise 9.11 (Hysteresis).
Besides the example of Eq. (9.43), a discontinuous phase transition also generically occurs when applying a symmetry-breaking field to systems that would otherwise have a continuous phase transition. For example, an Ising magnet at $T < T_c$ in an external magnetic field h is described by the Landau free energy $\mathcal{F}[m] = -\frac{r}{2}m^2 + \frac{u}{4}m^4 - hm$, with r and u real and positive. Whenever $h \neq 0$, there is a single global minimum of the free energy.
a. What is the value of the symmetry-breaking field h above which the global minimum is the only stable state? We will call this value h_{max}.
b. Imagine we initialise the system with $h = h_0 > h_{\text{max}}$. If we slowly change the strength of the field to $-h_0$, and then back to h_0, how does the magnetisation m evolve as a function of the field h?

Notice that going from $+h_0$ to $-h_0$ yields a different magnetisation curve $m(h)$ than going from $-h_0$ to $+h_0$. This phenomenon is known as **hysteresis**, and is a hallmark of metastable states surrounding a discontinuous phase transition.

9.4.2 Spinodal Decomposition

The supercooled symmetric state and the superheated symmetry-broken phases in the free energy of Eq. (9.43) are not present for all temperatures. As the sign of the second derivative of the free energy changes, each of these metastable states reaches its so-called limit of stability. The temperature at which this happens is called the **spinodal temperature**. Examining the shape of the free energy around a local minimum as the temperature evolves through the associated spinodal point, we find that it looks remarkably like the evolution of the global minimum in a continuous phase transition. Decreasing the temperature below the spinodal temperature while the system is in a supercooled state thus results in the same type of domain formation as discussed in Section 9.3, which in this case is referred to as the **spinodal decomposition** of the symmetric state. Like the evolution in a continuous phase transition, the spinodal decomposition is characterised by a diverging length scale and critical exponents. To distinguish these exponents from the ones occurring at a true continuous phase transition, they are sometimes called pseudocritical exponents.

Upon approaching the spinodal temperature, the metastable phase becomes increasingly fragile. The closer to the spinodal point, the easier it becomes to perturb the system and make it rapidly evolve into the stable phase. There is thus a practical superheating or supercooling limit, determined by dynamic properties, that prevents metastable configurations ever reaching the spinodal point in practice.

Exercise 9.12 (Spinodal).
Using the free energy of Eq. (9.43) with an additional source term $-h\psi$, show that upon approaching the spinodal temperature T_s from the high-temperature symmetric phase, the susceptibility $\chi = \left.\frac{\partial \psi}{\partial h}\right|_{h=0}$ diverges.

9.4.3 Homogeneous Nucleation

In practice, discontinuous phase transitions are triggered by the dynamical evolution of a small patch of stable phase within a bulk metastable state, in a process called **nucleation**. In the presence of a thermal bath or external noise, the order parameter can fluctuate and thus spontaneously create a small region of the stable phase within the bulk metastable configuration. The energy cost, or work W, required to generate such a bubble depends on its size and can be written as

$$W(r) = \sigma 4\pi r^2 + (\Delta\mathcal{F})\frac{4}{3}\pi r^3. \tag{9.46}$$

Here, r is the bubble's radius, $\Delta\mathcal{F} < 0$ is the free energy difference between the homogeneous stable and metastable phases, while $\sigma > 0$ is a parameter called the surface tension, which depends on the details of the interface between the two phases. We assume that the bubble is perfectly spherical and that the interface term depends only on the surface area.

Because of the energy cost of the interface, creating small bubbles generically requires net positive work. They can be spontaneously generated by thermal or external fluctuations, but quickly relax back to the bulk metastable phase. For bubbles with large radii, however, the free energy gain can overcome the surface tension. Consequently, there exists a critical nucleus with size r^*, whose creation requires maximal work:

$$r^* = \frac{2\sigma}{|\Delta\mathcal{F}|} \quad \text{with} \quad W(r^*) = \frac{16\pi\sigma^3}{3(\Delta\mathcal{F})^2}. \tag{9.47}$$

For a bubble with radius larger than r^*, the energy of the system will be reduced by increasing the bubble size, rather than decreasing it. This causes the process of **homogeneous nucleation**: the spontaneous formation of a nucleus larger than the critical size, which subsequently grows until the whole system is in the stable phase.

In the so-called **classical theory of nucleation**, nucleation is treated as an activated process with a nucleation rate J proportional to the Boltzmann weight of creating a critical nucleus:

$$J = J_0 e^{-W(r^*)/k_B T}. \tag{9.48}$$

The value of the prefactor J_0 depends on details of the phases and dynamics involved in the nucleation process. Ref. [44] provides an excellent in-depth introduction to the physics of metastable phases and homogeneous nucleation.

9.4.4 Heterogeneous Nucleation

The spontaneous appearance of critical bubbles due to thermal fluctuations is in practice quite rare as compared to nucleation of bubbles at the defects and surfaces that are present in any generic, finite system. This latter process is called **heterogeneous nucleation**, and it is easily observed, for example, when boiling water, as most vapor bubbles originate at the water–pot interface.

Heterogeneous nucleation is more common than the homogeneous variety because the work required to form a critical nucleus next to a surface or defect is decreased by their presence. For example, the work required to create a critical bubble that sticks to a surface with contact angle θ is given by [44]:

$$W(r^*) = \frac{16\pi\sigma^3}{3(\Delta\mathcal{F})^2} f(\theta). \tag{9.49}$$

Here, $f(\theta) = \frac{1}{4}(2 - 3\cos\theta - \cos^3\theta)$ is the geometric factor describing the reduction of interface area compared to the bubble volume. Since the nucleation rate of Eq. (9.48) is an exponential function of the work required to generate a critical bubble, even a small reduction of $W(r^*)$ due to the presence of defects or surfaces will cause first-order transitions to be dominated by heterogeneous nucleation in practice.

PART III

APPLICATIONS

Effective Field Theory 10

In this chapter, we discuss several of the most relevant aspects of spontaneous symmetry breaking in **effective field theory**. The methodology of effective field theory is central to modern particle physics and represents a large and still active field of research. We cannot possibly cover all of its facets here in a single chapter, nor hope to achieve the same depth and rigour that can be found in dedicated books like Refs. [32], [41], or [215]. Instead, we will outline in this chapter some of the methods that relate directly to spontaneous symmetry breaking and provide references to sources for further study. We also introduce several concepts that will be used in other contexts later, in particular in Chapters 11 and 13.

10.1 Constructing an Effective Field Theory

A continuum field theory, with continuous real space coordinates x and continuous fields $\phi(x)$, is mathematically well defined from infinitesimal to infinite distances and field values. But does that make sense physically? Indeed, time and again in the history of physics, theories have proven valid only in a limited domain, to be superseded by other theories that capture nature beyond that domain. The Coulomb potential of a point charge in classical physics, for example, diverges as the location of the point charge is approached, signalling the need for quantum mechanics to take over. This general pattern persists in quantum field theory: energy scales at very small distances can become very large, and we can never be sure that there won't be some new physics rearing its head at those scales.

Taking this point seriously, it is wise to always regard one's theory as having a limited domain of validity, and in particular a maximum energy scale beyond which the theory breaks down. It also means, however, that there is usually a domain of validity for theories that only consider physical effects up to some energy scale. This fact makes the separation of scales in physics enormously beneficial. High-energy physics can typically only have a non-qualitative, parametric influence on low-energy physics, so that to good approximation we need only concern ourselves with the latter. A well-known example is the Born–Oppenheimer approximation: an interacting system of heavy and light particles may be effectively described in terms of two subsystems that only influence one another through either their average or instantaneous properties. Similarly, field theorists often 'integrate out' heavy degrees of freedom from the path integral.

In many situations, we are interested only in a low-energy regime, which governs the long-range physics, and not in microscopic details. This is also typically the regime accessible to experiments. As a consequence, even if we are provided with a complete, microscopic theory valid up to very high energy scales, it is often more convenient to derive and work with an effective description of physics at low energies, which incorporates the influence of high-energy effects on the low-energy observables without having to keep track explicitly of high-energy degrees of freedom. This is the idea of formulating a **low-energy effective theory**.

The low-energy effective theory is especially relevant to systems with a spontaneously broken symmetry, for which Goldstone's theorem guarantees the existence of gapless Nambu–Goldstone modes. Since these modes persist to arbitrarily low energies, they are guaranteed to feature in any type of low-energy description. Moreover, in systems where all other excitations are gapped, the Nambu–Goldstone modes form the only constituents of the low-energy effective theory.

10.1.1 The Effective Action

Before turning to the construction of an explicit low-energy theory, we will give a brief general review of its central ingredients. Recall that in statistical physics the partition function $Z = \text{Tr}\ \exp(-\hat{H}/k_B T)$ captures the complete equilibrium physics. If you can actually perform the sum over all configurations, you have 'solved the system exactly'. This is equivalent to diagonalising the Hamiltonian in quantum mechanics.

Similarly, in a (relativistic) quantum field theory with action functional $S[\phi]$, the partition function in the path integral formulation can be defined as

$$Z = \int \mathcal{D}\phi\ \exp\left[\frac{\mathrm{i}}{\hbar}S[\phi]\right]. \tag{10.1}$$

Usually we are interested in n-point correlation functions, written in the path integral formalism as

$$\langle\phi_1(x_1)\cdots\phi_n(x_n)\rangle \equiv \frac{1}{Z}\int \mathcal{D}\phi\ \phi(x_1)\cdots\phi(x_n)\exp\left[\frac{\mathrm{i}}{\hbar}S[\phi]\right]. \tag{10.2}$$

These correlation functions can be calculated using a method that starts by adding an external, classical source term $J(x)$ to the action. The partition function now becomes a functional of this source:

$$Z[J] = \int \mathcal{D}\phi\ \exp\left[\frac{\mathrm{i}}{\hbar}S[\phi] + \frac{\mathrm{i}}{\hbar}\int \mathrm{d}^4x\ J(x)\phi(x)\right]. \tag{10.3}$$

The $Z[J]$ is often called the **generating functional** for correlation functions because any arbitrary n-point correlation function of the form of Eq. (10.2) can be expressed in terms of its functional derivatives in the limit of vanishing source strength:

$$\langle\phi_1(x_1)\cdots\phi_n(x_n)\rangle = \frac{1}{Z[J]}(-\mathrm{i}\hbar)^n\frac{\delta}{\delta J(x_1)}\cdots\frac{\delta}{\delta J(x_n)}Z[J]\bigg|_{J=0}. \tag{10.4}$$

There exist several more generating functionals that provide convenient ways of calculating particular quantities. One that is particularly interesting is the generator $W[J]$ of the connected correlation functions, which we briefly encountered before in Section 3.7. The two-point connected correlation function, for example, equals $\langle\phi_1(x_1)\phi_2(x_2)\rangle_{\text{connected}} \equiv \langle\phi_1(x_1)\phi_2(x_2)\rangle - \langle\phi_1(x_1)\rangle\langle\phi_2(x_2)\rangle$. The generator for connected correlation functions is defined by

$$W[J] = (-i\hbar)\log Z[J]. \tag{10.5}$$

With this definition, the connected n-point correlation functions become [35, 215]

$$\langle \phi_1(x_1) \cdots \phi_n(x_n) \rangle_{\text{connected}} = (-\mathrm{i}\hbar)^{n-1} \frac{\delta}{\delta J(x_1)} \cdots \frac{\delta}{\delta J(x_n)} W[J] \bigg|_{J=0}. \tag{10.6}$$

10.1.1.1 Quantum Effective Action

Given the generator $W[J(x)]$, one can define its Legendre transform $\Gamma[\varphi(x)]$ as

$$\Gamma[\varphi] = W[J] - \int \mathrm{d}^4 x J(x) \varphi(x). \tag{10.7}$$

In this expression, the newly introduced classical (c-number) field $\varphi(x)$ is taken to be the expectation value of the quantum field $\phi(x)$ entering the W generating functional: $\varphi(x) \equiv \langle \phi(x) \rangle = \delta W[J]/\delta J(x)$. Similarly, the value of the source term causing a particular configuration $\varphi(x)$ is found, using the chain rule, to be given by the functional derivative $-\delta\Gamma[\varphi]/\delta\varphi(x) = J(x)$. The field φ replacing J as the dynamical variable is analogous to the canonical momentum replacing velocity in the Legendre transform of a Hamiltonian to a Lagrangian.

The functional $\Gamma[\varphi]$ is known as the **quantum effective action**, or simply the **quantum action**. To see why, notice that the ground state expectation value of the field ϕ subject to the action $S[\phi]$ is realised in the absence of any external sources. It therefore corresponds to the configuration φ_0 that minimises the quantum action: $\delta\Gamma/\delta\varphi|_{\varphi=\varphi_0} = 0$. In this sense, the quantum action Γ plays a similar role as the usual action S, whose classical configuration is determined by the Euler–Lagrange equation $\delta S/\delta\phi = 0$. The difference is that the configuration φ_0 is the expectation value of the field ϕ in the exact ground state of the full quantum theory, rather than a configuration minimising only the classical potential.

To see how the quantum action can itself be used as a generating functional, we explicitly consider some of its derivatives. We already know that the first derivative $\delta\Gamma/\delta\varphi$ equals $-J(x)$. You will prove in Exercise 10.1 that the second derivative $\delta^2\Gamma/\delta\varphi(x)\delta\varphi(y)$ is equal to $G^{-1}(x,y)$, where G^{-1} is the inverse of the propagator $G(x,y) = -(i/\hbar)\langle\phi(x)\phi(y)\rangle_{\text{connected}}$ for the ϕ fields in the absence of any sources. Notice that this is the full propagator, including all connected diagrams, rather than only the classical part generated by the second derivative of the usual action.

To understand the meaning of higher derivatives of the quantum action, we can compare their expressions with those of the generating functional for connected diagrams. After some manipulations, we find, for example, the expression at third order:

$$\begin{aligned} \frac{\delta}{\delta J(x_1)} \frac{\delta}{\delta J(x_2)} \frac{\delta}{\delta J(x_3)} W[J] \bigg|_{J=0} &= - \int dy_1 dy_2 dy_3 G(x_1,y_1) G(x_2,y_2) G(x_3,y_3) \\ &\quad \times \frac{\delta}{\delta\varphi(y_1)} \frac{\delta}{\delta\varphi(y_2)} \frac{\delta}{\delta\varphi(y_3)} \Gamma[\varphi]. \end{aligned} \tag{10.8}$$

Here, we used $G(x,y) = -\delta^2 W[J]/\delta J(x)\delta J(y)|_{J=0} = -\delta\varphi(y)/\delta J(x)|_{J=0}$. From Eq. (10.6), we know that the left-hand side of Eq. (10.8) represents the sum of all connected diagrams with three external legs. The right-hand side on the other hand represents all diagrams ending in three fully summed two-point correlators (propagators), connected by the factor $\delta^3\Gamma/\delta\varphi^3$. That factor must then represent the sum of all one-particle irreducible (1PI) diagrams connecting to three legs, because any one-particle reducible contributions (that can be split in two by cutting a single line) will have already been included in the combination of irreducible diagrams and fully summed two-point correlators. This result readily generalises to higher derivatives, where $\Gamma[\varphi]$ is found to be the generator for general one-particle irreducible diagrams:

$$\langle \phi_1(x_1)\cdots\phi_n(x_n)\rangle_{1\text{PI}} = \frac{\mathrm{i}}{\hbar}\frac{\delta}{\delta\varphi(x_1)}\cdots\frac{\delta}{\delta\varphi(x_n)}\Gamma[\varphi], \qquad n \geq 3. \tag{10.9}$$

The power of the quantum action as a generating functional is especially apparent in scattering theory, where 1PI correlation functions are the prime objects of interest when calculating various cross sections (see for instance Refs. [35], [186], and [215]). These are quite complicated to calculate directly from $Z[J]$ or $W[J]$. Knowledge of $\Gamma[\varphi]$, on the other hand, allows straightforward identification of 1PI correlation functions as tree-level terms in its perturbation expansion. More generally, notice that any connected diagram can be constructed by concatenating 1PI diagrams, enabling the calculation of any physical response function from knowledge of the quantum effective action.

Exercise 10.1 (Inverse propagator from the quantum action).
To prove that the second derivative of the quantum action is proportional to the inverse propagator, we start by noticing that its definition in Eq. (10.7) yields the relation $-\delta\Gamma[\varphi]/\delta\varphi(x) = J(x)$.
a. Argue that this relation implies for the second functional derivative,

$$-\frac{\delta^2\Gamma[\varphi]}{\delta J(y)\delta\varphi(x)} = \delta(x-y). \tag{10.10}$$

b. Use the chain rule for functional derivatives and the relation between the generationing functional $W[J]$ and expectation values of the field ϕ to write this expression as

$$-\int \mathrm{d}^4 z\, \frac{\delta^2 W[J]}{\delta J(y)\delta J(z)}\frac{\delta^2\Gamma[\varphi]}{\delta\varphi(z)\delta\varphi(x)} = \delta(x-y). \tag{10.11}$$

c. Using the definition of the propagator as the connected two-point correlation function, and assuming translational invariance to write its Fourier transform as $G(y,z) \propto \int \mathrm{d}^4 p\, \mathrm{e}^{ip(y-z)}G(p)$, show that

$$\int \mathrm{d}^4 p \int \mathrm{d}^4 z\, \mathrm{e}^{ip(y-z)}G(p)\frac{\delta^2\Gamma[\varphi]}{\delta\varphi(z)\delta\varphi(x)} = \int \mathrm{d}^4 p\, \mathrm{e}^{-ip(x-y)}. \tag{10.12}$$

d. Finally, use translational invariance again to write the Fourier transform of $F(z,x) \equiv \delta^2\Gamma/(\delta\varphi(z)\delta\varphi(x))$ as $F(z,x) \propto \int \mathrm{d}^4 p\, \mathrm{e}^{ip(z-x)}F(p)$, to find

$$G(p)F(p) = 1\ \forall p. \tag{10.13}$$

This result shows that $F(p) = G^{-1}(p)$ and establishes the relation between the quantum action and the full propagator, including all connected diagrams.

Exercise 10.2 (Quantum action of a relativistic scalar field).
Consider a non-interacting real scalar relativistic field theory given by the action

$$S[\phi] = \int \mathrm{d}^d x \frac{1}{2}\left(-\partial_\mu\phi\partial^\mu\phi - m^2\phi^2\right). \tag{10.14}$$

a. Introduce a source field J as defined in Eq. (10.3), then Fourier transform and integrate out ϕ to find the explicit expression for the partition function:

$$Z[J] = \mathrm{e}^{-\frac{i}{2\hbar}\int \frac{\mathrm{d}^d p}{(2\pi)^d} J_p^* \frac{1}{p^2+m^2} J_p}. \tag{10.15}$$

b. Use the relation $\varphi(x) = \delta W[J]/\delta J(x)$ to show that $\varphi(x)$ can be written as

$$\varphi(x) = -\int \frac{\mathrm{d}^d p}{(2\pi)^d} \frac{e^{ipx}}{p^2+m^2} J_p^* \equiv -\int \mathrm{d}y\, G(x,y) J(y). \tag{10.16}$$

c. By inverting Eq. (10.16), write the source field as $J(x) = -\int \mathrm{d}y\, G^{-1}(x,y)\varphi(y)$.
d. Write out explicitly the quantum effective action $\Gamma[\varphi]$ and verify that its second derivative equals the inverse propagator $G^{-1}(x,y)$.

10.1.1.2 Thermal Effective Action

A quantity similar to the quantum effective action can be constructed by formulating the path integral appearing in the definition of the partition function of Eq. (10.1) in Euclidean time $\tau = it$, as we did before in Section 9.1.2:

$$Z = \mathrm{Tr}\, \exp(-\hat{H}/k_{\mathrm{B}}T) = \int \mathcal{D}\phi\, \exp\left[-\frac{1}{\hbar} S_{\mathrm{E}}[\phi]\right]. \tag{10.17}$$

Here, $S_{\mathrm{E}}[\phi] = \int_0^{\hbar\beta} \mathrm{d}\tau\, \mathcal{L}[\phi(\tau)]$ is the **Euclidean action**, describing a thermal quantum field theory with $\beta = 1/k_{\mathrm{B}}T$ the inverse temperature. We can again add a source term to the Euclidean action by writing

$$Z[J] = \int \mathcal{D}\phi\, \exp\left[-\frac{1}{\hbar} S_{\mathrm{E}}[\phi] - \frac{1}{\hbar}\int_0^{\hbar\beta} \mathrm{d}\tau \int \mathrm{d}^3x\, J(x,\tau)\phi(x,\tau)\right]. \tag{10.18}$$

As before, $Z[J]$ acts as a generating functional, this time for thermal correlation functions. The generator of connected thermal correlation functions is the free energy functional $\mathcal{F}[J] = -(1/\beta)\log Z[J]$, which is the thermal version of the generator $W[J]$ defined in Eq. (10.5). Through a Legendre transform, we find the **thermal effective action**:

$$\Gamma_{\mathrm{T}}[\varphi] = \hbar\beta\mathcal{F}[J] - \int_0^{\hbar\beta} \mathrm{d}\tau \int \mathrm{d}^3x\, J(x,\tau)\varphi(x,\tau). \tag{10.19}$$

The thermal effective action plays the same role as the quantum effective action of Eq. (10.7), but now for thermal quantum field theories. In addition, however, the thermal effective action $\Gamma_{\mathrm{T}}[\varphi]$ is a generalisation of the free energy. Whereas the usual free energy without source terms, $\mathcal{F}[J = 0]$, generates thermodynamic quantities such as entropy, specific heat, and so forth, the thermal effective action also yields correlation functions by writing the free energy as a functional of the field φ. We actually already encountered an example of the free energy written in this form before, when we discussed the Ginzburg–Landau free energy functional for charged superfluids in Chapter 8. Another example is the thermal effective action for the Higgs boson in a thermal universe, which we will encounter in Section 14.1.

10.1.1.3 Anomalies

An important difference between the usual action $S[\phi]$ and the quantum action $\Gamma[\varphi]$ is that the two do not necessarily have the same symmetries. At first sight, this is a surprising result. After all, in Chapter 1, we introduced symmetries as transformations on the fields that leave the action invariant. There, we silently assumed that the action determines the values of all physical observables, so that a symmetry of the action has no influence on any measurable quantities. The actual quantity predicting observable expectation values, however, is the partition function. In a quantum field theory, the summation over all possible configurations appearing in the partition function is written in terms of a path integral, and it is exactly this integration that can have symmetries different than the action itself. Since the quantum effective action derives from the partition function, its symmetries are constrained by the path integral as well as the original action. In particular, if there exists a symmetry of the usual action that is not a symmetry of the quantum action, the quantum field theory is said to be **anomalous**. Hand-waving, this is sometimes described as 'quantum corrections breaking the classical symmetry'. More precisely, whenever a symmetry of the action $S[\phi]$ is not a symmetry of the path integral measure $\mathcal{D}\phi$, an anomaly is present.

Given the existence of anomalies, you might wonder which action we should consider when talking about the symmetries of any physical system. To see this, recall that minimising the quantum effective action $\Gamma[\varphi]$ gives the exact ground state expectation values of the full quantum theory, while minimising $S[\phi]$ yields only the classical equations of motion. The solutions derived using the quantum action reflect the actual symmetries of the problem. The classical equations of motion are only an approximation to the solutions of the underlying quantum problem, and so are its symmetries. The presence of anomalies in a quantum field theory therefore does not constitute symmetry breaking, in the sense that solutions of the quantum effective action respect the symmetries of the full quantum problem.

A famous example of an anomaly occurring in particle physics is the **chiral anomaly**. As we will see in Section 11.2, chiral symmetry is responsible for the separate conservation of the number of left- and right-handed massless Dirac fermions. At the level of the action $S[\phi]$, the massless Dirac equation not only conserves the total number of left- and right-handed fermions $n_L + n_R$ but also their difference $n_L - n_R$. However, correctly calculating expectation values of correlation functions using the path integral indicates that some processes forbidden by chiral symmetry actually occur with a non-zero rate. This was first discovered using a one-loop calculation of the neutral pion decay, where a so-called triangular Feynman diagram shows that the difference $n_L - n_R$ is not conserved [2, 24]. The result of this calculation can be understood in terms of topological excitations of the Dirac vacuum called instantons. Alternatively, one can view the chiral anomaly as arising from the impossibility of regularising the massless Dirac equation [180]. Either way, the anomaly arises from the fact that the path integral measure is not invariant under the chiral symmetry.

The chiral anomaly also occurs in the context of topological condensed matter. Both in Weyl semimetals and topological insulators, it can be formulated in terms of topological surface states [54]. Finally, the chiral anomaly plays a role in theories of early universe baryogenesis, which we will discuss in Section 14.1.

For the study of spontaneous symmetry breaking, the realisation that anomalies may occur in quantum field theories signals the need to consider symmetries of the full quantum effective action, rather than those of the classical action.

10.1.1.4 Effective Potential

Going beyond the expectation values for field variables found by minimising the quantum effective action without source terms present, we can consider the generalisation to the case with source terms. A relation between the quantities Γ and φ and the original quantities S and ϕ is then given by the De Witt identity [47]:

$$\frac{\delta\Gamma[\varphi]}{\delta\varphi(x)} = \left\langle \frac{\delta S}{\delta\phi(x)} \right\rangle_{J=J[\varphi]}. \tag{10.20}$$

The quantum effective action can be expanded in powers of the field $\varphi(x)$ and its derivatives. In the special, but not unusual, case that the expectation value $\varphi(x)$ is constant in both space and time, all derivative terms in the expansion of Γ vanish, and we can write it as $\Gamma(\bar{\varphi}) = -\hbar \int d^4x\, V_{\text{eff}}(\bar{\varphi})$. Here, $\bar{\varphi}$ indicates the constant value of the field, of which the **effective potential** V_{eff} is simply a function (not a functional). If the ground state expectation value for ϕ is constant, its value lies in the minimum of this potential. We can similarly define the **thermal effective potential** starting from the thermal effective action.

Considering fluctuations around a constant configuration, the expansion of the quantum effective action can be written as

$$\Gamma[\varphi] = \int \mathrm{d}^4x \left[-V_{\text{eff}}(\varphi) + \frac{\hbar^2}{2m}(\partial_\mu\varphi)(\partial^\mu\varphi) + \ldots \right]. \tag{10.21}$$

This form of the effective action amounts to a gradient expansion about constant fields $\bar{\varphi}$. Disregarding all terms beyond the second one is called the **local potential approximation** (LPA) and considering it often forms the first approach in practical calculations. The same approximation for thermal field theories yields the Ginzburg–Landau free energy functional. In the LPA, the quantum equation of motion reduces to $\partial_\mu\partial^\mu\varphi \propto -\partial V_{\text{eff}}/\partial\varphi$, which again gives a direct analogy between the quantum effective action and the usual action S generating classical equations of motion. As before, however, the solutions to the quantum equation of motion represent the full quantum dynamics of the expectation value, rather than only its classical approximation.

10.1.2 Goldstone's Theorem from the Quantum Action

The relevance of the effective potential to spontaneous symmetry breaking stems from the observation that the quantum effective action $\Gamma[\varphi]$ correctly captures the symmetries of the full quantum problem [35, 41, 159, 215]. Recall that if the field ϕ transforms non-trivially under a symmetry operation, the symmetry is spontaneously broken if the ground state expectation value of ϕ is nonzero. Since φ equals the expectation value of ϕ whenever the quantum equations of motion are satisfied, the condition for symmetry breaking can directly be formulated as a requirement on the fields that extremise the quantum effective action.

For the simplest case of translationally invariant systems with constant $\varphi(x) = \bar{\varphi}$, the effective potential is simply a function (not a functional) of the variable $\bar{\varphi}$. The ground states of the system are then those with extremal effective potential, and if they have a nonzero value of $\bar{\varphi}$, the symmetry is broken spontaneously. That is, the requirement for spontaneous symmetry breaking to occur is

$$\frac{\mathrm{d}V_{\text{eff}}(\bar{\varphi})}{\mathrm{d}\bar{\varphi}} = 0 \qquad \text{for } \bar{\varphi} \neq 0. \tag{10.22}$$

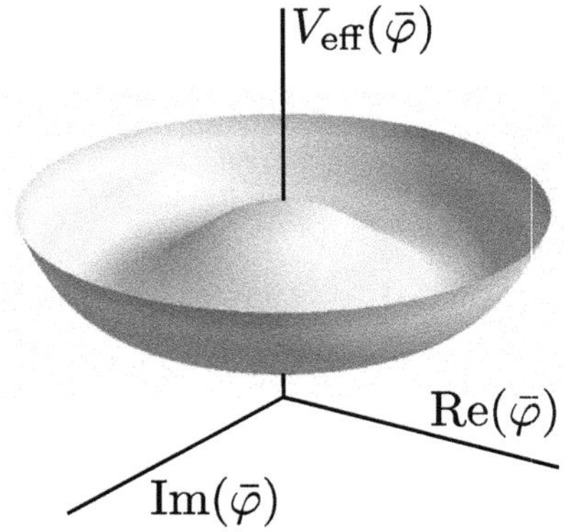

Figure 10.1 Effective potential for a scalar field spontaneously breaking $U(1)$ phase rotations.

This condition is central to the proof of Goldstone's theorem in terms of the quantum action, which you will encounter in Exercise 10.3.

To further explore the shape of $V_{\mathrm{eff}}(\varphi)$ in symmetry-broken systems, notice that because the quantum effective action has all symmetries of the full quantum problem, so does the effective potential in the presence of translational invariance. That is, if $\bar{\varphi}$ and $\bar{\varphi}'$ are related by a symmetry transformation, the values of the effective potential at these configurations must be equal. In particular, this also holds for the value of the effective potential at the minimum satisfying Eq. (10.22). For example, in the case of a single field spontaneously breaking $U(1)$ phase rotations, the effective potential at low temperatures will look like Figure 10.1. Any configuration along the bottom of the rim in this potential satisfies Eq. (10.22) and represents a possible symmetry-broken state. More generally, whenever a continuous global symmetry is spontaneously broken, an infinitesimal transformation on the symmetry-broken state characterised by $\bar{\varphi}$ and satisfying Eq. (10.22) takes it to a distinct configuration $\bar{\varphi} + \delta\bar{\varphi}$, which also satisfies Eq. (10.22) and has the same value for the effective potential.

Having reached this point, you might notice that the effective potential and quantum action bear great similarity to the free energy functionals encountered in the Ginzburg–Landau description of phase transitions. This is no coincidence, since we already saw in Eq. (10.19) that the thermal effective action expresses the free energy as a function of the field φ. Moreover, Eq. (10.22) indicates that the field $\bar{\varphi}$ can be used as an order parameter for the symmetry-breaking state, and V_{eff} becomes precisely the uniform and constant part of the Ginzburg–Landau free energy. This correspondence implies both that the quantum action can be used to derive coefficients for the Ginzburg–Landau description of any phase transition, and conversely, that the techniques developed for Ginzburg–Landau theory before the widespread use of quantum field theory can be employed to formulate approximations to the quantum effective action.

10.1.2.1 Excitations

Because the effective potential is constant along a path of $\bar{\varphi}$-values connected by continuous symmetry transformations, there is guaranteed to be a direction in which the potential $V_{\mathrm{eff}}(\bar{\varphi})$ is flat around any symmetry-broken state. Therefore, changing the field globally along that direction does not cost any energy. This result is an alternative formulation of the fact that the presence of a continuous symmetry ensures the existence of a continuous manifold of degenerate symmetry-broken states. Although global transformations within this manifold do not cost any energy, they should not be mistaken for Nambu–Goldstone modes.

The global transformations have strictly infinite wavelength. As argued at length in Section 4.1, Goldstone's theorem concerns bosonic modes approaching zero momentum, rather than strictly global transformations. Exercise 10.3 shows how the quantum effective action can be used to prove that a spontaneously broken continuous symmetry also implies the existence of modes at finite momenta, whose energy approaches zero as the momentum vanishes. Further applications of the quantum effective action in the context of spontaneous symmetry breaking are described in Project 21 and Section 10.3.

Exercise 10.3 (Goldstone's theorem from the quantum action).
Consider a system with a vectorial order parameter expectation value $\varphi(x)$, whose components are indicated by $\varphi_a(x)$. A global symmetry transformation may be defined by its action on the order parameter as $\varphi_a(x) \rightarrow \varphi_a(x) + (\delta_s\varphi)_a$. Because the transformation is a symmetry, the variation of the quantum effective action under this transformation must equal zero, so that $\delta\Gamma = \Gamma[\varphi(x) + \delta_s\varphi] - \Gamma[\varphi(x)] = 0$. In case the symmetry transformation is continuous, the vanishing of $\delta\Gamma$ can be written as

$$\sum_a \int \mathrm{d}^4x \, \frac{\delta\Gamma[\varphi]}{\delta\varphi_a(x)} (\delta_s\varphi)_a = 0. \tag{10.23}$$

Notice that the functional derivative $\delta\Gamma/\delta\varphi(x)$ produces a function of the initial state $\varphi(x)$ on which the transformation acts. The condition of Eq. (10.23) holds for any initial state.
a. For fields $\varphi(x)$ that are independent of x, the quantum effective action simplifies to the effective potential $V_{\text{eff}}(\varphi)$. Use this to show that Eq. (10.23) applied to the specific case of a global symmetry transformation on an initially homogeneous field $\varphi(x) = \bar{\varphi}$ yields

$$\sum_a \left. \frac{\partial V_{\text{eff}}(\varphi)}{\partial\varphi_a} \right|_{\varphi=\bar{\varphi}} (\delta_s\varphi)_a = 0. \tag{10.24}$$

b. The variation of the initial field $\bar{\varphi}$ can always be written in the form $(\delta_s\varphi)_a = \sum_b \Delta_{ab}\bar{\varphi}_b$ with Δ a matrix representation of the symmetry transformation. Show that taking the derivative $\partial/\partial\bar{\varphi}_c$ of Eq. (10.24) results in

$$\sum_{ab} \left. \frac{\partial^2 V_{\text{eff}}(\varphi)}{\partial\varphi_c \partial\varphi_a} \right|_{\varphi=\bar{\varphi}} \Delta_{ab}\bar{\varphi}_b + \sum_a \left. \frac{\partial V_{\text{eff}}(\varphi)}{\partial\varphi_a} \right|_{\varphi=\bar{\varphi}} \Delta_{ac} = 0. \tag{10.25}$$

The uniform ground states $\bar{\phi}$ of the model, including possible symmetry-broken ones, will sit at a minimum of the effective potential, so that

$$\left. \frac{\partial V_{\text{eff}}(\varphi)}{\partial\varphi_a} \right|_{\varphi=\bar{\phi}} = 0 \quad \text{for all } a. \tag{10.26}$$

For these ground states, we must therefore have

$$\sum_{ab} \left. \frac{\partial^2 V_{\text{eff}}(\varphi)}{\partial\varphi_c \partial\varphi_a} \right|_{\varphi=\bar{\phi}} \Delta_{ab}\bar{\phi}_b \equiv \sum_a m^2_{ca}(\bar{\phi})\,(\delta_s\bar{\phi})_a = 0. \tag{10.27}$$

c. Explain why a symmetric ground state must have $(\delta_s\bar{\phi})_a = 0$ for all a, and therefore automatically satisfies Eq. (10.27). Also argue that a symmetry-breaking ground state must have $(\delta_s\bar{\phi})_a \neq 0$ for at least one component and that this implies that the mass matrix $m^2(\bar{\phi})$ for any symmetry-breaking ground state $\bar{\phi}$ must have at least one zero eigenvalue.

Recall the result of Exercise 10.1, in which we showed that the inverse propagator $G^{-1}(x,y)$ for excitations on top of an initial state φ can be written in terms of the quantum effective action as

$$G^{-1}_{ab}(x-y) = \frac{\delta^2\Gamma[\varphi]}{\delta\varphi_a(x)\delta\varphi_b(y)}. \tag{10.28}$$

Here, we assumed translational invariance to write G^{-1} as function of the difference $x-y$ only.

d. Expressing the quantum effective action in the way of Eq. (10.21) show that the propagator for excitations on top of a homogeneous ground state can be written in momentum space as

$$\begin{aligned} G^{-1}_{ab}(p) &= -\left.\frac{\partial^2 V_{\text{eff}}(\varphi)}{\partial\varphi_a\partial\varphi_b}\right|_{\varphi=\bar{\phi}} + \mathcal{O}(p^2) \\ &= m^2_{ab}(\bar{\phi}) + \mathcal{O}(p^2). \end{aligned} \tag{10.29}$$

Notice that we silently assumed the theory not to have long-ranged interactions (mediated by gauge fields) in order for the Fourier transform underlying Eq. (10.29) to be well defined.

e. Explain how Eq. (10.29) implies there must be at least one well-defined mode whose inverse propagator approaches zero arbitrarily closely as p approaches zero.

Since the propagator $G(p)$ has a pole whenever its inverse is zero, Eq. (10.29) establishes Goldstone's theorem: a mode whose lifetime diverges as $p \to 0$ (i.e., energy p_0 and momentum $\vec{p}$ both approaching zero) necessarily exists in the presence of a spontaneously broken global symmetry.

10.1.3 Low-Energy Effective Theory

Having found that the quantum effective action is the object of interest for an effective low-energy description, the question remains how one can systematically obtain it. In fact, multiple approaches are possible, either starting from a known high-energy theory and finding out the effect of high-energy excitations on the low-energy degrees of freedom or starting from known symmetries and constraints and building up ever more detailed interactions between low and higher-energy degrees of freedom. We will briefly introduce typical approaches of both types.

10.1.3.1 Integrating Out Degrees of Freedom

The full path integral formulation for the partition function requires an integration over all degrees of freedom, including all the fields, with all their perturbations, at all frequencies and length scales. A simple intermediate step to consider is therefore to first integrate only part of the degrees of freedom. For example, in Section 14.1, we will be interested in the effective potential for the Higgs field, which in lowest order can be approximated by integrating out all other particles species (leptons, quarks, and vector bosons). Another example is the Gor'kov derivation of integrating out the fermionic degrees of freedom in a superconductor to obtain an effective theory in terms of the order parameter field [43].

This approach yields an effective description of the low-energy physics, but it is not the quantum effective action of Eq. (10.1.1) because it still contains microscopic degrees of freedom to be integrated over. For the Higgs field or superconductor, for example, even after integrating out some degrees of freedom, we are still left with a path integral over either the Higgs boson or the order parameter field.

10.1.3.2 Renormalisation Group

Instead of integrating out specific particles or fields, a more systematic way of arriving at a low-energy effective theory starting from the high-energy description was put forward by Wilson [223]. Suppose we have a momentum-space path integral $Z = \int \mathcal{D}\phi(p)\ \exp(\mathrm{i}S/\hbar)$, with an explicitly defined high-momentum cut-off Λ (which can be sent to infinity at the end of the calculation). We introduce a number b slightly larger than 1, such that we can separate the degrees of freedom into low-momentum $\phi_< = \{\phi(p < \Lambda/b)\}$ and high-momentum $\phi_> = \{\phi(\Lambda/b < p < \Lambda)\}$. The path integral can then be expressed as

$$Z = \int \mathcal{D}\phi_< \ \mathcal{D}\phi_> \ \mathrm{e}^{\frac{\mathrm{i}}{\hbar}(S_> + S_< + S_{\mathrm{int}})}. \tag{10.30}$$

Here, $S_>$ only depends on the high-momentum fields $\phi_>$, $S_<$ only depends on the low-momentum fields $\phi_<$, while S_{int} depends on both. We now want to carry out the path integral over $\phi_>$, integrating out the high-momentum degrees of freedom. On its own, $\int \mathcal{D}\phi_> \exp(\mathrm{i}S_>/\hbar)$ would just evaluate to some number, so the problem lies only in the term S_{int}. The main idea is that, if b is very close to 1, then the 'momentum shell' $\Lambda/b < p < \Lambda$ is very narrow, and the effects of S_{int} may be treated perturbatively [35, 58, 72, 230]. The result is a path integral over an action S' valid at scales $0 < p < \Lambda/b$, which looks like S, but with the values of its coefficients c_i' slightly modified from the values c_i in the original action. At that point, we can rescale the momentum cut-off back to its original value $\Lambda/b \to \Lambda$. This way we derived a new action with modified coupling constants but with the same momentum cut-off as before. This procedure can be repeated again and again.

Repeating the integration, the coefficients c_i become functions of the renormalisation parameter: $c_i = c_i(b)$, and we can track how they change as we integrate out more and more momentum shells. If $\partial_b c_i(b)$ is positive, the value of the coefficient becomes larger as we go to lower energy scales. If $\partial_b c_i(b)$ is negative, the coefficient shrinks. Finally, there may be a point where $\partial_b c_i(b) = 0$ for all couplings. At this point, the value of the coefficient c_i no longer changes under integrating out momentum shells, indicating a **fixed point**. This happens, for example, at a second-order phase transition where the correlation length ξ diverges and physics looks the same at all length scales, as discussed in Section 5.5. This procedure is called the momentum-shell or Wilsonian **renormalisation group** (RG), which was also discussed briefly in Sections 5.5 and 9.2. It constitutes a method for obtaining a low-energy effective theory in a controlled way, given any high-energy microscopic theory.

Spontaneous symmetry breaking only involves low-energy degrees of freedom. Determining whether or not a given high-energy system will spontaneously break a symmetry and what its physical consequences are, can therefore be done entirely within the renormalised low-energy effective theory. In particular, the renormalisation group flow does not allow the low-energy effective theory to break any symmetries that were present at high energies, thus guaranteeing that the symmetry-breaking pattern of the low-energy effective theory matches that of the full theory.

One exceptional circumstance to keep in mind, is that it is possible for the RG flow to generate low-energy terms that transform in the appropriate way under symmetry operations, but which cause the Lagrangian to change by a total derivative. Although the total derivative does not affect the (local) equations of motion, it can allow for (non-local) topological differences between ordered states. Whether such differences have measurable consequences depends on the boundary conditions imposed and may introduce anomalous contributions to conserved currents [35].

10.1.3.3 Relevance and Renormalisability

Originally, as the machinery of quantum field theory was first developed, terms appearing in a Lagrangian would be classified as **renormalisable** if the UV divergences they generated could be 'removed' or 'cancelled' by some renormalisation procedure, and **non-renormalisable** otherwise. In other words, non-renormalisable terms are problematic. Consequently, one stated that theories with only renormalisable terms are 'feasible', physical descriptions of nature, while non-renormalisable theories were to be discarded.

The perspective on renormalisation taken when constructing a low-energy effective theory seems at first sight almost entirely opposite [35, 163]. Since we are interested in the low-energy theory, it is considered not important what goes on at high energies. Terms with coefficients that decrease as we go to lower energies, corresponding to terms that increase towards higher energies and may diverge in the UV limit, are now called **irrelevant** since they contribute nothing to the low-energy region of interest. Conversely, in terms of what does contribute to the low energy physics, we need to only concern ourselves with the **relevant** terms that grow when going to lower energies and do not require renormalisation in the UV. This situation is summarised in Table 10.1.

The rationale in the effective theory approach is that irrelevant terms leading to UV divergences will be cancelled at those high-energy scales by some effects not yet taken into account. As long as this is possible in principle, the theory will be renormalisable. Because the high-energy contributions introduced in the renormalisation procedure do not contribute

Scaling dimension d and spacetime dimension D	Size as $E \to 0$	RG flow	Feasibility
$d < D$	Grows	Relevant	Superrenormalisable
$d = D$	Constant	Marginal	Strictly renormalisable
$d > D$	Falls	Irrelevant	Non-renormalisable

Table 10.1 Renormalisation nomenclature, following Ref. [163].

to the low-energy effective physics, there is no need to explicitly construct the renormalisation procedure. Identifying relevant terms in the action is thus complementary to establishing its renormalisability, and both are required for establishing the complete feasibility and features of a quantum field theory.

A first approach to determining the renormalisability and relevance of terms in the Lagrangian is by **power counting**. In natural units (where $\hbar = c = 1$), each dimensionful quantity corresponds to a certain power of mass dimension. In particular, a real scalar field ϕ in $D = 4$ has mass dimension $[\phi] = 1$ (see, e.g., [163]). The **scaling dimension** of a term is then the sum of the powers of mass dimension of fields in that term. For instance, with $[\phi] = 1$, the scaling dimension d of ϕ^ζ is given by $d = \zeta$. Coupling constants in a theory similarly have mass dimensions, which contribute to the scaling dimension of the associated terms in the Lagrangian. Comparing the scaling dimension d to the physical dimension D in which the theory is defined leads to Table 10.1, which indicates the relations between scaling dimensions, relevance, and renormalisability.

10.1.3.4 Effective Lagrangian by Construction

An alternative approach to the Wilsonian RG flow goes under the name of constructing the **effective Lagrangian**, which is also sometimes called the phenomenological Lagrangian or simply the effective field theory. A thorough treatment is given, for example, in Refs. [32] and [35]. The idea of constructing an effective Lagrangian rests on the observation that the low-energy theories obtained under RG flow obey all the same symmetries as the original, high-energy theory. Furthermore, when the symmetry in principle allows for terms in the Lagrangian that happen to be absent in the original high-energy theory, such terms are typically generated under the renormalisation group flow [32, 121]. We already saw that this emergence of new terms is possible in Section 8.6, where no odd term $\sim \phi^3$ was present in the original theory, but it appeared after integrating out contributions from other fields. Given these two facts, a good Ansatz for the low-energy effective theory is to simply write down the most general theory consistent with all the symmetries of the system. Even if the coefficients in such a theory have unknown values, a lot of information about its physical properties can be obtained, such as the number of degrees of freedom and the shapes of their dispersion relations.

In the study of spontaneous symmetry breaking, the construction of an effective Lagrangian is useful for determining the type and spectrum of NG modes. Goldstone's theorem ensures these are massless and feature in the effective theory at any arbitrarily low energy scale. To study the lowest-energy behaviour of NG modes, any energy scale in the system, and in particular any particle mass, can thus be considered to be large and irrelevant. In practice, there are often no massless particles other than the NG modes, unless some symmetry prevents the generation of mass by interactions under the RG flow. The low-energy spectrum then consists of NG modes only and is determined entirely by symmetry. Constructing the effective Lagrangian in such a case is the topic of Section 10.2. One particular consequence of symmetry on NG modes becomes apparent in this approach when we consider that in relativistic systems, Lorentz symmetry places strong constraints on the terms allowed in the effective Lagrangian, which are absent in non-relativistic systems. This directly leads to the distinction between type-A and type-B NG modes in the non-relativistic systems of Section 4.2.

Notice that the effective Lagrangian discussed in this section and the 1PI (quantum) effective action of the previous section are not the same. Because they are both effective descriptions of an original, high-energy action, however, they are related. In particular, a low-energy approximation to the quantum effective action is equivalent at the classical level to the Wilson action obtained by integrating out high-energy modes. Similarly, the quantum effective action can be used to show that renormalisation of high-energy (UV) divergences cannot be affected by the absence or presence of spontaneous symmetry breaking that dominates the effective Lagrangian [41]. Intuitively, this can be attributed to the fact that spontaneous symmetry breaking is a ground-state effect, and that systems excited to high energies or temperatures always have their symmetry restored.

10.2 The Coset-Space Construction

The construction of phenomenological effective field theories is especially useful for describing the physics of NG modes because they are guaranteed by Goldstone's theorem to be gapless, as shown in Section 4.1 and Exercise 10.3. In principle, other gapless modes may be present as well, and these can be systematically taken into account [35], but here we focus on the simplest case and consider all other particles to be massive. We can then choose any energy scale lower than the mass of the lightest massive particles in the system and consider an effective field theory in terms of NG physics only. From Section 10.1, we know that effective field theory should inherit the symmetries of the underlying high-energy theory and these are always given by the full symmetry group G, including both unbroken and spontaneously broken symmetries. By itself this condition is not sufficient to define the effective Lagrangian however because we do not yet know how the symmetry transformations will act on the low-energy, effective degrees of freedom parametrised by the fields in the effective field theory.

For example, consider a system with the effective potential depicted in Figure 10.1, which has a spontaneously broken rotational symmetry described by the complex field configuration $\varphi = \langle \phi \rangle$. In any symmetry-broken state along the bottom of the rim of V_{eff}, the expectation value φ has a non-zero amplitude determined by the interaction parameters in the theory. The value of the phase of φ, however, is chosen spontaneously. The effective Lagrangian will describe excitations about a particular symmetry-broken ground state. These consist of massive amplitude fluctuations taking the state φ in the radial direction of Figure 10.1 and massless phase fluctuations moving the state along the bottom of the rim. The energy spectrum of these modes will be identical starting from any particular value of the spontaneously chosen phase, but the same is not necessarily true for the definition of the modes themselves. If the spontaneously chosen phase is zero, an infinitesimal excitation of the field parallel to the imaginary axis in Figure 10.1 will be phase mode, but if the spontaneously chosen phase is $\pi/2$, an excitation in the same direction would be an amplitude mode. In other words, we need to establish how the definitions of modes exciting a symmetry-broken ground state change under symmetry transformations. In the current example, a convenient (but not unique) way to parametrise the symmetry-broken state is to write $\varphi = a_0 e^{i\theta_0}$, with excited states of the form $\varphi' = (a_0 + a)e^{i(\theta_0+\theta)}$. Here, a parametrises an amplitude mode and θ a phase mode. The symmetry transformation rotates the phase, and thus acts on the phase mode as $\theta \to \theta + \delta\theta$. This is a non-linear transformation, showing that in general, the broken symmetry does not simply act linearly on the fields, but transforms between them in potentially complicated, non-linear ways. We can therefore not simply resort to the standard linear representation theory of Section 1.7.3 to build up the effective Lagrangian.

It would be convenient if we could construct an effective field theory for the NG modes that is invariant under the full group G, using only the Lie group properties of G and the subgroup H of residual symmetries. In two seminal papers, Callan, Coleman, Wess, and Zumino (CCWZ) showed how to construct the most general effective Lagrangian given only these properties [37, 42]. A good textbook explanation can be found in [215], but here we follow the more modern treatment of [35]. As this heavily relies on the coset-space properties of G/H, introduced in Section 3.1, this is often referred to as the **coset-space construction**. In this chapter, we restrict ourselves to internal symmetries. The generalisation that includes spacetime symmetries is the topic of Section 13.1.

10.2.1 Symmetry Transformations

When we say that a system, described by a field theory, has a symmetry described by the transformation group G, we generally mean two things. First, the fields appearing in the theory transform under some linear representation of G, as described in Section 1.7.3. That is, the group elements $g \in G$ act on the collection of fields $\phi_a(x)$ indexed by a, as a linear transformation. The fields can therefore be described by representations of the transformation group.

Second, the Lagrangian constructed out of those fields and their derivatives is invariant under the transformations in G. This is what distinguishes a symmetry transformation from any other transformation of the fields. While the fields by themselves are usually not invariant under the symmetry transformations, combinations of them can be. For instance, ϕ_a may transform under rotations as $R_a^b\phi_b$, and thus not be invariant, while the combination $\phi_a^{\mathrm{T}}\phi_a$ is invariant under rotations, and is suitable for a term appearing in a rotation-symmetric Lagrangian.

In the traditional approach that we followed so far, the Lagrangian (or action, or Hamiltonian) comes first: it defines the actual system and its physical laws. We then observe which symmetries it possesses. In the effective Lagrangian, these roles are reversed. The symmetries are leading, and we construct the most general effective field theory that is invariant under those symmetries. To carry out that construction, we need to first identify fields that transform in some known way under the symmetry transformations $g \in G$. In other words, they should be described by representations of the symmetry group. Second, we need to construct terms in the Lagrangian from combinations of those fields that are invariant under all symmetries.

As alluded to in the example of Figure 10.1, when symmetries are spontaneously broken, the NG fields typically transform non-linearly under the broken symmetries. Those transformations can therefore not be described by an ordinary, linear representation of the group. The non-linear transformations are in general so complicated that these fields seem entirely unsuitable for constructing invariant Lagrangians. For example, if the symmetry acts in a non-linear way, a second-order interaction term coupling two fields may turn into an interaction between three entirely different fields after a symmetry transformation. Constructing an action that is invariant under such transformations then requires balancing the effects of the transformation on infinitely many terms.

Fortunately, we shall see shortly that there exist combinations of terms which transform as a linear representation of the unbroken symmetries H when acted upon by any symmetry transformation $g \in G$, including both the unbroken and broken ones. This takes us back into the known territory of constructing a theory from linear representations. We can then collect all possible combinations of these linear representations as usual and restrict the theory to contain only those that are invariant under H. As shown in the original CCWZ papers, and in Ref. [35], the theory obtained this way is also the most general Lagrangian that is invariant under G. Finding the combinations of fields on which broken symmetry transformations act as unbroken ones can be done using the coset-space properties introduced in Sections 1.7.1 and 3.1.

10.2.2 Constructing Invariant Terms

We start the search for invariant terms in the effective Lagrangian by considering a general symmetry group G, whose generators are separated into the unbroken ones T_i, indexed by i, j, k and the broken ones X_a, indexed by a, b, c. We know that the broken generators X_a will excite NG modes when acting on a symmetry-broken state. Formally, we can define the Lie-group-valued operator:

$$U(\pi) = \mathrm{e}^{\mathrm{i}\pi^a(x)X_a}. \tag{10.31}$$

The state $\Phi = U\chi$ obtained by acting with U on a ground state χ with a spontaneously broken symmetry, will contain NG modes parametrised by $\pi^a(x)$. As an example, consider again the specific case of a single field breaking $U(1)$ phase rotations depicted in Figure 10.1. Picking a particular ground state χ then corresponds to choosing a specific location along the bottom of the rim of the potential. The operator U transports this state around the rim if $\pi(x)$ is constant and corresponds to a local version of the symmetry transformation otherwise. The proof of Goldstone's theorem showed that such local symmetry transformations excite NG modes, so that U can be said to create the NG field configuration $\pi(x)$.

We would now like to construct a field theory for the NG fields $\pi^a(x)$. The central observation of the coset-space construction is that the effect of a general symmetry transformation $g \in G$ on the state Φ can be written as

$$g\Phi = gU(\pi)\chi \equiv \mathrm{e}^{\mathrm{i}\tilde{\pi}^a(x)X_a}\mathrm{e}^{\mathrm{i}\theta^i T_i}\chi = \mathrm{e}^{\mathrm{i}\tilde{\pi}^a(x)X_a}\chi = U(\tilde{\pi})\chi. \tag{10.32}$$

The second equality in this expression defines the fields $\tilde{\pi}^a(\pi, g)$ and $\theta^i(\pi, g)$ in terms of π. That the operator gU can always be expressed this way is guaranteed by the fact that we can write G in terms of left cosets of H, and H is generated by the T_i, as detailed, for example, in Refs. [35, 136]. That $\mathrm{e}^{\mathrm{i}\theta^i T_i}\chi = \chi$ in the third equality comes from the fact that χ was assumed to be a ground state, which is invariant under the action of the unbroken symmetry generators. Because $U(\tilde{\pi})$ still has the form of a Lie-group-valued operator for the broken symmetries, it generates NG modes, now parametrised by the fields $\tilde{\pi}$. This allows us to interpret the transformation g as acting on the NG modes, rather than the state:

$$g\colon \pi(x) \;\to\; \tilde{\pi}(x). \tag{10.33}$$

Here, we used the explicit notation $g\colon \pi \to \tilde{\pi}$ indicating that π transforms into $\tilde{\pi}$ under the action of g to avoid any suggestion that this is a usual, linear transformation. When formulating an effective field theory in terms of NG modes, without referring to the ground state explicitly, Eq. (10.33) describes how the NG modes appearing in the theory are affected by a symmetry transformation. From here on, we therefore refer to $\tilde{\pi}$ simply as the transformed NG modes. Because the definition of the transformed modes, Eq. (10.32), contains the ground state χ on both sides, we can write more generally

$$\begin{aligned} gU(\pi) &= \mathrm{e}^{\mathrm{i}\tilde{\pi}^a(x)X_a}\mathrm{e}^{\mathrm{i}\theta^i T_i}, \\ gU(\pi)h^{-1}(\pi,g) &= U(\tilde{\pi}). \end{aligned} \tag{10.34}$$

The second line implicitly defines the operator $h = \mathrm{e}^{\mathrm{i}\theta^i T_i}$ in the group H of unbroken symmetries, which will be central to the remainder of the coset-space construction.

10.2.2.1 Example: NG Mode Transformations for Broken SU(2)

Before continuing, we will consider an explicit example of the construction so far to the case of $SU(2)$ symmetry being broken down to $U(1)$. The subgroup of unbroken symmetries contains only the spin operator S_z as its generator. The broken generators consist of the S_x and S_y operators. The commutation relations between generators give rise to a set of structure constants, identified in Exercise 10.4.

Exercise 10.4 (Structure constants).
a. Recall the definition of the structure constants $f_{ab}{}^{c}$ of a Lie algebra, as given in Eq. (1.52). Using the fact that the T_i generate a subgroup of G (namely H) and assuming the structure constants are antisymmetric in all indices (which is the case for compact Lie groups, covering most cases of interest in physics), show that in general

$$[T_i, X_a] = \mathrm{i}\sum_b f_{ia}{}^{b} X_b. \tag{10.35}$$

In particular, notice that none of the unbroken generators T_j appear on the right-hand side.
b. Now use the usual commutation relations for spin operators to find the structure constants relevant to the case of $SU(2) \to U(1)$.

A general transformation $g \in SU(2)$ can now be written as

$$g = \mathrm{e}^{\mathrm{i}g^a X_a}\mathrm{e}^{\mathrm{i}g^i T_i} = \mathrm{e}^{\mathrm{i}(g^x S_x + g^y S_y)}\mathrm{e}^{\mathrm{i}g^z S_z}. \tag{10.36}$$

Combining this with $U = \mathrm{e}^{\mathrm{i}(\pi^x S_x + \pi^y S_y)}$, we can write Eq. (10.34) as

$$\begin{aligned} gU &= \mathrm{e}^{\mathrm{i}(g^x S_x + g^y S_y)}\mathrm{e}^{\mathrm{i}g^z S_z}\mathrm{e}^{\mathrm{i}(\pi^x S_x + \pi^y S_y)} \\ &= \left(1 + \mathrm{i}[\pi^x + g^x + g^z\pi^y]S_x + \mathrm{i}[\pi^y + g^y - g^z\pi^x]S_y\right) \\ &\quad \times \left(1 + \mathrm{i}\left[g^z + \frac{1}{2}(g^y\pi^x - g^x\pi^y)\right]S_z\right) + \mathcal{O}\left((g,\pi)^3\right). \end{aligned} \tag{10.37}$$

Here, we expanded the exponents for small values of both the g and π parameters, and commuted them to obtain the second line, retaining terms to quadratic order. Re-exponentiating the expression in the final line defines the transformed NG modes and the parameter θ introduced in Eq. (10.34) as

$$\begin{aligned}
\tilde{\pi}^x &= \pi^x + g^x + g^z\pi^y + \mathcal{O}\left((g,\pi)^3\right), \\
\tilde{\pi}^y &= \pi^y + g^y - g^z\pi^x + \mathcal{O}\left((g,\pi)^3\right), \\
\theta^z &= g^z + \frac{1}{2}\left(g^y\pi^x - g^x\pi^y\right) + \mathcal{O}\left((g,\pi)^3\right).
\end{aligned} \tag{10.38}$$

We will come back to this example after formulating the next steps.

10.2.2.2 The Maurer–Cartan Form

Having identified the way in which the NG modes transform, we will build up the effective field theory from suitably chosen combinations of the NG fields and their derivatives. In particular, we will look for combinations of fields on which any element of G acts in the same way as a linear transformation generated by some residual symmetry in H.

The identification of allowed terms in the effective field theory starts from the so-called **Maurer–Cartan form** $U^{-1}\partial_\mu U$. Here, $\partial_\mu = \partial/\partial x^\mu$ is the ordinary spacetime derivative that acts on the operators U via their dependence on the fields $\pi(x)$. We will show that the Maurer–Cartan form can be expressed in terms of fields that transform linearly under the residual symmetry transformations, and therefore, these fields can be used to construct our desired effective field theory. We start by using the fact that the Maurer–Cartan form takes values in the Lie algebra of G [32, 35], which is spanned by T_i and X_a. We can thus express it as

$$U^{-1}\partial_\mu U = \mathrm{i}(A^i_\mu(\pi)T_i + \mathrm{e}^a_\mu(\pi)X_a). \tag{10.39}$$

For convenience, we define $A_\mu(\pi) \equiv A^i_\mu(\pi)T_i$ and $e_\mu(\pi) = \mathrm{e}^a_\mu(\pi)X_a$. To see how these fields transform, we consider the Maurer–Cartan form associated with $U(\tilde{\pi}) = gU(\pi)h^{-1}(\pi, g)$. Simplifying notation by dropping the explicit π-dependencies in intermediate steps, we find

$$\begin{aligned}
U^{-1}(\tilde{\pi})\partial_\mu U(\tilde{\pi}) &= (gUh^{-1})^{-1}\partial_\mu(gUh^{-1}) \\
&= hU^{-1}g^{-1}g\left((\partial_\mu U)h^{-1} + U(\partial_\mu h^{-1})\right) \\
&= h(U^{-1}\partial_\mu U)h^{-1} - (\partial_\mu h)h^{-1} \\
&= h(U^{-1}(\pi)\partial_\mu U(\pi) - h^{-1}\partial_\mu h)h^{-1}.
\end{aligned} \tag{10.40}$$

Here, we used that g does not depend on π to go to the second line, and that $\partial_\mu(hh^{-1}) = 0$ in going from the second to the third line. Substituting the parametrisation of Eq. (10.39) into Eq. (10.40) for the Maurer–Cartan forms on both sides, and using the fact that h contains only unbroken transformations, we find the relations

$$\begin{aligned}
A_\mu(\tilde{\pi}) &= h\left(A_\mu(\pi) + \mathrm{i}h^{-1}(\partial_\mu h)\right)h^{-1}, \\
e_\mu(\tilde{\pi}) &= he_\mu(\pi)h^{-1}.
\end{aligned} \tag{10.41}$$

In the final line, it becomes clear that expressing the operator e_μ in terms of transformed fields $\tilde{\pi}$ has the same effect as transforming it under a linear representation $h \in H$. Because the transformation $\pi \to \tilde{\pi}$ corresponds to expressing operators in terms of NG modes on top of a transformed ground state, the effect on e_μ of any transformation $g \in G$ connecting distinct symmetry-broken ground states is found to be the same as that of an unbroken symmetry transformation. This is true not only for e_μ itself, but also for any function of e_μ. Products or functions of e_μ that are invariant under all transformations in H, are therefore also invariant under the transformations $\pi \to \tilde{\pi}$ generated by any element $g \in G$, containing both the broken and unbroken symmetries. In this way, writing a field theory in terms of H-invariant combinations of the fields e_μ, rather than directly in terms of π, ensures its invariance under the full symmetry group.

Besides the operator e_μ that always allows for the construction of invariant terms, the operator A_μ can also be used in specific circumstances. First of all, if the symmetry-breaking states happen to violate all symmetries of the group G, the only element in the residual symmetry group H will be the identity. In that case, both e_μ and A_μ are invariant under all transformations in G, and any function of them may appear in the effective field theory. A second, more subtle, effect arises when the residual symmetry group H is Abelian. In that case, it can be shown that the associated Maurer–Cartan form $h^{-1}\partial_\mu h$ equals precisely $i\partial_\mu \theta^j T_j$. That is, the term in Eq. (10.41) which prevents A_μ from transforming as a linear representation of H is a total derivative. Although an H-invariant Lagrangian containing A_μ is then not invariant under all symmetry transformations in the group G, the action typically will be, because total derivatives in the Lagrangian do not contribute to the action (barring topological effects dependent on boundary conditions). Notice that this argument allows A_μ to appear in the Lagrangian if it is invariant under all transformations $h \in H$, but not powers or functions of A_μ. As you will show in Exercise 10.5, this special circumstance actually occurs in the example of $SU(2) \to U(1)$ symmetry breaking, where it contributes a term that will be crucial to the classification of its NG modes in Section 10.3.

Exercise 10.5 (Invariant terms for broken SU(2)).

a. Returning to the example of $SU(2) \to U(1)$ symmetry breaking, show that its Maurer-Cartan form can be expanded as

$$U^{-1}\partial_\mu U = \mathrm{i}(\partial_\mu \pi^x)S_x + \mathrm{i}(\partial_\mu \pi^y)S_y + \frac{1}{2}\mathrm{i}[\pi^x\partial_\mu\pi^y - \pi^y\partial_\mu\pi^x]S_z + \mathcal{O}\left((\pi,\partial\pi)^3\right). \tag{10.42}$$

Comparing this expression to Eq. (10.39) yields the definitions

$$\begin{aligned} e^x_\mu &= \partial_\mu \pi^x + \mathcal{O}\left((\pi,\partial\pi)^3\right), \\ e^y_\mu &= \partial_\mu \pi^y + \mathcal{O}\left((\pi,\partial\pi)^3\right), \\ A^z_\mu &= \frac{1}{2}\left(\pi^x\partial_\mu\pi^y - \pi^y\partial_\mu\pi^x\right) + \mathcal{O}\left((\pi,\partial\pi)^3\right). \end{aligned} \tag{10.43}$$

b. Substitute the transformations identified in Eq. (10.38), and show that

$$e_\mu(\tilde{\pi}) = e_\mu + g^z\left((\partial_\mu\pi^y)S_x - (\partial_\mu\pi^x)S_y\right) + \mathcal{O}\left((g,\pi,\partial\pi)^3\right). \tag{10.44}$$

c. Show that $e_\mu(\tilde{\pi}) = h e_\mu(\pi) h^{-1}$ up to $\mathcal{O}\left((g,\pi,\partial\pi)^3\right)$.

d. Confirm that $A^z_\mu(\tilde{\pi}) - A^z_\mu(\pi) = -\partial_\mu \theta^z$ holds to leading order.

A noticeable feature of the coset-space construction is that the definition $U = e^{i\pi^a X_a}$ implies that the Maurer–Cartan form $U^{-1}\partial_\mu U$ and hence the operators e_μ, are proportional to $\partial_\mu \pi(x)$. This can be seen directly from its definition:

$$\begin{aligned} U^{-1}\partial_\mu U &= e^{-i\pi^a X_a}\partial_\mu e^{i\pi^b X_b} \\ &= i(\partial_\mu \pi^b)\left(e^{-i\pi^a X_a} X_b e^{i\pi^c X_c}\right). \end{aligned} \tag{10.45}$$

Combined with Eq. (10.39), this expression implies that also the fields e^a_μ contain a derivative of π as an overall prefactor. As will be discussed in more detail in the next section and in Project 21, this is a generic feature for NG modes, which signals the fact that they become non-interacting in the long wavelength limit.

10.2.2.3 The Covariant Derivative

Before formulating the final form of the effective Lagrangian, a last difficulty to overcome is the fact that $h(\pi, g)$ introduced in Eq. (10.34) depends on the spacetime coordinates x_μ through its dependence on the fields θ^i, and thus π. Because of this, the derivative $\partial_\nu e_\mu$ does not transform in the same way as e_μ itself: $\partial_\nu e_\mu(\tilde{\pi}) \neq h(\pi)(\partial_\nu e_\mu(\pi))h^{-1}(\pi)$. This prevents us from using derivatives of e_μ in the construction of the effective Lagrangian.

The lack of a proper derivative can be remedied, however, by using Eqs. (10.40)– (10.41) to define the **covariant derivative** $\mathcal{D}_\mu = \partial_\mu + iA^i_\mu(\pi)T_i$. It can then be straightforwardly confirmed using the transformations of Eq. (10.41) that

$$\tilde{\mathcal{D}}_\nu e_\mu(\tilde{\pi}) = h(\pi)\mathcal{D}_\nu e_\mu(\pi)h^{-1}(\pi). \tag{10.46}$$

Here, the notation $\tilde{\mathcal{D}}_\nu$ indicates that the field A_μ in the definition of the covariant derivative is a function of $\tilde{\pi}$ rather than π. This is what makes it a covariant derivative, whose definition varies along with the ground state around which the fields are defined.

Exercise 10.6 (The covariant derivative to leading order).
a. Show that for the specific example of $SU(2) \to U(1)$ symmetry breaking,

$$\mathcal{D}_\nu e_\mu = \partial_\nu e_\mu + \mathcal{O}\left((g, \pi, \partial\pi, \partial^2\pi)^3\right). \tag{10.47}$$

b. Use this to confirm that

$$\begin{aligned} \tilde{\mathcal{D}}_\nu e_\mu(\tilde{\pi}) &= h(\mathcal{D}_\nu e_\mu)h^{-1} \\ &= \mathcal{D}_\nu e_\mu + g^z\left((\partial_\nu\partial_\mu\pi^y)S_x - (\partial_\nu\partial_\mu\pi^x)S_y\right) + \mathcal{O}\left((g, \pi, \partial\pi, \partial^2\pi)^3\right). \end{aligned} \tag{10.48}$$

A shift between symmetry-broken ground states $\pi \to \tilde{\pi}$ generated by any element $g \in G$, is shown in Eq. (10.46) to result in both the field e_μ and its covariant derivative $\mathcal{D}_\nu e_\mu$ transforming in the same way as they would under an unbroken symmetry transformation. This implies that any combination of the field e_μ and its covariant derivatives that is invariant under the unbroken symmetries in H, is also automatically invariant under any transformation $g \in G$, including the broken symmetries. We have thus found a way of constructing combinations of the NG fields $\pi(x)$ and their derivatives which are invariant under the full group G, despite the NG fields transforming in a complicated, non-linear way under the broken-symmetry transformations.

Because the operators e_μ transform linearly, we can treat them just as we did in the standard linear representation theory of Section 1.7.3. That is, we are looking for the Lagrangian associated with an excited state of the from $e^{ie_\mu(\pi)}\chi$. This Lagrangian should be such that its value is unaffected by the transformation $\pi \to \tilde{\pi}$, and can be built from the components e^a_μ and their covariant derivatives, which parametrise the excited state:

$$\begin{aligned}&\mathcal{L}(e^a_\mu(\tilde{\pi}),\, e^a_\nu(\tilde{\pi})e^b_\mu(\tilde{\pi}),\, \tilde{\mathcal{D}}_\nu e^a_\mu(\tilde{\pi}), \ldots)\\ &= \mathcal{L}(e^a_\mu(\pi),\, e^a_\nu(\pi)e^b_\mu(\pi),\, \mathcal{D}_\nu e^a_\mu(\pi), \ldots).\end{aligned} \tag{10.49}$$

It is proven in the original CCWZ papers and in [35] that the effective Lagrangian constructed this way is also the most general effective Lagrangian possible.

As an explicit example, we return once more to the case of $SU(2) \to U(1)$ symmetry breaking. Using the result of Exercises 10.5 and 10.6 the building blocks for the effective Lagrangian in this case are found to be, up to $\mathcal{O}\left((\pi, \partial\pi)^3\right)$:

$$\begin{aligned}e^a_\mu &= \partial_\mu \pi^a,\\ \mathcal{D}_\nu e^a_\mu(\pi) &= \partial_\nu \partial_\mu \pi^a,\\ A^z_\mu &= \pi^x \partial_\mu \pi^y - \pi^y \partial_\mu \pi^x.\end{aligned} \tag{10.50}$$

Notice that because of the transformation properties in Eq. (10.41), the effective Lagrangian may contain any combination of the components e^a_μ and $\mathcal{D}_\nu e^a_\mu$, whereas A^z_μ may only appear in isolated form, and is then guaranteed to add only a total derivative to $\mathcal{L}_{\text{eff}}$ under the transformation $\pi \to \tilde{\pi}$. To find combinations of e^x_μ and e^y_μ that are left invariant under the same transformation, we can use the fact that the residual symmetry in this specific case causes a rotation between components of the S_a operators. The energy of excitations can thus not depend on their direction in this plane, and only the combination $(e^x_\mu)^2 + (e^y_\mu)^2$ is allowed to appear in the effective Lagrangian. You can check explicitly using the results of Exercise 10.5 that indeed $(\partial_\mu \tilde{\pi}^x)^2 + (\partial_\mu \tilde{\pi}^y)^2 = (\partial_\mu \pi^x)^2 + (\partial_\mu \pi^y)^2 + \mathcal{O}(g^2)$, while other combinations of the components of e_μ are not left invariant under the transformation. We thus find that the effective Lagrangian for the case of $SU(2) \to U(1)$ symmetry breaking should be of the form

$$\mathcal{L}_{\text{eff}} = \beta_{ab}\pi^a \partial_\mu \pi^b + \gamma_{ab}(\partial_\mu \pi^a)(\partial_\mu \pi^b) + \ldots. \tag{10.51}$$

The first term comes from A^z_μ, so that β_{ab} must be an antisymmetric matrix, while γ_{ab} in the second term must be proportional to the identity matrix. In general, no sum over μ is implied in this expression, and all derivatives may have distinct coefficients (so that β_{ab} and γ_{ab} depend on μ). If additional symmetries in spacetime are present, these will impose further constraints. You will show in Exercise 10.7, for example, that isotropy of space implies $\beta_{ab} = 0$ for all spatial derivatives in the first term. Similarly, the second term reduces to $\gamma \sum_{a,\mu} \partial_\mu \pi^a \partial^\mu \pi^a$ if Lorentz invariance is imposed.

10.2.2.4 Summary

After this long discussion, let us summarise the central points of the coset-space construction for the effective Lagrangian: We started from the observation that broken symmetry transformations are parametrised by the NG fields π^a and excited in the ground state χ by the operator $U = e^{i\pi^a(x)X_a}$. These fields map one-to-one to elements of the coset space G/H. We then noticed that acting with any transformation $g \in G$ on the excited state $U\chi$ results in a reparametrisation $\pi \to \tilde{\pi}$ of the fields. The transformation $\pi \to \tilde{\pi}$ is in general non-linear, preventing us from directly writing the effective Lagrangian in terms of the π fields. Next, we introduced the Maurer–Cartan form, which can be used to identify combinations $e_\mu(\pi)$ of the broken symmetries that transform under any element $g \in G$ as a linear representation of H. Moreover, the Maurer–Cartan form also allows the construction of a covariant derivative $\mathcal{D}_\nu e_\mu$ that transforms in the same way. Finally, with the fields e_μ and their covariant derivatives transforming as linear representations of H, we identify combinations of their components that are invariant under all unbroken symmetries $h \in H$. The most general form of the effective Lagrangian contains only such terms (and possibly A^i_μ), and describes all allowed NG fields and their interactions.

We will return to this derivation in Section 13.1, where it is generalised to include broken (external) spacetime symmetries.

10.3 NG Modes in the Effective Lagrangian

One of the most consequential aspects of the coset-space construction is the result following from Eq. (10.45), that the fields $e_\mu(\pi)$ in terms of which the effective field theory is formulated, are proportional to a derivative $\partial_\mu \pi$ of a Nambu–Goldstone mode. This implies that all terms in the effective Lagrangian containing NG modes must contain at least one such derivative, while terms like $\pi_a \pi^a$ cannot occur, because they transform non-linearly under G. The prefactor to the derivative in any individual term, however, may contain arbitrary functions of the fields. The forms of these functions may be derived explicitly using the construction of the previous section, but many of their properties can also be established directly by demanding that the Lagrangian be invariant under its full symmetry.

Focussing on the long-wavelength physics that dominates at low energies, we can, for example, write the effective Lagrangian containing only NG modes in terms of a gradient expansion. As long as the theory is Lorentz invariant, it must then be of the form

$$\mathcal{L}_{\text{eff}} = C_{ab}(\pi)\partial_\mu \pi^a \partial^\mu \pi^b + \mathcal{O}\left((\partial\pi)^4\right). \tag{10.52}$$

Here, C_{ab} may be constant or a functional of the fields, but does not depend on the field derivatives. The relativistic notation makes it obvious that terms with an odd number of derivatives cannot be brought into scalar form and therefore cannot appear in the Lagrangian. The functionals C_{ab}, moreover, have to be such that invariance of the theory under the full symmetry group G (containing Lorentz transformations, as well as additional broken and unbroken symmetries) is obtained.

Shifting to systems without Lorentz invariance, such as non-relativistic phases of matter, also terms with odd powers of the derivatives become allowed, and the most general form of the effective Lagrangian becomes

$$\begin{aligned}\mathcal{L}_{\text{eff}} = &\, B_a(\pi)\partial_t \pi^a + B'_{a,r}(\pi)\partial_r \pi^a + C_{ab}(\pi)\partial_t \pi^a \partial_t \pi^b \\ &+ C'_{ab,r}(\pi)\partial_t \pi^a \partial_r \pi^b + C''_{ab,r,r'}(\pi)\partial_r \pi^a \partial_{r'} \pi^b + \mathcal{O}\left((\partial\pi)^3\right).\end{aligned} \tag{10.53}$$

Here, the subscripts r and r' refer to spatial coordinates and should be summed over. Again, the coefficients need to be such that overall invariance of the Lagrangian under its symmetries is guaranteed. One common simplification occurs for isotropic systems that maintain invariance under spatial rotations despite the broken Lorentz invariance. As you will show in Exercise 10.7, no odd powers of the spatial derivative are allowed in that case, and the effective theory becomes

$$\mathcal{L}_{\text{eff}} = B_a(\pi)\dot{\pi}^a + C_{ab}(\pi)\dot{\pi}^a\dot{\pi}^b + C''_{ab}(\pi)\nabla\pi^a \cdot \nabla\pi^b + \mathcal{O}\left((\partial\pi)^3\right). \tag{10.54}$$

Here, the gradients are over spatial coordinates only, and the time derivative of NG modes is indicated by dotted fields. This general form of the effective low-energy theory for NG modes in isotropic non-relativistic systems suffices to fully characterise possible types of NG modes and their dispersions, as well as establish the (non-)vanishing of their scattering amplitudes in the long-wavelength limit.

Exercise 10.7 (Isotropic effective field theory).
Consider a system with unbroken rotational symmetry, so that it is isotropic.
a. Confirm explicitly that a term with a single gradient is not invariant under a two-fold (meaning: 180°) rotation along any axis.
b. Use the effect of general rotations on spatial vectors to argue that no terms with an odd number of spatial derivatives (gradients) can appear in the effective Lagrangian.

The spatial rotation symmetry does not restrict terms with temporal derivatives in the effective theory. We will, however, still ignore any term that can be written as a total time derivative.

c. Explain why terms of the form $\partial_t(f(\pi))$, with f an arbitrary functional of the fields π, can be neglected in the effective Lagrangian.
Hint: Consider the effect of such a term on the action.

10.3.1 Types of NG Modes

In Chapter 4, we made a distinction between type-A and type-B NG modes. Both are gapless scalar modes, but the former typically feature a linear dispersion relation, while the latter usually have quadratic dispersion and generically come with a gapped partner mode. Furthermore, we stated the generic low-energy effective Lagrangian for NG modes in Eq. (4.9), and we anticipated the fact that the term linear in time derivatives has a prefactor proportional to the so-called Watanabe–Brauner matrix. This was the central ingredient required to explain the different dispersion relations of different types of NG modes. Now that we derived the effective Lagrangian for NG modes in Eq. (10.54), we are in a position to prove these earlier assertions and discuss the differences between type-A and type-B NG modes in more detail.

10.3.1.1 Canonically Conjugate Fields

Following [207] and [32], we first explore the possibility that two broken symmetry generators excite the same NG mode. If the two symmetries are generated by Q_a with $a \in \{1, 2\}$, we can always choose coordinates such that the corresponding fields π_a will behave under an infinitesimal symmetry transformation as $\pi^a \to \pi^a + \delta\pi^a$, with $\delta\pi^a$ an infinitesimal constant. This is the non-linear transformation discussed in detail in the beginning of Section 10.2. For now, focus only on the first term of the generic NG mode effective Lagrangian in Eq. (10.54), so that $\mathcal{L}_{\text{eff}} = B_a(\pi)\dot{\pi}^a + \ldots$. Expanding the prefactor in powers of the fields yields

$$\mathcal{L}_{\text{eff}} = \left(\alpha_a + \beta_{ab}\pi^b\right)\dot{\pi}^a + \mathcal{O}\left((\pi, \partial\pi)^2\right). \tag{10.55}$$

We discussed an example of this effective Lagrangian for the specific case of $SU(2) \to U(1)$ symmetry breaking in Section 10.2.2. Here, we proceed to formulate generic properties for them, applicable to all symmetry-breaking patterns.

In fact, the terms proportional to α_a as well as the symmetric part of $\beta_{ab}\pi^b\dot{\pi}^a$ may be neglected, as they correspond to total time derivatives, which you argued in Exercise 10.7 do not contribute to the action. A direct application of Eq. (1.13) then shows the Noether charge densities associated with the two symmetries to be

$$J_a^0 = \frac{\partial\mathcal{L}_{\text{eff}}}{\partial\dot{\pi}^a} = \beta_{ab}\pi^b + \mathcal{O}\left(\pi^2, \partial\pi\right). \tag{10.56}$$

Here, we used the fact that $\delta\pi_a$ is a constant to set the variation $\Delta_s\pi$ appearing in Eq. (1.13) to one.

Applying an infinitesimal symmetry transformation to the fields in Eq. (10.56) shows that the Noether charge densities transform as $J_a^0 \to J_a^0 + \beta_{ab}\delta\pi^b + \ldots$. On the other hand, we can also consider the Noether charge to be the expectation value of the operator for the observable conserved charge. The effect of an infinitesimal symmetry transformation can then be expressed as

$$\left\langle \hat{J}_a^0 \right\rangle \to \left\langle e^{i\delta\pi^b\hat{Q}_b}\hat{J}_a^0 e^{-i\delta\pi^c\hat{Q}_c}\right\rangle = \left\langle \hat{J}_a^0\right\rangle + i\left\langle \delta\pi^b\left[\hat{Q}_b,\hat{J}_a^0\right]\right\rangle + \mathcal{O}\left(\delta\pi^2\right). \tag{10.57}$$

Comparing the two expressions for the transformation of the Noether charge density shows that the coefficients in the effective Lagrangian are directly related to the conserved charges:

$$\beta_{ab} = i\left\langle\left[\hat{Q}_b,\hat{J}_a^0\right]\right\rangle = i\left\langle\left[\hat{Q}_b,\hat{Q}_a\right]\right\rangle/V. \tag{10.58}$$

In the final equality, we invoked translational invariance to write the Noether charge density in terms of the spacetime volume V and the total charge operator. Notice that although all expressions for the Noether charge and its variation in preceding expressions were leading orders in an expansion, the final result for the relation between the coefficients and charges is exact.

Exercise 10.8 (Coefficients in the effective Lagrangian).
Show that the tensor $T_{ab} = \left\langle\left[\hat{A}_a,\hat{B}_b\right]\right\rangle$ with $\hat{A}$ and $\hat{B}$ Hermitian operators has only purely imaginary elements.

Combined with the fact that $\left\langle\left[\hat{A}_a,\hat{A}_b\right]\right\rangle = -\left\langle\left[\hat{A}_b,\hat{A}_a\right]\right\rangle$, this result implies that β_{ab} must be a real, antisymmetric tensor.

Exercise 10.8 shows that β_{ab} must be a real, antisymmetric tensor. This confirms that diagonal elements of the form $\pi^1\dot{\pi}^1$ or $\pi^2\dot{\pi}^2$ do not appear in the effective Lagrangian. Moreover, off-diagonal elements always come in pairs:

$$\mathcal{L}_{\text{eff}} = \beta\left(\pi^1\dot{\pi}^2 - \pi^2\dot{\pi}^1\right) + \mathcal{O}\left((\pi, \partial\pi)^2\right). \tag{10.59}$$

This expression for the effective Lagrangian is precisely of the canonical form $L = p\dot{q} - H$, with $H = q\dot{p} + \ldots$ consistent with the Hamilton equations. In other words, the two symmetry transformations π^1 and π^2 represent conjugate variables. Together, they constitute a single degree of freedom, corresponding to a single NG mode, that can be excited by local symmetry transformations of either charge.

The general principle of multiple symmetries not being independent is present already in the simplest example of $SU(2) \rightarrow U(1)$ symmetry breaking in a ferromagnet. The unbroken $SU(2)$ symmetry is described by three generators S_x, S_y, and S_z, two of which are broken in the $U(1)$ phase. Taking the broken symmetries to be S_x and S_y, Eq. (10.58) shows that the coupling β_{ab} is proportional to $\langle S_z \rangle$. This means that if an excitation on top of the symmetry-broken state with $\langle S_x \rangle = \langle S_y \rangle = 0$ introduces some non-zero magnetisation along the x-direction, this magnetisation will precess in the order parameter field $\langle S_z \rangle$, and generate magnetisation in the y-direction as well. The overall effect of an excitation in either of these directions is thus the same. Notice how this differs from the situation in an antiferromagnet. Although it has the same $SU(2) \rightarrow U(1)$ symmetry-breaking pattern, the absence of a total magnetisation in the z-direction in its ordered state allows the excitations associated with S_x and S_y to remain independent.

10.3.1.2 Counting NG Modes

Returning to the general form of the effective Lagrangian containing only NG modes at low energies, we use the results so far to write it as

$$\mathcal{L}_{\text{eff}} = \frac{1}{2}\beta_{ab}\left(\pi^a\dot{\pi}^b - \pi^b\dot{\pi}^a\right) + C_{ab}(\pi)\dot{\pi}^a\dot{\pi}^b + C''_{ab}(\pi)\vec{\nabla}\pi^a \cdot \vec{\nabla}\pi^b + \ldots. \tag{10.60}$$

For each field π^a corresponding to a broken symmetry, Eq. (10.58) can be used to check whether or not terms linear in $\dot{\pi}^a$ appear in the effective Lagrangian. If so, it is part of a canonically conjugate pair. Both fields in such a pair excite the same NG mode, which defines it as a type-B NG mode. As discussed in Section 4.2, they also give rise to a gapped partner mode, whose gap size is determined by the value of the β_{ab} coupling constant. All fields for which there is no term linear in time derivatives in the effective Lagrangian correspond to independent excitations and are called type-A NG modes.

As argued in Section 4.2, Fourier transforming $\mathcal{L}_{\text{eff}}$ will introduce a frequency ω for every time derivative and a momentum k for every gradient, so that type-B modes generically have a quadratic dispersion, while the dispersion of type-A modes is typically linear. This rule is not absolute, however, because the coefficients C''_{ab} may be zero, in which case type-A modes may also obtain a quadratic dispersion.

To count the number of type-A and type-B modes in any given symmetry-breaking pattern, it suffices to compute the Watanabe–Brauner matrix β_{ab}. As argued before, this matrix is real and antisymmetric, and you showed in Exercise 4.1 that there exist a choice of coordinates in which it appears as

$$\tilde{\beta} = O\beta O^{-1} = \begin{pmatrix} M_1 & & & & & \\ & \ddots & & & & \\ & & M_m & & & \\ & & & 0 & & \\ & & & & \ddots & \\ & & & & & 0 \end{pmatrix}, \qquad M_i = \begin{pmatrix} 0 & \lambda_i \\ -\lambda_i & 0 \end{pmatrix}. \tag{10.61}$$

Here, O is an orthogonal matrix implementing the basis rotation, and λ_i are real. The blocks M_i correspond to the pairs of canonically conjugate fields, with each pair exciting a single type-B NG mode and its associated gapped partner mode. On the other hand, each row (or column) containing only zeroes corresponds to a field exciting a single type-A mode. Since the total size of the matrix β is equal to the number of broken symmetry generators $\dim G/H$, we thus find the number of different types of NG modes to equal

$$n_{\rm A} = \dim G/H - \operatorname{rank}\beta, \qquad n_{\rm B} = \frac{1}{2}\operatorname{rank}\beta. \tag{10.62}$$

This reproduces the results announced in Eq. (4.8) and establishes the general counting rule for NG modes.

10.3.2 Adler Zeroes

The effective field theory of the previous section contained only NG modes. Considering more general situations, in which other modes can also be excited at low energies, the effective Lagrangian may, for example, contain terms like $\pi\partial_\mu\pi\bar{\psi}\psi$ or $\eta\partial_\mu\pi\partial^\mu\pi$, in which the NG field parametrised by π couples to either fermionic modes ψ or bosonic modes η. In such cases, the fact that π is generated by a broken symmetry may lead to constraints beyond the usual ones implied by unbroken symmetries of the effective Lagrangian.

For broken Abelian symmetry transformations, for example, the symmetry-broken state and NG fields can always be parametrised by a coordinate along the minimum in the effective potential, as described in Section 10.1.2. In that parametrisation, the effect of the broken symmetry transformation on the NG fields is to shift them by a constant: $\pi \to \pi + \delta\pi$. For the effective Lagrangian to be invariant under such shifts, it must be proportional to derivatives of the fields, as we have seen in the derivation of the coset-space construction in Section 10.2.2. In fact, all NG modes associated with Abelian symmetries that occur in the effective Lagrangian do so with derivatives.

After a Fourier transform, the derivative of an NG field appears as the field multiplied by energy or momentum. The appearance of only derivatives of an NG field thus has the immediate consequence that all interactions involving those NG modes seem to vanish in the limit that their energy and momentum goes to zero. In the strict infinite-wavelength limit, this is not a surprising result. As discussed several time before, NG modes in that limit correspond to transitions between symmetry-related ground states, which do not affect any internal processes. The presence of the derivative also implies, however, that the strength of interactions involving a NG mode vanishes smoothly as the energy of that mode approaches zero.

Although this result is straightforwardly found in the particular parametrisation aligning NG fields with the minimum of the effective potential, it is far from trivial in any other basis. As you will discover in Project 21, working in a more generic basis typically allows for many non-vanishing interaction terms to appear in the effective Lagrangian. Since observable scattering amplitudes should not depend on the basis in which they are computed, the sum of all interactions contributing to a particular scattering process still has to be zero, but this cancellation cannot be seen directly from the form of the Lagrangian in a generic basis. The vanishing scattering amplitudes involving NG modes in the limit that their energy approaches zero are known as **Adler zeroes**.

The appearance of Adler zeroes indicates that low-energy NG modes are weakly interacting. Intuitively, the suppression of scattering amplitudes can be thought of as causing the lifetime of NG modes to become infinite in the long wavelength limit, showing that the NG modes are well-defined, stable quasi-particles even in the presence of other fields. Despite this seemingly general argument, however, the vanishing of interactions for low-energy NG modes is not an absolute rule, and there are several exceptions to the Adler zero principle.

10.3.2.1 Exception: Interactions with Other Fields

Although interactions between the NG fields π and low-energy fermionic fields ψ of the type $\bar{\psi}\gamma_\mu\psi\partial_\mu\pi$ contain only derivatives of the NG field, they may nonetheless contribute to non-vanishing scattering amplitudes. To see this, consider the Feynman diagram of Figure 10.2, which contributes to the scattering amplitude of an NG mode off the fermionic field. Here, the propagator for the NG mode is indicated by the dashed line, while the fermionic propagator corresponds to the solid line. This diagram is proportional to $p \cdot qD(q)G(k)G(k+q)G(k+q-p)D(p)$, with $D(q)$ and $G(k)$ the bosonic and fermionic propagators, and the momenta p and q appearing due to the derivatives in the Lagrangian. Even if the momentum q of the incoming NG mode is taken to zero, this diagram can yield a non-zero contribution to the scattering amplitude if $G(k+q)$ has a singularity in that limit. Without showing it explicitly, we note that this may happen if the limit $q \to 0$ takes the virtual fermion onto its mass-shell, as described in detail, for example, in Ref. [32].

That diagrams such as the one in Figure 10.2 may be non-zero, is not in contradiction with the earlier intuition of Adler zeroes being connected to the infinite lifetime of NG modes. The lifetime is determined by the dressing of an NG mode with virtual excitations. The scattering amplitude to which the diagram contributes, on the other hand, is one between an NG mode and a real fermion already present in the initial state.

10.3.2.2 Exception: Interactions between NG Fields

A second situation in which the Adler zero principle may be violated occurs in the low-energy effective field theory for NG modes associated with non-Abelian symmetry. The

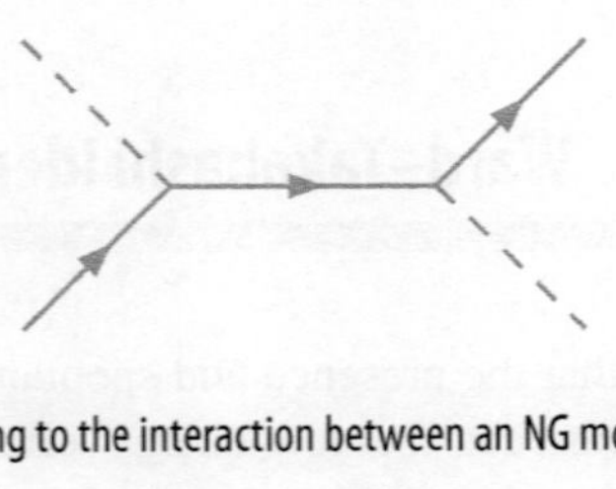

Figure 10.2

One of the Feynman diagrams contributing to the interaction between an NG mode (dashed line) and a fermionic field (solid line).

coset-space construction of Section 10.2 can then be used to find the form of the effective Lagrangian. Writing the symmetry transformations as $U = e^{i\pi^a X_a}$ as before, the fields e^a_μ in terms of which $\mathcal{L}_{\text{eff}}$ is defined can be expanded in powers of the π^a field as $e^a_\mu = \partial_\mu \pi^a - f_{bc}{}^a \pi^c \partial_\mu \pi^b + \ldots$ [32]. Composing allowed terms in the effective Lagrangian such as $g_{ab} e^a_\mu e^b_\mu$ may now yield terms containing products of field operators and their derivatives. In contrast to broken Abelian symmetries, the NG fields associated with non-Abelian symmetry thus do not always occur only with derivatives. Because the interactions are then not proportional to the momenta of all involved NG modes, it is possible for scattering amplitudes to remain non-zero in the long wavelength limit.

Exercise 10.9 (Violation to the Adler zero principle).
As a concrete example of a situation in which the Adler zero principle may be violated, consider the completely broken spin rotation symmetry $SU(2) \to 1$. This situation occurs, for example, in canted antiferromagnets, with S_x, S_y, and S_z all representing broken symmetry generators.

a. Find the structure constants $f_{ab}{}^c$ defined by the commutation relations:

$$[S_a, S_b] = \mathrm{i} \sum_c f_{ab}{}^c S_c. \tag{10.63}$$

b. Assume that the leading term in the effective Lagrangian is of the form $\mathcal{L}_{\text{eff}} \simeq g\delta_{ab} e^a_\mu e^b_\mu$ with $e^a_\mu \simeq \partial_\mu \pi^a - f_{bc}{}^a \pi^c \partial_\mu \pi^b$ and show that all interactions involving three NG fields cancel.

c. Show that the interaction vertex for a π^y and a π^z mode to scatter into a π^y and a π^z mode remains non-zero in the limit of vanishing momentum for one of the incoming fields.

d. Use conservation of energy and momentum, as well as Fermi's golden rule, to argue that despite there being non-vanishing amplitudes for scattering events involving zero-momentum π^z modes, the lifetime of these modes (which is proportional to the inverse of the total scattering rate) still diverges in the long-wavelength limit.

That interactions involving zero-momentum NG modes do not necessarily vanish for non-Abelian broken symmetries, can be intuitively understood as a consequence of the broken global symmetry transformations not being independent. In the case of broken $SU(2)$ symmetry discussed in Exercise 10.9, for example, a simultaneous rotation around the x and y axes can also be written as a rotation around the x and z axes. Different combinations of modes thus become fundamentally indistinguishable. As alluded to in the final part of the exercise, however, this does not limit the diverging lifetime of NG modes in the limit of vanishing momentum.

10.4 Ward–Takahashi Identities

The previous sections showed that the presence and spontaneous breakdown of symmetries allow us to systematically construct the low-energy effective Lagrangian for Nambu–Goldstone modes and their interactions with other low-energy degrees of freedom. In this section, symmetries are shown to also imply constraints on the dynamics of fields in general, at any energy scale. The most well-known example of such a constraint is Noether's theorem, which guarantees that if the action describing any physical system has a symmetry, there exists a corresponding conserved quantity. As described in Chapter 1, the conservation law takes the form of a continuity equation $\partial_\nu j^\nu = 0$ in terms of the local Noether charge density.

Further constraints on the propagators and interaction vertices describing the dynamics of quantum fields can be deduced from the continuity equation because the Noether charge density $j^\nu(x)$ is built from these fields. Such constraints are known as **Ward–Takahashi identities**, and they complement the somewhat similar constraints that Noether's second theorem, discussed in Section 1.6.3, may impose. The derivation of the Ward–Takahashi identities itself builds on Noether's trick, introduced in Exercise 1.3.

To derive the Ward–Takahashi identities, we start from a fully general action $S[\phi_a]$, with the index a labelling different fields, and the associated partition function $Z = \int \mathcal{D}\phi\, \mathrm{e}^{\mathrm{i}S/\hbar}$. As explained in Section 10.1.1, it is useful to also include a source term, $Z = \int \mathcal{D}\phi\, \mathrm{e}^{\mathrm{i}(S+\int dx\, J^a\phi_a)/\hbar}$, so that we can formulate relations between propagators by taking derivatives with respect to J^a before letting all J^a tend to zero. Recall that the effect of a general infinitesimal transformation of the fields, regardless of whether it is a symmetry, can be written to linear order as

$$\begin{aligned}
\phi_a &\to \phi'_a = \phi_a + \delta\phi_a, \\
\mathcal{L} &\to \mathcal{L}' = \mathcal{L} + \frac{\partial\mathcal{L}}{\partial\phi_a}\delta\phi_a + \frac{\partial\mathcal{L}}{\partial(\partial_\nu\phi_a)}\delta(\partial_\nu\phi_a), \\
Z &\to Z' = \int \mathcal{D}\phi'\, \mathrm{e}^{\mathrm{i}S[\phi']/\hbar} = \int \mathcal{D}\phi'\, \mathrm{e}^{\mathrm{i}\int dx\,(\mathcal{L}'+J^a\phi'_a)/\hbar}.
\end{aligned} \tag{10.64}$$

In the final line, the fields ϕ' appear only as the fields being integrated over in the functional integral for the partition function. Their redefinition does not affect the result of the functional integral, just as renaming the integration variable in a usual integral does not affect its value. Therefore, $Z' = Z$, and we obtain the relation

$$\begin{aligned}
Z = Z' &= \int \mathcal{D}\phi'\, \mathrm{e}^{\mathrm{i}\int dx\,(\mathcal{L}'+J^a\phi'_a)/\hbar} \\
&= \int \mathcal{D}\phi\, \det\left(\frac{\delta\phi'}{\delta\phi}\right) \mathrm{e}^{\mathrm{i}\int dx\,(\mathcal{L}'+J^a\phi'_a)/\hbar} \\
&\approx \int \mathcal{D}\phi\, \det\left(\frac{\delta\phi'}{\delta\phi}\right) \mathrm{e}^{\mathrm{i}\int dx\,(\mathcal{L}+J^a\phi_a)/\hbar} \\
&\quad \times \left(1 + \mathrm{i}\int dx \left[\frac{\partial\mathcal{L}}{\partial\phi_a}\delta\phi_a + \frac{\partial\mathcal{L}}{\partial(\partial_\nu\phi_a)}\delta(\partial_\nu\phi_a) + J^a\delta\phi_a\right]\right).
\end{aligned} \tag{10.65}$$

Here, we kept only terms up to linear order in the transformation. Notice the argument of the determinant is a functional derivative rather than a ratio of variations.

The term with the determinant appears from the transformation of the fields in the measure of the functional integral. If it is not equal to one, this indicates there is an obstruction to regularising the path integral in a way that is consistent with its symmetry, and the action is said to contain an **anomaly**, as discussed in Section 10.1.1. Although anomalies do occur in various models of physics, we will ignore them in the present discussion and focus on the situation without anomalies from here on. In that case, the first term in Eq. (10.65) is just Z in its original form, so that the full equation becomes

$$Z = Z + \mathrm{i}\int \mathcal{D}\phi\, \mathrm{e}^{i\int dx\,(\mathcal{L}+J^a\phi_a)/\hbar} \int dx \left[\frac{\partial\mathcal{L}}{\partial\phi_a}\delta\phi_a + \frac{\partial\mathcal{L}}{\partial(\partial_\nu\phi_a)}\delta(\partial_\nu\phi_a) + J^a\delta\phi_a\right]. \quad (10.66)$$

Using the definition of the expectation value in its path integral formulation, we thus find

$$\begin{aligned}\left\langle\int dx \left[\frac{\partial\mathcal{L}}{\partial\phi_a}\delta\phi_a + \frac{\partial\mathcal{L}}{\partial(\partial_\nu\phi_a)}\delta(\partial_\nu\phi_a) + J^a\delta\phi_a\right]\right\rangle &= 0\\ \Rightarrow \quad \left\langle\int dx \left[-\partial_\nu\frac{\partial\mathcal{L}}{\partial(\partial_\nu\phi_a)}\delta\phi_a + J^a\delta\phi_a\right]\right\rangle &= 0. \end{aligned} \quad (10.67)$$

The final line follows from using the chain rule and assuming that ϕ_a obeys the Euler–Lagrange equations, as we did in in Chapter 1 when deriving Noether's theorem.

In the derivation so far, we made no assumptions on the form of $\delta\phi_a$, so that Eq. (10.67) has to hold for $\delta\phi_a$ being any arbitrary function of position. Now, we specialise to the case of $\delta\phi_a$ being proportional to a global symmetry transformation. That is, we consider $\delta\phi_a(x) = \epsilon(x)\delta_s\phi_a$, with $\delta_s\phi_a$ such that the transformation $\phi_a \to \phi_a + \delta_s\phi_a$ is a global symmetry. Notice that it could either be a global symmetry by itself, or the global part of a local gauge freedom, which we saw in Chapter 8 is also always a global symmetry. Because the symmetry is global, $\delta_s\phi_a$ does not depend on position directly, but it may depend on the fields ϕ_a and their derivatives.

Recall from the discussion in Chapter 1 that a symmetry transformation leaves the action invariant, so that the Lagrangian density is left invariant up to at most a total derivative: $\mathcal{L}' - \mathcal{L} = \partial_\nu K^\nu$. Noether's theorem then guarantees the existence of a conserved current j defined by

$$\partial_\nu\left(\frac{\partial\mathcal{L}}{\partial\left(\partial_\nu\phi_a\right)}\delta_s\phi_a - K^\nu\right) \equiv \alpha\partial_\nu j^\nu = 0. \quad (10.68)$$

Here, α parametrises the transformation $\delta_s\phi_a$, and we define $\Delta_s\phi_a = \delta_s\phi_a/\alpha$ as in Chapter 1. The conserved current j in this expression is unrelated to the source strength J in the equations above, which is introduced only to simplify definitions of observables and must be taken to zero at the end of every calculation.

Inserting the definition of the Noether current into Eq. (10.67) yields

$$\begin{aligned}\left\langle\int dx\left(-\partial_\nu K^\nu - \alpha\partial_\nu j^\nu + J^a\delta_s\phi_a\right)\epsilon(x)\right\rangle &= 0\\ \Rightarrow \quad \left\langle\partial_\nu j^\nu\right\rangle &= \left\langle J^a\Delta_s\phi_a - \partial_\nu K^\nu/\alpha\right\rangle. \end{aligned} \quad (10.69)$$

The final expression follows from the fact that the first line should hold for any function $\epsilon(x)$, so that the integrand must vanish at every position individually. From this expression, we can derive a series of relations between the Noether current and correlation functions, known as Ward–Takahashi identities. The most powerful results are found for global transformations that not only leave the action invariant, but also the Lagrangian. This implies $\partial_\nu K^\nu = 0$ and in that case Eq. (10.69) yields

$$\int \mathcal{D}\phi\, \mathrm{e}^{i\int dx\,(\mathcal{L}+J^a\phi_a)/\hbar}\left(\partial_\nu j^\nu - J^a\Delta_s\phi_a\right) = 0. \quad (10.70)$$

Because of the source terms, this expression is in fact a generating functional for a whole series of Ward–Takahashi identities. These can be found by taking functional derivatives with respect to J^a before letting them tend to zero. The most basic expression, setting $J^a \to 0$ without taking any derivatives, yields $\partial_\nu \langle j^\nu \rangle = 0$. This describes the fact that the expectation value of the Noether current is conserved, as already implied by the combination of Noether's theorem and the absence of anomalies. More instructive identities appear when including functional derivatives with respect to the sources. For example,

$$\frac{\delta}{\delta J^b(x_b)} \frac{\delta}{\delta J^c(x_c)} \int \mathcal{D}\phi \, \mathrm{e}^{i \int dx\, (\mathcal{L}+J^a\phi_a)/\hbar} \left(\partial_\nu j^\nu - J^a \Delta_s \phi_a\right)\Big|_{J=0} = 0$$
$$\Rightarrow \quad \left\langle \phi_b(x_b)\phi_c(x_c)\partial_\nu j^\nu(x)\right\rangle = i \left\langle \phi_c(x_c)\delta(x-x_b)\Delta_s\phi_b + \phi_b(x_b)\delta(x-x_c)\Delta_s\phi_c\right\rangle . \quad (10.71)$$

Other Ward–Takahashi identities can be generated by taking functional derivatives with respect to different sources and different numbers of sources. In the present form, these Ward–Takahashi identities apply to fields ϕ_a obeying the Euler–Lagrange equations (i.e., they are 'on-shell'). Keeping the form of the equation in the first line of Eq. (10.67), however, similar identities can be derived that also hold off-shell.

10.4.1 Example: The Three-Point Interaction Vertex

To see the implications of the Ward–Takahashi identities, consider the specific example of a single field ϕ that is symmetric under the global $U(1)$ phase rotation associated with number conservation. In this case, the infinitesimal symmetry transformation corresponds to $\phi \to \mathrm{e}^{i\alpha}\phi \approx \phi + i\alpha\phi$, so that we define $\Delta_s\phi = i\phi$. For consistency, we should also consider the conjugate field $\bar{\phi}$, with $\Delta_s\bar{\phi} = -i\bar{\phi}$. Considering the source term $J\delta\phi + \bar{J}\delta\bar{\phi}$, then gives the Ward–Takahashi identity:

$$\partial_\nu \left\langle \phi(x_b)\bar{\phi}(x_c) j^\nu(x)\right\rangle = \left(\delta(x-x_c) - \delta(x-x_b)\right)\left\langle \phi(x_b)\bar{\phi}(x_c)\right\rangle . \quad (10.72)$$

The expectation value on the right-hand side of this expression is a real-space propagator $G(x_b, x_c)$ for the field ϕ. The expectation value on the left is a three-point Green's function $G^\nu(x, x_b, x_c) = \left\langle \phi(x_b)\bar{\phi}(x_c) j^\nu(x)\right\rangle$, which can be used to define the three-point interaction vertex $\Gamma^\nu(x, x', x'')$ by writing

$$G^\nu(x, x_b, x_c) = \int dx' \int dx''\, G(x_b, x')\Gamma^\nu(x, x', x'')G(x'', x_c). \quad (10.73)$$

The three-point vertex describes the scattering of a field excitation by the Noether current. Assuming translational symmetry, all propagators, and vertices must conserve momentum, so that the Fourier transform of the three-point propagator can be written as $G^\nu(k'-k, k, k') = G(k)\Gamma^\nu(k'-k, k, k')G(k')$. The Ward–Takahashi identity of Eq. (10.72) then yields a relation between the propagators and interaction vertex:

$$iq_\nu G^\nu(q, k, k+q) = G(k+q) - G(k)$$
$$\Leftrightarrow \quad iq_\nu \Gamma^\nu(q, k, k+q) = G^{-1}(k) - G^{-1}(k+q). \quad (10.74)$$

Here, we wrote $k' = k + q$. This expression constrains the allowed form of the interaction vertex appearing in any theory with global $U(1)$ phase symmetry (or $U(1)$ gauge freedom) by relating it to its single-particle propagators.

A particularly powerful implication can be obtained by considering the limit of vanishing q, in which Eq. (10.74) takes on the differential form

$$i\Gamma^{\nu}(0,k,k) = \frac{\partial}{\partial k^{\nu}} G^{-1}(k). \tag{10.75}$$

This expression is general and holds both for interacting and free theories. Writing Γ_0 and G_0 for the free interaction vertex and propagator, and defining the self-energy Σ by writing $G^{-1} = G_0^{-1} + \Sigma$, Eq. (10.75) reduces to

$$i\Gamma^{\nu}(0,k,k) = i\Gamma_0^{\nu}(0,k,k) + \frac{\partial}{\partial k^{\nu}} \Sigma(k). \tag{10.76}$$

All effects of the interaction, or quantum corrections, on the vertex $\Gamma^{\nu}(0,k,k)$ thus come from the quantum corrections of the propagator itself. This fact is used for constraining the construction of interacting theories in particle physics, but is also used, for example, in numerical dynamical mean field theory calculations, which employ the momentum-independence of the self energy in infinite dimensions to obtain a situation in which quantum corrections to the interaction vertex vanish identically.

11 Particle Physics

Symmetry and symmetry breaking play a central role in our understanding of fundamental particles and their interactions. The relativistic Lorentz symmetry of spacetime, as discussed in Section 11.1, restricts what type of particles can exist and how they should be represented. The approximate chiral symmetry of quark doublets and triplets described in Section 11.2 inspired the systematic ordering of the vast collection of subatomic particles in terms of possible symmetry-breaking processes, which culminated in the formulation of the **Standard Model** of fundamental forces and particles.

One of the key ingredients of the Standard Model is the spontaneous symmetry breaking of a global $SU(2) \times U(1)$ symmetry to a global $U(1)_{\mathrm{em}}$ subgroup. This so-called **electroweak symmetry breaking** is discussed in Section 11.3. It occurs in the presence of associated gauge fields and will therefore lead to an Anderson–Higgs mechanism, which we introduced in Chapter 8 in the context of $U(1)$ gauge freedom. In fact, because of its importance to fundamental particle physics, electroweak symmetry breaking is often colloquially referred to simply as the Higgs effect and the associated amplitude mode as the Higgs boson.

As already discussed in Sections 1.6 and 8.1, gauge invariance can never be broken, and it is exactly the inviolable gauge invariance of the Standard Model that allows the Anderson–Higgs mechanism to arise in electroweak symmetry breaking. Because of this key role played by gauge freedom in the Standard Model, we briefly discuss pure gauge theories (not coupled to matter) in Section 11.4.

We conclude this chapter with a few remarks on $\mathcal{CPT}$-symmetry in Section 11.5, and on possible further unification efforts in Section 11.6. Particle physics is a very large and active field of physics, which this chapter cannot, and does not attempt to, provide an exhaustive overview of. Instead, we focus on highlighting the essential role of symmetry breaking in some of the central components of particle physics.

11.1 Spacetime Symmetries

As explained in Section 1.5.4, an internal symmetry refers to properties of fields and observables other than their spacetime coordinate. An example is the $U(1)$ phase symmetry associated with particle number conservation. Global internal symmetries lead to conserved Noether currents and can be spontaneously broken. In local internal symmetries, the symmetry transformations acting on the internal degrees of freedom depend on the spatial (and possibly temporal) coordinates, but the coordinates themselves do not transform. A prominent example is the local $U(1)$-transformation of the vector potential $A_\mu(x)$ in electromagnetism. As discussed in Section 2.3, Elitzur's theorem ensures that local symmetries cannot be spontaneously broken, although Chapter 8 showed that their global part may be broken.

External symmetries act foremost on spacetime itself. The transformation of quantities that depend on spacetime, such as fields and operators, then follows from the transformation of space and time coordinates. A global spacetime transformation leads to a global symmetry transformation of physical fields.

In particle physics, internal symmetries provide relations between particles and, upon introducing the associated gauge fields, can describe the three fundamental forces. The external symmetries of spacetime, on the other hand, determine the properties of individual particles, such as their spin and exchange statistics. In particular, we will see in Section 11.1.3 that every fundamental particle is defined by an irreducible representation of the Poincaré group.

11.1.1 Spacetime Transformations

Following the usual relativistic nomenclature, we call a point in spacetime, $x^\mu = (x^0, x^i) = (ct, \mathbf{x})$, an event. A **spacetime transformation** then sends each event in spacetime to another event:

$$x^\mu \to x'^\mu. \tag{11.1}$$

This can be formally written as $x' = Tx$, with T the transformation operator acting on the event x.

Exercise 11.1 (Spacetime transformation group).
Argue that the transformations of Eq. (11.1) form a group.

A field $\phi(x)$ is a quantity that depends on spacetime. The effect of the spacetime transformation of Eq. (11.1) on the field ϕ is defined by

$$\phi'(x') = \phi(x). \tag{11.2}$$

In words, the value of the transformed field ϕ' at position x' in the transformed coordinate system is defined to equal the value of ϕ at the position x in the original coordinate system that transforms to x' under the spacetime transformation. Using the relation $x' = Tx$, we can express the transformed field in terms of the original coordinates:

$$\begin{aligned} \phi'(Tx) &= \phi(x), \\ \phi'(x) &= \phi(T^{-1}x). \end{aligned} \tag{11.3}$$

The transformation of vector and tensor fields is derived in the same way. Eq. (11.3) can thus be treated like an internal symmetry, defined by its action on the field variables.

In quantum mechanics, symmetry operations can generally be represented by unitary operators. We can then associate each spacetime transformation T with a unitary operator $\hat{U}(T)$ acting on states $|\psi(x)\rangle$ that depend on spacetime:

$$|\psi'(x)\rangle \equiv \hat{U}(T)|\psi(x)\rangle = |\psi(T^{-1}x)\rangle. \tag{11.4}$$

In other words, the transformations $\hat{U}(T)$ form a **representation** of the spacetime symmetries.

Exercise 11.2 (Spacetime transformation operator group).
Show that the unitary operators $U(T)$ indeed satisfy the requirements of a group.

As a result of the state transformations, operators $\hat{O}(x)$ that depend on spacetime should transform as

$$\begin{aligned} U(T)\hat{O}(x)|\phi(x)\rangle &= \hat{O}'(x)|\phi'(x)\rangle \\ \Rightarrow \quad \hat{O}'(x) &= \hat{U}(T)\hat{O}(x)\hat{U}^{-1}(T). \end{aligned} \tag{11.5}$$

11.1.2 Poincaré Group

The group of all symmetries of Minkowski space is called the **Poincaré group**. It is a Lie group, and therefore, its generators form a Lie algebra. In addition to translations and reflections, the Poincaré group consists of **Lorentz transformations**: spatial rotations and Lorentz boosts. The boosts can be regarded as hyperbolic rotations that mix space and time coordinates. We will see in Section 13.2 that Lorentz boosts are broken in any medium with non-zero energy density. Here, we instead focus on the generators of the Lorentz transformations in vacuum.

11.1.2.1 The Proper Lorentz Group

Three-dimensional rotations are collected in the rotation group $SO(3)$ and can be parametrised by three Euler angles. A rotation over an angle ϕ around the z-axis, for example, is given by

$$x \to x\cos\phi \; - y\sin\phi \,, \qquad y \to x\sin\phi \; + y\cos\phi \,, \qquad z \to z. \tag{11.6}$$

These relations can also be expressed in terms of a rotation matrix $R_z(\phi) = \exp(-\mathrm{i}\phi J_z)$ acting on the vector $(x, y, z)^T$. The generators J_i for rotations around the axes labelled by $i \in \{x, y, z\}$ are then defined as

$$J_x = \begin{pmatrix} 0 & 0 & 0 \\ 0 & 0 & -\mathrm{i} \\ 0 & \mathrm{i} & 0 \end{pmatrix}, \qquad J_y = \begin{pmatrix} 0 & 0 & \mathrm{i} \\ 0 & 0 & 0 \\ -\mathrm{i} & 0 & 0 \end{pmatrix}, \qquad J_z = \begin{pmatrix} 0 & -\mathrm{i} & 0 \\ \mathrm{i} & 0 & 0 \\ 0 & 0 & 0 \end{pmatrix}. \tag{11.7}$$

Note that these rotations leave the distance r between a point (x, y, z) and the origin invariant.

Exercise 11.3 (Rotation Lie algebra).
Verify that the generators for rotations satisfy the commutation relation

$$[J_a, J_b] = \mathrm{i}\epsilon_{abc} J_c. \tag{11.8}$$

Note that this defines the structure constants of Eq. (1.52) for the rotation group.

Complementing the spatial rotations, **Lorentz boosts** describe a transformation between two inertial frames that differ by a velocity $\mathbf{v}$. For instance, if $\mathbf{v}$ is parallel to the z-axis, the spacetime coordinates transform as

$$ct \to \gamma(ct - \beta x), \qquad x \to \gamma(-\beta ct + x), \qquad y \to y, \qquad z \to z. \tag{11.9}$$

Here, c is the speed of light, $\beta = v/c$, and $\gamma = 1/\sqrt{1-\beta^2}$ is the Lorentz factor. Writing $\cosh\zeta = \gamma$ and $\sinh\zeta = \gamma\beta$, the transformation becomes

$$ct \to ct\cosh\zeta - x\sinh\zeta, \qquad x \to -ct\sinh\zeta + x\cosh\zeta, \qquad y \to y, \qquad z \to z. \tag{11.10}$$

This has a very similar form to Eq. (11.6), with the trigonometric functions replaced by hyperbolic ones. We can therefore view the Lorentz boost as a hyperbolic rotation in the tx-plane. In analogy to the spatial rotations, generators K_i for Lorentz boosts along each of the spatial axes $i \in \{x, y, z\}$ can be defined by

$$K_x = \begin{pmatrix} 0 & -\mathrm{i} & 0 & 0 \\ -\mathrm{i} & 0 & 0 & 0 \\ 0 & 0 & 0 & 0 \\ 0 & 0 & 0 & 0 \end{pmatrix}, \quad K_y = \begin{pmatrix} 0 & 0 & -\mathrm{i} & 0 \\ 0 & 0 & 0 & 0 \\ -\mathrm{i} & 0 & 0 & 0 \\ 0 & 0 & 0 & 0 \end{pmatrix}, \quad K_z = \begin{pmatrix} 0 & 0 & 0 & -\mathrm{i} \\ 0 & 0 & 0 & 0 \\ 0 & 0 & 0 & 0 \\ -\mathrm{i} & 0 & 0 & 0 \end{pmatrix}. \tag{11.11}$$

For example, a Lorentz boost along the x-direction is given by acting with $R_x(\zeta) = \exp(-i\zeta K_x)$ on the four-vector $(ct, x, y, z)^T$.

In the non-relativistic limit $v \ll c$, or equivalently $\beta \ll 1$ (but $\beta c = v$), while also $\gamma \approx 1$. The only non-trivial transformation in Eq. (11.9) is then $x \to -vt + x$. This is often referred to as a **Galilean boost** or Galilean transformation.

Exercise 11.4 (Lorentz boosts non-closure).
Show that the Lorentz boosts do not by themselves form a group, by calculating the total transformation resulting from two non-parallel boosts.

The Lorentz boosts together with the rotations form a Lie group, called the **proper Lorentz group**, denoted by $SO(1,3)$. It has six generators, consisting of the three rotations J_a given by Eq. (11.7), and the three boosts K_a defined in Eq. (11.11). The commutation relations between all six generators are

$$[J_a, J_b] = \mathrm{i}\epsilon_{abc}J_c, \qquad [J_a, K_b] = \mathrm{i}\epsilon_{abc}K_c, \qquad [K_a, K_b] = -\mathrm{i}\epsilon_{abc}J_c. \tag{11.12}$$

We can collect these six generators in an antisymmetric tensor $M^{\mu\nu}$ through the definitions $J_a = -\frac{1}{2}\epsilon_{abc}M^{bc}$ and $K_a = -M^{0a}$. The Lie-algebra relations are then summarised in the single expression

$$[M^{\alpha\beta}, M^{\gamma\delta}] = -\mathrm{i}\left(\eta^{\alpha\gamma}M^{\beta\delta} - \eta^{\alpha\delta}M^{\beta\gamma} - \eta^{\beta\gamma}M^{\alpha\delta} + \eta^{\beta\delta}M^{\alpha\gamma}\right). \tag{11.13}$$

Here $\eta^{\mu\nu} = \mathrm{diag}(-1, +1, +1, +1)$ is the (inverse) Minkowski metric tensor.

Exercise 11.5 (Lorentz group tensor generators).
Verify that Eq. (11.13) is equivalent to Eq. (11.12).

Given a four-vector x^μ, a general Lorentz transformation can be written as

$$x^{\mu\prime} = \Lambda^\mu{}_\nu x^\nu. \tag{11.14}$$

Here, $\Lambda^{\mu}{}_{\nu}$ is a 4×4-matrix, generated by the three rotations and three boosts. The fact that all Lorentz transformations $\Lambda^{\mu}{}_{\nu}$ are generated by the rotations and boosts means that we can represent them using six parameters $\vec{\theta}$ and $\vec{\zeta}$. One possible, but not unique, way to do this is to write

$$\Lambda = \exp\left[-\mathrm{i}\vec{\theta}\cdot\vec{J} - \mathrm{i}\vec{\zeta}\cdot\vec{K}\right]. \tag{11.15}$$

Because spacetime is not Euclidean, the matrix Λ is not orthogonal, but instead satisfies $\eta\Lambda^{\mathrm{T}}\eta = \Lambda^{-1}$, or $(\Lambda^{-1})^{\mu}{}_{\nu} = \Lambda_{\nu}{}^{\mu} = \eta_{\nu\kappa}\eta^{\mu\lambda}\Lambda^{\kappa}{}_{\lambda}$, with $\eta_{\mu\nu}$ again the Minkowski metric tensor. These relations imply that the interval $s = x_{\mu}x^{\mu}$ between an event $x^{\mu} = (ct, x, y, z)$ and the origin is left invariant by the elements of the Lorentz group.

Exercise 11.6 (Lorentz group structure).
a. Argue (don't calculate) that the Lorentz transformations $\Lambda^{\mu}{}_{\nu}$ acting on spacetime form a group.
b. The action of Lorentz transformations on quantum mechanical states is given by unitary representations $\hat{U}(\Lambda)$ defined in analogy to Eq. (11.4). Show that the unitary operators $\hat{U}(\Lambda)$ satisfy the requirements of a group.

11.1.2.2 Translation Symmetries

Besides Lorentz transformations, the possible spacetime symmetries include translations, both in time and space. Transformations of this kind leave the interval $\Delta s = |x'^{\mu} - x^{\mu}|$ between any two events x, x' invariant. Unlike the transformations, we encountered so far, translations are not linear, in the sense that there is no matrix M such that $x'^{\mu} = M^{\mu}{}_{\nu}x^{\nu}$. Nevertheless, translations together with rotations and boosts form a group called the **proper Poincaré group**, which is denoted by $ISO(1,3)$ (the full Poincaré group $IO(1,3)$ also includes the discrete transformations discussed in Section 11.1.2.3).

As we saw, for example, in Section 1.4.4, the Hamiltonian H is the generator of time translations, while the momentum operators $\mathbf{P}$ generate spatial translations. These can be collected in the four-momentum $P^{\mu} = (H/c, \mathbf{P})$. Each four-momentum commutes with any other, but their commutation relations with rotations and Lorentz boosts are not trivial. The full Lie algebra of the Poincaré group therefore consists of the ten generators $\mathbf{J}$, $\mathbf{K}$, $\mathbf{P}$, and H. Their commutation relations are

$$\begin{aligned}
&[J_a, J_b] = \mathrm{i}\epsilon_{abc}J_c, && [J_a, K_b] = \mathrm{i}\epsilon_{abc}K_c, && [K_a, K_b] = -\mathrm{i}\epsilon_{abc}J_c,\\
&[J_a, P_b] = \mathrm{i}\epsilon_{abc}P_c, && [K_a, P_b] = \mathrm{i}\delta_{ab}H/c, && [K_a, H] = \mathrm{i}cP_a,\\
&[J_a, H] = 0, && [P_a, H] = 0, && [P_a, P_b] = 0.
\end{aligned} \tag{11.16}$$

Using the tensor $M^{\mu\nu}$ defined above Eq. (11.13), this can be condensed into

$$\begin{aligned}
&[M^{\alpha\beta}, M^{\gamma\delta}] = -\mathrm{i}\left(\eta^{\alpha\gamma}M^{\beta\delta} - \eta^{\alpha\delta}M^{\beta\gamma} - \eta^{\beta\gamma}M^{\alpha\delta} + \eta^{\beta\delta}M^{\alpha\gamma}\right),\\
&[M^{\alpha\beta}, P^{\gamma}] = \mathrm{i}\left(\eta^{\alpha\gamma}P^{\beta} - \eta^{\gamma\beta}P^{\alpha}\right),\\
&[P^{\mu}, P^{\nu}] = 0.
\end{aligned} \tag{11.17}$$

The second line indicates that the Poincaré group is not simply a direct product of the Lorentz group and spacetime translations. Instead it is a so-called semidirect product with Lorentz transformations acting on the translations, which is written as $IO(1,3) = \mathbb{R}^{1,3} \rtimes O(1,3)$. In particular, Lorentz transformations and spacetime translations in general do not commute.

11.1.2.3 Discrete Symmetries

Complementing the continuous Lorentz transformations and translations are the discrete spacetime transformations, which consist of reflections in both spatial and temporal coordinates. They are called **parity** $\mathcal{P}$ and **time reversal** $\mathcal{T}$, respectively, and can be described by the matrices:

$$\mathcal{P} = \begin{pmatrix} 1 & & & \\ & -1 & & \\ & & -1 & \\ & & & -1 \end{pmatrix}, \qquad \mathcal{T} = \begin{pmatrix} -1 & & & \\ & 1 & & \\ & & 1 & \\ & & & 1 \end{pmatrix}. \tag{11.18}$$

Both of these form groups isomorphic to the two-element cyclic group C_2 (or $\mathbb{Z}_2$). They also commute with each other.

Spacetime transformations that include reflections are called **improper**, to distinguish them from the regular, or **proper** rotations, boosts, and translations discussed in Sections 11.1.2.1 and 11.1.2.2. Proper and improper transformations can be distinguished by their determinants. All continuous Lorentz transformations have positive determinant equal to one, while the discrete spacetime transformations have negative determinant -1. The continuous Lorentz transformations and the two discrete transformations together form the **improper Lorentz group** $O(1,3)$, of which the proper Lorentz group $SO(1,3)$ is a subgroup. In fact, the group $O(1,3)$ can be divided into four $SO(1,3)$-pieces, each labelled by one of four possible combinations of the ± 1 eigenvalues of parity and time reversal.

Parity and Chirality

As we will argue in Section 11.1.3, fundamental particles are eigenstates of the Poincaré transformations, which include in particular the discrete spacetime reflections. States and observable properties of fundamental particles that obtain a minus sign under spatial reflection $\mathbf{x} \rightarrow -\mathbf{x}$ are said to be 'odd' under the parity transformation. In contrast, 'even' states and observables are left invariant by the parity transformation. The names used for observables labelled by their behaviour under parity transformations are listed in Table 11.1.

The most common scalar observables, such as mass, are even under parity. However, odd scalar observables, called pseudoscalars, do exist. Similarly, many vectorial quantities, such as position and momentum, are odd under parity, but observables like angular momentum and magnetisation are even. Vector quantities that are even under parity are called axial vectors.

Within the proper Lorentz group, rotations are even under parity whereas boosts are odd:

$$\mathcal{P} K_i \mathcal{P} = -K_i, \qquad \mathcal{P} J_i \mathcal{P} = J_i. \tag{11.19}$$

We can use this observation to introduce an alternative description of the group, in terms of so called left-handed (L_i) and right-handed (R_i) generators:

$$L_i = \frac{1}{2}(J_i - iK_i), \qquad R_i = \frac{1}{2}(J_i + iK_i). \tag{11.20}$$

Name	Rank	Parity	Example
Scalar	0	+1	Energy, mass
Pseudoscalar	0	−1	Helicity
Vector	1	−1	Position **x**, momentum **p**
Axial vector	1	+1	Angular momentum **L**, magnetisation

Table 11.1 Parity classification of physical properties.

Each of these **chiral** operators forms an $SU(2)$ angular momentum algebra, but the left- and right-handed generators commute with one another:

$$[L_a, L_b] = \mathrm{i}\epsilon_{abc}L_c, \qquad [R_a, R_b] = \mathrm{i}\epsilon_{abc}R_c, \qquad [L_a, R_b] = 0. \tag{11.21}$$

Applying a parity transformation exchanges left- and right-handed operators:

$$\mathcal{P}L_i\mathcal{P} = R_i, \qquad \mathcal{P}R_i\mathcal{P} = L_i. \tag{11.22}$$

This rewriting of the generators allows us to denote the proper Lorentz group as the product of a left- and right-handed rotational group $SU(2)_L \times SU(2)_R$. Instead of labelling physical states with their parity eigenvalues, we then can also define them using their chirality.

Time Reversal

Just as we did for Lorentz transformations and translations in Eq. (11.3), or equivalently in Eq. (11.5), parity transformations in quantum mechanics can be implemented by a unitary operator. Time reversal, on the other hand, cannot. As it turns out, time reversal is instead implemented by an anti-unitary operator, satisfying $\hat{\mathcal{T}}^{-1} = -\hat{\mathcal{T}}^{\dagger}$. To see that this results in the behaviour we would expect from time reversal, consider the four-momentum operator $\hat{P}^{\mu}$. If time reversal were implemented unitarily, following Eq. (11.5), we would find $\hat{\mathcal{T}}\hat{P}^{\mu}\hat{\mathcal{T}}^{-1} = \mathcal{T}^{\mu}{}_{\nu}\hat{P}^{\nu}$ with $\mathcal{T}^{\mu}{}_{\nu}$ given by Eq. (11.18). This would mean, for example, that time reversal leaves $\hat{P}^{x}$ unaffected, in contrast to the classical expectation that momentum should be proportional to velocity, which is odd under time reversal. Moreover, it would also imply that $\hat{H} = \hat{P}^0/c$ would satisfy $\hat{\mathcal{T}}\hat{H}\hat{\mathcal{T}}^{-1} = -\hat{H}$. For every eigenstate of a time reversal symmetric Hamiltonian with positive energy, there would then also be one with negative energy, and the Hamiltonian cannot be bounded from below. Implementing time reversal as a non-unitary transformation, on the other hand, yields $\hat{\mathcal{T}}\hat{P}^{\mu}\hat{\mathcal{T}}^{-1} = -\mathcal{T}^{\mu}{}_{\nu}\hat{P}^{\nu}$. This leads to momentum being odd and the Hamiltonian being even under time reversal, according to physical expectation.

Anti-unitary operators have some unfamiliar properties as compared to the usual unitary transformations. For example, the fact that position operators are even under time reversal while momentum is odd implies that it implements complex conjugation:

$$\hat{\mathcal{T}}\mathrm{i}\hat{\mathcal{T}}^{-1} = \hat{\mathcal{T}}[\hat{X}, \hat{P}^x/\hbar]\hat{\mathcal{T}}^{-1} = [\hat{X}, -\hat{P}^x/\hbar] = -\mathrm{i}. \tag{11.23}$$

Put more generally, the anti-unitary operator $\hat{\mathcal{T}}$ is also an anti-linear operator, which satisfies $\langle\hat{\mathcal{T}}\psi|\hat{\mathcal{T}}\psi'\rangle = \langle\psi|\psi'\rangle^*$.

Exercise 11.7 (Time reversal and spacetime translations).
Write spacetime translations as $\hat{T}(a_\mu) = \exp^{-ia_\mu \hat{P}^\mu}$ and calculate $\hat{\mathcal{T}}\hat{T}(a_\mu)\hat{\mathcal{T}}^{-1}$ to show that spatial translations are even under time reversal, while time translations are odd. Use the anti-unitarity of $\hat{\mathcal{T}}$.

Because $\hat{\mathcal{T}}$ is anti-unitary, we do not necessarily have $\hat{\mathcal{T}}^2 = 1$ in every representation, even though obviously $-(-t) = t$. Instead $\mathcal{T}^2$ can be either $+1$ or -1, with the minus sign applying to representations of the Lorentz group associated with particles of half-integer spin, that is, fermions. This is related the fact that a fermionic wave function changes sign under 2π-rotations, as we will discuss in the context of the spin–statistics relation in Section 11.1.3.

For representations of the Lorentz group with $\hat{\mathcal{T}}^2 = 1$, associated with bosons, the eigenvalues of $\hat{\mathcal{T}}$ are ± 1. We can then speak of time-reversal even or time-reversal odd eigenstates, like we did for parity. This situation also applies often to states of non-relativistic quantum mechanics that do not consist of fundamental particles. For instance, the spin operators in a Heisenberg magnet like Eq. (2.17) are taken to be time-reversal odd.

Motion reverses direction under time reversal, and this leads to the fact that all momenta, angular momenta, currents, magnetic dipole moments, and so forth are time-reversal odd. Hypothetical elementary magnetic monopoles would be particles that are odd under time reversal by themselves, without reverting any motion. Such particles feature in several proposals for a grand unified theory (which combine the strong and electroweak interactions, as we will see in Section 11.6). They have been experimentally sought after, but have not been found to date.

11.1.3 Fundamental Particles

A minimal requirement for defining a **fundamental particle** is that it transforms as an irreducible representation of the Poincaré group. The reason is that, to the best of our experimental knowledge, Poincaré transformations are (locally) symmetries of the empty universe. Stable, static excitations of fields in empty spacetime must be eigenstates of its symmetry transformations, and therefore transform as a representation of its isometries. Moreover, elementary particles must correspond to irreducible representations, because if it were reducible, an excitation could be decomposed into particles obeying irreducible representations. One consequence of the requirement for fundamental particles to transform as an irreducible representation is that their properties include, at minimum, the distinct representation labels of the symmetry group. This implies, for example, that fundamental particles in empty spacetime occur as eigenstates (and have definite values) of energy, momentum, and spin.

Spacetime translations form an Abelian subgroup of the Poincaré group. It is a central result of group theory that complex irreducible representations of Abelian groups are one-dimensional. Therefore, the eigenstates of spacetime translations are simply complex scalar functions, labelled by energy and momentum eigenvalues E and $\mathbf{p}$. In contrast, the Lorentz group of rotations and boosts is highly non-trivial, and the representation theory of the Lorentz group is of central importance in particle physics. Good treatments are found, for example, in the textbooks of Refs. [186] and [214].

On of the key properties of irreducible representations of the Poincaré group is their behaviour under a 2π rotation. Simple scalar and vector fields, for example, are symmetric under such rotations. But **spinor** representations are also possible, with states reversing sign under a 2π rotation. Similarly, higher-dimensional irreducible tensor representations are possible as well. The metric tensor $g_{\mu\nu}$, for instance, is a (0,2)-tensor in a 10-dimensional irreducible representation. In the general theory of relativity, the metric tensor is itself a physical field, and its elementary excitations may be conjectured to be tensorial particles called gravitons.

The **spin–statistics relation** indicates that fundamental particles with integer spin (0,1,...) are bosons, while fundamental particles with half-integer spin ($\frac{1}{2}, \frac{3}{2}, \ldots$) must be fermions [151]. Notice that this relation assumes that spacetime is four-dimensional and that Lorentz invariance is unbroken. Non-fundamental particles, such as the quasiparticles featuring in condensed matter systems with broken Lorentz symmetry, therefore need not obey this relation. Indeed, without Lorentz symmetry it is possible to encounter seemingly exotic excitations like 'spinless fermions', which are quasiparticle with fermion statistics but spin zero [219]. Moreover, in two dimensions one can have quantum statistics distinct from just that of bosons and fermions. Quasiparticles called anyons, for example, obtain a nontrivial complex phase rather than just a factor ± 1 upon 2π rotations. A prominent example of such a quasiparticle is the Majorana fermion proposed to occur in both topological superconductors [26] and the fractional quantum Hall effect [12, 60]. As long as we consider only fundamental particles in the Lorentz invariant vacuum, the irreducible spin representations of those particles are listed in Table 11.2.

11.1.3.1 Spinors

To describe spin-$\frac{1}{2}$ fermionic fundamental particles, it is convenient to introduce four-component vectors called **spinors**. The transformation of spinors under elements of the Lorentz group can be neatly expressed using **Dirac gamma matrices**. They are a set of four matrices γ^μ, which satisfy anticommutation relations known as the Clifford algebra $\{\gamma^\mu, \gamma^\nu\} = 2\eta^{\mu\nu}$. The generators of the Lorentz group $M^{\alpha\beta}$, introduced in Eq. (11.13), can be expressed in terms of gamma matrices as

$$M^{\mu\nu} = \frac{i}{4}\left[\gamma^\mu, \gamma^\nu\right]. \tag{11.24}$$

You will use these in Exercise 11.8 to show that the Dirac gamma matrices form a representation of the Lorentz group.

Name	Spin	Example
Scalar	0	Higgs boson
Spinor	$\frac{1}{2}$	Electron, quark
Vector	1	Photon, W- and Z-boson
Tensor	2	Graviton

Table 11.2 Spins of fundamental particles.

Exercise 11.8 (Dirac matrix representation of Lorentz group).
a. Recall that we defined the Minkowski metric as $\eta^{\mu\nu} = \mathrm{diag}(-1, +1, +1, +1)$. Write down explicitly four matrices γ^μ that satisfy the commutation relations $\{\gamma^\mu, \gamma^\nu\} = 2\eta^{\mu\nu}$.
b. Recall that the generators of rotations are $J_a = -\frac{1}{2}\epsilon_{abc}M^{bc}$, and the generators of Lorentz boosts are $K_a = -M^{0a}$. Write down the rotation and boost generators using the relation $M^{\mu\nu} = \frac{\mathrm{i}}{4}\left[\gamma^\mu, \gamma^\nu\right]$ and show that they satisfy the algebra given in Eq. (11.12).

Because the Lorentz group contains left- and right-handed subgroups, $SU(2)_L$ and $SU(2)_R$, spinor representations can be split into two (left- and right-handed) components. One way to identify these components is by using the so-called fifth gamma matrix, defined as

$$\gamma_5 \equiv -\mathrm{i}\gamma^0\gamma^1\gamma^2\gamma^3. \tag{11.25}$$

This matrix anticommutes with all other gamma matrices and has eigenvalues ± 1. It can be used to define the projection operators:

$$P_L = \frac{1}{2}(1 + \gamma_5), \qquad P_R = \frac{1}{2}(1 - \gamma_5). \tag{11.26}$$

You can verify that the left- and right-handed generators L_i, and R_i defined in Eq. (11.20) act solely within either subspace projected on by P_L and P_R. Particles with definite chirality will thus be eigenstates of γ_5, with eigenvalues corresponding to their handedness. More generally, a spinor ψ can be split into a left-handed component $\psi_\mathrm{L} = \frac{1}{2}(1 + \gamma_5)\psi$ and a right-handed component $\psi_\mathrm{R} = \frac{1}{2}(1 - \gamma_5)\psi$.

As noted in Eq. (11.22), the parity transformation flips the sense of handedness. In terms of Dirac gamma matrices, the parity transformation may be written as $\mathcal{P} = i\gamma^0$. Moreover, the gamma matrices transform as a vector, with $\mathcal{P}\gamma^i\mathcal{P} = -\gamma^i$ for $i = 1, 2, 3$. At the same time, the combination $\gamma_5\gamma^i$ does not change sign under the parity transformation, $\mathcal{P}\gamma_5\gamma^i\mathcal{P} = \gamma_5\gamma^i$, and is therefore an axial vector according to the classification of Table 11.1.

11.2 Chiral Symmetry Breaking

Focussing for now on massless fermions, the Dirac Lagrangian for four-component spinors $\psi(x)$ is given by

$$\mathcal{L}_\mathrm{Dirac} = -\overline{\psi}\gamma^\mu\partial_\mu\psi. \tag{11.27}$$

Here γ^μ are the Dirac gamma matrices and $\overline{\psi} \equiv \psi^\dagger\gamma^0$ is the Dirac adjoint of the spinor ψ. In addition to the usual $U(1)$ symmetry $\psi \to \mathrm{e}^{\mathrm{i}\theta}\psi$, this Lagrangian has an additional **chiral symmetry**:

$$\psi \to \mathrm{e}^{\mathrm{i}\theta^\mathrm{A}\gamma_5}\psi. \tag{11.28}$$

As a result, there are two conserved Noether currents, consisting of the usual vector current V^μ associated with particle number conservation and the additional **axial current** associated with the chiral symmetry:

$$V^\mu = \mathrm{i}\overline{\psi}\gamma^\mu\psi, \qquad A^\mu = \mathrm{i}\overline{\psi}\gamma^\mu\gamma_5\psi. \tag{11.29}$$

Using the projectors of Eq. (11.26), we can split the Dirac spinors into left- and right-handed components χ_L and χ_R, defined by

$$\chi_L = P_L\psi, \qquad \chi_R = P_R\psi. \tag{11.30}$$

Using these, the Dirac Lagrangian can be written as:

$$\mathcal{L} = -\overline{\chi}_L\gamma^\mu\partial_\mu\chi_L - \overline{\chi}_R\gamma^\mu\partial_\mu\chi_R. \tag{11.31}$$

This formulation of the Lagrangian makes explicit that the fields χ_L and χ_R both have a global $U(1)$ phase rotation symmetry, and thus that the model for massless Dirac spinors separately conserves the number of left- and right-handed fermions. This is consistent with the vector and axial Noether currents being conserved, because their Noether charges can be written as $\mathrm{i}V^0 = n_L + n_R$ and $\mathrm{i}A^0 = n_L - n_R$. These thus conserve both the total number of fermions, and the difference between the number of left- and right-handed particles. The combination of independent symmetries for the left- and right-handed sectors is often written as $U(1)_L \times U(1)_R$.

11.2.1 *SU*(2) Chiral Doublets

In the massless Dirac Lagrangian of Eq. (11.27) we could write the four-component spinor as a doublet of two-component spinors because of the chiral symmetry, as we did explicitly in Eq. (11.30). The converse is also true: any time a model contains a doublet of fields, there exists a chiral transformation connecting them. Such a model will have a chiral symmetry unless it contains an explicit symmetry-breaking term.

Consider, for example, a complex scalar doublet $\Phi = \begin{pmatrix}\phi_1 & \phi_2\end{pmatrix}^{\mathrm{T}}$ where ϕ_1 and ϕ_2 are complex scalar fields, with the Lagrangian:

$$\mathcal{L} = -\frac{1}{2}(\partial_\mu\Phi^\dagger)(\partial^\mu\Phi) - \frac{1}{2}r\Phi^\dagger\Phi - \frac{1}{4}u(\Phi^\dagger\Phi)^2. \tag{11.32}$$

This Lagrangian has an $SU(2)$ symmetry associated with transformation of the type $\Phi \to L\Phi$ with $L \in SU(2)$. You will find in Exercise 11.9 that it is also invariant under a second 'right-handed' $SU(2)$ transformation. The total symmetry group is thus $SU(2)_L \times SU(2)_R$. The presence of separate left- and right-handed symmetries is a manifestation of chiral symmetry.

Exercise 11.9 (SU(2) chiral symmetry breaking).
In this exercise, we will explore explicitly the chiral symmetry of the model in Eq. (11.32).
a. Show that the Lagrangian is invariant under $\Phi \to L\Phi$, with $L \in SU(2)$ a unitary 2×2 matrix with determinant 1.
b. Show that for any coefficients r_1 and r_2 obeying $r_1^* r_1 + r_2^* r_2 = 1$, the Lagrangian is also invariant under the transformation:

$$\phi_1 \to r_2\phi_1 + r_1\phi_2^*, \qquad \phi_2 \to -r_1\phi_1^* + r_2\phi_2. \tag{11.33}$$

We can use the components of Φ to define a $U(2)$-matrix $\breve{\Phi}$ and similarly use the coefficients r_1 and r_2 to define an $SU(2)$-matrix R, given by

$$\breve{\Phi} = \begin{pmatrix} \phi_2^* & \phi_1 \\ -\phi_1^* & \phi_2 \end{pmatrix}, \qquad R = \begin{pmatrix} r_2 & -r_1 \\ r_1^* & r_2^* \end{pmatrix}. \tag{11.34}$$

We can then use $\Phi^\dagger \Phi = \frac{1}{2}\mathrm{Tr}\ \breve{\Phi}^\dagger \breve{\Phi}$ to express the Lagrangian in terms of $\breve{\Phi}$. This shows explicitly that the Lagrangian is invariant under any transformation of the form

$$\breve{\Phi} \to L\breve{\Phi}R^\dagger. \tag{11.35}$$

Because the matrices L and R in this expression can be chosen independently, the Lagrangian is in fact invariant separately under multiplication of $\breve{\Phi}$ with an $SU(2)$ matrix from either the left or the right. This is the **chiral symmetry** of the model. The Lagrangian is also invariant under global $U(1)$ phase rotations of the form $\Phi \to e^{i\alpha}\Phi$, and the full symmetry group is $SU(2)_\mathrm{L} \times SU(2)_\mathrm{R} \times U(1)$. We will ignore the $U(1)$ symmetry in this exercise.

The groups $SU(2)_\mathrm{L}$ and $SU(2)_\mathrm{R}$ are generated by the matrices Q_a^L and Q_a^R, which satisfy the usual $SU(2)$-relations (as derived, for example, in Exercise 1.7) $[Q_a^\mathrm{L}, Q_b^\mathrm{L}] = i\epsilon_{abc}Q_c^\mathrm{L}$, $[Q_a^\mathrm{R}, Q_b^\mathrm{R}] = i\epsilon_{abc}Q_c^\mathrm{R}$, and $[Q_a^\mathrm{L}, Q_b^\mathrm{R}] = 0$. We use these to define vector and axial charges as: $Q_a^\mathrm{V} = Q_a^\mathrm{L} + Q_a^\mathrm{R}$ and $Q_a^\mathrm{A} = Q_a^\mathrm{L} - Q_a^\mathrm{R}$.
c. Show that the vector and axial charges satisfy the algebra relations:

$$\begin{aligned}
[Q_a^\mathrm{V}, Q_b^\mathrm{V}] &= i\epsilon_{abc}Q_c^\mathrm{V}, \\
[Q_a^\mathrm{A}, Q_b^\mathrm{A}] &= i\epsilon_{abc}Q_c^\mathrm{V}, \\
[Q_a^\mathrm{V}, Q_b^\mathrm{A}] &= i\epsilon_{abc}Q_c^\mathrm{A}.
\end{aligned} \tag{11.36}$$

This shows the Q_a^V generate a subgroup of $SU(2)_\mathrm{L} \times SU(2)_\mathrm{R}$, but the Q_a^A do not.

If $r < 0$ and $u > 0$ in Eq. (11.32), we know from Section 5.2 that the potential has a minimum at $\langle\Phi\rangle = \begin{pmatrix} 0 & v \end{pmatrix}^\mathrm{T}$ with $v \in \mathbb{R}$. Equivalently, we can say there is a minimum at $\langle\breve{\Phi}\rangle = \mathrm{diag}(v, v)$.
d. Show that a pure vector transformation ($L = R$ in Eq. (11.35)) leaves the expectation value $\langle\breve{\Phi}\rangle$ invariant, while a purely axial transformation ($L = R^\dagger$ in Eq. (11.35)) does not.

This result shows that realising a specific value for v in $\langle\breve{\Phi}\rangle$ corresponds to the spontaneous breaking of the symmetry of the model from $SU(2)_\mathrm{L} \times SU(2)_\mathrm{R}$ to the diagonal subgroup $SU(2)_{\mathrm{L+R}}$ generated by the vector charges Q_a^V. Since there are three broken generators Q_a^A, we expect three type-A NG modes. The Lagrangian expressed in terms of $\breve{\Phi}$, with the chiral symmetry of Eq. (11.35) is the same as the one describing the Higgs field before coupling to any other degrees of freedom, as we will see in Eq. (11.58) of Section 11.3. Note that once we include coupling to gauge fields in that section, however, the three NG modes become massive.

The $SU(2)$ chiral symmetry in theories containing a fermion doublet may be spontaneously broken. As an example, consider a doublet consisting of the up and down quark. Because these are the lightest quarks, we may approximate them as having zero mass. In that case, the Dirac Lagrangian becomes

$$\mathcal{L} = -\bar{q}\gamma^\mu \partial_\mu q + \ldots . \tag{11.37}$$

Here, chiral doublet q is given by $q = \begin{pmatrix} u & d \end{pmatrix}$, with u and d the four-component spinors describing up and down quark states. This means that q itself is an eight-component spinor. The Lagrangian is invariant under the general $SU(2) \times SU(2)$ transformations given by

$$\begin{pmatrix} u \\ d \end{pmatrix} \rightarrow \exp\left(\mathrm{i}\vec{\theta}^V \cdot \vec{\tau} + \mathrm{i}\gamma_5 \vec{\theta}^A \cdot \vec{\tau}\right) \begin{pmatrix} u \\ d \end{pmatrix}. \tag{11.38}$$

In this expression, the component τ_i of the vector $\vec{\tau}$ are the three Pauli matrices, which act on the two components of q, while γ_5 is the usual fifth gamma matrix acting on the u and d spinors. The Noether currents associated with these transformations again contain a vector and an axial component, one for each generator of the $SU(2)$ transformations:

$$\vec{V}^\mu = \mathrm{i}\bar{q}\gamma^\mu \vec{\tau} q, \qquad \vec{A}^\mu = \mathrm{i}\bar{q}\gamma^\mu \gamma_5 \vec{\tau} q. \tag{11.39}$$

Noether's theorem guarantees that both currents are conserved, $\partial_\mu \vec{V}^\mu = \partial_\mu \vec{A}^\mu = 0$.

Nambu–Goldstone modes associated with the spontaneous breakdown of the usual, vectorial $SU(2)$ symmetry can be excited by the vector charge. On the other hand, if the chiral symmetry is broken, the corresponding Nambu–Goldstone mode will carry an axial Noether charge proportional to $\vec{A}^0$. In either case, Eq. (11.38) shows that the excited state can be expressed in terms of the combinations $\bar{u}d$, $\bar{u}u$, $\bar{d}d$, and $\bar{d}u$. Such states of bound quark and an antiquark components are called **mesons**. In particular, bound states of up and down quarks are known as pi mesons or **pions**.

11.2.2 Approximate Symmetry Breaking

Historically, the idea that pions are Nambu–Goldstone modes, corresponding to the spontaneous breaking of a chiral symmetry between up and down quarks, was a crucial step in recognising the potential for symmetry to guide the construction of a Standard Model of particle physics, for which Nambu received the 2008 Nobel Prize. Its success pointed Weinberg, Salam and Glashow (who received the 1979 Nobel Prize) to propose symmetry breaking as the principle uniting the weak and electromagnetic forces into one electroweak force, which we will discuss in Section 11.3.

There is an obvious flaw with the identification of pions as Nambu–Goldstone modes, however. Pions are not massless. Moreover, up and down quarks are not massless either, which means their chiral symmetry is explicitly broken. If we assume that the up and down quark have the same mass $m_{u/d}$, we can write the symmetry-breaking term as

$$\mathcal{L}_{\text{mass}} = -m_{u/d}\bar{q}q. \tag{11.40}$$

While proving Noether's theorem in Chapter 1, we wrote a general variation of the Lagrangian in the form of Eq. (1.10). Applying this to variations $(\delta\mathcal{L})_\alpha$ of the Lagrangian in Eq. (11.37), under the transformation $q \rightarrow e^{i\gamma_5 \theta^A_\alpha \tau^\alpha} q$ (no summation over α), we find $(\delta\mathcal{L})_\alpha = \partial_\mu A^\mu_\alpha$. If the transformation is a symmetry, we have $(\delta\mathcal{L})_\alpha = 0$, and the current A^μ_α is conserved. However, with the mass term $\mathcal{L}_{\text{mass}}$ included in the Lagrangian, the axial current is no longer conserved:

$$\partial_\mu A^\mu_\alpha = 2im_{u,d}\bar{q}\gamma_5 \tau_\alpha q. \tag{11.41}$$

This observation is colloquially known as the partial conservation of the axial current (PCAC).

It is sometimes said that the massive pion arises from spontaneously breaking an **approximate symmetry**. This should not be taken literally, because with the mass term already explicitly breaking chiral symmetry, there is nothing left of it that can be spontaneously broken. Instead, the phrasing 'spontaneously broken approximate symmetry' is intended to suggest that in the limit where the explicit symmetry-breaking term vanishes, the restored symmetry would be spontaneously broken. Moreover, it also suggests that the explicit symmetry-breaking term is weak, so that many properties of the system will be close to those expected of the idealised system with a spontaneously broken symmetry. In particular, apart from its small mass, the pion closely resembles a massless Nambu–Goldstone mode, and as mentioned in Section 4.3.1, it is often called a **pseudo Goldstone boson** or **pseudo NG mode**. Note that pions are also pseudoscalars in the sense of Table 11.1, but that the 'pseudo' in pseudoscalar has a completely different meaning from the 'pseudo' in pseudo NG mode.

The phenomenology of pseudo NG modes is general, and they arise whenever a system has an approximate symmetry that would be spontaneously broken in the symmetric limit. We can derive the behaviour of their small masses using the effective potential of Section 10.1. Consider a system with fields ϕ_a and multiple continuous transformations $\phi_a^{(\alpha)} \to \phi_a + \delta\phi_a^{(\alpha)}$, labelled by α. We can always write the transformations as $\delta\phi_a^{(\alpha)} \equiv \sum_b t_{ab}^{(\alpha)}\phi_b$, so that $t^{(\alpha)}$ are the generators of the transformations. Having an approximate symmetry means the effective potential consists of a symmetric term V_0, which is invariant under the transformations $t^{(\alpha)}$, as well as a small symmetry-breaking term λV_1, which is not:

$$V_{\text{eff}}(\phi) = V_0(\phi) + \lambda V_1(\phi). \tag{11.42}$$

The ground state configuration corresponds to a field $\overline{\phi}$ minimising V_{eff}, so that $\partial V_{\text{eff}}/\partial\phi_a|_{\phi=\overline{\phi}} = 0$ for all components a. Treating λ as a small parameter, we can expand the ground state in orders of λ, and write $\overline{\phi} = \overline{\phi}_0 + \lambda\overline{\phi}_1 + \mathcal{O}(\lambda^2)$. At zeroth order, the minimisation condition gives a constraint on $\overline{\phi}_0$, namely,

$$\left.\frac{\partial V_0(\phi)}{\partial\phi_a}\right|_{\phi=\overline{\phi}_0} = 0. \tag{11.43}$$

If the system is in a state that breaks the symmetry of V_0 , so that $\overline{\phi}_0 \neq 0$, there is a degeneracy of possible solutions to Eq. (11.43). All of these solutions have the same amplitude $|\overline{\phi}_0| \equiv v$, and they are connected by symmetry transformations.

Continuing the expansion, at first order in λ the minimisation condition yields

$$\sum_b \left.\frac{\partial^2 V_0(\phi)}{\partial\phi_a\partial\phi_b}\right|_{\phi=\overline{\phi}_0} \overline{\phi}_{1,b} + \left.\frac{\partial V_1(\phi)}{\partial\phi_a}\right|_{\phi=\overline{\phi}_0} = 0. \tag{11.44}$$

You might recognise the second-order derivative of the potential in this expression as the so-called mass matrix defined in Eq. (10.27) of Exercise 10.3. Given that $\overline{\phi}_0$ breaks a symmetry of V_0, the theory with $\lambda = 0$ should have NG modes. As you showed in Exercise 10.3, this means that the mass matrix satisfies the condition:

$$\sum_a \left.\frac{\partial^2 V_0(\phi)}{\partial\phi_a\partial\phi_b}\right|_{\phi=\overline{\phi}_0} (\delta\overline{\phi}_0)_a^{(\alpha)} \equiv \sum_a (m_0^2)_{ab}(\delta\overline{\phi}_0)_a^{(\alpha)} = 0. \tag{11.45}$$

Here, $(\delta\overline{\phi}_0)_a^{(\alpha)} = \sum_b t_{ab}^{(\alpha)} \phi_b|_{\phi=\overline{\phi}_0}$ indicate the variations of the field $\overline{\phi}_0$ under the symmetry transformation t^α. As in Exercise 10.3, the fact that $(\delta\overline{\phi}_0)_a^{(\alpha)} \neq 0$ for any broken symmetry labelled by α, implies that the matrix $(m_0^2)_{ab}$ associated with V_0 must have a zero eigenvalue whenever a symmetry is broken, and hence that there must exist a massless NG mode if we only consider the symmetric part V_0 of the effective potential.

Multiplying Eq. (11.44) with $(\delta\overline{\phi}_0)_a^{(\alpha)}$ and summing over a yields the so-called **vacuum alignment condition**:

$$\sum_a \frac{\partial V_1(\phi)}{\partial \phi_a}\bigg|_{\phi=\overline{\phi}_0} (\delta\overline{\phi}_0)_a^{(\alpha)} = 0. \tag{11.46}$$

This shows that the symmetry-breaking term V_1 in the effective potential induces an extra condition on the zeroth order solution $\overline{\phi}_0$. It forces the solution to point in the specific direction preferred by the symmetry-breaking perturbation, rather than picking any of the other directions that were degenerate solutions in the absence of the perturbation. This is visualised in Figure 11.1 for the example of a complex scalar field with $U(1)$ phase rotation symmetry.

Excitations on top of the aligned ground state in the presence of V_1 are no longer massless Nambu–Goldstone modes. We can find their mass by explicitly expanding the mass matrix associated with the full potential $V_{\text{eff}} = V_0 + \lambda V_1$, which is given by $m_{ab}^2 = \partial^2 V(\phi)/\partial\phi_a\partial\phi_b|_{\phi=\overline{\phi}}$. At zeroth order, the mass matrix is just the m_0^2 introduced in Eq. (11.45), which has zero eigenvalues corresponding to massless Nambu–Goldstone modes for each symmetry in V_0 that is spontaneously broken. The first-order correction to the mass matrix becomes

$$(m_1^2)_{ab} = \sum_c \frac{\partial^3 V_0(\phi)}{\partial\phi_a\partial\phi_b\partial\phi_c}\bigg|_{\phi=\overline{\phi}_0} \overline{\phi}_{1,c} + \frac{\partial^2 V_1(\phi)}{\partial\phi_a\partial\phi_b}\bigg|_{\phi=\overline{\phi}_0}. \tag{11.47}$$

We are interested in the fate of the excitations that were massless Nambu–Goldstone modes in the absence of the perturbation V_1. For each broken symmetry transformation $t^{(\alpha)}$,

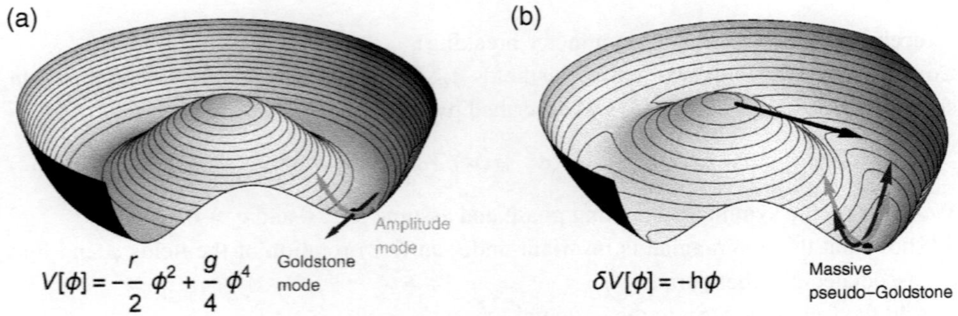

Figure 11.1

Approximate symmetry breaking in a complex scalar field. (a) The potential $V[\phi] = -r|\phi|^2/2 + g|\phi|^4/4$ with an exact symmetry. The symmetry-broken state has a massless Nambu–Goldstone mode corresponding to excitations around the rim of the potential. (b) The symmetric potential perturbed by a small symmetry-breaking term $\delta V[\phi] = -(h\phi^* + h^*\phi)$, with the phase of h visualised by the direction of the black arrow emanating from the origin. The favouring of one of the previously degenerate ground states by the perturbation is known as vacuum alignment. As a result of the perturbation, the rim of the potential is canted, and excitation along it are massive pseudo NG modes.

such a mode exists and is proportional to $(\delta\overline{\phi}_0)_a^{(\alpha)}$, which is an eigenvector of the mass matrix m_0^2. Normalising these eigenvectors by writing t for the norm of the eigenvector $(\delta\overline{\phi}_0)_a^{(\alpha)}$, we can construct the mass matrix for just the pseudo NG modes as

$$M_{\alpha\beta}^2 = |t|^{-2} \sum_{ab} (\delta\overline{\phi}_0)_a^{(\alpha)} (m^2)_{ab} (\delta\overline{\phi}_0)_b^{(\beta)}. \tag{11.48}$$

The indices α and β in this expression run over the broken symmetry generators only. Using the definition $(\delta\overline{\phi}_0)_a^{(\alpha)} = \sum_m t_{mb}^{(\alpha)} \overline{\phi}_{0,b}$ and expanding in lowest order in λ yields

$$M_{\alpha\beta}^2 = |t|^{-2} \sum_{ab,mn} t_{ma}^{(\alpha)} t_{nb}^{(\beta)} \overline{\phi}_{0,a} \overline{\phi}_{0,b} (m_1^2)_{ab}. \tag{11.49}$$

Here, we used the fact that m_0^2 is zero for these particular excitations.

The final step is to insert the definition of $(m_1^2)_{ab}$ from Eq. (11.47). The third-order derivative in this definition can be rewritten explicitly as the derivative of $\frac{\partial^2 V(\phi)}{\partial\phi_a \partial\phi_b}$. Using Eq. (11.44), the third-order derivative term can therefore be simplified to something that only depends on V_1:

$$\frac{1}{|t|^2} \sum_{mnab} t_{ma}^{(\alpha)} t_{nb}^{(\beta)} \overline{\phi}_{0,b} \left[\frac{\partial \overline{\phi}_{0,a}}{\partial \overline{\phi}_{0,n}} \left. \frac{\partial V_1(\phi)}{\partial \phi_m} \right|_{\phi=\overline{\phi}_0} - \frac{\partial}{\partial \overline{\phi}_{0,n}} \left(\overline{\phi}_{0,a} \left. \frac{\partial V_1(\phi)}{\partial \phi_m} \right|_{\phi=\overline{\phi}_0} \right) \right]. \tag{11.50}$$

The final term in this expression vanishes because of the vacuum alignment condition of Eq. (11.46), while the first term can be rewritten to express the mass matrix in the form

$$M_{\alpha\beta}^2 = \sum_m \frac{(t^{(\alpha)} t^{(\beta)} \overline{\phi}_0)_m}{|t|^2} \left. \frac{\partial V_1(\phi)}{\partial \phi_m} \right|_{\phi=\overline{\phi}_0} + \sum_{mn} \frac{(t^{(\alpha)} \overline{\phi}_0)_m (t^{(\beta)} \overline{\phi}_0)_n}{|t|^2} \left. \frac{\partial^2 V_1(\phi)}{\partial \phi_m \partial \phi_n} \right|_{\phi=\overline{\phi}_0}. \tag{11.51}$$

If the perturbation λV_1 breaks the symmetries of V_0, this matrix will have non-zero eigenvalues. The squared mass of the pseudo NG mode is then proportional to the symmetry-breaking potential, or in other words, the square of the pion mass must be proportional to the quark mass.

Exercise 11.10 (Approximate symmetry breaking).
Consider a theory with two real scalar fields ϕ_a, $a = 1, 2$, with a potential of the form shown in the left of Figure 11.1 and described by

$$V_0[\phi_a] = \frac{\mu}{2}(\phi_1^2 + \phi_2^2) + \frac{g}{4}(\phi_1^2 + \phi_2^2)^2. \tag{11.52}$$

We focus on the symmetry-breaking phase and assume $\mu < 0$ and $g > 0$.
a. Show that this Lagrangian is invariant under an $O(2)$ rotation of the fields ϕ and find the associated Noether current.
b. Add the following term to the potential,

$$V_1[\phi_a] = -h\phi_1, \tag{11.53}$$

with $h > 0$ and show that the total potential $V = V_0 + V_1$ is no longer symmetric under $O(2)$ rotations of the fields. Also show that the Noether current you derived in (a) is no longer conserved.
c. Find the position $\overline{\phi}$ of the minimum of the potential V to first order in h. To do so, write the solution as $\overline{\phi} = \overline{\phi}^{(0)} + h\overline{\phi}^{(1)}$ and expand $\partial V/\partial\phi_a$ to first order in h.

d. Calculate the mass matrix at the minimum of the total potential, $m^2_{ab} = \left.\frac{\partial^2 V}{\partial\phi_a\partial\phi_b}\right|_{\phi=\bar{\phi}}$, to first order in h. Show that there is a massless Nambu–Goldstone mode when $h = 0$ and indicate what the mass of that same mode is for h not equal to zero.

The idea that there is an approximate chiral $SU(2)$ symmetry between up and down quarks that are nearly massless can be extended to higher-order chiral symmetry groups. The most prominent application is the approximate $SU(3) \times SU(3)$ chiral symmetry of the strong interactions, which combines the up, down, and strange quarks into a triplet. The breaking of this approximate symmetry then yields an octet of pseudo Goldstone mesons: in addition to the pions, we find kaons and eta mesons.

11.3 Electroweak Symmetry Breaking

We introduced symmetry breaking in the presence of gauge fields in Section 8.1, for the specific example of $U(1)$ symmetry and $U(1)$ gauge fields. The physics discussed there plays an essential role throughout particle physics, but typically in the context of more involved non-Abelian symmetry groups. A primary example is the model for the **electroweak force**, which is precisely such a non-Abelian gauge theory.

11.3.1 Yang–Mills Theory

To describe gauge invariance and symmetry breaking for the electroweak force, we start from the general Lagrangian for a two-component scalar field:

$$\mathcal{L} = -\frac{1}{2}\left|\partial_\mu\phi\right|^2 - \frac{r}{2}\phi^\dagger\phi - \frac{u}{4}(\phi^\dagger\phi)^2. \tag{11.54}$$

Here, $\phi(x)$ is the two-component vector $\phi(x) \equiv \begin{pmatrix}\phi_1(x) & \phi_2(x)\end{pmatrix}^T$, with $\phi_{1,2}(x)$ complex scalars. As discussed in Section 1.7, this Lagrangian is invariant under a global $SU(2)$ rotation of the fields $\phi_{1,2}(x)$. In addition, there is also a global $U(1)$ rotation that changes the complex phase of both fields at the same time. Therefore, Eq. (11.54) has a global $SU(2) \times U(1)$ symmetry, which can be characterised by four parameters α^a ($a = 1, 2, 3$) and β:

$$\phi \to e^{i\alpha^a\sigma^a} e^{i\beta/2}\phi. \tag{11.55}$$

The σ^a in this expression are Pauli matrices, and the factor 1/2 is introduced for later convenience. If the coefficient of the quadratic term in the Lagrangian is negative, $r < 0$, spontaneous symmetry breaking occurs. The ground-state expectation value for the field ϕ is found by solving the Euler–Lagrange equations, which yields $\langle\phi^\dagger\rangle\langle\phi\rangle = \frac{|r|}{u} \equiv v^2$. Choosing a specific configuration that obeys this condition breaks the $SU(2) \times U(1)$ symmetry down to a $U(1)$ subgroup.

Exercise 11.11 ($SU(2) \times U(1)$ symmetry breaking).
a. Verify that Eq. (11.54) is invariant under the global transformation of Eq. (11.55).
b. Show that $\phi_2(x)$ is unaffected by the transformation of Eq. (11.55) for the specific choice of parameters $\alpha^3 = \beta$ and $\alpha^{1,2} = 0$.
c. Show that for $r < 0$, solving the Euler–Lagrange equations yields the condition $\langle\phi^\dagger\rangle\langle\phi\rangle = \frac{|r|}{u} \equiv v^2$.
d. Consider the ground state $\bar{\phi} = \begin{pmatrix} 0 \\ v \end{pmatrix}$, with v real. Write $\phi = \bar{\phi} + \delta\phi$ and expand the Lagrangian in small $\delta\phi$.
e. Identify which fields are Nambu–Goldstone modes, and which is the amplitude mode. How many Nambu–Goldstone modes you find?
f. After symmetry breaking, which choice of parameters in the transformation of Eq. (11.55) still leave the fields invariant? In other words, what is the set of unbroken symmetry transformations? Consequently, what is the coset space of inequivalent symmetry-broken states?

For each global symmetry of the Lagrangian in Eq. (11.54), we can introduce a gauge field. Traditionally, the gauge field associated with the $SU(2)$ symmetry is denoted A^a_μ, while B_μ is used for the $U(1)$ gauge field. The covariant derivative for the two-component field ϕ can then be defined as

$$D_\mu\phi = \left(\partial_\mu - \mathrm{i}gA^a_\mu\sigma^a - \mathrm{i}\tfrac{1}{2}g'B_\mu\right)\phi. \tag{11.56}$$

Here, g is the coupling constant or charge associated with the $SU(2)$ symmetry, which is called the **weak isospin**, while g' is the coupling constant for the $U(1)$ symmetry and is called the **weak hypercharge**. The gauge transformations that leave the covariant derivative invariant are obtained by making the transformations of Eq. (11.55) local:

$$\begin{aligned} \phi(x) &\rightarrow \mathrm{e}^{\mathrm{i}\alpha^a(x)\sigma^a}\mathrm{e}^{\mathrm{i}\beta(x)/2}\phi(x), \\ A^a_\mu(x) &\rightarrow A^a_\mu(x) + \frac{1}{g}\partial_\mu\alpha^a(x), \\ B_\mu(x) &\rightarrow B_\mu(x) + \frac{1}{g'}\partial_\mu\beta(x). \end{aligned} \tag{11.57}$$

Exercise 11.12 ($SU(2) \times U(1)$ gauge transformation).
Verify that $|D_\mu\phi|^2$ is invariant under the gauge transformation of Eq. (11.57).

The Lagrangian that describes the dynamics of the gauge fields A^a and B, as well as their minimal coupling to the matter fields ϕ, is called **Yang–Mills theory**:

$$\begin{aligned} \mathcal{L}_{YM} = &-\frac{1}{4}\left(\partial_\mu A^a_\nu - \partial_\nu A^a_\mu + g\epsilon_{abc}A^b_\mu A^c_\nu\right)^2 - \frac{1}{4}\left(\partial_\mu B_\nu - \partial_\nu B_\mu\right)^2 \\ &-\frac{1}{2}\left|D_\mu\phi\right|^2 - \frac{r}{2}\phi^\dagger\phi - \frac{u}{4}(\phi^\dagger\phi)^2. \end{aligned} \tag{11.58}$$

The ϵ_{abc} in this expression is the completely antisymmetric Levi-Civita tensor. The first line of Eq. (11.58) describes the energy of the gauge field and is the non-Abelian generalisation of the Lagrangian for electric and magnetic fields, $\mathbf{E}^2 - \mathbf{B}^2$, in the usual theory of electromagnetism.

To count the physical degrees of freedom in the Lagrangian of Eq. (11.58), first note that there are four massless gauge bosons, described by the fields A^a and B. Each is a massless vector field constrained by a gauge freedom, similar to the electromagnetic vector potential in electromagnetism. These fields carry two propagating, physical degrees of freedom each [49, 71]. On top of that, the two-component complex scalar field ϕ has a total of four degrees of freedom. The full $SU(2) \times U(1)$ theory therefore includes a grand total of 12 propagating degrees of freedom. These degrees of freedom are present in both the symmetric and symmetry-broken phase. As we will see shortly, the Anderson–Higgs mechanism will yield three massive vector bosons in the symmetry-breaking phase, with 3 degrees of freedom each, complemented by one massless vector boson with 2 degrees of freedom, and one massive real scalar boson with a single degree of freedom, leading to the same total of 12 physical degrees of freedom.

11.3.2 Massive *W* and *Z* Bosons

The presence of gauge freedom in the Yang–Mills theory of Eq. (11.58) means that breaking the global $SU(2) \times U(1)$ symmetry will cause the three massless Nambu–Goldstone modes you found in Exercise 11.11 for the theory without gauge fields, to become massive. This is the Anderson–Higgs mechanism introduced in Section 8.3, and it causes the carriers of the weak force, the W and Z bosons, to be massive. Moreover, because the W and Z bosons are massive, the weak force is short-ranged, as a result of a non-Abelian version of the Meissner effect.

To see how the massive W and Z bosons arise from global $SU(2)$ symmetry breaking, consider Eq. (11.58) with a negative coefficient $r < 0$ for the term quadratic in matter fields. The scalar matter field ϕ then obtains a non-zero expectation value, which is commonly chosen to be

$$\langle \phi \rangle = \begin{pmatrix} 0 \\ v \end{pmatrix}. \tag{11.59}$$

We also choose $v = \sqrt{|r|/u}$ to be real and positive. The first choice, of a symmetry-breaking ground state with the ϕ_1 component zero, corresponds to picking a particular direction for the $SU(2)$ order parameter, while the second choice, to have v real and positive, is a gauge fix that is commonly called the **unitary gauge**. It is similar to the unitary gauge fix used in Section 8.2 in the context of the Abelian Anderson–Higgs mechanism.

Having chosen a ground state configuration that spontaneously breaks the global $SU(2) \times U(1)$ symmetry down to $U(1)$, the Anderson–Higgs mechanism comes into play. That is, in the presence of the broken symmetry, the covariant derivatives in the kinetic term of the ϕ-field result in mass terms for the A^a_μ and B_μ gauge fields. To see the effect, we can use a mean-field approximation and replace the field ϕ with its expectation value. This yields two separate mass terms for the gauge fields:

$$\mathcal{L} = -\frac{1}{8} v^2 \left[g^2 \left| A^1_\mu - \mathrm{i} A^2_\mu \right|^2 + \left(g A^3_\mu - g' B_\mu \right)^2 \right] + \ldots . \tag{11.60}$$

The first term combines the two real components A^1 and A^2 into a single complex massive boson. It has two independent degrees of freedom, which we call the W bosons. They are defined in terms of the A^a_μ fields as $W^\pm_\mu = (A^1_\mu \mp iA^2_\mu)/\sqrt{2}$, so that the W mass term becomes

$$\mathcal{L}_W = -\frac{v^2 g^2}{4} W^+_\mu W^-_\mu \equiv -\frac{1}{2} m_W^2 \left(W^+_\mu W^-_\mu + W^-_\mu W^+_\mu \right). \tag{11.61}$$

The mass of the W bosons is thus finally given by $m_W = vg/2$.

The second term of Eq. (11.60) indicates a specific real, linear combination of the components A^3_μ and B_μ that obtains a mass. This suggests that there also exists an orthogonal combination of those components that remains massless. To describe both the massive and massless field, we can introduce a transformation of the form

$$\begin{pmatrix} Z_\mu \\ A_\mu \end{pmatrix} = \begin{pmatrix} \cos\theta_w & -\sin\theta_w \\ \sin\theta_w & \cos\theta_w \end{pmatrix} \begin{pmatrix} A^3_\mu \\ B_\mu \end{pmatrix}. \tag{11.62}$$

Here, θ_w is called the **electroweak mixing angle** or sometimes the **Weinberg angle**. Choosing the value of the mixing angle such that Z_μ becomes the specific combination of fields appearing in Eq. (11.60), we have identified two different gauge fields originating from A^3_μ and B_μ. There is the massive Z boson, as well as a massless photon A_μ (not to be confused with the A^a_μ gauge fields associated with the unbroken $SU(2)$ symmetry). Comparing Eq. (11.62) with the Lagrangian of Eq. (11.60), we find that the electroweak mixing angle is given by

$$\cos\theta_w = \frac{g}{\sqrt{g^2 + g'^2}}, \qquad \sin\theta_w = \frac{g'}{\sqrt{g^2 + g'^2}}. \tag{11.63}$$

Written in terms of the Z boson, the mass term becomes

$$\mathcal{L}_Z = -\frac{1}{2}\frac{v^2(g^2 + g'^2)}{4} Z^2_\mu \equiv -\frac{1}{2} m_Z^2 Z^2_\mu. \tag{11.64}$$

The mass of the Z boson is thus equal to $m_Z = \frac{v}{2}\sqrt{g^2 + g'^2}$. Comparing this to the W, boson mass provides a direct relation between the electroweak mixing angle and the ratio of bosonic masses:

$$\cos\theta_w = \frac{m_W}{m_Z}. \tag{11.65}$$

The experimentally determined masses of the W and Z bosons are $m_W = 80.38$ GeV and $m_Z = 91.19$ GeV, yielding a mixing angle that is approximately $\theta_w = 28°$. This is often expressed in terms of $\sin^2(\theta_w) = 1 - (m_W/m_Z)^2$, which then approximately equals $\sin^2(\theta_w) \approx 0.2229$.

11.3.3 The Residual $U(1)$ Symmetry: Electromagnetism

Having broken the global $SU(2) \times U(1)$ symmetry, a residual unbroken $U(1)$ global symmetry remains. Within the unitary gauge in Eq. (11.59) and our choice of ground state expectation value for the Higgs field ϕ, the residual symmetry corresponds to rotations of the phase of the first component in the ϕ-doublet. In other words, it corresponds to transformations with parameters $\alpha^3 = \beta$ in Eq. (11.55).

The local version of this transformation, with $\alpha^3(x) = \beta(x) \equiv \alpha(x)$, is part of the gauge freedom, which remains regardless of whether the global symmetry $SU(2) \times U(1)$ is broken. Using Eq. (11.57), it can be seen to act only on the first component of ϕ and the massless boson $A_\mu(x)$:

$$\begin{aligned} \phi_1(x) &\to e^{i\alpha(x)}\phi_1(x), \\ A_\mu(x) &\to A_\mu(x) + \left(\frac{\sin\theta_w}{g} + \frac{\cos\theta_w}{g'}\right)\partial_\mu\alpha(x). \end{aligned} \tag{11.66}$$

Comparing this expression to the usual electromagnetic $U(1)$ gauge transformation in Eq. (1.49), we see that the massless A_μ boson transforms under $U(1)$ gauge transformations exactly in the way expected for the electromagnetic photon, if we identify $-1/(\sin\theta_w/g + \cos\theta_w/g')$ with the electromagnetic charge e. Using Eq. (11.63) to rewrite this yields

$$g = \frac{-e}{\sin\theta_w}, \qquad g' = \frac{-e}{\cos\theta_w}. \tag{11.67}$$

This provides a relation between on the one hand the elementary electromagnetic charge e and the mixing angle θ_w, and on the other the weak isospin g and weak hypercharge g' of the $SU(2) \times U(1)$ theory.

Notice that the W bosons transform non-trivially under the residual $U(1)$ symmetry, and that they therefore represent particles that are charged under the electromagnetic interaction. Also, the three transformations contained in Eq. (11.57) besides the electromagnetic gauge transformation of Eq. (11.66), remain gauge freedoms of the theory, regardless of any symmetries being spontaneously broken. These may affect the representation of the bosonic modes, which we defined in Eq. (11.62) for a particular gauge fix. For example, the gauge transformation with $\alpha^3(x) = -\beta(x)$ mixes the Z boson and the phase mode of the $\phi_2(x)$ field.

11.3.4 The Higgs Boson

So far, we identified three massive bosons (W^+, W^-, and Z) in the symmetry-breaking phase of the Yang–Mills theory, each with three physical degrees of freedom, as well as a massless photon (A) with two physical, propagating degrees of freedom. Since the theory before symmetry breaking had twelve components, one is left over. This is the massive amplitude mode of the Higgs field, which is commonly referred to as 'the' Higgs boson.

Ironically, the Higgs boson itself has little to do with the gauge fields and gauge invariance that define the Anderson–Higgs mechanism. An amplitude mode naturally appears whenever the amplitude of the order parameter represents a physical quantity whose size can vary. This includes all cases of a global symmetry being broken in relativistic quantum field theories. For the Yang–Mills theory, we can write the field ϕ as a sum of its ground-state expectation value and a real-valued bosonic field $H(x)$ affecting its local amplitude:

$$\phi(x) = \begin{pmatrix} 0 \\ v + H(x) \end{pmatrix}. \tag{11.68}$$

With this, the Lagrangian for the Higgs boson becomes

$$\mathcal{L}_H = -\frac{1}{2}|D_\mu H|^2 - \frac{1}{2}(2|r|)H^2 - \sqrt{|r|u}H^3 - \frac{u}{4}H^4. \tag{11.69}$$

The mass of the Higgs boson is then given by the parameter $m_H^2 = 2|r|$ in the symmetry-broken phase.

The core of the electroweak theory was constructed in the 1970s, alongside the discovery of the electroweak force carriers W and Z. The relation between the mixing angle and the different particle masses suggested they were related by the breaking of an $SU(2)$ symmetry, which prompted a search for the associated amplitude mode. After years of effort, a Higgs boson was experimentally identified in 2013, with a mass of $m_H = 125$ GeV. This is consistent with the scalar Higgs boson of Eq. (11.69), but it does not by itself rule out the possibility of more involved structures, like a Higgs doublet. The search for possible evidence of particles beyond the scalar Higgs boson thus continues.

11.3.5 Fermion Mass

Besides the bosons discussed in Section 11.3.4, the Standard Model also contains fermions. These are represented by four-component Dirac spinors $\psi(x)$ as described by the Dirac Lagrangian:

$$\mathcal{L}_{\text{Dirac}} = -\overline{\psi}\gamma^{\mu}\partial_{\mu}\psi. \tag{11.70}$$

Here, γ^{μ} are the Dirac gamma matrices described in Section 11.1.3. Because the Lagrangian does not contain a mass term, we can use the chiral symmetry of Section 11.2 to split the spinors in separate left- and right-handed components. These fermions will be minimally coupled to the gauge fields of the electroweak interaction, and the global part of the (broken) electroweak $SU(2)$ transformations thus also acts on the fermions. In fact, each type of left-handed lepton or quark in the Standard Model combines with another to form a doublet that transforms under the global $SU(2)$ symmetry in the same way as the two-component Higgs field ϕ. For example, the doublet E_{L} combining the left-handed electron and electron neutrino, or the doublet Q_{L} combining left-handed up and down quarks, can be written as

$$E_{\text{L}} = \begin{pmatrix} \nu_e \\ e \end{pmatrix}_{\text{L}}, \qquad Q_{\text{L}} = \begin{pmatrix} u \\ d \end{pmatrix}_{\text{L}}. \tag{11.71}$$

Notice that the electroweak $SU(2)$ symmetry acts only on left-handed fermions. The right-handed fermions are unaffected by the global $SU(2)$ transformations. In fact, right-handed neutrinos have not yet been observed, and the quest for them is an ongoing research effort.

How the electroweak $SU(2) \times U(1)$ gauge transformation affects the different fermion fields depends on their charges. The charge associated with the B_{μ} field is called the **hypercharge** and is represented by the letter Y. The charges associated with the A^a_{μ} fields are labelled T^a, which are necessarily representations of $SU(2)$. With this, the covariant derivative for the fermion fields reads

$$D_{\mu} = \partial_{\mu} - igA^a_{\mu}T^a - ig'YB_{\mu}. \tag{11.72}$$

Comparing this to Eq. (11.56) shows that the charges indeed appear as the coupling constants determining the strength of the interactions between matter and gauge fields. Notice the distinction between the conserved Noether charges in Section 1.3.4, which correspond to the expectation value of a symmetry generator, and the elementary charges of fundamental particles appearing here as coupling constants in the interaction between gauge and matter fields.

These are of course related, and global $U(1)$ phase rotations in the electroweak symmetry corresponds to conservation of the total electromagnetic charge made up out of the discrete elementary charges carried by individual particles.

Rewriting the A and B fields in terms of the W and Z bosons relevant for broken electroweak symmetry yields

$$D_\mu = \partial_\mu - \mathrm{i}\frac{g}{\sqrt{2}}(W_\mu^+ T^+ + W_\mu^- T^-) - \mathrm{i}\frac{1}{\sqrt{g^2+g'^2}} Z_\mu (g^2 T^3 - g'^2 Y) - \mathrm{i}eA_\mu(T^3 + Y). \tag{11.73}$$

From this, we see that the eigenvalues of $Q \equiv T^3 + Y$ are the elementary charges (in units of e) associated with electromagnetism, as this matrix couples the fermions to the electromagnetic vector potential A_μ. For example, for the left-handed leptons appearing in the Standard Model, it turns out that $Y = -1/2$ and $T^3 = \frac{1}{2}\begin{pmatrix} 1 & 0 \\ 0 & -1 \end{pmatrix}$. The doublet E_L is then an eigenvector of $Q = T^3 + Y$, and the covariant derivative causes separate couplings of the electron and electron neutrino to the electromagnetic gauge field. Acting with Q on the doublet, we see that in the presence of electroweak symmetry breaking, the neutrino has no electromagnetic charge, whereas the electron has electromagnetic charge -1. Similarly, left-handed quarks turn out to have the same T^3, but $Y = 1/6$, so that the up quark has electromagnetic charge $2/3$ and the down-quark $-1/3$.

Although combining electrons and electron neutrinos into a doublet explains how these particles interact with the electroweak gauge fields, at first sight it does not seem to allow for the experimentally observed fermion masses. We cannot, for example, add an explicit mass terms of the type $-m\bar{e}_\mathrm{L}e_\mathrm{L}$ to the Dirac Lagrangian, as that would violate the global $SU(2)$ symmetry. Moreover, if electrons and neutrinos are part of the same doublet, there would be no reason to include such a mass term for the electron component, but not for the neutrinos. The resolution to this conundrum is again provided for by electroweak symmetry breaking. The global $SU(2)$ symmetry does allow a so-called **Yukawa coupling** between both left- and right-handed fermions and the Higgs field. For the electron and electron neutrino, this has the form

$$\mathcal{L}_{\phi e} = -\lambda_e \overline{\begin{pmatrix} \nu_e \\ e \end{pmatrix}}_L \begin{pmatrix} \phi_1 \\ \phi_2 \end{pmatrix} e_R. \tag{11.74}$$

Whenever the Higgs field has a spontaneous ground state expectation value of the form of Eq. (11.59), the Yukawa coupling looks like a mass term for the electrons only:

$$\mathcal{L}_{\phi e} = -m_e\, \bar{e}_\mathrm{L} e_\mathrm{R}, \qquad m_e = \lambda_e v. \tag{11.75}$$

This coupling requires experimental determination of the same number of parameter values as an explicit fermionic mass term, with the electron-Higgs Yukawa coupling λ_e appearing instead of the electron mass m_e. However, it does provide for an elegant mathematical structure explaining the relation between electrons, neutrinos, and the Higgs field. Because of the combination of symmetry breaking and Yukawa coupling, it is often said colloquially that 'the Higgs field give particles mass'. In fact, the global $SU(2)$ symmetry breaking introduces masses in two distinct ways. The first is the Anderson–Higgs mechanism, which provides mass for the W and Z bosons. The second is the Yukawa coupling between the Higgs field and the fermions. While in both cases masses arise due to the breaking of global $SU(2)$ symmetry, the mass generation mechanisms are unrelated in all other regards.

More detailed discussion of various aspects of electroweak symmetry breaking can be found in Refs. [159] and [215].

11.4 Gauge Confinement

In the Yang–Mills theory of Eq. (11.58), the electroweak interaction is mediated by non-Abelian gauge fields, whose coupling to the Higgs field in the symmetry-breaking phase causes the Anderson–Higgs mechanism. As it turns out, the non-Abelian gauge fields by themselves, without additional scalar gauge or spinor matter fields, are already capable of hosting distinct phases. The **pure gauge theories** describing them have a surprising connection to the Higgs phase of spontaneously broken global $SU(2)$ symmetry.

In this context, it is conventional to include the coupling constant in the definition of the gauge field itself, $gA^a_\mu \to A^a_\mu$, so that the Yang–Mills Lagrangian without matter or scalar gauge fields can be written as

$$\mathcal{L}_{\text{YM}} = -\frac{1}{4g^2} G^a_{\mu\nu} G^{\mu\nu,a}. \tag{11.76}$$

Here, $G^a_{\mu\nu} \equiv \partial_\mu A^a_\nu - \partial_\nu A^a_\mu + \epsilon_{abc} A^b_\mu A^c_\nu$ is the non-Abelian field strength tensor.

Although the pure gauge theory contains only gauge fields, which are themselves not observable, it does describe the behaviour of physical observables corresponding to gauge-invariant functions of the fields. For example, we can consider a generalisation of the magnetic flux $\Phi = \oint \vec{A} \cdot d\vec{l}$ constructed from the vector potential in electromagnetism. In the context of non-Abelian theories, this becomes a **Wilson loop**, defined by

$$W_C = \mathcal{P} \exp \left[\mathrm{i} \oint_C A^a_\mu \sigma^a dx^\mu \right]. \tag{11.77}$$

Here, $\mathcal{P}$ is the path-ordering operator, and σ^a are the Pauli matrices, which form a representation of the $SU(2)$ gauge freedom. As long as the contour C is closed, the Wilson loop matrix W_C is invariant under the gauge transformation $A^a_\mu(x) \to A^a_\mu(x) + \partial_\mu \alpha^a(x)$.

To understand the physical significance of the Wilson loop, we can add a source term to the Lagrangian density, by writing $\mathcal{L} = \mathcal{L}_{\text{YM}} - A^a_\mu J^{\mu,a}$. Rather than using this expression as a generating functional for correlation functions of the A^a_μ field, as we did in Sections 10.1.1 and 10.4, we now interpret $J^{\mu,a}$ as an externally applied current, which acts as a source for the A^a_μ field. This is in direct analogy, for example, to how we describe the electromagnetic vector potential in the presence of a given electric charge configuration. Consider now the particular case of a current $J^{\mu,a}$ that is non-zero only along some spacetime loop C, and which has unit amplitude along that loop. We can then interpret the expectation value of the Wilson loop in terms of the partition function:

$$\left\langle \mathcal{P} \exp \left[\mathrm{i} \oint_C A^a_\mu \sigma^a dx^\mu \right] \right\rangle = \frac{Z[J]}{Z[0]}. \tag{11.78}$$

Here, $Z[J]$ is the partition function for the theory with current J, and $Z[0]$ that for the theory without any currents present.

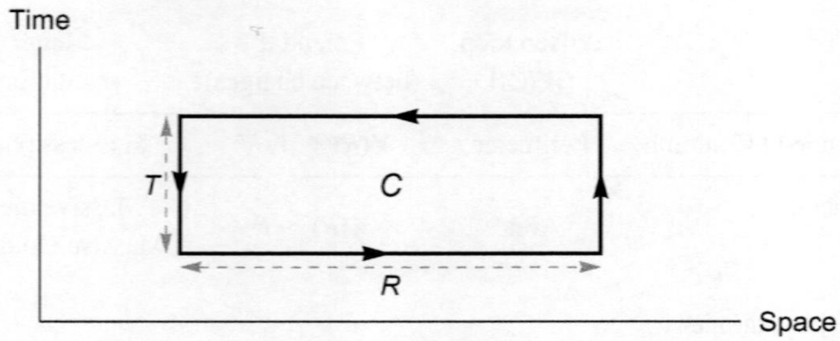

Figure 11.2

A rectangular contour in spacetime, which extends over a distance R in space and a distance T in time. Its spacetime area is $A = RT$, and the perimeter $P = 2(T + R)$. The scaling of a Wilson loop defined along this contour with either its area or perimeter, determines whether charges are confined or deconfined.

We then specialise further and consider the specific current along the contour C shown in Figure 11.2, consisting of a rectangle in spacetime extending a distance R in space and T in time. On any time slice cutting the loop, there is a charge $+1$ on the right edge of C, while a charge -1 sits at its left edge. In the case of T being much larger than R, most of the loop current consist of a static charge configuration, and in the limit $T \to \infty$ only this static part remains. In that limit, we can use the relation between the partition function and the free energy of a system in equilibrium to interpret the ratio of partition functions as $Z[J]/Z[0] = \mathrm{e}^{-\beta(F[J]-F[0])}$. Here, F is the free energy of a hypothetical four-dimensional classical system at temperature $1/\beta$. Using the correspondence between classical equilibrium physics in d dimensions, and quantum dynamics in $d-1$ dimensions [95], the classical free energy divided by temperature, βF, corresponds to the quantum ground state energy times time, $E_0 T/\hbar$. Using this slight detour, we can thus finally identify the expectation value of the Wilson loop as

$$\left\langle \mathcal{P} \exp\left[\mathrm{i} \oint_C A^a_\mu \sigma^a dx^\mu \right] \right\rangle \approx \mathrm{e}^{-\frac{T}{\hbar}(E_0[J]-E_0[0])}. \tag{11.79}$$

Notice that T here is the temporal extent of the rectangular contour C and that the correspondence becomes exact in the $T \to \infty$ limit. Because the charge configuration for $T \gg R$ is nearly static, the energy difference $E[J] - E[0]$ is almost exclusively a potential energy $V(R)$, which corresponds to the energy required to keep two opposite charges a distance R apart. The physical significance of the rectangular Wilson loop is thus that its expectation value in the pure gauge theory determines the potential energy that will be experienced by an externally imposed, static configuration of charges interacting through the gauge field:

$$\langle W(C) \rangle \propto \exp\left[-V(R)T\right]. \tag{11.80}$$

The rectangular contour in the Wilson loop can be characterised by the two parameters T and R, or equivalently, by the area $A = RT$ and perimeter $P = 2(R + T)$ of the loop. The latter have the advantage that they are geometric quantities that can be generalised to other shapes of loops. Expanding the potential $V(R)$, the leading contribution to the Wilson loop expectation value can then be argued to be of the form [95]

$$\langle W(C) \rangle \propto \exp\left[-(\sigma A + \mu P + \ldots)\right]. \tag{11.81}$$

	Wilson loop $\langle W(\mathcal{C})\rangle$	Potential (between charges)	Matter excitations
Deconfined / Coulomb	Perimeter	$V(R) \sim 1/R^n$	Massless photon
Confinement Higgs	Area	$V(R) \sim R$	Massive meson Massive Goldstone

Table 11.3 Phases of gauge theories.

Here, μ and σ are expansion coefficients. This form of $\langle W(C)\rangle$ suggests the existence of two distinct physical regimes. If $\sigma = 0$, the perimeter constitutes the leading term in the exponent, and in the limit $T \to \infty$ the potential energy required to keep charges a distance R apart vanishes, $V(R) \to 0$. In fact, higher orders in the expansion typically yield corrections to this of the form $V(R) \propto 1/R^n$, but these still vanish in the limit of large separation between charges. This situation is familiar, for example, from the **Coulomb phase** describing electromagnetic charges whose interactions are suppressed with distance.

A second regime arises if $\sigma \neq 0$, and the leading term in Eq. (11.81) is proportional to the area. In this case, the asymptotic limit of the potential energy keeps increasing linearly with distance:

$$V(R) \sim \sigma R. \tag{11.82}$$

This regime is called the **confinement phase** and was first discovered in the context of quantum chromodynamics (QCD) describing the strong nuclear force. As the potential grows with distance, pulling charges apart becomes harder the further they are separated. As a results, it is impossible to fully separate charges in this phase, and this is the reason that, for example, quarks are never observed in isolation. Equivalently, this can be interpreted by saying there is a 'string tension' σ connecting opposite charges. The properties of the two regimes are summarised in Table 11.3.

Generalising to non-Abelian pure gauge theories with an $SU(N)$ gauge freedom, it can be shown that these are always in a confining phase in dimensions $D > 2$ [64]. What happens if you couple these non-Abelian gauge fields to various form of bosonic or fermionic matter is the subject of ongoing research. The behaviour of the Wilson loops depends strongly on the type of matter fields added to the theory, and besides direct applications in particle physics, this also allows for the study of possible deconfinement transitions of emergent gauge fields in condensed matter contexts.

One particular case of interest arises when a scalar or 'Higgs' field is added that transforms under the same symmetry group as the gauge fields. The Lagrangian of Eq. (11.76) is then extended to

$$\mathcal{L}_{\mathrm{YM}} = -\frac{1}{4g^2} G^a_{\mu\nu} G^{\mu\nu,a} - \frac{\gamma}{2} |D_\mu \phi|^2 + \dots. \tag{11.83}$$

Here, γ parametrises the coupling strength between the Higgs field and the gauge bosons, and the dots include a potential for the Higgs field.

In the confinement phase, any physical excitations of the Higgs field must involve charge-neutral pairs of charges. Insisting that the excitations are also gauge invariant implies that they can be written in the form

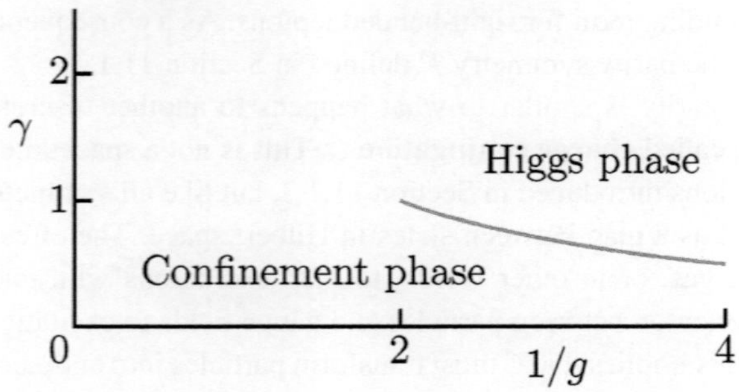

Figure 11.3

Phase diagram for the non-Abelian gauge theory of Eq. (11.83). The Higgs phase is smoothly connected to the confinement phase at low values of the gauge field coupling $1/g$, while a discontinuous transition between the two phases exist at large coupling.

$$\phi(\mathbf{r})\left(\mathcal{P}\exp\left[\mathrm{i}\int_{\mathbf{0}}^{\mathbf{r}} A_\mu^a T^a dx^\mu\right]\right)\phi^\dagger(\mathbf{0}). \tag{11.84}$$

In analogy to the colour-charge neutral excitations of QCD, such excitations are often called **mesons**. At the same time, if the Higgs field obtains a non-zero ground state expectation value because the global $SU(N)$ symmetry is spontaneously broken, the operator of Eq. (11.84) describes the massive gauge boson resulting from the Anderson–Higgs mechanism. Remarkably, it was suggested in the context of lattice gauge theories that these two limits, of the theory being in a confinement phase of the gauge field or in a Higgs phase with massive gauge bosons, can be smoothly connected [53]. In other words, these two phases are found to correspond to the same physical state. In several theories of this type a parameter regime has been identified in which a discontinuous transition separates the Higgs and confinement phases, but like in the liquid–gas transition of Project 17, there is also a regime that smoothly interpolates between the two phases. This is schematically shown in Figure 11.3.

Although the extension of the original arguments relating the Higgs and confinement phase for lattice gauge theories has not been rigorously proven to also apply to continuum models yet, the result that these phases may be smoothly connected at all is surprising. After all, the confinement phase does not possess any obvious broken symmetry, while the Higgs phase does results from the spontaneous breakdown of a global symmetry. How to resolve this seeming contradiction is an open question. Perhaps there is a broken symmetry hiding in the confinement phase, or a phase transition hiding from the arguments connecting the two phases.

11.5 CPT Symmetry

One of the surprising things about the electroweak force is that it only affects left-handed fermions. For example, the interactions between W bosons and the left-handed electrons and electron neutrinos is described by

$$\mathcal{L} = \mathrm{i}\frac{g}{\sqrt{2}}\left(\overline{\nu}_L\gamma^\mu e_L W_\mu^+ + \overline{e}_L\gamma^\mu \nu_L W_\mu^-\right). \tag{11.85}$$

There is no corresponding term for right-handed leptons. As a consequence, the electroweak interactions **violate** the parity symmetry $\mathcal{P}$ defined in Section 11.1.2.

The violation of parity is similar to what happens to another discrete symmetry in the electroweak theory, called **charge conjugation** $\mathcal{C}$. This is not a spacetime symmetry like the discrete transformations introduced in Section 11.1.2, but like all symmetry transformations, it can be understood as a map between states in Hilbert space. The effect of $\mathcal{C}$ is to change the signs of all charges, or in other words, to map excitations with one sign of coupling constant for the interaction between particles and gauge fields to excitations interacting with the opposite sign. This implies that $\mathcal{C}$ must transform particles into antiparticles. For example, the antiparticle of a left-handed electron is a right-handed anti-electron, so $\mathcal{C}$ interchanges e_L and $\overline{e}_R$. It is clear that the electroweak interaction of Eq. (11.85) also violates charge conjugation. However, the combination of $\mathcal{C}$ and $\mathcal{P}$ is a symmetry of Eq. (11.85), because its combined effect maps the two terms in $\mathcal{L}$ onto each other:

$$\overline{\nu}_L\gamma^\mu e_L W^+_\mu \underset{\mathcal{C}}{\rightarrow} \overline{e}_R\gamma^\mu \nu_R W^-_\mu \underset{\mathcal{P}}{\rightarrow} \overline{e}_L\gamma^\mu \nu_L W^-_\mu . \tag{11.86}$$

The overall electroweak theory is therefore said to be $\mathcal{CP}$-conserving.

The story does not end there, however. Cronin and Fitch discovered in 1964 that kaon decay is not invariant under the combination of charge conjugation and parity, and thus presents an example $\mathcal{CP}$-violation. Since kaons are bound states of quarks and antiquarks, the observed $\mathcal{CP}$-violation must occur in the electroweak interaction of quarks. The general form of the coupling between quarks allowed by the electroweak $SU(2)$ gauge freedom, similar to the interaction of Eq. (11.85) between electrons and neutrinos, is given by

$$\mathcal{L} = \mathrm{i}\frac{g}{\sqrt{2}}\left(V_{ij}\overline{u}^i_L\gamma^\mu d^j_L W^+_\mu + V^*_{ij}\overline{d}^j_L\gamma^\mu u^i_L W^-_\mu\right). \tag{11.87}$$

Here, u_L is the triplet of left-handed up, charm, and top quarks, while d_L is the triplet of left-handed down, strange, and bottom quarks. The off-diagonal elements of the **Cabibbo–Kobayashi–Maskawa (CKM) matrix** V_{ij} allow the weak interaction to mix different quark families. Moreover, the combination of $\mathcal{C}$ and $\mathcal{P}$ transformations maps terms of the interaction onto each other only if the CKM matrix is purely real. Experimentally, this is found not to be the case, and therefore the electroweak interactions for quarks violates $\mathcal{CP}$.

Finally, the charge and parity transformations $\mathcal{C}$ and $\mathcal{P}$ are tightly related to the remaining discrete spacetime symmetry, time-reversal $\mathcal{T}$. Parity and time-reversal together change all spacetime coordinates, while charge conjugation makes particles into antiparticles. Intuitively, we can think of antiparticles somewhat like as particles travelling back in time with opposite momentum. In that sense, the effect of $\mathcal{PT}$ is then the same as that of $\mathcal{C}$, and we may expect that the combination $\mathcal{CPT}$ must always be a symmetry. Indeed, the so-called **$\mathcal{CPT}$ theorem** shows that any Lorentz-invariant theory must respect $\mathcal{CPT}$. Only when Lorentz symmetry is violated, can one violate $\mathcal{CPT}$. Whether or not our universe has spontaneously broken $\mathcal{CPT}$-symmetry is an active field of research.

11.6 Grand Unification

The Standard Model unifies the weak nuclear force and electromagnetism into the electroweak force, by recognising that they are mediated by gauge fields associated with the same $SU(2) \times U(1)$ gauge freedom. This was in part inspired by the work of Maxwell, who showed that electric and magnetic forces are part of the same electromagnetic interaction. Continuing in this fashion, it is natural to wonder whether all forces of nature can be unified into a single framework. The next step towards that goal would be a **Grand Unified Theory** (GUT) that unifies the electroweak and the strong nuclear forces [215].

The strong nuclear force is described by **quantum chromodynamics** (QCD), which is an $SU(3)$-symmetric Yang–Mills theory with massless gauge bosons called **gluons**. Inspired by the success of the electroweak theory, the idea of grand unification is to find a large symmetry group that contains both the $SU(3)$ symmetry of chromodynamics and the spontaneously broken $SU(2) \times U(1)$ symmetry of electroweak theory. The most straightforward proposal of this kind is to consider the group:

$$SU(5) \supset SU(3) \times SU(2) \times U(1). \tag{11.88}$$

The so-called Georgi–Glashow model of grand unification is based on $SU(5)$ symmetry. It posits that at some very high energy scale, the global $SU(5)$ symmetry is spontaneously broken down to $SU(3) \times SU(2) \times U(1)$. This would require an extension of the Higgs field to a five-component scalar, and it would lead to various new gauge bosons that become massive due to the Anderson–Higgs mechanism of the $SU(5)$ breaking. Another detectable prediction of some GUTs is the possibility of proton decay [106].

There is no strong experimental indication yet that any form of grand unification is actually realised in nature. There are, however, some results that may point in its direction. Most notable is the observation of **neutrino oscillations**, which indicates that neutrinos propagate in superpositions of flavour eigenstates. The mixing between flavours can be described by a matrix akin to the CKM matrix of Eq. (11.87), indicating the existence of interactions between differently flavoured neutrinos. Such interactions are not part of the Standard Model, but could possibly be part of a grand unified theory [85]. Similarly, the Standard Model does not provide an explanation for why the hypercharges of leptons and quarks are perfect ratios, rather than generic real numbers, which might perhaps be explained by them originating from a theory based on a higher symmetry group [101].

Finally, in the desire to unify all fundamental forces, gravity remains. A hypothetical theory that somehow unifies the electroweak, strong, and gravitational forces into a single set of gauge fields, whose associated global symmetry is unbroken at sufficiently high energies, is sometimes referred to as a 'theory of everything' [213]. At the time of writing, no signatures of such unification have been observed in any experiments. That many theoretical physicists nonetheless pursue it, is owing to the temptation of interpreting all of the richness of fundamental particle physics as arising from consecutive levels of spontaneous symmetry breaking starting from a single, universal symmetry.

12 Fermionic Condensates

The Ginzburg–Landau theory for charged superfluids introduced in Section 8.1 explained how bosonic particles called Cooper pairs can become phase coherent and form a superconducting state at sufficiently low temperatures. In contrast to the neutral superfluids of Section 2.5, the fact that Cooper pairs are electrically charged allows charged superfluids to carry an electric (super)current. Moreover, the coupling of the charges to the electromagnetic field results in an Anderson–Higgs mechanism of Section 8.3, giving mass to photons inside the superconductor (the Meissner effect). A second distinction between real-life superconductors and neutral superfluids, which we have not touched on so far, is that in superconducting materials there really are no bosons that can condense into a charged superfluid state. Superconductors arise from the interacting electron fluid existing in metals, and electrons are fermions!

The aim of this chapter is therefore to discuss **fermionic symmetry breaking**. When we introduced the order parameter operator in Section 3.2, we tacitly assumed it to be bosonic, in the sense that local order parameter operators acting on different spatial positions commute rather than anticommute. At first sight, this seems to exclude fermions from obtaining any kind of ground state expectation value and forming an ordered state. As we will see, however, interactions between fermionic particles can allow them to combine into nearly bosonic objects (a process dubbed 'pairing'), which can subsequently form a condensate that in almost all aspects mimics the by-now familiar bosonic theory of spontaneous symmetry breaking.

The archetypal example for symmetry breaking in fermionic systems is that of a **BCS superconductor**. Here, the acronym BCS stands for Bardeen–Cooper–Schrieffer, who explained in 1957 how the observed phenomenology of superconducting materials arises from ordering among pairs of electrons. There are many other examples of fermionic condensates. Some, like the superfluid phases of helium-3, or quark condensate contributions to the QCD vacuum, are close relatives of BCS superconductors. For others, such as the so-called spin and charge density waves, it is perhaps not immediately obvious that their order parameters arise from fermion pairing. Nevertheless, all of them adhere to the basic outlines of BCS theory that we discuss in this chapter. We will skip over many aspects that are covered at length in excellent textbooks [192] and instead focus on the aspects of symmetry breaking, the order parameter, and the specific characteristics of fermionic condensates that set them apart from the bosonic superfluids of Section 2.5 and the charged superfluids of Section 8.1.

12.1 The Fermionic Coherent State

At first sight, the fermionic nature of electrons seems to make it impossible to form anything like a superfluid state. Certainly, a simple accumulation of particles of the form $(\hat{c}^\dagger)^N |\text{vac}\rangle$

with $\hat{c}^\dagger$ a fermion creation operator is out of the question, due to the Pauli exclusion principle. But even a Bose–Einstein condensate breaking $U(1)$ phase rotation symmetry, as discussed in Section 2.5, is not possible for fermions in any straightforward way. To see why, recall that the Bose–Einstein condensate arises from interacting bosons, whose collective, $k = 0$, Hamiltonian is given by

$$\hat{H}_0 = -\mu \hat{a}^\dagger \hat{a} - \lambda \sqrt{V} \left(\hat{a}^\dagger e^{i\phi} + \hat{a}\, e^{-i\phi} \right). \tag{12.1}$$

Here, $\hat{a}^\dagger$ creates a boson with wave number $k = 0$, and the term proportional to λ is the symmetry-breaking field. The classical state describing the broken symmetry in a Bose–Einstein condensate without quantum corrections, is an eigenstate of the symmetry-breaking field. It should thus be an eigenstate of the annihilation operator $\hat{a}$, or more precisely a right eigenstate of $\hat{a}$ and left eigenstate of $\hat{a}^\dagger$. We called this a **coherent state** in Eq. (2.38), and as you will show in Exercise 12.1, it is a superposition of states with different numbers of particles:

$$|\phi\rangle = \mathrm{e}^{-\frac{|\phi|^2}{2}} \sum_n \frac{\left(\phi \hat{a}^\dagger\right)^n}{n!} |\mathrm{vac}\rangle. \tag{12.2}$$

Here, $\hat{a}^\dagger$ is the creation operator for a boson, and ϕ is a complex number indicating the eigenvalue of the annihilation operator, $\hat{a}\,|\phi\rangle = \phi\,|\phi\rangle$.

Exercise 12.1 (Eigenstates of the annihilation operators).
Eigenstates $|\phi\rangle$ of the bosonic annihilation operator $\hat{a}$ obey the eigenvalue equation $\phi\,|\phi\rangle = \hat{a}\,|\phi\rangle$.
a. Write the state $|\phi\rangle$ in the form $|\phi\rangle = \sum_n C_n \frac{(\hat{a}^\dagger)^n}{\sqrt{n!}} |\mathrm{vac}\rangle$ and then use the bosonic commutator and the fact that $\hat{a}\,|\mathrm{vac}\rangle = 0$ to show that the eigenvalue equation can be written as

$$\phi\,|\phi\rangle = \sum_n \sqrt{n+1}\, C_{n+1} \frac{\left(\hat{a}^\dagger\right)^n}{\sqrt{(n)!}} |\mathrm{vac}\rangle. \tag{12.3}$$

b. Comparing this result to the original eigenvalue equation yields a recursion relation for the coefficients C_n. Ignoring normalisation and arbitrarily choosing $C_0 = 1$, show that the coherent state $|\phi\rangle$ equals

$$|\phi\rangle = e^{\phi \hat{a}^\dagger} |\mathrm{vac}\rangle. \tag{12.4}$$

Exercise 12.2 (Eigenstates of the creation operators).
a. Show that the bosonic creation operator $\hat{a}^\dagger$ cannot have an eigenstate solving the equation $\hat{a}^\dagger\,|\theta\rangle = \theta\,|\theta\rangle$.
b. The creation operator does have a left eigenstate, solving the equation $\langle\theta|\,\hat{a}^\dagger = \theta\,\langle\theta|$. What is it, and what is the corresponding eigenvalue?
Hint: No calculations are required to answer these questions.

Having found that the eigenstates of the bosonic annihilation operator constitute the classical symmetry-breaking state of the Bose–Einstein condensate, it is tempting to try and repeat the analysis for a fermionic annihilation operator. In analogy to Eq. (12.1), we may expect the phase rotation symmetry of free fermions with $\hat{H} \propto \hat{c}^\dagger \hat{c}$ to be broken by a perturbation of the form $\hat{c}^\dagger e^{i\phi} + \hat{c}\, e^{-i\phi}$, with $\hat{c}$ a fermionic annihilation operator. If we can find a fermionic coherent state obeying $\hat{c}\,|\psi\rangle = \psi\,|\psi\rangle$, this would then be the classical symmetry-breaking state for a fermionic Bose–Einstein condensate. The fermionic eigenvalue equation, however, immediately presents us with a problem. If the state $|\psi\rangle$ is an eigenstate of the annihilation operator $\hat{c}$, then we should also be able to apply two fermionic annihilation operators to it:

$$\hat{c}\hat{c}\,|\psi\rangle = \psi^2\,|\psi\rangle\,. \tag{12.5}$$

Because the Pauli exclusion principle guarantees $\hat{c}\hat{c} = 0$, this equation can only hold if ψ squares to zero. More generally, if we allow coherent states of multiple types of fermions defined by $\hat{c}_x\,|\psi\rangle = \psi_x\,|\psi\rangle$, the same reasoning would imply that the numbers ψ_x need to anticommute, so that $\{\psi_x, \psi_{x'}\} = 0$.

In fact, a set of non-commuting numbers can be constructed in a mathematically consistent way, and they are called **Grassmann variables**. They are much like normal numbers, except that they are defined as antisymmetric under exchange: $\psi_x\psi_y = -\psi_y\psi_x$, which holds for all values of x and y. Using them, the fermionic coherent state can be expressed in terms of the Grassman variable ψ as

$$|\psi\rangle = e^{-\sum_x \psi_x \hat{c}_x^\dagger}\,|\text{vac}\rangle\,. \tag{12.6}$$

The fermionic coherent state $|\{\psi_x\}\rangle$ plays a central role in the path integral formulation of fermionic quantum field theories. For our present purpose of identifying a classical state that is an eigenstate of $\hat{c}_x$, however, the fermionic coherent state falls short. The eigenvalue of the fermionic annihilation operator is a Grassmann variable, and therefore certainly not the expectation value of any observable quantity. In particular, it is not a phase variable labelling different ways of breaking phase rotation symmetry. Although a very useful, and often used, mathematical construction, the fermionic coherent state is not a physical state that can result from spontaneous symmetry breaking.

12.1.1 Pairs of Fermions

We thus seem to arrive at the conclusion that there simply is no fermionic analogue to the bosonic coherent state or Bose–Einstein condensate. Yet, we know from experiment that superconductors behave precisely according to the Ginzburg–Landau theory of Chapter 8 for charged superfluids. It turns out that the superconducting state side-steps the impossibility of constructing a symmetry-breaking coherent state of fermions, by combining electrons into so-called **Cooper pairs** before condensing them. As we will see, this means that electrons in the superconducting state do not exist on their own, but always appear in pairs of the form $\hat{a}_k^\dagger \equiv \hat{c}_{k,\uparrow}^\dagger \hat{c}_{-k,\downarrow}^\dagger$. These Cooper pairs contain two electrons each, and as a single, combined unit the pair behaves much like a boson. In particular, exchanging two pairs causes the wave function to pick up two simultaneous minus signs, which cancel. Distinct Cooper pairs thus obey bosonic commutation relations:

$$\left[\hat{a}^\dagger_k, \hat{a}^\dagger_{k'}\right] = 0, \qquad \left[\hat{a}^\dagger_k, \hat{a}_{k'}\right] = 0, \qquad \forall k' \neq k. \tag{12.7}$$

Notice, however, that although the wave function for two Cooper pairs is bosonic in the sense that it is symmetric under the exchange of the pairs, the individual pairs still contain electrons, which must obey the Pauli exclusion principle. It is therefore still impossible to create a product state of two Cooper pairs with exactly the same quantum numbers, and $\hat{a}^\dagger_k \hat{a}^\dagger_k = 0$. Because of this constraint, bound pairs of fermions are often referred to as **hard-core bosons**.

The hard-core nature of Cooper pairs prevents us from defining a coherent state of the form $e^{\alpha \hat{a}^\dagger_k} |\text{vac}\rangle$, in direct analogy to the wave function for a Bose–Einstein condensate. Expanding the exponent in this expression would yield only two terms, in contrast to the infinitely many terms appearing for true bosons. Heisenberg's uncertainty principle guarantees that it is impossible to create a state with broken phase rotation symmetry without superposing infinitely many states of different particle number, and the state $e^{\alpha \hat{a}^\dagger_k} |\text{vac}\rangle$ is thus not a symmetry-breaking state.

Although we assigned a quantum number k to the Cooper pair creation operator $\hat{a}^\dagger_k$, the pair as a whole actually has zero momentum, since it consist of one fermion with momentum k and one with $-k$. We can therefore make a superposition of Cooper pairs that all have zero total momentum, but which consist of fermion pairs with different opposing momenta. To see how this helps, consider a superposition of Cooper pairs with different quantum numbers, such as $\sqrt{1/2}(\hat{a}^\dagger_k + \hat{a}^\dagger_{k'})$. Unlike the single Cooper pair, it is possible to create a product state of two such superposed Cooper pairs:

$$\frac{1}{2}\left(\hat{a}^\dagger_k + \hat{a}^\dagger_{k'}\right)^2 |\text{vac}\rangle = \hat{a}^\dagger_k \hat{a}^\dagger_{k'} |\text{vac}\rangle \neq 0. \tag{12.8}$$

Higher powers of the superposed creation operators do still vanish, but the hard core constraint is clearly weaker for the two-state superposition than it was for the single Cooper pair state. Adding more components to the superposition weakens the constraint further. This can be continued until finally, in the thermodynamic limit, the state $\sum_k \alpha_k \hat{a}^\dagger_k$ becomes truly bosonic. In that limit, it is then possible to construct a coherent state of the form

$$|\text{BCS}\rangle = e^{\sum_k \alpha_k \hat{a}^\dagger_k} |\text{vac}\rangle\,. \tag{12.9}$$

This state has all the same properties as the bosonic Bose–Einstein condensate, and it is this state that we will find in Section 12.2.2 to be the symmetry-breaking classical state realised in superconductors. Moreover, because the Cooper pairs are charged, its macroscopic properties are precisely those described in Chapter 8 in terms of the Ginzburg–Landau theory for a charged superfluid.

The condensate of superposed Cooper pairs allows for a superconducting state to be defined, and thus explains the appearance of broken phase-rotation symmetry, the Anderson–Higgs mechanism, the Josephson effect, and the emergence of dissipationless supercurrents. Its form, however, leaves us with a fundamental problem: electrons are negatively charged particles, which interact through purely repulsive Coulomb forces, and at first sight do not seem to have any reason for attracting one another and forming effectively bosonic pairs. In fact, the attraction between electrons in superconductors, and many other types of fermionic condensates, does not come from a direct Coulomb interaction. Instead, it is an effective interaction that emerges at sufficiently low energies from the exchange of virtual excitations. In the case of superconductors, the best-understood way to achieve an attractive, or pairing, interaction, is through the coupling between electrons and phonons (crystal lattice vibrations).

12.1.2 The Pairing Interaction

To see how the presence of phonons may give rise to emergent attractive interactions between electrons, it suffices to consider a single band of electrons, interacting with so-called Einstein phonons whose energy $\hbar\Omega$ is constant for all momenta:

$$\hat{H} = \sum_{\vec{k},\sigma} \left(\epsilon(\vec{k}) - \mu\right) \hat{c}^{\dagger}_{\vec{k},\sigma} \hat{c}_{\vec{k},\sigma} + \sum_{\vec{q}} \hbar\Omega \hat{b}^{\dagger}_{\vec{q}} \hat{b}_{\vec{q}} + \frac{1}{\sqrt{N}} \sum_{\vec{k},\vec{q},\sigma} g \hat{c}^{\dagger}_{\vec{k}+\vec{q},\sigma} \hat{c}_{\vec{k},\sigma} \left(\hat{b}_{\vec{q}} + \hat{b}^{\dagger}_{-\vec{q}}\right). \tag{12.10}$$

The operator $\hat{c}^{\dagger}_{\vec{k},\sigma}$ with $\sigma \in \{\uparrow, \downarrow\}$ creates an electron whose spin is aligned or anti-aligned with the z-axis, while μ is the chemical potential of the electrons with energies $\epsilon(\vec{k})$. Notice that unlike the bosonic chemical potential in Section 2.5, the fermionic potential μ in Eq. (12.10) is not necessarily negative. Its value indicates the inflection point of the Fermi–Dirac distribution separating occupied and unoccupied electronic states. The second line can be interpreted as describing the absorption or emission of phonons by electrons, while conserving total momentum. For simplicity, we assume the amplitude g for this electron-phonon coupling to be positive and not to have any momentum dependence. In Section 12.3.2, we will consider more general cases.

In the limit of temperature being low compared to the phonon energy, any quasiparticle excitations (electrons or holes) in the system will have excitation energies $|\epsilon(\vec{k}) - \mu|$ that are small compared to $\hbar\Omega$. If, moreover, the electron–phonon coupling is weak in the sense that $g \ll \hbar\Omega$, we can use a perturbation theory in powers of $g/(\hbar\Omega)$ to formulate a low-energy effective theory. This can be done by integrating out the phonons in a Lagrangian formulation. You can also formulate an effective Hamiltonian from Eq. (12.10) by dividing the Hilbert space into so-called Hubbard sectors and applying strong-coupling perturbation theory [160, 192]. Either way, we end up with an effective theory of the form

$$\hat{H}_{\text{eff}} = \sum_{\vec{k},\sigma} \left(\epsilon(\vec{k}) - \mu\right) \hat{c}^{\dagger}_{\vec{k},\sigma} \hat{c}_{\vec{k},\sigma} + \hat{H}^{(2)} + O\left(g^3/(\hbar\Omega)^2\right). \tag{12.11}$$

The lowest order contribution in this Hamiltonian is just the dispersion relation of non-interacting electrons. The first-order contribution to the perturbation expansion vanishes, while the second-order contribution $\hat{H}^{(2)}$ equals

$$\begin{aligned} \hat{H}^{(2)} &= -\frac{g^2}{\hbar\Omega} \frac{1}{N} \sum_{\vec{k},\sigma} \sum_{\vec{k}',\sigma'} \sum_{\vec{q}} \hat{c}^{\dagger}_{\vec{k}+\vec{q},\sigma} \hat{c}_{\vec{k},\sigma} \hat{c}^{\dagger}_{\vec{k}'-\vec{q},\sigma'} \hat{c}_{\vec{k}',\sigma'} \\ &= -\frac{g^2}{\hbar\Omega} \sum_{\vec{q}} \hat{\rho}_{\vec{q}} \hat{\rho}_{-\vec{q}}. \end{aligned} \tag{12.12}$$

The operator $\hat{\rho}_{\vec{q}}$ in the final line is the Fourier transform of the real space electron density operator. Its eigenstates are waves of electron density with $\vec{q}$ labelling the propagation vector.

Exercise 12.3 (The density operator).
Show that $\hat{\rho}(\vec{q}) \equiv \frac{1}{\sqrt{N}} \sum_k \hat{c}^\dagger_{\vec{k}} \hat{c}_{\vec{k}+\vec{q}}$ is the Fourier transform of the local electron charge density $\hat{\rho}(\vec{x}) = \hat{c}^\dagger_{\vec{x}} \hat{c}_{\vec{x}}$.

In the low-energy effective theory described by $\hat{H}_{\text{eff}}$, the phonon degrees of freedom have disappeared and the non-interacting electrons are replaced by interacting ones. The interaction of Eq. (12.12), moreover, is attractive. It lowers the energy whenever an electronic density wave with momentum $\vec{q}$ is accompanied by a similar density wave with momentum $-\vec{q}$. The qualitative effect is that electrons can lower their energy by organising into momentum-space pairs.

The physical understanding of how the pairing interaction arises is that the electron–phonon coupling term in the original Hamiltonian allows electrons to gain interaction energy by exciting or absorbing phonons. When the phonon energy is high, such phonon excitations are suppressed. Quantum corrections to the completely non-interacting limit can then be described in terms of virtual phonons being exchanged between electrons with opposite momenta. The ability of such electron pairs to gain energy results in a preference for any occupied electronic states to always be accompanied by occupied states with opposite momentum.

Notice that in this derivation, the fact that $\hat{b}^\dagger_{\vec{q}}$ creates a phonon rather than another type of bosonic excitation, did not play any particular role. Although many superconducting materials in nature contain Cooper pairs bound by phonon exchange, we will see in Section 12.3.2 that other types occur as well.

12.1.3 Particle–Hole Pairing

The mechanism of interactions between fermions and bosons leading to a low-energy effective theory featuring a pairing interaction is not particular to electrons combining into Cooper pairs. In slightly different circumstances, the attractive interaction can be mediated between electrons and holes instead, leading to bound electron–hole pairs. In fact, many of the known electronically ordered states can be described in terms of electron–hole pairing.

Consider, for example, the operator for the spin of an electron, which is given by

$$\hat{\mathbf{S}}_{\vec{r}} = \hat{c}^\dagger_{\vec{r},\alpha} \, \vec{\sigma}_{\alpha\beta} \, \hat{c}_{\vec{r},\beta} \,. \tag{12.13}$$

Spins in a material can magnetically order, as we saw in the context of the Heisenberg antiferromagnet in Section 2.4. If the non-zero expectation value $\langle \hat{\mathbf{S}}_{\vec{r}} \rangle$ in a magnetically ordered material arises from its electron spins, the order can be considered a condensation of electron–hole pairs because Eq. (12.13) expresses the spin operator as an electron–hole pair creation operator. In general, whenever spontaneous symmetry breaking leads to a non-zero electron–hole expectation value of the form $\langle \hat{c}^\dagger_{\vec{k},\alpha} \, \vec{\sigma}_{\alpha\beta} \, \hat{c}_{\vec{k}+\vec{Q},\beta} \rangle$, the resulting phase is called a **spin-density wave** with ordering wavevector $\vec{Q}$.

Similarly, electron–hole pairing can lead to a charge modulation that breaks translational symmetries. A state with a non-zero expectation value $\langle \hat{c}^\dagger_{\vec{k},\sigma} \, \hat{c}_{\vec{k}+\vec{Q},\sigma} \rangle$ is called a **charge density**

wave. In systems with more involved spin or orbital configurations, other forms of electron–hole pairing can occur. This includes electron–hole pairs with the electron and hole having different orbitals, spin, band index, valley, and so forth.

The electron–hole pairing can result, as in the case of Cooper pairing, from basically any type of interaction. It can be caused by the exchange of phonons, through electromagnetic excitations or direct Coulomb interactions, or even combinations of those. Similarly, a condensate consisting of paired quarks and anti-quarks may arise from the exchange of gluons in approximate theories of quantum chromodynamics. The main difference between the electron–hole and electron–electron pairing is that electron–hole pairs carry no electric charge and therefore have no direct coupling to gauge fields.

12.2 BCS Theory of Superconductivity

Focussing again on superconductivity, and anticipating the emergence of Cooper pairs, a more convenient way of writing Eq. (12.12) is by grouping creation and annihilation operators:

$$\hat{H}^{(2)} \approx -\frac{g^2}{\hbar\Omega}\frac{1}{N}\sum_{\vec{k},\sigma}\sum_{\vec{k}',\sigma'}\sum_{\vec{q}} \hat{c}^{\dagger}_{\vec{k}+\vec{q},\sigma}\,\hat{c}^{\dagger}_{\vec{k}'-\vec{q},\sigma'}\,\hat{c}_{\vec{k}',\sigma'}\,\hat{c}_{\vec{k},\sigma}. \tag{12.14}$$

In this expression, we neglected the contribution from the anticommutator of $\hat{c}_{\vec{k}}$ and $\hat{c}^{\dagger}_{\vec{k}'-\vec{q}}$, since it is only non-zero for $\vec{k} = \vec{k}' - \vec{q}$, and therefore has a relative contribution of only order $1/N$ to the total sum. In the thermodynamic limit, this may be ignored. Written this way, the interaction term may be interpreted as describing processes in which a pair of electrons, with momenta $\vec{k}$ and $\vec{k}'$, is annihilated, while a new pair is created with momenta $\vec{k} + \vec{q}$ and $\vec{k}' - \vec{q}$.

At this point, we should recall that $\hat{H}^{(2)}$ is part of a low-energy effective theory, which should include only electronic states with energies well below the phonon energy $\hbar\Omega$. In particular, this means that the energy needed to annihilate one pair of electrons, and create another, should be small compared to the phonon energy:

$$\epsilon(\vec{k} + \vec{q}) + \epsilon(\vec{k}' - \vec{q}) - \epsilon(\vec{k}') - \epsilon(\vec{k}) \ll \hbar\Omega. \tag{12.15}$$

At low temperatures, this implies that the sum in Eq. (12.14) can only involve electron or hole states within $O(\hbar\Omega)$ from the Fermi energy. Since $\hbar\Omega$ is typically much smaller than E_{F}, the inequality then places a stringent constraint on the momenta of paired electrons that may appear in the sum. As shown in Figure 12.1, it is practically impossible for two states close to the Fermi surface to gain momenta $\vec{q}$ and $-\vec{q}$, and still end up in states that are also close to the Fermi surface. That is, say you start with an electron at $\vec{k}$ close to the Fermi surface, and choose $\vec{q}$ such that $\vec{k} + \vec{q}$ is again close to a Fermi momentum vector. Then there are generically only very few vectors $\vec{k}'$ close to the Fermi surface for which $\vec{k}' - \vec{q}$ is also close to a Fermi momentum. In the limit of $\hbar\Omega \ll E_{\mathrm{F}}$, practically the only way in which this **kinematic constraint** can be satisfied is if the momenta of the paired electrons are precisely opposite.

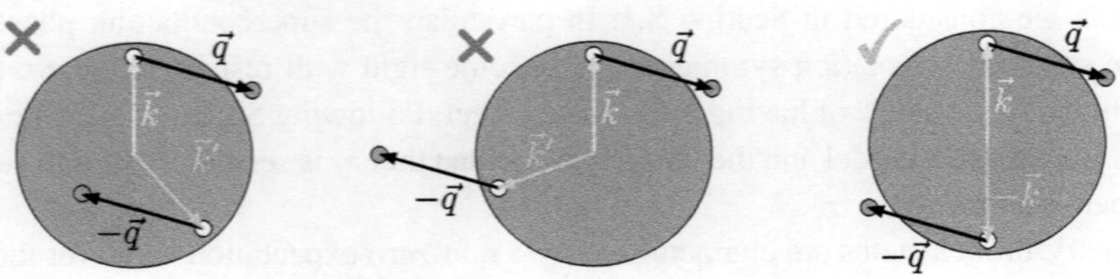

Figure 12.1

The kinematic constraint on scattering processes in Eq. (12.14). An electron close to E_F gaining momentum $\vec{q}$ and ending up in a state that is close to E_F, can only be accompanied by an electron losing the same momentum $\vec{q}$ if both the initial and final state of the second electron are likewise close to the Fermi surface. The only pairs for which these constraints can be satisfied for more than $\mathcal{O}(1)$ phonon momenta, have opposite electronic momentum.

Exercise 12.4 (The kinematic constraint on pairing).
Consider a spherical Fermi surface and explain why the number of states a pair of particles with $\vec{k}+\vec{k}' \neq 0$ can scatter into is negligible compared to the number of states a pair with vanishing centre of mass momentum ($\vec{k}+\vec{k}' = 0$) may end up in.
Assume the ordering of energy scales $|\epsilon(\vec{k}) - E_{\mathrm{F}}| \ll \hbar\Omega \ll E_{\mathrm{F}}$.

In the discussion so far, we did not pay much attention to the spins of the electrons making up the Cooper pair. We will discuss a range of possible symmetry-based constraints on preferred spin configurations in Section 12.3.2. For now, consider the simplest possible situation in which there is an kinetic energy cost for the total (spin) angular momentum of each Cooper pair. The lowest energy configuration will then consist of spin-singlet Cooper pairs, in which the electron spins are necessarily anti-aligned. We then finally end up with the effective low-energy model:

$$\hat{H}_{\text{eff}} = \sum_{\vec{k},\sigma} \left(\epsilon(\vec{k}) - \mu\right) \hat{c}^{\dagger}_{\vec{k},\sigma} \hat{c}_{\vec{k},\sigma} - V {\sum_{\vec{k},\vec{k}'}}' \hat{c}^{\dagger}_{\vec{k}',\uparrow} \hat{c}^{\dagger}_{-\vec{k}',\downarrow} \hat{c}_{-\vec{k},\downarrow} \hat{c}_{\vec{k},\uparrow}. \tag{12.16}$$

Here, $V = 2g^2/(\hbar\Omega)$, and the prime on the second summation indicates that the sum includes only those momenta for which both $|\epsilon_{\vec{k}} - E_{\mathrm{F}}|$ and $|\epsilon_{\vec{k}'} - E_{\mathrm{F}}|$ are small compared to the phonon energy. Precisely where to place this ultraviolet cut-off is not well defined in the formalism developed here. As we will see shortly, however, the results of the theory are rather insensitive to this choice, at least for small $g/(\hbar\Omega)$. Notice that the final term can also be written in terms of the Cooper pair creation and annihilation operators defined in Section 12.1.1, as $-V \sum'_{\vec{k},\vec{k}'} \hat{a}^{\dagger}_{\vec{k}} \hat{a}_{\vec{k}'}$. Equation (12.16) defines the **Bardeen–Cooper–Schrieffer (BCS) model** for superconductivity, which yields a good description of conventional, weak-coupling superconductors.

12.2.1 The BCS Order Parameter

Based on the universality of spontaneous symmetry breaking, we expect the low-energy BCS model to behave qualitatively the same as the Ginzburg–Landau theory of charged bosons that we considered in Section 8.1. In particular, the superconducting phase should break the same phase-rotation symmetry and become rigid with respect to electromagnetic perturbations, in the sense of having a Meissner effect. Following Section 5.3.2, the connection between the BCS model and the Ginzburg–Landau theory is seen most straightforwardly using mean-field theory.

Symmetry-broken states are characterised by a non-zero expectation value for their order parameter operators, which can therefore be used as the mean-field expectation value whose fluctuations are assumed to be small in mean-field theory. Neglecting these fluctuations is then appropriate whenever relative variations in the order parameter are small. This is the same requirement as the Ginzburg criterion of Section 5.4.2, which defines the range of applicability for Ginzburg–Landau theory.

Since we expect the superconducting state to resemble a Bose–Einstein condensate of Cooper pairs, a natural choice for the order parameter would be

$$V\sum_{\vec{k}}\langle \hat{c}^{\dagger}_{\vec{k},\uparrow}\hat{c}^{\dagger}_{-\vec{k},\downarrow}\rangle = \Delta e^{-i\phi}, \qquad V\sum_{\vec{k}}\langle \hat{c}_{-\vec{k},\downarrow}\hat{c}_{\vec{k},\uparrow}\rangle = \Delta e^{i\phi}. \tag{12.17}$$

This can be compared directly to the order parameter $\langle\hat{a}\rangle$ for a Bose–Einstein condensate, with $\sum_{\vec{k}}\hat{c}_{-\vec{k},\downarrow}\hat{c}_{\vec{k},\uparrow}$ in place of the bosonic annihilation operator, as explained in Section 12.1.1. Notice that we followed convention by including the interaction strength V in the definition of the order parameter. The amplitude Δ therefore has units of energy, and we will see shortly that it corresponds to a directly observable quantity. In addition, whenever Δ is non-zero, there exists a phase of the order parameter, ϕ. Using Eq. (12.17), it is clear that ϕ corresponds to the average phase of all Cooper pairs. The superconducting state with a well-defined phase ϕ thus corresponds to configuration breaking the global $U(1)$ phase-rotation symmetry of the BCS Hamiltonian.

As phase and number are canonically conjugate variables, breaking phase-rotation symmetry is only possible in a superposition of states with different particle numbers. This can also be seen directly from the expectation value of two annihilation operators in Eq. (12.17) becoming non-zero in the ordered state. As discussed in Section 2.5.2 for the case of Bose–Einstein condensates, violating conservation of particle number requires coupling to a coherent source of particles that acts as a symmetry-breaking field. The coupling required to stabilise the symmetry-broken state may be exceedingly weak in real-life, macroscopic superconductors, but it cannot be zero. For a piece of superconducting material, a weak interaction with a second superconductor could act as the source of particle fluctuations, and the value of their relative phases is then observable through the Josephson effect of Section 3.5. Electrons inside an isolated piece of superconducting material that do not participate in the formation of Cooper pairs can in principle fulfill the same role, although the gauge freedom of Section 8.2 prevents us from experimentally verifying the realisation of a stable, symmetry-breaking value of the phase ϕ without invoking a second superconductor.

Notice that the mean-field expectation value $\Delta e^{i\phi}$ is also an order parameter in the sense of Section 3.2. Its amplitude is zero in the symmetric state and non-zero in any symmetry-breaking configuration, while its phase is an element of the coset space labelling all inequivalent ways of breaking the symmetry. Moreover, the definition of the order parameter as the expectation value of a bilinear product of fermionic creation and annihilation operators is general for fermionic condensates. Density waves, for example, combine particle and hole operators, with an order parameter of the form $\sum_{\vec{k},\sigma}\langle \hat{c}^{\dagger}_{\vec{k},\sigma}\hat{c}_{\vec{k}+\vec{q},\sigma}\rangle$ signifying the presence of a charge density wave with wave number $\vec{q}$, or the combination $\sum_{\vec{k},\sigma}\langle(-1)^\sigma \hat{c}^\dagger_{\vec{k},\sigma}\hat{c}_{\vec{k}+\vec{q},\sigma}\rangle$ for the simplest kind of spin-density wave. In direct analogy to the BCS theory, these types of order can be thought of as Bose–Einstein condensates of particle–hole pairs, in which the pairs behave like bosons when averaged over a macroscopic number of ways of realising the same bosonic, centre-of-mass momentum.

Using the mean fields defined in Eq. (12.17) and discarding fluctuations as in Section 5.3.2, the mean-field Hamiltonian for the BCS model of Eq. (12.16) becomes

$$\hat{H}^{\rm MF}_{\rm BCS} = \sum_{\vec{k}}\Big[\left(\epsilon(\vec{k})-\mu\right)\left(\hat{c}^\dagger_{\vec{k},\uparrow}\hat{c}_{\vec{k},\uparrow}+\hat{c}^\dagger_{\vec{k},\downarrow}\hat{c}_{\vec{k},\downarrow}\right)$$
$$-\Delta e^{i\phi}\hat{c}^\dagger_{\vec{k},\uparrow}\hat{c}^\dagger_{-\vec{k},\downarrow}-\Delta e^{-i\phi}\hat{c}_{-\vec{k},\downarrow}\hat{c}_{\vec{k},\uparrow}\Big]+\frac{|\Delta|^2}{V}. \tag{12.18}$$

Notice that throughout the formulation of this mean-field Hamiltonian, we implicitly assumed that there are no electromagnetic fields present, and that we may therefore choose the vector potential $\vec{A}$ to be zero, and ignore it altogether in writing $\hat{H}^{\rm MF}_{\rm BCS}$. As emphasised in Chapter 8, however, the vector potential should in fact be an integral part of the superconducting order parameter. As it turns out, it is difficult to construct a manifestly gauge-invariant weak-coupling theory for BCS superconductivity. We will thus proceed by working in the gauge fix $\vec{A}=0$ and make sure at the end of the calculation that our results are gauge invariant upon re-introducing electromagnetic fields.

Writing Eq. (12.18) in a Lagrangian formulation and integrating out the electrons would yield an action in terms of only the order parameter field, which corresponds directly to the Ginzburg–Landau theory for charged superfluids introduced in Section 8.1. Sticking to the Hamiltonian formalism instead, we can minimise the ground state energy of $\hat{H}^{\rm MF}_{\rm BCS}$ with respect to the order parameter $\Delta e^{i\phi}$. This will yield a set of saddle point equations, whose solution in turn gives the equilibrium value of the order parameter.

Exercise 12.5 (Global phase rotations).
Show explicitly that $\hat{H}^{\rm MF}_{\rm BCS}$ is invariant under global rotations of the electron phases, assuming that the order parameter rotates as well. Notice that the BCS Hamiltonian is not invariant under any local (momentum-dependent) phase transformations. Explain why it really ought to be and trace back at what point in the derivation of the BCS model we implicitly chose a gauge fix.

12.2.2 The Superconducting Ground State

To diagonalise $\hat{H}^{\rm MF}_{\rm BCS}$ and find its ground state energy, we need a Bogoliubov transformation. As in our discussion of antiferromagnetism in Chapter 6, the transformation contains a superposition of particle and hole operators:

$$\begin{aligned}\hat{c}_{\vec{k},\uparrow} &= u^*_{\vec{k}}\hat{\gamma}_{\vec{k},0}+v_{\vec{k}}\hat{\gamma}^\dagger_{\vec{k},1}, & \hat{\gamma}_{\vec{k},0} &= u_{\vec{k}}\hat{c}_{\vec{k},\uparrow}-v_{\vec{k}}\hat{c}^\dagger_{-\vec{k},\downarrow},\\ \hat{c}^\dagger_{-\vec{k},\downarrow} &= -v^*_{\vec{k}}\hat{\gamma}_{\vec{k},0}+u_{\vec{k}}\hat{\gamma}^\dagger_{\vec{k},1}, & \hat{\gamma}^\dagger_{\vec{k},1} &= v^*_{\vec{k}}\hat{c}_{\vec{k},\uparrow}+u_{\vec{k}}*\hat{c}^\dagger_{-\vec{k},\downarrow}.\end{aligned} \tag{12.19}$$

Here, the spin labels of the original electrons are replaced by a 0 or 1 index on the transformed particles because the new particles are a superposition of electrons and holes with spin up and down. The transformed particles thus do not have definite values of spin or charge. They are known as **Bogoliubov quasiparticles**. To ensure that they are fermions, with all the usual fermionic anticommutation relations, we should demand the coefficients to obey $|u_{\vec{k}}|^2 + |v_{\vec{k}}|^2 = 1$. We also assumed $\epsilon(\vec{k}) = \epsilon(-\vec{k})$ as well as $u_{\vec{k}} = u_{-\vec{k}}$ and $v_{\vec{k}} = v_{-\vec{k}}$.

Exercise 12.6 (Bogoliubov quasiparticles).
Check that the transformed particles in Eq. (12.19) obey the usual fermionic anti-commutation relations.

Writing the mean-field Hamiltonian in terms of the transformed particles and choosing the values of the coefficients $u_{\vec{k}}$ and $v_{\vec{k}}$ such that it is diagonal yields the final form:

$$\hat{H}^{\mathrm{MF}}_{\mathrm{BCS}} = \frac{|\Delta|^2}{V} + \sum_{\vec{k}} \left[\left(\epsilon(\vec{k}) - \mu \right) - E_{\vec{k}} \right] + \sum_{\vec{k}} E_{\vec{k}} \left(\hat{\gamma}^{\dagger}_{\vec{k},0} \hat{\gamma}_{\vec{k},0} + \hat{\gamma}^{\dagger}_{\vec{k},1} \hat{\gamma}_{\vec{k},1} \right). \tag{12.20}$$

Here, we introduced $E_{\vec{k}} = \sqrt{\left(\epsilon(\vec{k}) - \mu\right)^2 + |\Delta|^2}$.

Exercise 12.7 (Diagonalising the BCS Hamiltonian).
Find expressions for the coefficients $u_{\vec{k}}$ and $v_{\vec{k}}$ that result in the diagonal form of Eq. (12.20) for the BSC Hamiltonian.

The diagonalised mean-field Hamiltonian looks just like a model for non-interacting fermions in a band with dispersion $E_{\vec{k}}$. However, there is one major difference with the usual electronic Hamiltonians. In the mean-field BCS Hamiltonian, the number of electrons was not a conserved quantity. To see how many of the Bogoliubov quasiparticle states in the bands of Eq. (12.20) are occupied in the ground state, we should minimise the expectation value of the mean-field Hamiltonian. We thus occupy states with negative energies and leave states with positive energies unoccupied. Since $E_{\vec{k}}$ is positive for all $\vec{k}$, the ground state of $\hat{H}^{\mathrm{MF}}_{\mathrm{BCS}}$ does not have any Bogoliubov quasiparticles occupied. To see that this really is the correct way of filling up quasiparticle states, you can consider the limit $\Delta \to 0$ and confirm that the energy of the state without any Bogoliubov quasiparticles coincides with the ground state energy of the non-interacting Fermi gas.

As Δ becomes non-zero, the ground state energy in Eq. (12.20) is lowered by the gain in interaction energy associated with the emerging, ordered, superconducting phase. The Bogoliubov quasiparticles describe fermionic excitations on top of the superconducting condensate, which is literally the quasiparticle vacuum. We can use this observation to construct the electronic wave function $|\mathrm{BCS}\rangle$ of the superconducting ground state in our weak coupling description:

$$\begin{cases} \hat{\gamma}_{\vec{k},0} |\mathrm{BCS}\rangle = \left(u_{\vec{k}} \hat{c}_{\vec{k},\uparrow} - v_{\vec{k}} \hat{c}^{\dagger}_{-\vec{k}\downarrow} \right) |\mathrm{BCS}\rangle = 0 & \forall \vec{k}, \\ \hat{\gamma}_{\vec{k},1} |\mathrm{BCS}\rangle = \left(v_{\vec{k}} \hat{c}^{\dagger}_{\vec{k},\uparrow} + u_{\vec{k}} \hat{c}_{-\vec{k},\downarrow} \right) |\mathrm{BCS}\rangle = 0 & \forall \vec{k}. \end{cases} \tag{12.21}$$

It is straightforward to verify that these equations imply the famous expression for the **BCS wave function**:

$$|\mathrm{BCS}\rangle = \prod_{\vec{k}} \left(u_{\vec{k}} + v_{\vec{k}} \hat{c}^{\dagger}_{\vec{k},\uparrow} \hat{c}^{\dagger}_{-\vec{k},\downarrow} \right) |\mathrm{vac}\rangle . \tag{12.22}$$

Here, $|\mathrm{vac}\rangle$ is the vacuum of the original, non-transformed electrons, and the product runs over all momenta, both below and above the Fermi momentum of the unperturbed model.

Exercise 12.8 (The BCS wave function).
a. Check that the BCS wave function defines a vacuum for the Bogoliubov quasiparticles.
b. Check that the BCS wave function in Eq. (12.22) is normalised.
c. Use the BCS wave function in Eq. (12.22) to evaluate the expectation value $\sum_{\vec{k}} \langle \hat{c}_{-\vec{k},\downarrow} \hat{c}_{\vec{k},\uparrow} \rangle$ and write a self-consistency condition for the gap amplitude $|\Delta|$. Show that any choice for the phase ϕ gives a self-consistent solution. Argue why this must be the case based on the symmetry of the mean-field Hamiltonian.

As expected from the fact that the superconductor is a charged superfluid, the BCS wave function in Eq. (12.22) is a product state that is similar to a Bose–Einstein condensate. This can be made more explicit by writing the BCS wave function as

$$|\mathrm{BCS}\rangle = \left(\prod_{\vec{k}} u_{\vec{k}} \right) e^{\sum_k \frac{v_{\vec{k}}}{u_{\vec{k}}} \hat{c}^{\dagger}_{\vec{k},\uparrow} \hat{c}^{\dagger}_{-\vec{k},\downarrow}} |\mathrm{vac}\rangle . \tag{12.23}$$

As anticipated in Section 12.1.1, this expression is of the same form as that of the coherent state wave function of a Bose–Einstein condensate, $|\mathrm{BEC}\rangle = e^{-|\alpha|^2/2} e^{\alpha \hat{a}^{\dagger}} |\mathrm{vac}\rangle$, with the bosonic creation operator defined as $\hat{a}^{\dagger} = \sum_k \frac{v_{\vec{k}}}{u_{\vec{k}}} \hat{c}^{\dagger}_{\vec{k},\uparrow} \hat{c}^{\dagger}_{-\vec{k},\downarrow}$. These are the bosons that condense in the superconducting state, in a coherent state whose phase coincides with that of the mean-field order parameter.

Because the BCS mean-field Hamiltonian of Eq. (12.18) does not conserve particle number, its eigenstates will be superpositions of states with different numbers of electrons. In Exercise 12.9, you will compare the extent of particle-number uncertainty in the BCS and BEC wave functions. Besides this quantitative comparison, however, there is also a fundamental difference between the two cases. The superconducting order parameter $\langle V \sum_{\vec{k}} \hat{a}^{\dagger}_{\vec{k}} \rangle$ written in terms of the Cooper pair creation operators breaks the global phase-rotation symmetry $\hat{a}^{\dagger}_{\vec{k}} \to \mathrm{e}^{\mathrm{i}\varphi} \hat{a}^{\dagger}_{\vec{k}}$. However, the Cooper pairs condensing in the BCS state are not really bosons and consist of two electrons each. Written explicitly in terms of electrons, the order parameter $\langle V \sum_{\vec{k}} \hat{c}^{\dagger}_{\vec{k},\uparrow} \hat{c}^{\dagger}_{-\vec{k},\downarrow} \rangle$ is seen to be invariant under the transformation $\hat{c}^{\dagger}_{\vec{k},\sigma} \to \mathrm{e}^{\mathrm{i}\pi} \hat{c}^{\dagger}_{\vec{k},\sigma}$. The BCS ground state therefore does not fully break the global phase rotation symmetry of electrons, but rather reduces it from $U(1)$ to $\mathbb{Z}_2$. This situation is reminiscent of crystals breaking continuous spatial translation symmetry down to discrete lattice translations, and the interpretation is similar as well. The broken spatial invariance in crystals allowed for a localised ground state consisting of a superposition of different momentum states. At the same time, the remnant discrete translation symmetry ensured that momentum is still conserved in crystals up to multiples of the reciprocal lattice vector. In the same way, the BCS superconductor superposes states with different numbers of Cooper pairs and thus different numbers of electrons, but the remnant $\mathbb{Z}_2$ symmetry guarantees that the electron number is conserved up to multiples of two. That is, changes in the Cooper pair number always increase or decrease the electron number by a multiple of two, and therefore the parity of the number of electrons in the superconducting state is a conserved quantity.

The fact that the bilinear fermionic order parameter has a remnant symmetry is generic to all fermionic condensates. In density waves with an order parameter proportional to $\langle\sum_{\vec{k}}\hat{c}^\dagger_{\vec{k}}\hat{c}_{\vec{k}+\vec{q}}\rangle$, for example, transformations of the type $\hat{c}^\dagger_{\vec{k}} \to e^{i\vec{a}\cdot\vec{k}}\hat{c}^\dagger_{\vec{k}}$ are broken except for values of $\vec{a}$ such that $\vec{a}\cdot\vec{q}$ is a multiple of 2π. Written in terms of the bosonic operator $\hat{a}^\dagger_{\vec{q}} = \sum_{\vec{k}}\hat{c}^\dagger_{\vec{k}}\hat{c}_{\vec{k}+\vec{q}}$, on the other hand, the symmetry $\hat{a}^\dagger_{\vec{q}} \to e^{i\vec{a}\cdot\vec{q}}\hat{c}^\dagger_{\vec{q}}$ is fully broken for any non-zero value of $\vec{a}$. In real space, this corresponds to the overall density wave localising at a precise position, while the translational symmetry of the electrons is only reduced to a discrete subset and still allows for the electronic momentum to be conserved up to multiples of an emergent reciprocal lattice vector.

Exercise 12.9 (Number fluctuations).
In this exercise, we investigate the extent to which particle number is uncertain in the BCS ground state. For simplicity, and without loss of generality, we consider a situation in which the phase of the order parameter is zero and write the BCS ground state as

$$|\mathrm{BCS}\rangle = \prod_k \left[\cos(\theta_k) - \sin(\theta_k)\hat{c}^\dagger_{k,\uparrow}\hat{c}^\dagger_{-k,\downarrow}\right]|\mathrm{vac}\rangle\,. \tag{12.24}$$

a. The operator for the total particle number is given by $\hat{N} = \sum_{k,\sigma}\hat{c}^\dagger_{k,\sigma}\hat{c}_{k,\sigma}$. Show that the mean particle number in the BCS ground state is equal to $\langle\hat{N}\rangle = 2\sum_k \sin^2(\theta_k)$.
b. Show that the (absolute) variance is

$$\begin{aligned}(\delta N)^2 &\equiv \langle\mathrm{BCS}|\hat{N}^2|\mathrm{BCS}\rangle - \left(\langle\mathrm{BCS}|\hat{N}|\mathrm{BCS}\rangle\right)^2 \\ &= 4\sum_k \cos^2(\theta_k)\sin^2(\theta_k).\end{aligned} \tag{12.25}$$

c. Using the explicit expressions for $\cos(\theta_k)$ and $\sin(\theta_k)$ from Exercise 12.8, give a rough estimate of the relative standard deviation $\frac{\delta N}{\langle N\rangle}$ in terms of the gap Δ, for small Δ.
d. What is the fate of the absolute variance and the relative standard deviation of the particle number in the thermodynamic limit? How do you think they compare to the absolute variance and relative standard deviation of the position of a table, which spontaneously breaks translational symmetry?
e. Repeat the analysis for a Bose–Einstein condensate with ground state wave function $|\mathrm{BEC}\rangle = e^{-|\Delta|^2/2}e^{\Delta\hat{a}^\dagger}|\mathrm{vac}\rangle$ and particle number operator $\hat{N} = \hat{a}^\dagger\hat{a}$.

Notice the difference indicated by your final result between the variances in the number of bosons and fermions. The relative standard deviation in the boson or Cooper pair number vanishes in the limit of a macroscopic average number of bosons participating in the condensate. The relative fluctuations in electron number, on the other hand, vanish in the thermodynamic limit regardless of the order parameter amplitude.

12.2.3 The Superconducting Gap

Although we identified the ground state of the BCS mean-field model in Eq. (12.22), we did not yet show that the order parameter can actually be non-zero in any self-consistent solution. This can be done by minimising the ground state energy of $\hat{H}^{\text{MF}}_{\text{BCS}}$ and solving the resulting saddle point equations. The procedure is detailed in various textbooks [160, 192] and finally results in an expression for the self-consistent mean-field value of the order parameter:

$$|\Delta| = \frac{\hbar\Omega}{\sinh(1/VN(0))} \approx 2\hbar\Omega e^{-1/VN(0)}. \tag{12.26}$$

Here, $N(0)$ is the electronic density of states at the Fermi level of the unperturbed metal. Notice that no features of the electronic dispersion $\epsilon(\vec{k})$ other than its non-zero density of states $N(0)$ enter the expression for the order parameter. Superconductivity is therefore predicted to arise in any metal as it is cooled to zero temperature, regardless of its electronic structure or dimensionality. In practice, however, superconductivity is often pre-empted by either the emergence of different types of order or the destruction of the metallic state itself. Barring these, the exponential dependence of the superconducting order parameter on the interaction strength V indicates that there will be a finite superconducting order parameter, for any non-zero strength of the pairing interaction, regardless how weak it is. The magnitude of the order parameter, however, becomes very small for small $VN(0)$. The prefactor in the equation for the order parameter is the phonon frequency, playing the role of an ultraviolet cut-off scale. Since changes of the prefactor are easily overtaken by changes in the exponent, it only has a minor influence on the gap amplitude. In fact, this is one of the reasons that we could be sloppy with our definition of the cut-off energy in Eq. (12.16). Even a small numerical uncertainty in the value of the factor $VN(0)$ can make relatively large changes in the choice of the upper limit on the energy integrals undetectable.

The amplitude Δ of the superconducting order parameter can be given physical meaning by realising that it determines the size of the gap in the energy spectrum of the superconductor. This follows immediately from the diagonal form of $\hat{H}^{\text{MF}}_{\text{BCS}}$ in Eq. (12.20), which shows that excitations on top of the superconducting ground state created by $\hat{\gamma}^{\dagger}_{\vec{k}}$ cost at least an energy $E_{\vec{k}=\vec{k}_{\text{F}}} = \Delta$ to make. For this reason, Δ is commonly referred to as 'the BCS order parameter', or the BCS gap, and Eq. (12.26) as the **BCS gap equation**.

The identification of the BCS order parameter with the energy cost for creating fermionic excitations indicates an important difference between the purely bosonic charged superfluid of Section 8.1 and the fermionic condensate. The order in the former could only be destroyed by the proliferation of phase fluctuations, either in the form of (gapped) Nambu–Goldstone modes or vortices. In the fermionic condensate, on the other hand, the bosonic particles can be destroyed themselves. Perturbations or thermal fluctuations with energies in excess of the BCS gap will excite Bogoliubov quasiparticles, and in doing so break apart Cooper pairs. The superconducting order is then destroyed by a direct suppression of the order parameter amplitude, rather than by averaging over a disordered phase field.

The way in which superconductivity melts in actual superconducting materials depends on the distinction between type-I and type-II superconductors introduced in Section 8.4. Amplitude melting is typical for strong type-I superconductors, whereas phase-melting is the most prominent mechanism in strong type-II materials. For other fermionic condensates with bilinear order parameter operators, either phase melting or amplitude melting may dominate, depending on the strength and structure of the pairing interaction, the coherence length, and other material-dependent quantities.

12.3 Gap Symmetry, Triplet Pairs, and High-T_c Superconductors

Although the BCS Hamiltonian can be used to understand the superconducting phases in many materials, there are several ways in which it may be insufficient for others. In those cases, the BCS theory may be extended by going beyond the mean-field approximation, or by lifting some of the assumptions on the symmetry of the order parameter or the nature of the interactions between electrons.

12.3.1 Strong Coupling Superconductivity

In the BCS gap equation in Eq. (12.26), the strength of the superconducting interactions is characterised by the combination of the potential $V = 2g^2/(\hbar\Omega)$ and the electronic density of states $N(0)$. Their product defines a dimensionless electron–phonon coupling strength $\lambda \equiv VN(0)$. Whenever λ becomes of order one or larger, we speak of **strong coupling superconductivity**. It is still possible to understand these systems in terms of a perturbative expansion akin to the BCS model, but this becomes technically more cumbersome. It is most conveniently done as a diagrammatic expansion in the context of quantum field theory and is then known as **Migdal–Eliashberg theory**. Unlike BCS theory, this method includes the frequency dependence of the electron–phonon coupling. It allows for a detailed and quantitative description of a great number of superconducting materials, including the paradigmatic strong coupling superconductor lead.

Despite the strength of the dimensionless coupling, the same hierarchy of energy scales as in BCS theory is present also in strong coupling superconductors. The gap amplitude $|\Delta|$ is much smaller than the phonon energy scale $\hbar\Omega$, which in turn is much smaller than the Fermi energy E_F:

$$|\Delta| \ll \hbar\Omega \ll E_F. \tag{12.27}$$

In lead, for example, the coupling strength can be as large as $\lambda = 1.9$, but still the superconducting transition occurs at $T_c = 7.2$ K, while the phonon energy approximated by the Debye frequency and expressed as temperature is $\Theta_D \approx 105$ K, and the Fermi energy is as large as $E_F = 9.5$ eV $\approx 110{,}000$ K. Because of this, Migdal–Eliashberg superconductors are not much different from BCS superconductors, other than their more involved theoretical description.

12.3.2 Anisotropic Order Parameters

Going beyond just the strength of the electron–phonon coupling, also its form in Eq. (12.10) is greatly simplified and will not necessarily be a good approximation for any particular material. In general, the coupling g will depend on both the incoming and outgoing momenta (as well as possibly other degrees of freedom) of electrons scattering off phonons. Keeping the momentum dependence of the coupling explicit, the effective low-energy theory becomes of the form

$$\hat{H}_{\text{eff}} = \sum_{\vec{k},\sigma} \left(\epsilon(\vec{k}) - \mu\right) \hat{c}^{\dagger}_{\vec{k},\sigma} \hat{c}_{\vec{k},\sigma} - \sum_{\vec{k},\vec{k}'}{}' V_{\vec{k},\vec{k}'} \hat{c}^{\dagger}_{\vec{k}',\uparrow} \hat{c}^{\dagger}_{-\vec{k}',\downarrow} \hat{c}_{-\vec{k},\downarrow} \hat{c}_{\vec{k},\uparrow} . \tag{12.28}$$

Here, the momentum-dependence of $V_{\vec{k},\vec{k}'}$ is inherited directly from that of the electon–phonon coupling. Introducing the order parameter $\Delta_{\vec{k}} = \sum_{\vec{k}'} V_{\vec{k},\vec{k}'} O_{\vec{k}'}$ with $O_{\vec{k}} = \langle \hat{c}_{-\vec{k},\downarrow} \hat{c}_{\vec{k},\uparrow} \rangle$, the mean-field Hamiltonian for the BCS model can be written as

$$\hat{H}^{\text{MF}}_{\text{BCS}} = \sum_{\vec{k}} \Delta^*_{\vec{k}} O_{\vec{k}} + \sum_{\vec{k}} \Big\{ \epsilon(\vec{k}) \left(\hat{c}^{\dagger}_{\vec{k},\uparrow} \hat{c}_{\vec{k},\uparrow} + \hat{c}^{\dagger}_{\vec{k},\downarrow} \hat{c}_{\vec{k},\downarrow} \right)$$
$$- \Delta^*_{\vec{k}} \hat{c}_{-\vec{k},\downarrow} \hat{c}_{\vec{k},\uparrow} - \Delta_{\vec{k}} \hat{c}^{\dagger}_{\vec{k},\uparrow} \hat{c}^{\dagger}_{-\vec{k},\downarrow} \Big\} . \tag{12.29}$$

Here, we made the usual assumptions that $\epsilon(\vec{k}) = \epsilon(-\vec{k})$, and $V_{\vec{k},\vec{k}'} = V_{\vec{k}',\vec{k}}$.

The mean-field Hamiltonian can be diagonalised using a Bogoliubov transformation. For ease of notation, we choose the phase ϕ of the order parameter to be zero. The minimisation of the mean field ground state energy then yields the anisotropic gap equation:

$$\Delta_{\vec{k}'} = -\frac{1}{2} \sum_{\vec{k}} V_{\vec{k}',\vec{k}} \frac{\Delta_{\vec{k}}}{\sqrt{(\epsilon(\vec{k}) - \mu)^2 + \Delta^2_{\vec{k}}}} . \tag{12.30}$$

The self-consistent solutions of this equation determine the variation of the order parameter amplitude in k-space. The case of a uniform and attractive interaction $V_{\vec{k}',\vec{k}} = V > 0$ leads to the standard BCS model and an isotropic or **s-wave** gap $\Delta_{\vec{k}} = \Delta$. In the presence of more general interactions $V_{\vec{k}',\vec{k}}$, different types of gap function may emerge, and additional symmetries may be spontaneously broken.

For example, in real materials, the Fermi surface is typically far from isotropic, even if the electronic dispersion consists of just a single band of nearly free electrons. The coupling between electronic states can mirror such anisotropies, resulting in the order parameter $\Delta_{\vec{k}}$ varying with direction. A sketch of the order parameter for such **anisotropic s-wave** superconductivity is shown in Figure 12.2(a). This type of superconductivity has been proposed to occur, for example, in certain heavy-fermion compounds [189].

In other materials, for example, those consisting of multiple pockets that are not isotropically distributed over the Brillouin zone, the interaction between electrons may end up being attractive within a single pocket, but repulsive between different ones. In such cases, the order parameter may have a different sign for electrons on pockets in different locations, even if the sign does not vary with direction in the Brillouin zone. This corresponds to a so-called **s+− symmetry** for the superconducting gap (sketched in Figure 12.2(b)), which has been suggested to arise in some pnictide (iron-based) superconductors.

Notice that although the gap is not momentum-independent in these examples, its variations do not break any symmetries of the Hamiltonian. The shape of the Fermi surface, and therefore the momentum dependence of the dispersion $\epsilon(k)$, in Figure 12.2(a), for example, has only twofold rotational symmetry. The gap function $\Delta_{\vec{k}}$ has the same symmetry. Likewise, the Fermi surface in Figure 12.2(b) has fourfold rotational symmetry, as does the associated gap function. It is also possible, however, that spatial symmetries of the Hamiltonian are broken spontaneously by the formation of a gap function with a lower symmetry than the Fermi surface.

This can occur, for example, when the electron–electron interactions are repulsive, $V_{\vec{k}',\vec{k}} > 0$. Somewhat counterintuitively, these can still cause Cooper pairing. Consider, for

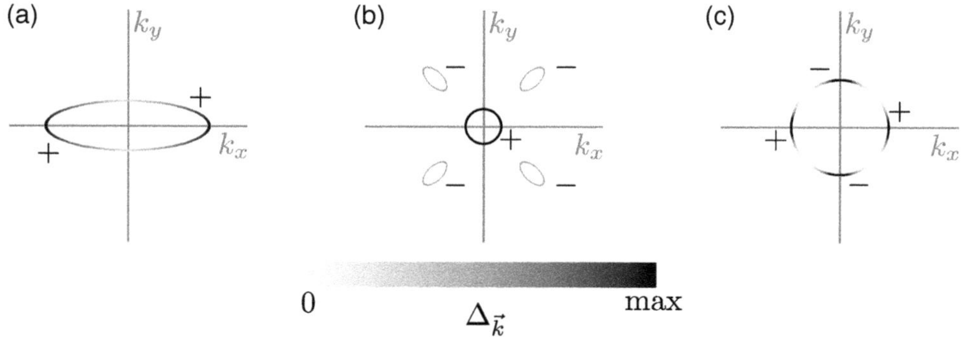

Figure 12.2 Sketches of anisotropic superconducting order parameters. Panel (a) shows a Fermi surface with an anisotropic s-wave symmetric gap indicated by the changing shade along the surface. In this case, the anisotropy could be inherited directly from the Fermi surface anisotropy. The order parameter in panel (b) has s+− symmetry, as indicated by the signs on the different pockets. This situation may arise, for example, due to differences between interactions within and between pocket. Panel (c) indicates d-wave symmetry, which may occur if the sign of the interactions between electrons becomes repulsive.

example, a situation in which $V_{\vec{k}',\vec{k}}$ is large and positive for $\vec{k}' \perp \vec{k}$ but small and positive for parallel momenta. Close inspection of Eq. (12.30) shows there may still be a solution, as long as the order parameter changes sign as a function of momentum, as sketched in Figure 12.2(c). The resulting **d-wave** symmetry implies a vanishing superconducting order parameter along directions in the Brillouin zone separating order parameters with different sign. Notice that the interaction $V_{\vec{k}',\vec{k}}$ in this case depends on the momentum difference, but that the BCS Hamiltonian in Eq. (12.28) with a spherical Fermi surface is still invariant under rotations. The d-wave configuration in this case spontaneously breaks both the global $U(1)$ phase rotation symmetry, and the $SO(3)$ symmetry associated with spatial rotations. Superconductivity with a d-wave order parameter is believed to occur in cuprate high-T_c superconductors and in some heavy fermion superconductors. The Fermi surface in those materials is not spherically symmetric, however, and the rotational symmetry is broken explicitly rather than spontaneously. The Nambu–Goldstone modes associated with the broken $SO(3)$, corresponding to oscillations of the direction in which superconductivity vanishes, are therefore typically gapped and not observed in these materials.

Additional symmetries can be broken by superposing d-wave or other anisotropic configurations. For example, naively adding two d-wave states of the type shown in the right of Figure 12.2, but rotated over an angle with respect to one another will yield just another, rotated d-wave state. However, if the two components are superposed with a relative phase difference, the resulting state breaks time reversal invariance as well as the spatial and phase rotation symmetries.

In general, the phase variations of the gap function in d-wave superconductors can be experimentally confirmed using the Josephson effect. For example, if you take an annulus of superconducting material in the d-wave state of Figure 12.2, and cut out one quarter of it, then the sign of the gap will be opposite on the two cuts. Now consider putting a quarter-annulus of known s-wave superconducting material in the opening, leaving two weak links at the cuts. The phase difference between the d-wave and s-wave materials will be $\Delta\phi = \phi_d - \phi_s$ at one cut, and $\Delta\phi = \phi_s - (\phi_d + \pi)$ at the other. Following the discussion of Section 3.5, a Josephson current proportional to $\sin(\Delta\phi)$ will arise at each of the cuts, and since $\sin(\phi_s - \phi_d - \pi)) = \sin(\phi_d - \phi_s)$, a macroscopic current will run around the annulus. Detecting the dependence of this loop current on the phase difference $\Delta\phi$ gives direct experimental evidence for the d-wave symmetry of the superconducting gap function.

Exercise 12.10 (Momentum-dependent pairing).
Isotropic or s-wave superconductors are common among materials in which the effective interactions are mediated solely by phonons. This is less usual in unconventional superconductors, where for instance spin-wave-like excitations can mediate strongly $\vec{k}$-dependent effective interactions.
a. Consider an interaction term of the BCS kind, but keeping the momentum dependence explicit:

$$\hat{H} = \sum_{\vec{k},\sigma} \epsilon(\vec{k})\hat{c}^{\dagger}_{\vec{k},\sigma}\hat{c}_{\vec{k},\sigma} - \sum_{\vec{k},\vec{k}'} V_{\vec{k},\vec{k}'}\hat{c}^{\dagger}_{\vec{k}',\uparrow}\hat{c}^{\dagger}_{-\vec{k}',\downarrow}\hat{c}_{-\vec{k},\downarrow}\hat{c}_{\vec{k},\uparrow}. \tag{12.31}$$

Define $\Delta_{\vec{k}} = \sum_{\vec{k}'} V_{\vec{k},\vec{k}'} O_{\vec{k}'}$ and show that the gap equation becomes

$$\Delta_{\vec{k}'} = -\frac{1}{2}\sum_{\vec{k}} V_{\vec{k}',\vec{k}} \frac{\Delta_{\vec{k}}}{\sqrt{(\epsilon(\vec{k}) - \mu)^2 + \Delta^2_{\vec{k}}}}. \tag{12.32}$$

The Fermi surface of electrons hopping between the nearest atoms of a square lattice is precisely square at half-filling, and therefore far from isotropic. Consider a situation in which the Fermi surface anisotropy causes the pairing interaction to be strongly attractive along the $(1, 0)$ and $(0, 1)$ directions in the Brillouin zone, while it is repulsive along the $(1, 1)$ and $(1, -1)$ directions.
b. Sketch how the gap function $\Delta_{\vec{k}}$ evolves around the Fermi surface, both for the case in which $\Delta_{\vec{k}}$ has the same symmetry as the Fermi surface, and for the case in which $\Delta_{\vec{k}}$ has a lower symmetry but is still even under reflections in the $k_x = 0$ and $k_y = 0$ axes. The former situation is an example of anisotropic s-wave symmetry, while the latter has d-wave symmetry.

12.3.3 Broken Spin-Rotation Symmetry

In all forms of pairing interaction we considered so far, we implicitly assumed that the spins of the electrons bound into a Cooper pair form a singlet state. Because the singlet state is odd under the exchange of its two spins, the spatial or 'orbital' part of the Cooper pair wave function needs to be even in order to satisfy the fermionic exchange statistics of the electrons making up the pair. All s- and d-wave symmetric order parameters we considered above are invariant under the transformation $\vec{k} \to -\vec{k}$, and thus have even spatial wave functions. All superconductors with these types of order parameters must therefore contain spin singlet pairs. In fact, the link between the parity of the order parameter and the spin structure of the Cooper pair does not have to be imposed separately, but is hard-wired into the mean-field equation. This can be seen by considering

$$\hat{H}_{\text{pairing}} = -\sum_{\vec{k}} \Delta_{\vec{k}} \hat{c}^{\dagger}_{\vec{k},\uparrow} \hat{c}^{\dagger}_{-\vec{k},\downarrow}$$
$$= \begin{cases} -\sum_{\vec{k}>0} \Delta_{\vec{k}} \left(\hat{c}^{\dagger}_{\vec{k},\uparrow} \hat{c}^{\dagger}_{-\vec{k},\downarrow} - \hat{c}^{\dagger}_{\vec{k},\downarrow} \hat{c}^{\dagger}_{-\vec{k},\uparrow} \right) & \text{if } \Delta_{\vec{k}} = \Delta_{-\vec{k}}, \\ -\sum_{\vec{k}>0} \Delta_{\vec{k}} \left(\hat{c}^{\dagger}_{\vec{k},\uparrow} \hat{c}^{\dagger}_{-\vec{k},\downarrow} + \hat{c}^{\dagger}_{\vec{k},\downarrow} \hat{c}^{\dagger}_{-\vec{k},\uparrow} \right) & \text{if } \Delta_{\vec{k}} = -\Delta_{-\vec{k}}. \end{cases} \tag{12.33}$$

Self-consistent solutions of the mean field equations can thus exist only for combinations of order parameters and spin states that obey the correct fermionic exchange statistics. For completeness, notice that the same-spin pairing interaction $\sum_{\vec{k}} \Delta_{\vec{k}} \hat{c}^{\dagger}_{\vec{k},\uparrow} \hat{c}^{\dagger}_{-\vec{k},\uparrow}$ is non-zero only if $\Delta_{\vec{k}} = -\Delta_{-\vec{k}}$ also.

Considering interactions beyond electron–phonon coupling, states with non-zero spin angular momentum can be favoured over the spin-singlet configuration. **Superfluid ^{3}He**, for example, is believed to consist of pairs of two ^{3}He atoms (or, more precisely, two quasiparticles associated with the ^{3}He Fermi liquid). The effective interaction between quasiparticles in superfluid ^{3}He arises from the exchange part of the Coulomb interaction, rather than the phonon exchange gluing electrons together in the BCS superconductor. The electromagnetic exchange interaction favours alignment of spins on different quasiparticles [204].

The ferromagnetic nature of the exchange interaction may be understood from the fact that aligned spins enforce a node in the antisymmetric wave function of two interacting fermions, which reduces the repulsive Coulomb energy compared to a symmetric configuration because of the increased average distance between charges. Alternatively, it can be interpreted as the magnetic field produced by one particle causing a Zeeman splitting of the spin states on another particle, which in turn produces a Zeeman field for the first particle. This exchange of magnetic fields or spin fluctuations is closely analogous to the exchange of phonons or displacement fields in the usual Cooper pair formation, except that it favours the alignment of spins. In fermionic superfluids like ^{3}He, the order parameter may thus be expected to be of odd parity, and the spins arranged in a triplet configuration.

The most general form of the pairing interaction in a BCS-type Hamiltonian can be written as

$$\hat{H}_{\text{pairing}} = -\sum_{\vec{k},\vec{k}'} \sum_{\sigma_1,\sigma_2} V_{\vec{k},\vec{k}'} \hat{c}^{\dagger}_{\vec{k}',\sigma_1} \hat{c}^{\dagger}_{-\vec{k}',\sigma_2} \hat{c}_{-\vec{k},\sigma_2} \hat{c}_{\vec{k},\sigma_1}. \tag{12.34}$$

Splitting the interaction strength $V \equiv V^+ + V^-$ into a term V^+ that is even and a term V^- that is odd under both $\vec{k} \to -\vec{k}$ and $\vec{k}' \to -\vec{k}'$, this becomes

$$\begin{aligned} \hat{H}_{\text{pairing}} = & -\sum_{\vec{k},\vec{k}'} V^-_{\vec{k},\vec{k}'} \left(\hat{c}^{\dagger}_{\vec{k}',\uparrow} \hat{c}^{\dagger}_{-\vec{k}',\uparrow} \hat{c}_{-\vec{k},\uparrow} \hat{c}_{\vec{k},\uparrow} + \hat{c}^{\dagger}_{\vec{k}',\downarrow} \hat{c}^{\dagger}_{-\vec{k}',\downarrow} \hat{c}_{-\vec{k},\downarrow} \hat{c}_{\vec{k},\downarrow} \right) \\ & + \frac{1}{2} V^-_{\vec{k},\vec{k}'} \left(\hat{c}^{\dagger}_{\vec{k}',\uparrow} \hat{c}^{\dagger}_{-\vec{k}',\downarrow} + \hat{c}^{\dagger}_{\vec{k}',\downarrow} \hat{c}^{\dagger}_{-\vec{k}',\uparrow} \right) \left(\hat{c}_{-\vec{k},\downarrow} \hat{c}_{\vec{k},\uparrow} + \hat{c}_{-\vec{k},\uparrow} \hat{c}_{\vec{k},\downarrow} \right) \\ & + \frac{1}{2} V^+_{\vec{k},\vec{k}'} \left(\hat{c}^{\dagger}_{\vec{k}',\uparrow} \hat{c}^{\dagger}_{-\vec{k}',\downarrow} - \hat{c}^{\dagger}_{\vec{k}',\downarrow} \hat{c}^{\dagger}_{-\vec{k}',\uparrow} \right) \left(\hat{c}_{-\vec{k},\downarrow} \hat{c}_{\vec{k},\uparrow} - \hat{c}_{-\vec{k},\uparrow} \hat{c}_{\vec{k},\downarrow} \right). \end{aligned} \tag{12.35}$$

In situations where the interactions between quasiparticles favour alignment of spins, V^- dominates and Cooper pairs preferentially form in triplet configurations. The mean-field order parameter then consists of three components, for the three ways of pairing into spin-triplet states. These can be combined into a matrix:

$$\Delta_{\sigma_1,\sigma_2}(\vec{k}) = \sum_{\vec{k}'} V^-_{\vec{k},\vec{k}'} \langle \hat{c}^{\dagger}_{\vec{k}',\sigma_1} \hat{c}^{\dagger}_{-\vec{k}',\sigma_2} \rangle. \tag{12.36}$$

Here, we defined $\Delta_{\uparrow,\downarrow} = \Delta_{\downarrow,\uparrow}$, so that $\Delta_{\sigma_1,\sigma_2}$ is specified by three components. Notice that the order parameters are defined here with respect to an arbitrarily chosen spin-z axis. Depending on the relative values of the order parameters, the spin rotational symmetry of the full Hamiltonian is spontaneously broken. This is sometimes colloquially referred to as the Cooper pairs obtaining a magnetisation. Additionally, the odd spatial wave function of the Cooper pairs has an orientation in real space, and spontaneously breaks spatial, or 'orbital', rotation symmetry. In the absence of spin–orbit coupling the orbital and spin polarisation are independent. The expectation values in each of the three spin-resolved order parameters may then be further decomposed into terms with different values of the angular momentum l^z. As you will confirm in Exercise 12.11, a full p-wave triplet superfluid state thus simultaneously breaks the $SO(3)$ symmetry of spatial rotations, the $SU(2)$ symmetry of spin rotations, and the $U(1)$ symmetry of phase rotations.

The possibility of 'dressing' the order parameter with additional degrees of freedom and additional broken symmetries is not specific for superconductors and occurs generically for all fermionic condensates. In charge density waves, for example, the particle and hole operators combined in the order parameter expectation value may refer to specific atomic orbitals or spin states. The ordered configuration can then break spatial or spin-rotational symmetries on top of the spontaneously broken translational invariance. Moreover, just as in superconductors, superposing multiple such anisotropic density waves can yield yet more spontaneously broken symmetries, including spatial inversion and time reversal invariance.

Exercise 12.11 (The Cooper pair wave function).
In analogy to Section 12.1.1, we can define hard-core boson creation operators $\hat{a}^\dagger_{k,\sigma}$ for spin triplet Cooper pairs by writing

$$\begin{aligned}
\hat{a}^\dagger_{\vec{k},1} &= \hat{c}^\dagger_{\vec{k},\uparrow}\hat{c}^\dagger_{-\vec{k},\uparrow}, \\
\hat{a}^\dagger_{\vec{k},0} &= \sqrt{1/2}(\hat{c}^\dagger_{\vec{k},\uparrow}\hat{c}^\dagger_{-\vec{k},\downarrow} + \hat{c}^\dagger_{\vec{k},\downarrow}\hat{c}^\dagger_{-\vec{k},\uparrow}), \\
\hat{a}^\dagger_{\vec{k},-1} &= \hat{c}^\dagger_{\vec{k},\downarrow}\hat{c}^\dagger_{-\vec{k},\downarrow}.
\end{aligned} \tag{12.37}$$

We can additionally define Cooper pair spin operators of the form $\hat{S}^j(\vec{k}) = \sum_{\sigma_1,\sigma_2}\hat{a}^\dagger_{\vec{k},\sigma_1} S^j_{\sigma_1,\sigma_2}\hat{a}_{\vec{k},\sigma_2}$, with $S^j_{\sigma_1,\sigma_2}$ a 3×3 matrix, and $j \in \{x,y,z\}$.

a. Defining $S^z_{\sigma_1,\sigma_2} = \mathrm{diag}(1,0,-1)$, give expressions for $S^x_{\sigma_1,\sigma_2}$ and $S^y_{\sigma_1,\sigma_2}$ and show that the matrices S^j generate an $SU(2)$ algebra.

b. Consider only the part of the general pairing interaction of Eq. (12.35) proportional to V^- and write it in the form

$$\hat{H}^-_{\text{pairing}} = -\sum_{\vec{k},\vec{k}'}\sum_{\sigma_1,\sigma_2} V^-_{\vec{k},\vec{k}'}\hat{a}^\dagger_{\vec{k}',\sigma_1} h_{\sigma_1,\sigma_2'}\hat{a}_{\vec{k},\sigma_2}. \tag{12.38}$$

Show that h_{σ_1,σ_2} commutes with all three $S^j_{\sigma_1,\sigma_2}$ matrices.

c. Consider a mean-field superconducting state characterised by the expectation values $O_{\vec{k},j} = \langle\hat{a}^\dagger_{\vec{k},j}\rangle$. Focussing on the specific case that $O_{\vec{k},j}$ is real and of the form $O_{\vec{k},j} = f(\vec{k})o_j$ explain why a generic configuration with three distinct components o_j breaks the $SU(2)$ Cooper pair spin rotation symmetry down to $U(1)$.

Notice that although o_j has three components, these are not independent superconducting order parameters. The hard-core bosons $\hat{a}^{\dagger}_{\vec{k},j}$ really consist of electrons, and the full Hamiltonian in terms of electrons does not have separate $U(1)$ phase-rotation symmetries for $j = 1, 0$, and -1. The generalised form of the gap equation in Eq. (12.32) including spins will therefore couple the different o_j components. Nevertheless, a mean-field ground state described by non-zero values of o_j breaks both the $SU(2)$ spin rotation and the superconducting $U(1)$ phase rotation symmetries.

Next, consider the spatial part of the mean-field expectation value. Expanding $f(\vec{k})$ in spherical harmonics, we can write it as $f(\vec{k}) = r(k) \sum_{l,m} f_{lm} Y^l_m(\theta, \phi)$, where $k = |\vec{k}|$, while θ and ϕ are the Euler angles describing the direction of $\vec{k}$.

d. Explain why components of $f(\vec{k})$ with even values of l do not contribute to the mean-field expectation value $\langle \hat{H}^{-}_{\text{pairing}} \rangle$.

e. Give an argument for why the components with the lowest allowed value of l contribute the most to the mean-field expectation value $\langle \hat{H}^{-}_{\text{pairing}} \rangle$.

Considering only the $l = 1$ contributions to $f(\vec{k})$, and assuming them to be real, the mean-field expectation value is characterised by the three parameters f_{1m}.

f. Assuming $V^{-}_{\vec{k}\vec{k}'}$ depends only on the scalar product $\vec{k} \cdot \vec{k}'$, and not on the individual momenta, argue that $\hat{H}^{-}_{\text{pairing}}$ is invariant under global rotations of the Cooper pair momentum $\hat{a}^{\dagger}_{\vec{k},\sigma} \to \hat{a}^{\dagger}_{R\vec{k},\sigma}$, with R a rotation operator.

g. Explain why a generic mean-field configuration with three distinct components f_{1m} breaks the $SO(3) \sim SU(2)$ spatial rotation symmetry down to $U(1)$.

In the absence of spin–orbit coupling, spin and spatial rotations are independent, and there is no reason for the mean-field expectation value to factorise, and its more general form is $O_{\vec{k},j} = r(k) \sum_m g_{jm} Y^1_m(\theta, \phi)$. Because $\langle \hat{a}^{\dagger}_{\vec{k},j} \rangle$ may be complex, g_{jm} contains 18 real parameters, which together fully describe the superconducting state. As before, however, notice that Cooper pairs really consist of electrons. This causes the 18 components to be coupled in the saddle point equations, and not be independent. Nevertheless, the mean-field ground state of $\hat{H}^{-}_{\text{pairing}}$ without spin–orbit coupling described by these 18 parameters realises the symmetry-breaking pattern $SO(3) \times SU(2) \times U(1) \to U(1) \times U(1)$.

12.3.4 Helium-3

In practice, only a small subset of all possible symmetry-breaking patterns arises in proposed p-wave triplet superfluids or superconductors. ^{3}He is probably the best understood triplet superfluid and has a rich phase diagram shown in Figure 12.3 [204]. Each different phase corresponds to a different choice of triplet-order parameter.

Arguably, the simplest form is known as the **A_1-phase** of superfluid ^{3}He, which arises under the application of an external magnetic field. Since the field explicitly breaks spin rotation symmetry, pairs that have a magnetisation aligned with the magnetic field are preferred. Writing the order parameter in the form of Eq. (12.36), only the component $\Delta_{\uparrow,\uparrow}$ (with S^z defined parallel to the magnetic field) will be non-zero. In terms of the orbital decomposition

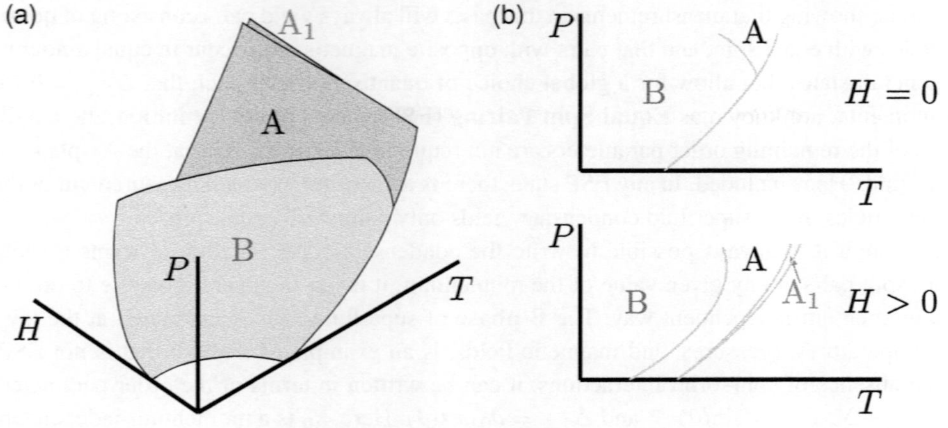

Figure 12.3

Schematic phase diagram of superfluid ^{3}He. Panel (a) shows the phase diagram as a function of temperature T, pressure P, and applied magnetic field H. The phases indicated all occur on the low-temperature side of the corresponding surfaces. Panel (b) indicates slices through the three-dimensional phase diagram, at fixed values of the magnetic field. In all cases, ^{3}He is a normal fluid at the highest temperatures.

of the pair wave function, it turns out the A_1-phase also contains only a single component, given by the $l = 1$, $m = 1$ spherical harmonic. The momentum dependence of the wave function is thus $Y_{11}(\vec{k}) \propto \sin(\theta)e^{i\varphi}$, where θ and φ are the polar and azimuthal Euler angles describing the direction of $\vec{k}$. Notice that k^z here is defined to lie parallel to the orbital angular momentum $\vec{l}$, which in the absence of spin–orbit coupling is independent of the spin quantisation axis. Because of the spatial part, the order parameter amplitude vanishes when $\sin(\theta) = 0$, in directions parallel or anti-parallel to the orbital angular momentum. The nodes in the gap render the superfluid state spatially anisotropic, on top of its magnetic anisotropy due to the equal, non-zero magnetisation of all Cooper pairs. Without spin–orbit coupling, the direction of $\vec{l}$ and thus the spatial directions in which the order parameter has nodes is independent of the applied magnetic field and spontaneously breaks rotational symmetry. Because spin rotations are explicitly broken by the applied field, the pattern of spontaneous symmetry breaking in this case is $SO(3) \times U(1) \to U(1)$.

A very similar state is realised in the **A-phase** of superfluid 3He of superfluid ^{3}He, upon removing the external magnetic field. The spatial part of the order parameter is not affected in the transition from the A_1 to the A-phase, and in particular the nodes at orientations of $\vec{k}$ parallel or anti-parallel to the orbital angular momentum persist. The spin decomposition of the order parameter, however, adjusts to the fact that there is no longer a preferred magnetisation direction, and in the A-phase only the component $\Delta_{\uparrow,\downarrow}$ is non-zero. Without spin–orbit coupling, the spontaneously broken symmetries are then $SO(3) \times SU(2) \times U(1) \to U(1) \times U(1)$.

Choosing an alternative spin-quantisation axis, the same phase can also be written in terms of the alternative set of order parameters $\Delta_{\uparrow,\downarrow} = 0$ and $\Delta_{\uparrow,\uparrow} = e^{i\phi}\Delta_{\downarrow,\downarrow} > 0$. Notice that the relative phase ϕ is fully determined by the choice of quantisation axis, so that the A-phase is still characterised by only a single amplitude and (overall) phase for its order parameters. The amplitudes $|\Delta_{\uparrow,\uparrow}|$ and $|\Delta_{\downarrow,\downarrow}|$ can be made equal by a momentum-independent choice of the spin axis, showing that measurements in this basis will always yield pairs consisting of quasiparticles with equal spin, and that pairs with opposite magnetisation occur in equal amounts. In general, states that allow for a global choice of quantisation axis such that $\Delta_{\uparrow,\downarrow} = 0$ for all momenta, are known as **Equal Spin Pairing** (ESP) states. In this definition, the amplitudes of the remaining order parameters are not required to be equal, so that the A_1-phase of superfluid ^{3}He is included. In any ESP state, there is a direction in which measurement of the quasiparticles in the superfluid condensate yields only pairs with equal spin.

Although it is always possible to write the condensate wave function in terms of only equal spin pairs at any given value of the momentum, it is not in general possible to do this in a momentum-independent way. The **B-phase** of superfluid ^{3}He, encountered at the lowest temperatures, pressures, and magnetic fields, is an example of a phase that is not ESP. In the absence of spin–orbit interactions, it can be written in terms of the order parameters $\Delta_{\downarrow,\downarrow} = -\Delta^*_{\uparrow,\uparrow} = d_0 \sin(\theta) e^{i\varphi}$ and $\Delta_{\uparrow,\downarrow} = d_0 \cos(\theta)$. Here, d_0 is a momentum-independent constant while θ and φ again denote the polar and azimuthal Euler angles for the direction of $\vec{k}$ relative to $\vec{l}$. This phase has no nodes in any direction, and in fact it features an isotropic total gap in the excitation spectrum proportional to $\sum_{\sigma_1,\sigma_2} |\Delta_{\sigma_1,\sigma_2}|^2$. Despite the isotropy of the gap, the direction dependence of the order parameter decomposition into $\Delta_{\sigma_1,\sigma_2}$ components breaks the spatial rotational symmetry, in addition to the spin and phase rotations.

Both the A- and B-phases are examples of so-called **unitary phases** of the triplet superfluid. The name comes from the observation that the 2×2 matrix with elements $\Delta_{\sigma_1,\sigma_2}$ is unitary (up to normalisation) in these phases. Historically, the customary way of describing unitary phases is in terms of the so-called **d-vector**, whose components are defined as

$$\begin{aligned} d_x &= -\Delta_{\uparrow,\uparrow} + \Delta_{\downarrow,\downarrow}, \\ d_y &= -i\left(\Delta_{\uparrow,\uparrow} + \Delta_{\downarrow,\downarrow}\right), \\ d_z &= \Delta_{\uparrow,\downarrow} + \Delta_{\downarrow,\uparrow}. \end{aligned} \tag{12.39}$$

In terms of the d-vector, unitary phases are ones for which all components of $\vec{d}$ are real, up to an overall total phase shared by all components of the order parameter. In fact, all phases of the triplet superfluid could be described in terms of a complex-valued d-vector, but it is used mostly for unitary phases where it has the physical interpretation that $\vec{d}$ points in the direction of zero spin projection.

The requirement for the d-vector to be real implies that in unitary phases the spin-up and spin-down components of the order parameter always have equal weight: $|\Delta_{\downarrow,\downarrow}| = |\Delta_{\uparrow,\uparrow}|$. In contrast, ESP phases are defined in terms of the d-vector by the requirement that there is a momentum-independent choice of spin-quantisation axis in which d_z vanishes. Of the phases we encountered so far, the A_1-phase is ESP but not unitary, the A-phase is both, and the B-phase is unitary but not ESP.

12.3.4.1 Spin–Orbit Coupling

In all phases considered so far, we assumed the spin and orbital sectors of the order parameter to be independent. In particular, this implies that global rotations of only the spins or only the orbital wave function do not affect the energy, so that $SU(2)$ spin rotation and $SO(3)$ spatial rotation symmetries can be separately broken. In real superfluids, however, there are often weak interactions that couple the spin to the orbital degree of freedom. In ^{3}He, for example, the electromagnetic interaction between two nuclear dipole moments is invariant under the simultaneous rotation of both spin and orbital orientations, but not under individual rotations of either. The dipole interaction is very weak compared to other forces and has a mostly negligible effect in normal fluids. In superfluid phases, however, the spin and orbital orientations of individual atoms are related to one another by the superfluid order parameter. That is, rotating the spin or d-vector on a single atom causes an excitation of the ordered state unless all spins or d-vectors in the entire condensate are rotated in the same way. This is an incarnation of the rigidity of the superfluid order discussed in Section 4.6. It amplifies the effect of the dipole interactions, since aligning the spin and orbital orientation in a favourable way on a single atom now simultaneously leads to all other atoms finding similarly favourable alignment. This effect is sometimes called 'spontaneously broken spin–orbit symmetry' (SBSOS), although the spin–orbit coupling is actually explicit.

The effect of the dipole interactions is different for the different phases of ^{3}He. In the A-phase it forces $\vec{d}$ to be parallel to $\vec{l}$, so that combined spin–orbital transformations are the only rotation symmetry to be broken spontaneously. Applying a small magnetic field will break this symmetry explicitly and align $\vec{d}$ perpendicular to the field, so that only the $U(1)$ phase rotations are broken spontaneously. In the A_1-phase, the d-vector is complex valued, and its direction loses meaning. The dipole interactions, however, still prefer $\vec{l}$ to be perpendicular to the magnetisation direction of Cooper pairs, which in turn align with the applied field. Again, only the $U(1)$ phase symmetry is then spontaneously broken.

Finally, for the B-phase, the most symmetric combination of spin and orbital components is given by the state $\vec{d} = d_0\vec{k}/|k|$ with coinciding choices for the spin and orbital quantisation axes. In this configuration, the combined angular momentum of orbital and spin, $\vec{J} = \vec{l} + \vec{S}$, is zero for each Cooper pair, and the state is fully isotropic. In fact, the gap in the energy spectrum depends only on $|d|$, so that it is indistinguishable from the gap of an s-wave superfluid. However, this particular relation between spins and orbitals does not minimise the dipole interactions. Instead, the spin–orbit symmetry breaking prefers a uniform rotation of all d-vectors in this state over an angle of $\cos^{-1}(-1/4)$ around a spontaneously chosen axis. The resulting state is still isotropic with respect to the spin and orbital sectors individually, but spontaneously breaks the $SO(3)$ rotational symmetry of their relative orientation.

12.3.4.2 Excitations

One of the main conceptual differences between the triplet and singlet superfluids is the number of internal degrees of freedom carried by the Cooper pairs. In the singlet superfluid, Cooper pairs have only a phase, and a spatial variation of the phase corresponds directly to a supercurrent running along its gradient. The triplet superfluid, on the other hand, has both spin and orbital components in addition to its phase. Having different spatial variations of the phase for different spin or orbital components will lead to spin or orbital supercurrents that do not necessarily transport any electric charge. Moreover, in addition to variations of the phase, the spin directions characterised by $\vec{d}$, the orbital angular momentum $\vec{l}$, and even the angle between them may be spatially modulated. Configurations with such spatially varying order parameter components are in general called **superfluid textures**.

At long wavelengths, in an otherwise uniform background, textures in phase, spin, or orbit, may correspond to the Nambu–Goldstone modes of the superfluid state, if the particular state considered spontaneously breaks the corresponding symmetries. On the other hand, textures may also be introduced explicitly into the ground state, for example, by boundary conditions. For example, in the A-phase of superfluid ^{3}He it has been observed that $\vec{l}$ is always perpendicular to the walls of any container. Yet other textures in triplet superfluids correspond to topological defects, such as vortices in the superfluid phase, in $\vec{l}$, in $\vec{d}$, or in their relative angle. A special case of these can again be found in the A-phase of superfluid ^{3}He , where a 4π rotation of the superfluid phase around a circle does not necessarily lead to a core of vanishing superfluidity at the centre of the circle. Instead, the circulation of the phase can be undone by smoothly turning $\vec{l}$ over an angle of π as the centre is approached. The resulting state is in the A-phase everywhere and has constant total order-parameter amplitude, but for every individual value of $\vec{k}$ the superfluid goes from zero to non-zero rotation as a function of radial distance. This **coreless vortex** is not a true topological defect, because it does not have a point of enforced vanishing order-parameter amplitude. Closely related effects also arise in true defects of the triplet superfluid, however. For example, even if a 2π rotation of the superfluid phase enforces the appearance of a core within the A- or B-phase of ^{3}He, the state inside that core may be a different type of triplet superfluid, rather than reverting to the normal fluid state. These types of structures make the study of textures and defects in superfluids with internal degrees of freedom much richer than their plain, singlet counterparts. More detailed descriptions of them and other elements of triplet superfluidity can be found in Refs. [111, 204].

12.3.5 High-Temperature Superconductors

Looking even further out and going beyond materials that can be described by BCS theory or any of its enhanced versions employing strong coupling, alternative pairing interactions, or internal degrees of freedom, there are the cuprate **high-T_c superconductors**. These materials are in a class of their own, and they present a fundamental conundrum. If they are heated above their superconducting transition temperatures, the resulting metallic state is not a Fermi liquid. In fact, its nature is still poorly understood, and for lack of a better name it is referred to simply as a **non-Fermi liquid** or **strange metal**. Whereas the Fermi liquid is built from non-interacting, fermionic quasiparticles at low energies, the non-Fermi liquid seems not to contain electron-like quasiparticles at all, and hence no fermions to make Cooper pairs out of. It is at present not clear what exists instead, let alone how one can address the superconducting instability of such a mysterious state. This is widely regarded as one of the fundamental open problems in modern condensed matter physics.

13 Spacetime Symmetry Breaking

Every solid object, and even every liquid, in our everyday world breaks spacetime symmetries. None of them are invariant under all of the translations, rotations, and boosts that we typically assume to be symmetries of our universe. Indeed, the first example of spontaneous symmetry breaking encountered in this book was that of the harmonic crystal, in Section 2.3. Despite the comforting familiarity of solid objects and liquids, however, the way in which spontaneous symmetry breaking unfolds for them is mathematically more involved than the breaking of symmetries unrelated to spacetime. For example, it has long been noted that the excitation spectrum of crystalline solids contains phonons due to translation symmetry breaking, but that rotational Nambu–Goldstone modes appear to be absent, even though rotational symmetry is also spontaneously broken. This puzzle was finally solved by results on spacetime-symmetry breaking obtained over the past two decades or so.

As stressed in Section 1.5.4, there is no fundamental difference between the breaking of spacetime (external) and other (internal) symmetries. The mathematical challenges associated with the symmetry transformation affecting spacetime coordinates rather than properties of fields within spacetime, however, stopped us from including broken Lorentz invariance in the discussion of Section 10.2. We extend that discussion to spacetime symmetries in Section 13.1. We then give explicit examples of systems with broken Lorentz boosts (Section 13.2), followed by an overview of the properties of relativistic superfluids (Section 13.3), solids and liquids (Section 13.4), liquid crystals (Section 13.5) and time crystals (Section 13.6). The breaking of discrete, or CPT, spacetime symmetries was briefly addressed in Section 11.5, and will not be revisited here.

The systematic study of the spontaneous breaking of Poincaré symmetry (introduced in Section 11.1.2) is still an active and rapidly evolving field of research. It is included in some recent treatises [32, 136, 154], and interested readers may also refer to some of the original research literature [5, 45, 141, 143, 145, 183, 184, 185].

13.1 Dependent Symmetry Generators

The fact that broken spacetime symmetries do not always excite independent NG modes, such as a crystal having phonons but no rotational NG modes originates in the Lie-algebra structure of the symmetry generators. The underlying mathematics was established some time ago [149, 203], but requires techniques beyond the scope of this book. A recent alternative approach is given in Ref. [32]. We will review some of the central results, without attempting to be thorough or complete.

13.1.1 Coset-Space Construction for Spacetime Symmetries

As discussed in Section 10.2, an effective Lagrangian for NG modes can be constructed from knowledge of their transformation properties under the full group G of broken and unbroken symmetries. The starting point for that construction was the realisation that broken symmetries act non-linearly on the coset or order parameter space G/H, and hence on the NG fields. Mathematically, if X_a are broken generators, then an element of G/H can always be parametrised as $U = \mathrm{e}^{\mathrm{i}\pi^a(x)X_a}$. The action of any symmetry transformation g in the full group G on an element of the coset space G/H can then be written in the form of Eq. (10.34):

$$gU = \mathrm{e}^{\mathrm{i}\tilde{\pi}^a(x)X_a}\mathrm{e}^{\mathrm{i}\theta^i T_i}. \tag{13.1}$$

Here, T_i are the unbroken generators. Because the T_i act trivially on the coset space, Eq. (13.1) can be interpreted as defining a transformation of the NG fields $\pi^a(x) \to \tilde{\pi}^a(x)$ under elements g of the full symmetry group G.

When spacetime symmetries are involved, these steps in the construction are complicated by the fact that also unbroken translation symmetry acts non-linearly on the coset space. This stems from the fact that position itself transforms non-linearly under translations: $x^\mu \to x^\mu + \alpha^\mu$. One way to include this effect on positions in the coset-space construction is to parametrise elements of the coset G/H as

$$U = \mathrm{e}^{\mathrm{i}x^\mu \bar{P}_\mu}\mathrm{e}^{\mathrm{i}\pi^a(x)X_a}. \tag{13.2}$$

Here, we divided the group of unbroken symmetries H into the unbroken translation generators $\bar{P}_\mu$, and a set of remaining unbroken operators T_i, which span the group H_0 of unbroken symmetries other than translations. The unbroken translations $\bar{P}_\mu$ are included in the parametrisation of U. The remaining unbroken operators will be factored out as an unbroken symmetry transformation $h \in H_0$ as before. The action of $g \in G$ on U is then given by

$$gU = \mathrm{e}^{\mathrm{i}\tilde{x}^\mu(x)\bar{P}_\mu}\mathrm{e}^{\mathrm{i}\tilde{\pi}^a(x)X_a}\mathrm{e}^{\mathrm{i}\theta^i T_i} = U(\tilde{x}, \tilde{\pi})h(x, \pi, g). \tag{13.3}$$

This defines the transformed fields $\tilde{x}(x, \pi, g)$ and $\tilde{\pi}(x, \pi, g)$, and also $h(x, \pi, g) \in H_0$.

From here, it is possible to carry through a derivation similar to that of Section 10.2.2. The result will again be an effective Lagrangian in terms of the NG-fields and their covariant derivatives [149, 154].

13.1.2 Inverse Higgs Constraints

For internal symmetries, we saw in Section 4.2 that NG fields correspond to independent NG modes only if the commutators of their generators have a vanishing ground state expectation value. When we include spacetime symmetries, there can be additional constraints determining whether or not two generators excite the same NG mode. Some intuition for this can be gained by considering spatial translations and rotations. While on a global level, a rotation is clearly different from a translation, at any given point the local effect of an infinitesimal rotation is a translation in the direction transverse to the line connecting the point to the centre of rotation. Locally, rotations and translations are therefore indistinguishable, see

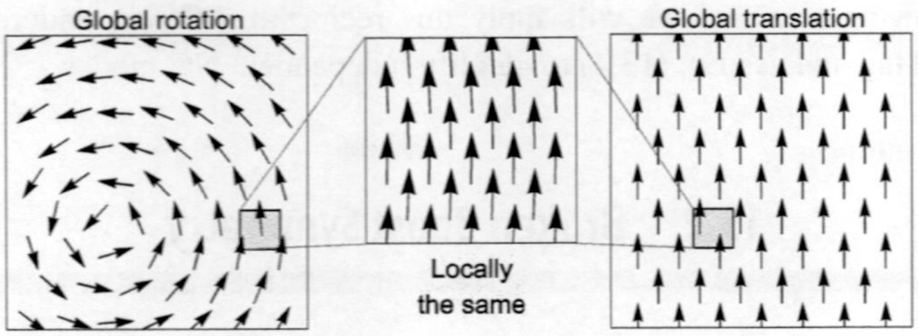

Figure 13.1

Locally, rotations and translations are indistinguishable. This puts a constraint on the corresponding NG modes when both rotation and translation symmetry are broken.

Figure 13.1. Because NG modes are excited by local versions of a global symmetry transformation, as seen in Section 4.1, this yields a constraint on which NG fields can be considered independent modes.

Exercise 13.1 (Infinitesimal rotations and translations).
a. Give an expression for how a 3-vector **x** transforms under a rotation around the z-axis (generated by J_z in Eq. (11.7)).
b. Consider an infinitesimal rotation by neglecting terms of second order in the rotation angle and show that it is locally equivalent to an infinitesimal translation.

Without proving it, a general recipe for recognising which broken spacetime generators excite independent modes can be formulated in terms of their commutation relations with the unbroken translations $\bar{P}_\mu$ appearing in Eq. (13.2). In general, we saw in Exercise 10.4 that such commutators are of the form

$$[\bar{P}_\mu, X_a] = if_{\mu a}{}^{b} X_b. \tag{13.4}$$

For any given broken generator X_a on the left-hand side, we inspect which broken generators X_b appear on the right-hand side with non-zero structure factor $f_{\mu a}{}^{b}$. Now, since H_0 is unbroken, the broken generators transform in a representation of H_0 and can therefore be grouped into multiplets (if there is any $h \in H_0$ such that $hX_a h^{-1} = X_b$, then X_a and X_b belong to the same multiplet). If there are generators X_b on the right-hand side, which do not belong to the same multiplet as X_a under the action of H_0, then the fields π_a and π_b excited by them are not independent modes. One way to see this is that for the symmetry-broken state Ψ, there is a non-trivial solution to the equation [45]:

$$\left(\pi^a(x)X_a + \pi^b(x)X_b\right)\Psi = 0 \qquad \text{no sum over } a \text{ or } b. \tag{13.5}$$

This result can be used to eliminate an extraneous mode when carrying out the coset-space construction. For instance, for a pair of generators X_a and X_b satisfying Eq. (13.4) and belonging to different multiplets, we can eliminate π_b from the effective low-energy theory by imposing the constraint $f^{\mu}{}_{a}{}^{b}\mathcal{D}_\mu \pi_b = 0$, where $\mathcal{D}_\mu$ is the covariant derivative defined in Section (10.2.2) [154]. Such constraints have become known as **inverse Higgs constraints** [79].

In this following sections, we will apply this recipe to different patterns of Poincaré symmetry breaking and use Eq. (13.4) to identify independent NG modes.

13.2 Broken Boost Symmetry

Lorentz symmetry appears to be a symmetry of our universe, in the sense that we expect the laws of nature applied to fundamental particles in empty spacetime to be invariant under Lorentz transformations. Indeed, the development of quantum field theory sprouted from the requirement of such relativistic invariance. In any physical context containing a medium with non-zero mass density, however, the medium defines a preferred frame and breaks Lorentz boosts.

To see this, we can consider the energy–momentum tensor $T^{\mu\nu}$, which characterises the distribution of matter in space. In general, the energy–momentum tensor determines the curvature of spacetime according to the Einstein equations, while in Eq. (1.36) we identified it with the Noether current for spacetime translations. The most symmetric medium possible for which we can consider the energy–momentum tensor is the **perfect fluid** as defined in the context of cosmology as a fully homogeneous and isotropic (i.e., invariant under translations and rotations) form of matter. It is characterised by its mass density ρ and pressure p, and in its rest frame, it yields an energy–momentum tensor of the form

$$T^{\mu\nu} = \begin{pmatrix} c^2\rho & & & \\ & p & & \\ & & p & \\ & & & p \end{pmatrix}. \tag{13.6}$$

In a general frame defined by a four-velocity field $u^\mu = \gamma(c, \mathbf{u})$ with respect to the rest frame, the energy–momentum tensor of the perfect fluid is given by $T^{\mu\nu} = (\rho + p/c^2)u^\mu u^\nu + p\eta^{\mu\nu}$ [38].

The Lorentz transformations $\Lambda^\mu{}_\nu$ defined in Section 11.1.2 act on the energy–momentum tensor as

$$T^{\mu\nu} \to \Lambda^\mu{}_\kappa \Lambda^\nu{}_\lambda T^{\kappa\lambda}. \tag{13.7}$$

You will show in Exercise 13.2 that the energy–momentum tensor of Eq. (13.6) is not invariant under boosts. In other words, even the perfect fluid breaks Lorentz invariance.

Exercise 13.2 (Broken Lorentz boosts).
a. Show that the generator K_x defined in Eq. (11.11) generates the Lorentz boost:

$$(\Lambda_x)^\mu{}_\nu = \begin{pmatrix} \cosh\zeta & -\sinh\zeta & 0 & 0 \\ -\sinh\zeta & \cosh\zeta & 0 & 0 \\ 0 & 0 & 1 & 0 \\ 0 & 0 & 0 & 1 \end{pmatrix}. \tag{13.8}$$

Analogous expressions hold for boosts in the y- and z-directions.

b. Show that the Lorentz boost Λ_x over a general angle ζ does not leave the energy–momentum tensor of Eq. (13.6) invariant.
c. Show that a perfect fluid with negative pressure equal to precisely $p = -c^2\rho$ is invariant under the Lorentz boost Λ_x. In fact, it is invariant under all Lorentz transformations.

Generalising the results of Exercise 13.2, we see that the presence of any form of matter or energy, even if it is maximally symmetric like the perfect fluid, breaks Lorentz boosts. The only exception is the one mentioned at the end of Exercise 13.2, a perfect fluid with a particular, negative pressure. For that fine-tuned pressure, the energy–momentum tensor happens to be proportional to the metric and equals $c^2\rho\eta^{\mu\nu}$. Such an energy–momentum tensor would correspond to a Lorentz-invariant scalar field $\Phi(x)$ having the expectation value $\langle|\Phi|\rangle = c^2\rho$. This suggests that any field theory in which Lorentz invariance is to be preserved after (internal) symmetry breaking can involve at most expectation values of scalar fields. For this reason, the Higgs boson is expected to be a Lorentz scalar. We will encounter the very unusual situation of matter fields with negative pressure again in Section 14.2, in the context of cosmology and inflation.

If we assume that any fundamental laws of nature are strictly Lorentz invariant and that there is nothing with broken Lorentz invariance in the initial state of the universe, then the breaking of Lorentz symmetry must be spontaneous. As we will see in Section 14.2.3, however, the idea that boosts, rotations, or translations can be broken spontaneously in the early universe is problematic. On the other hand, if Lorentz symmetry in the universe is broken explicitly, it must be broken very weakly. After all, at large scales we do not observe any consequences of Lorentz symmetry being broken, such as the speed of light deviating from c, up to extremely low bounds [46].

As we saw in Section 4.3, NG modes can still be found in systems with a weakly but explicitly broken symmetry, albeit with small masses. That NG modes associated with boost symmetry are not observed in our universe is due to the fact that all known states of matter breaking Lorentz boosts happen to also break additional spacetime symmetries. Through the inverse Higgs constraints of Eq. (13.4) the would-be boost NG modes are removed as independent degrees of freedom. Notice that although not known in our universe, hypothetical states breaking only Lorentz boosts can formally be defined [5, 45].

Exercise 13.3 (Lorentz symmetry Noether currents).
In this exercise, we will derive the Noether currents associated with Lorentz symmetry, closely following the discussion for translations in Section 1.4.4.
a. Consider a Lorentz transformation $\Lambda^{\mu}{}_{\nu}$ acting on coordinates x^{ν} and define $\omega^{\mu}{}_{\nu}$ to be the infinitesimal version of the transformation. Give an expression for the variation $\delta_{\mathrm{S}} x^{\mu}$ of the coordinates under the infinitesimal transformation.
b. Argue that $\omega^{\mu\nu}$ is antisymmetric in $\mu \leftrightarrow \nu$, using the fact that the metric is invariant under Lorentz transformations: $\Lambda^{\mu}{}_{\nu}\Lambda^{\kappa}{}_{\lambda}\eta^{\nu\lambda} = \eta^{\mu\kappa}$.
c. As in Section 1.4.4, we know that a field $\phi_a(x^{\mu})$ will transform into $\phi_a\left((\Lambda^{-1})^{\mu}{}_{\nu}x^{\nu}\right)$ under the Lorentz transformation. Use this to find the variation $\delta_{\mathrm{S}}\phi_a(x)$ of the field.
d. Substitute the expressions for $\delta_{\mathrm{S}}\phi_a$ and $\delta_{\mathrm{S}} x^{\mu}$ into Eq. (1.33). Then apply the Euler–Lagrange equations and write $\delta\mathcal{L}$ as a total derivative. Hint: Use the anti-symmetry of $\omega^{\mu\nu}$ to argue that $\omega^{\mu}{}_{\nu}x^{\nu}\partial_{\mu}\mathcal{L} = \omega^{\mu}{}_{\nu}\partial_{\mu}(x^{\nu}\mathcal{L})$.

e. Find the Noether current $(j_\Lambda)^\mu$ associated with the Lorentz transformation Λ. Show that it can be written in terms of the energy–momentum tensor of Eq. (1.36) as $(j_\Lambda)^\mu = \omega^\nu{}_\lambda x^\lambda T^\mu{}_\nu$.

You have thus shown that the Noether current associated with boosts is proportional to that for spacetime translations.

13.3 Relativistic Superfluids

A **relativistic superfluid** is arguably the simplest form of matter, as it ostensibly breaks only the $U(1)$ phase symmetry associated with particle number conservation, and nothing else. As a possible phenomenological Lagrangian describing the relativistic superfluid, we write

$$\mathcal{L} = -|\partial_\mu \Phi|^2 - m^2|\Phi|^2 - U|\Phi|^4. \tag{13.9}$$

The superfluid is described by a single complex scalar field $\Phi(x) = |\Phi(x)|e^{i\phi(x)}$, with $\Phi^\dagger(x)\Phi(x) = \rho_s(x)$ the superfluid density, and the phase $\phi(x)$ becoming long-range correlated in the symmetry-broken state. Comparing this to the description of Bose–Einstein condensation and non-relativistic superfluids in Section 2.5, we can interpret the field $\Phi(x)$ as the order parameter $\langle \hat{b}_0 \rangle$ in a coherent state of the form of Eq. (2.38). In that interpretation, the parameter m^2 originates from the mass of the bosons making up the superfluid (although m is not directly equal to the boson mass), while the quartic term with $U > 0$ stems from the short-range repulsions of Eq. (2.36).

The relativistic Lagrangian of Eq. (13.9) may also be compared to the Landau free energy of Eq. (5.6) describing the breaking of $U(1)$ symmetry in a non-relativistic setting and is a direct relativistic generalisation of it. The Lagrangian is invariant under $U(1)$-transformations of the form $\Phi(x) \to \Phi(x)e^{i\alpha}$, for which we derived the Noether current in Eq. (1.22). The Noether charge generating the symmetry is

$$Q = \int d^3x \, i\Phi^*(\partial_t \Phi) - i(\partial_t \Phi^*)\Phi, \tag{13.10}$$

which differs from its non-relativistic counterpart $\int \rho_s(x)$. For $m^2 < 0$, the phase rotation symmetry will be spontaneously broken.

Exercise 13.4 (Superfluid Noether currents).
In addition to phase rotations, the relativistic superfluid is also invariant under spacetime translations $x^\mu \to x^\mu + \alpha^\mu$.
Following the steps of Section 1.4, formulate expressions for the Noether charges $\int T^{00}$ and T^{i0} associated with the spacetime translations.

The five hydrodynamic modes in a relativistic superfluid follow from these four Noether charges associated with translations, together with Q [183].

A superfluid with non-zero superfluid density $|\Phi|^2$ can only be formed from a non-zero density of bosons. As in Section 2.5, it is thus crucial to include a chemical potential for the bosons. This must be a potential for the charge Q of Eq. (13.10), and we should consequently consider the grand-canonical partition function $Z(\mu, \beta) = \text{Tr}\, \exp(-(H - \mu Q)/\beta)$. Carefully deriving the Lagrangian in the relativistic theory implies the gradient term changes as [108]

$$-|\partial_\mu \Phi|^2 \to |(\partial_t - \mathrm{i}\mu)\Phi|^2 - |\nabla\Phi|^2. \tag{13.11}$$

So there is a shift $\partial_t \to \partial_t - \mathrm{i}\mu$. This, in turn, shows that we can consider the equilibrium state as the ground state of a modified Hamiltonian:

$$\bar{H} = H - \mu Q. \tag{13.12}$$

The Heisenberg time evolution of operators, however, is still generated by H. In terms of H, the ground state has the form [32, 178]

$$\Phi(x, t) = \Phi_0(x)\mathrm{e}^{\mathrm{i}\mu t}. \tag{13.13}$$

Here, $\Phi_0(x)$ is the ground state at zero chemical potential. This state is time dependent and breaks time translations spontaneously, due to the presence of the chemical potential.

Equation (13.13) holds for any relativistic fluid at finite density. For a superfluid, the charge Q is furthermore spontaneously broken. The system is still in a ground state of $\bar{H}$ in the thermodynamic limit, as explained in Section 3.3. But by inverting Eq. (13.12) to $H = \bar{H} + \mu Q$, this implies that H is also spontaneously broken. This is a complementary way of understanding the time translation symmetry-breaking manifest in Eq. (13.13) [142]. The notation $\bar{H}$ in Eq. (13.12) is used to connect to the formalism developed in Section 13.1.1: it indicates the unbroken time translation operator [142].

The finite density of bosons breaks Lorentz invariance as well. Indeed, it can be shown that the superfluid is a perfect fluid with its energy–momentum tensor given by Eq. (13.6). Its density and pressure are related by the thermodynamic identity $(c^2\rho + p) = \mu$ [184], which we saw in Exercise 13.2 breaks boost symmetry for any non-zero value of μ. Just as for time translations, this symmetry breaking can be regarded as spontaneous. Alternatively, the same Lagrangian equation with Eq. (13.11) can be taken as fundamental, where the chemical potential breaks Lorentz invariance explicitly. What follows then holds for pseudo NG-modes with a small gap, like we did for chiral symmetry breaking in Section 11.2.2.

Summarising, the relativistic superfluid breaks Lorentz boosts K_a and $U(1)$ phase symmetry Q, but preserves $\bar{H}$, momentum P_a, and rotations J_a in the Poincaré group. Following the recipe in Section 13.1.2, we need to consider the commutators of the broken generators with the unbroken spacetime translations $\bar{P}_a$. In particular, we see that

$$[\bar{P}_a, K_b] = -\frac{\mathrm{i}}{c}\delta_{ab} H = -\frac{\mathrm{i}}{c}\delta_{ab}(\bar{H} + \mu Q). \tag{13.14}$$

Here, we used the Poincaré group commutators of Eq. (11.16). Because Q generates a spontaneously broken symmetry, this expression satisfies the conditions for which an inverse Higgs constraint should be implemented. None of the boosts K_b therefore excite independent NG modes, massless or otherwise, and the low-energy spectrum consists only of the single NG mode associated with fluctuations in the superfluid phase.

13.4 Solids and Liquids

The most common states of matter from a human perspective are solids and liquids. Crystalline solids break continuous translation symmetries to a discrete subgroup. In fact, this symmetry-breaking pattern defines the crystal to such an extent that the representation theories of finite space groups are used in crystallography to distinguish and classify crystalline materials. As we have seen, broken translations also necessarily imply broken rotations. On the other hand, perfectly homogeneous and isotropic liquids preserve both translations and rotations, but they break boosts due to their non-zero mass density. Both solids and fluids are observed to host longitudinal phonons, while only solids additionally allow for transverse phonon excitations.

To understand these observations from a symmetry-breaking perspective, we need to apply the inverse Higgs constraints of Eq. (13.4), which in turn requires a field-theoretical description of solids and liquids. For this purpose, we introduce internal coordinates $\Phi^I(\mathbf{x})$, with $I \in \{x, y, z\}$, which are comoving with the medium. They can be regarded as the positions of volume elements, as well as a continuum version of the atom positions X in Eq. (1.40) and in Section 2.3.

The Lagrangian is obtained as a Legendre transform of the continuum version of Eq. (2.7) and the kinetic term is of the form $\mathcal{L} \sim (\partial_\mu \Phi^I(x))^2$. The action is invariant under the usual (external) spacetime translations $x^\mu \to x^\mu - \alpha^\mu$ generated by P^μ as introduced in Section 1.4.4. It is furthermore invariant under the (internal) global translations $\Phi^I(x) \to \Phi^I(x) + \alpha^I$. These internal translations can be thought of as moving the medium with respect to the background spacetime (along with all other media in the universe). We write $\mathcal{P}^I$ for the generators of this symmetry. The different effects of the internal and external translations on a generic configuration $\Phi^I(\mathbf{x})$ of the medium are shown in Figure 13.2.

The Lagrangian is also invariant under rotations J^a and boosts K^a, and thus in fact under the entire Poincaré group. And finally, it is also invariant under internal rotations $\Phi^I(x) \to R^{IJ}\Phi_J(x)$ for orthogonal matrices R, for which we write the generators as $\mathcal{J}^I$. The internal symmetry transformations $\mathcal{P}^I$ and $\mathcal{J}^I$ satisfy the commutations relations of the Euclidean group:

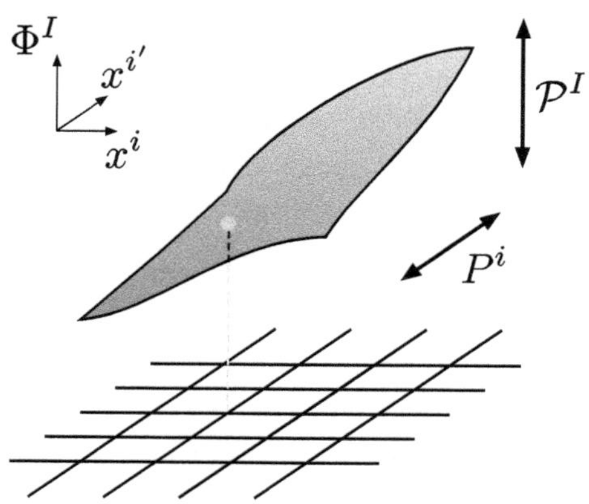

Figure 13.2 The effects of internal translations generated by $\mathcal{P}^I$ and external translations generated by P^i on the field representation $\Phi^I(x^i)$ of a continuous medium.

$$[\mathcal{P}^I, \mathcal{P}^J] = 0, \qquad [\mathcal{J}^I, \mathcal{P}^J] = \mathrm{i}\epsilon^{IJK}\mathcal{P}_K, \qquad [\mathcal{J}^I, \mathcal{J}^J] = \mathrm{i}\epsilon^{IJK}\mathcal{J}_K. \tag{13.15}$$

As they are internal symmetries, they commute with the Poincaré group of the symmetries of the background spacetime.

In the equilibrium state of a liquid or solid, each volume element is at a certain equilibrium position with respect to the background spacetime. There is a gauge freedom in choosing the internal coordinates to describe the equilibrium, and a natural choice is given by

$$\langle \Phi^I(x) \rangle = \delta^I{}_j x^j. \tag{13.16}$$

Clearly, this state is not invariant under spatial translations $x^i \to x^i - \alpha^i$, and in fact there is no choice of internal coordinates in which it would be. But there is a combined symmetry given by $x^i \to x^i - \alpha^i$ and $\Phi^I(x) \to \Phi^I(x) + \alpha^I$ under which the equilibrium state is invariant. So while P^i and $\mathcal{P}^I$ are both broken, their combination $\bar{P}^i = P^i + \mathcal{P}^i$ is not. The equilibrium state also breaks background and internal rotations J^i and $\mathcal{J}^I$, while their combination $\bar{J}^i = J^i + \mathcal{J}^i$ is unbroken. Finally, Lorentz boosts K^a are broken as they are for any medium with finite density.

Exercise 13.5 (Translation and rotation in co-moving coordinates).
a. Show by explicit calculation that the equilibrium state with $\Phi^I(x) = \delta^I{}_j x^j$ is invariant under the combined transformation:

$$x^i \to x^i - \alpha^i, \qquad \Phi^I(x) \to \Phi^I(x) + \alpha^I. \tag{13.17}$$

b. Similarly show that it is invariant under the combined transformations

$$x^i \to R^{ij} x_j, \qquad \Phi^I(x) \to (R^{\mathrm{T}})^{IJ} \Phi_J(x). \tag{13.18}$$

In summary, the symmetry generators to consider when describing solids or liquids are [143, 145]:

$$\text{Unbroken} \begin{cases} H, \\ \bar{P}^i = P^i + \mathcal{P}^i, \\ \bar{J}^i = J^i + \mathcal{J}^i, \end{cases} \qquad \text{Broken} \begin{cases} \mathcal{P}^i, \\ \mathcal{J}^i, \\ K^a. \end{cases} \tag{13.19}$$

There are nine broken generators, so nine possible NG modes. But, using the commutators of Eqs. (11.16) and (13.15), we find that [145]

$$\begin{aligned} [H, K^a] &= \mathrm{i}cP^a = \mathrm{i}c\bar{P}^a - \mathrm{i}c\mathcal{P}^a, \\ [\bar{P}^i, \mathcal{J}^j] &= \mathrm{i}\epsilon^{ijk}\mathcal{P}_k. \end{aligned} \tag{13.20}$$

Following the recipe of Section 13.1.2, we see that the broken generators K^a and $\mathcal{J}^i$ do not generate independent NG modes. We are therefore left with only three physical NG modes, caused by the broken translations, which we call phonons.

13.4.1 Solids

The translation symmetry in a crystalline solid is broken from the continuous group $\mathbb{R}^3$ to a discrete subgroup $\mathbb{Z}^3$ of discrete lattice translations. The order parameter should reflect the remnant symmetry, as discussed in Section 3.2, and the coset space in this case is $\mathbb{R}^3/\mathbb{Z}^3 \simeq U(1) \times U(1) \times U(1)$. A possible order parameter for a crystal breaking translations down to a discrete subgroup is then given by [19]

$$\mathcal{O}_{\mathbf{q}}(x) = \mathrm{e}^{2\pi \mathrm{i} \Phi^I(x) q_I} \rho(x). \tag{13.21}$$

Here, $\mathbf{q}$ is a reciprocal lattice vector for the crystalline lattice. Linearly independent reciprocal lattice vectors define independent local order parameters $\mathcal{O}_{\mathbf{q}}(x)$ associated with the breakdown of translation symmetry in different real-space directions. The amplitude $\rho(x)$ is the density of the medium, which will be uniform in the fluid phase, but non-homogeneous within a unit cell of the crystalline solid. The average of $\mathcal{O}_{\mathbf{q}}(x)$ over one or multiple unit cells vanishes for $\Phi^I(x^i) = \delta^I{}_i x^i$ and $\rho(x)$ constant, while it is non-zero in the crystalline state. Moreover, the averaged order parameter is a complex number for each independent direction in the solid, whose phase maps in a one-to-one fashion onto the quotient set G/H classifying broken symmetry states. The definition of Eq. (13.21) therefore satisfies all requirements introduced in Section 3.2 for local order parameters.

Notice that although the crystal breaks continuous translation symmetry, the remnant discrete translation invariance still allows for the definition of momentum eigenstates within a Brillouin zone. These suffice to define propagating long-wavelength modes, and in particular, they may be used to define the Nambu–Goldstone modes.

13.4.2 Liquids

Unlike solids, liquids are homogeneous and isotropic and therefore do not break translation or rotation symmetries. Nevertheless, they do break Lorentz boosts, whose associated NG modes should, according to the inverse Higgs constraint of Eq. (13.20), coincide with those for broken translation symmetry. In practice, however, both longitudinal and transverse phonons can be excited in solids, while liquids carry only longitudinal modes. In fact, the excitation energy required to create transverse strain in a solid is often used in practice as the defining characteristic distinguishing it from liquids, which have no resistance to shear deformations. It thus seems like the liquid misses $D-1$ transverse phonon modes that would be expected based on its broken boost symmetry.

This paradox cannot be resolved from the symmetry analysis alone. Instead, we need to also consider the low-energy effective theory for the NG modes. Since phonons are quantised sound or displacement waves, they can be expressed directly in terms of the position fields as $\delta\Phi^I(x) = \Phi^I(x) - \delta^I_i x^i$ [143, 145]. Constructing the effective Lagrangian from those fields, the kinetic term should be of the form $C_{IJ}\partial_\mu \delta\Phi^I \partial^\mu \delta\Phi^J$. Because the effective Lagrangian must be symmetric under the full group G of broken and unbroken symmetries, C_{IJ} should be such that the Lagrangian is isotropic. The diagonalised form of this expression then describes the dispersion of phonons.

We should realise that the liquid state has more symmetry than the solid. Because it is isotropic and homogeneous at all scales, the position field in equilibrium is invariant under all transformations that do not affect its local density, including shear displacements. The same shear modes as in a solid may thus be excited, but because they leave the state invariant, they cannot cost any energy. In terms of the effective theory, this means that after diagonalising the kinetic term, we will find the dispersion law $\omega = 0$ for shear modes in a liquid, and they do not manifest as propagating NG modes [145].

13.5 Liquid Crystals

In terms of symmetry, **liquid crystals** fall between solids and liquids. There are many different types of liquid crystal, but all are liquid in the sense of having at least some unbroken continuous translation symmetry, while they are like crystals in the sense that they do break at least some rotational symmetry [39]. A classic example is given by **nematic liquid crystals**, which often consist of essentially classical rod-like, elongated molecules that flow freely as a fluid, but which are all oriented along the same preferred direction.

As compared to a true fluid, the symmetry of a nematic liquid crystal is reduced from the full $SO(3)$ rotational symmetry to $SO(2) \rtimes \mathbb{Z}_2$. Here, the unbroken $SO(2)$ corresponds to rotations around the preferred direction of the molecules, while the $\mathbb{Z}_2$ indicates the inversion symmetry of individual molecules, which do not have a distinct head or tail. The orientational ordering in nematic liquid crystals is employed, for example, in liquid-crystal displays. Since rotational symmetry is broken, one would expect the appearance of rotational NG modes in a nematic phase. In practice, however, these modes are easily dissipated by the hydrodynamic motion of the fluid and are strongly overdamped [39].

Another interesting example of a liquid crystalline phase is that of a **smectic liquid crystal** in three dimensions. Like the nematics, classical smectic liquid crystals consist of elongated molecules with a common orientation. In the smectic phase, however, the molecules are arranged in periodically stacked layers. Translational symmetry within each layer is unbroken and the molecules flow as a liquid throughout each layer, but because molecules mostly do not travel between layers, translational symmetry in the direction perpendicular to the layers is broken. Since the orientation of the molecules is generally unrelated to the orientation of the layers, rotational symmetry is broken completely.

To see how many NG modes may be expected from this pattern of broken symmetries, we use the inverse Higgs constraint. We know from Eq. (11.16) that the commutator between the generators of translations and rotations is given by $[P_a, J_b] = \mathrm{i}\epsilon_{abc}P_c$. Two of the broken rotations, with rotation axes in the layer, are subject to an inverse Higgs constraint, since their commutator with one of the unbroken translations within the layers yields the generator of broken translations. But rotations around an axis perpendicular to the liquid layers suffer no constraint, and a corresponding rotational NG mode is present. The total spectrum of the smectic thus consist of a phonon mode corresponding to the broken translation, and the rotational NG mode of broken and unconstrained rotations in the liquid plane [21]. Like in nematics, in classical smectics, this mode is overdamped.

While the phase transition from a liquid to a solid is discontinuous and breaks rotations and translations simultaneously, it may be envisaged that using a liquid crystal intermediate state, the breaking of symmetries can occur in two steps. Heating a solid at low temperature, the translational order would then first be melted, and translation symmetry restored, without affecting the rotational order. This would result in a liquid crystal phase. Only at higher temperatures, the rotational order may then melt, and rotational symmetry be restored, to finally reach the true liquid phase. It has been suggested that such a scenario may be realised in two-dimensional crystalline order with hexagonal symmetry, if the melting is of the BKT-type discussed in Section 7.4 so that it results from the unbinding of topological defects [67, 140, 226]. In the crystalline solid, the topological defects related to translational order are called **dislocations**, and the first stage in the two-step melting process is therefore called dislocation-mediated melting. It leads to a complete restoration of translation symmetry while retaining the sixfold discrete rotation symmetry. The transition turns out to be continuous, and the resulting phase is a so-called **hexatic liquid crystal**. The topological defects related to rotational symmetry are called **disclinations**. They are confined in a crystalline solid, but deconfined in liquid crystals, and they may therefore appear and proliferate in the hexatic phase as temperature is raised further. The phase transition from the hexatic to a liquid phase may then be viewed as disclination-mediated melting.

Going beyond these classical examples, any quantum form of matter that breaks rotations but has at least some translational symmetry left can be called a **quantum liquid crystal**. In the quantum setting, numerical simulations of dislocation-mediated melting show that the rotational NG modes may be undamped at sufficiently low energies [21, 22, 227]. Finally, NG modes associated with the boosts that are broken by the presence of non-zero density are subject to inverse Higgs constraints in all liquid crystal phases and do not yield independent NG modes.

13.6 Time Crystals

After discussing phases with broken translations, rotations, and Lorentz boosts, you may wonder if there are also phases that break the one remaining continuous Poincaré symmetry, time translations. Just as a crystalline solid is a medium in which translational symmetry is broken down to a discrete subgroup, a **time crystal** could be defined as a systems in which time translation symmetry is broken down to a discrete subgroup.

The possibility of realising time crystals through spontaneous symmetry breaking has been proposed theoretically [222], but it was soon realised that this is fundamentally impossible in any equilibrium state of matter [34, 211]. The no-go theorems, however, do not exclude the possibility of time translation symmetry being broken out of equilibrium. In fact, there are plenty of trivial examples of broken time translations. The earth going around the sun, for example, breaks continuous time translation to the discrete subgroup of shifting time by an integer number of years. This is not the intended meaning of a time crystal, however, just like a row of equally spaced marbles does not constitute a spatial crystal. The essential distinction is that the earth's orbit does not have the stability discussed in Section 3.7, as even small perturbations of, for example, its angular velocity may send it spiralling into the sun. Moreover, it lacks the rigidity of Section 4.6 in the sense that local forces acting on the earth do not cause global changes to its orbit.

More non-trivial situations may occur in the non-linear dynamics of driven dissipative systems and have been observed, for example, in active elastic metamaterials [202]. The dynamics of these may converge to an attractive limit cycle, whose periodicity is independent of the driving frequency. Moreover, the instantaneous position (i.e., the phase along the limit cycle) may be chosen spontaneously, as it can have a diverging sensitivity to the initial, static configuration from which the driving commences. Finally, it is stable and rigid, in the sense that local and instantaneous perturbations may be dissipated, returning the dynamics to its limit cycle.

A second class of examples arising in driven dissipative systems is that of a **discrete time crystals**, in which recurring dynamics emerges with a period equal to an integer multiple of the driving period. The rigidity in these systems is manifested by the fact that perturbations in the driving frequency do not affect the periodicity of the emergent dynamics. Such discrete time-crystalline order has been reported, for example, in trapped ions driven by periodic laser pulses [229], as well as in diamond spin impurities driven by microwave radiation [40]. A more detailed discussion can be found in Ref. [51].

14 Cosmology

Cosmology is the study of the universe as a whole, encompassing all of space and time. Our current understanding of the history of the universe is firmly based on concepts of spontaneous symmetry breaking. Here we will highlight just a few of them, starting in Section 14.1 with the observation that the universe used to be much denser, hotter, and more symmetric than it is now, necessitating a sequence of phase transitions associated with symmetries of particle physics in order to evolve from the early universe to its present configuration. In Section 14.2, we zoom in on the large-scale structure observed in our universe, and the symmetry breaking inherent in the theory of inflation that explains its emergence. Finally, we discuss the appearance and nature of cosmological defects in Section 14.3.

More detailed discussions of these topics, including cosmological physics beyond the role of symmetry breaking, can be found, for example, in Refs. [132, 216].

14.1 Phase Transitions of the Early Universe

When observations on the grand scale of our universe as a whole started to appear in the early twentieth century, three major discoveries were made. First, we realised that the universe looks roughly the same in all directions and may thus be assumed to be isotropic on sufficiently large scales. Second, it also appears to have roughly the same content around every point in space and is therefore homogeneous when sufficiently course grained. And finally, it was found that everything in the universe moves away from us with ever increasing speed, which implies that our universe is expanding.

A metric for the expanding universe can be formulated by assuming it is filled with a **perfect fluid** (introduced in Section 13.2), which guarantees that it will be homogeneous and isotropic. This yields the Friedmann–Lemaître–Robertson–Walker metric:

$$\mathrm{d}s^2 = -c^2\mathrm{d}t^2 + R^2(t)\left(\frac{\mathrm{d}\chi^2}{1-k\chi^2} + \chi^2\mathrm{d}\psi^2\right). \tag{14.1}$$

Here, k is an index indicating whether we consider a universe with positive ($k = 1$) overall curvature, a flat universe ($k = 0$), or a negatively curved ($k = -1$) case. Notice that the spatial part of the metric is multiplied by an overall **scale factor** $R(t)$, which grows as a function of time. The dimensionless coordinate χ, meanwhile, is the co-moving distance or coordinate separation. In a visualisation of the universe as an expanding grid of spacetime points, as depicted in Figure 14.1, the dimensionless co-moving distance represents the number of links connecting two points. For static objects inside the expanding universe (vertices of the grid in Figure 14.1), the dimensionless co-moving distance between them stays constant while the dimensionful size of the interval separating them, $R(t)\chi$, grows. The scale factor in the metric of Eq. (14.1) thus describes the expansion of an isotropic and homogeneous universe.

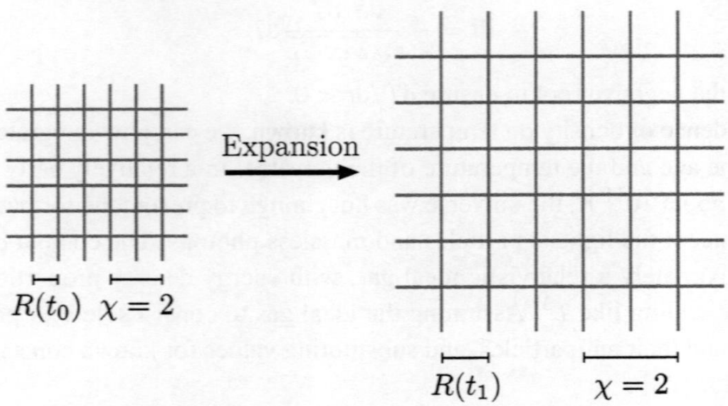

Figure 14.1

In an expanding universe, two points maintain a constant co-moving distance χ as the scale factor $R(t)$ increases.

A relativistic model for the expansion of the universe as a whole is obtained by inserting the metric in the relativistic Einstein equations, which results in the Friedmann equation:

$$H^2 + \frac{kc^2}{R^2} = \frac{8\pi G}{3}\rho. \tag{14.2}$$

Here, ρ is the energy density of the universe, G is the gravitational constant, and $H \equiv (\mathrm{d}R/\mathrm{d}t)/R$ is a rate called the **Hubble parameter**, which characterises the expanding scale of the universe.

It should be emphasised that the observed expansion of the universe implies a violation of Lorentz symmetry. After all, the value of the scale factor can be used to distinguish between different epochs in the evolution of the universe. In other words, time translation and time reversal symmetry are broken. Nevertheless, effective descriptions of physics in the universe that hold locally, at short time scales over which the expansion of the universe may be ignored, may still be expected to possess full Lorentz symmetry. The role of broken time-translation symmetry will be discussed further in Section 14.2, in the context of inflation.

14.1.1 A Hot Early Universe

As the universe is currently expanding, it must have been smaller before. Assuming conservation of energy and using standard thermodynamic arguments, this means it must have also been denser and hotter. We can estimate how hot it was at any given moment by assuming the expansion to be adiabatic, so that the entropy per unit co-moving volume, S/χ^3, is constant. Written in terms of the observed entropy density $s(t) = S/(R(t)\chi)^3$, this means that $s(t)R^3(t)$ is constant. Taking the time derivative yields a relation between the Hubble parameter and the entropy density:

$$H = -\frac{1}{3s}\frac{\mathrm{d}s}{\mathrm{d}T}\frac{\mathrm{d}T}{\mathrm{d}t}. \tag{14.3}$$

Notice that in this equation, H, s, and T all depend on time. Combining this with Eq. (14.2) and focussing on a period in which the universe is already sufficiently large for the curvature term kc^2/R^2 to be neglected, yields

$$dt = -\frac{dS/dT}{s\sqrt{24\pi G\rho}}dT. \tag{14.4}$$

Here, we used the negative root to ensure $dT/dt < 0$.

If the dependence of density on temperature is known, we can now integrate to find a relation between the age and the temperature of the universe. In a relatively early period, with a temperature of about 10^{11} K, the universe was hot enough to prevent the formation of any stable structure, except the lightest particles and massless photons. The content of the universe was then approximately a relativistic ideal gas, with energy density proportional to T^4 and entropy density scaling like T^3. Assuming the ideal gas to consist solely of photons, neutrinos, electrons, and their antiparticles, and substituting values for known constants, Eq. (14.4) becomes [29, 75, 123]

$$\frac{t - t_0}{\text{sec}} \approx \left(\frac{T}{10^{10}\text{K}}\right)^{-2} \approx \left(\frac{T}{\text{MeV}}\right)^{-2}. \tag{14.5}$$

Here, t_0 is a constant of integration, which is small compared to t for the period of interest. At later times, when the temperature drops below approximately 10^9 K, electrons and positrons get sufficient opportunity to mutually annihilate, and the corresponding change in the number of degrees of freedom causes Δt to slightly increase to about 2 seconds. Even later, for temperatures below about 10^6 K, heavier particles appear in non-negligible numbers, and Eq. (14.5) breaks down.

Throughout the hot, a dense early period described by Eq. (14.5), several of the symmetry-breaking transitions of the Standard Model introduced in Chapter 11 are expected to have taken place [29, 75, 123]. Going back in time, and up in temperature, from 10^{11} K, we first encounter the **QCD confinement** transition at temperature $T \sim 170$ MeV $\sim 10^{12}$ K and time $t \sim 10^{-5}$ s. At this point, the strong nuclear force causes a transition between confined and deconfined phases of colour-charged particles (quarks). At low temperatures, these are bound together into baryons such as protons and neutrons, and mesons such as pions and kaons. Above the transition temperatures, these particles fall apart into quarks and gluons, which form a quark-gluon plasma. The confinement transition was the subject of Section 11.4.

Continuing backwards in time, the **electroweak symmetry breaking** transition appears at $T \sim 100$ GeV $\sim 10^{15}$ K, at time $t \sim 10^{-12}$ s. Below this temperature, the Higgs field had a non-zero expectation value, corresponding to a mass for the Z and W bosons, decoupling electromagnetic photons from the weak force carriers. This symmetry-breaking transition from $SU(2) \times U(1)$ to $U(1)$ was introduced in Section 11.3, and its application to the evolution of the universe will be discussed later in this section.

At even higher temperatures of $T \sim 10^{16}$ GeV $\sim 10^{29}$ K, we go beyond the applicability of the Standard Model, perhaps encountering a transition related to **grand unification**. At this energy, the strong and electroweak interactions have similar strengths and a higher symmetry group relating them may start to show, as anticipated in Section 11.6. Beyond even this scale, there are indications that the extremely early universe went through a phase of exponential expansion known as inflation, which we will encounter in Section 14.2.

14.1.2 The Electroweak Transition

Around the time of the electroweak transition in the early universe, at $t \sim 10^{-12}$ seconds, the fields affected by electroweak interactions were at thermal equilibrium. In this context, it means that the relaxation time τ defined in Section 9.2, describing the time scale over which disturbances in the fields persist, was much shorter than the cooling time H^{-1} due to the expansion of the universe. Any perturbations could then return to equilibrium before the temperature changed significantly. Because of this, we can use the equilibrium techniques of Chapters 5 and 9 to describe the electroweak phase transition.

A convenient formulation can be given in terms of the of Section 10.1. To define this for the Higgs field ϕ, we start from the Yang–Mills theory of Eq. (11.58) describing the electroweak interactions between the Higgs field and the electroweak force carriers. We can take the bare potential for the Higgs field in that theory as the zeroth-order contribution to the thermal effective potential V_T, and write

$$V_\mathrm{T}(\phi)^{(0)} = -\frac{m_\mathrm{H}^2}{4}\phi^2 + \frac{u}{4}\phi^4. \tag{14.6}$$

Here, we expressed the parameter r appearing in Eq. (11.58) explicitly in terms of the Higgs mass as $m_\mathrm{H}^2 = -2r$.

Higher-order terms in the thermal effective potential arise from interactions of the Higgs field with both vector bosons (W and Z) and fermions (quarks). Formally, these terms can be found by 'integrating out' the vector bosons and the fermions from the full interacting theory. However, the leading, **one-loop correction** to the bare Higgs potential can also be constructed based on intuitive arguments. Namely, if the value of the Higgs field ϕ is non-zero, we know from Section 11.3 that the vector bosons acquire a mass $m_\mathrm{B} = \frac{g}{2}\phi$, while the fermions obtain a mass $m_\mathrm{F} = \lambda\phi$. To include the effect of this on the thermal effective potential, we can add a free energy to it for free bosons and fermions with masses proportional to the Higgs field ϕ. For free bosons with dispersion $\omega_k^2 \sim k^2 + m^2$, the free energy $f = -T \log Z$ can be expanded in the high temperature limit $T \gg m$ as

$$f_\mathrm{B}(T) = -\frac{\pi^2}{90}T^4 + \frac{1}{24}m^2T^2 - \frac{1}{12\pi}m^3T + \dots. \tag{14.7}$$

Notice that this expression contains a cubic term in terms of the mass $m \propto \phi$, similar to the way a cubic term arose upon integrating out the photon field in the description of the superconducting phase transition in Section 8.6. As was the case there, we may expect a third order term in V_T to render the transition discontinuous.

Because the electroweak Lagrangian contains one Z boson and two W bosons, each of which has three degrees of freedom, the one-loop correction to the bare Higgs potential contains nine times the contribution of Eq. (14.7) with $m = \frac{g}{2}\phi$. In addition, there will be a contribution from the quarks, which may be constructed from the free energy density for massive fermion fields:

$$f_\mathrm{F}(T) = -\frac{7}{8}\frac{\pi^2}{90}T^4 + \frac{1}{48}m^2T^2 + \dots. \tag{14.8}$$

The fermionic one-loop correction will be dominated by the heaviest particles, so we neglect the lighter quarks and focus exclusively on the top quark. It is a colour triplet of four-component spinors, giving a total of 12 times the contribution of Eq. (14.8). Instead of writing the top quark mass $m = \lambda_\mathrm{t}\phi$ in terms of the Yukawa coupling λ_t, it is customarily expressed using $m = m_\mathrm{t}\phi/v$ where m_t is the observed mass of the top quark and v the ground state expectation value of the Higgs field in the present, symmetry-broken phase.

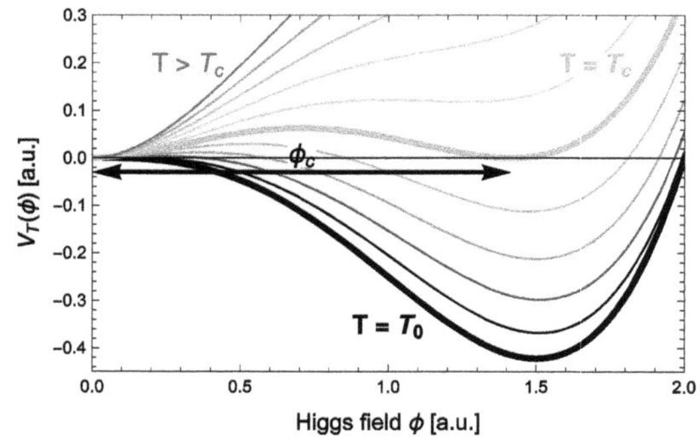

Figure 14.2 The thermal effective potential of Eq. (14.9), visualised in arbitrary units. The presence of the ϕ^3 term due to the interactions between the Higgs field and vector bosons, causes the global minimum of the potential to change discontinuously at the critical temperature T_c given by Eq. (14.11). At the spinodal temperature T_0 the symmetric ground state becomes unstable.

All together this yields the thermal effective potential for the Higgs field:

$$V_{\mathrm{T}}(\phi) = \left(\frac{3g^2}{32} + \frac{m_{\mathrm{t}}^2}{4v^2}\right)\left(T^2 - T_0^2\right)\phi^2 - \frac{3g^3}{32\pi}T\phi^3 + \frac{u}{4}\phi^4. \tag{14.9}$$

Here, we introduced the temperature scale T_0, given by

$$T_0 = \frac{m_{\mathrm{H}}}{\sqrt{\frac{3g^2}{8} + \frac{m_{\mathrm{t}}^2}{v^2}}}. \tag{14.10}$$

Substituting the known values of the weak coupling constant and the top quark mass, we find that $T_0 \approx 1.3m_{\mathrm{H}}$. We thus expect the phase transition to happen at a temperature scale comparable to the mass of the Higgs boson. Notice that Eq. (14.9) takes into account only the simplest, one-loop corrections to the bare potential. More accurate calculations also include the Higgs self-energy as well as higher-loop corrections. The qualitative behaviour of the thermal effective potential, however, is captured already by the one-loop theory.

14.1.2.1 Thermal Crossover

As visualised in Figure 14.2, the presence of the cubic term in the thermal effective potential causes a discontinuous change in the position of its global minimum as a function of temperature. Writing the Higgs potential as $V_{\mathrm{T}}(\phi) = r'(T^2 - T_0^2)\phi^2 - aT\phi^3 + \frac{u}{4}\phi^4$, its global minimum changes away from zero at the critical temperature defined by

$$T_{\mathrm{c}} = \frac{T_0}{\sqrt{1 - \frac{a^2}{r'u}}}. \tag{14.11}$$

At temperatures just below T_{c}, the minimum lies at $\phi_{\mathrm{c}} = (2a/u)T_{\mathrm{c}}$, indicating that the electroweak transition in the early universe should be a discontinuous phase transition.

The evolution of any observables associated to the ordering, however, is influenced by the presence of thermal fluctuations as well as the position of the global minimum. Following Section 6.2, the amplitude of the thermal fluctuations $\delta\phi = \sqrt{\langle\phi^2\rangle}$ of the Higgs field can be written in the high temperature limit as

$$\delta\phi^2 = \int \frac{d^3k}{(2\pi)^3} \frac{1}{2\omega_k} n_B(\omega_k) \approx \frac{T^2}{24}. \tag{14.12}$$

If these thermal fluctuations are larger at T_c than half of the expected jump ϕ_c in the Higgs field amplitude, the local configuration of the Higgs field may freely fluctuate between the metastable minimum at $\phi = 0$ and the symmetry breaking minimum at ϕ_c. The discontinuous phase transition will then appear as a smooth **thermal crossover**, in which the value of any local observable evolves smoothly across the transition temperature. This is the thermal equivalent of the zero-temperature Higgs-confinement crossover that was discussed in Section 11.4.

A truly discontinuous jump in thermal averages of local observables will be observed only if the fluctuations are sufficiently small:

$$\frac{\phi_c}{2} > \delta\phi = \frac{T_c}{\sqrt{24}}. \tag{14.13}$$

We can use Eq. (14.11) to relate ϕ_c to the parameter u, and Eq. (14.6) to write $u = m_H^2/(2v^2)$, in terms of the observed mass of the Higgs boson m_H. The electroweak phase transition can then only be discontinuous if the Higgs boson mass is sufficiently small:

$$m_H < 75\,\text{GeV}. \tag{14.14}$$

The experimental observation of the Higgs boson mass being $m_H = 125$ GeV thus implies the electroweak transition in the early universe was in fact a smooth crossover [84].

14.1.3 Matter–Antimatter Symmetry Breaking

A manifestly asymmetric property of our observable universe is the abundance of matter over antimatter. The ratio of antimatter to matter is of the order 10^{-10} [187] and is known as the **baryon asymmetry**. Its origin is still unknown.

Because particles and their antiparticles have the same mass and energy within the Standard Model, their numbers would be expected to be the same in equilibrium. We might then wonder if the generation of more matter than antimatter in the early universe was a matter of spontaneous symmetry breaking. There are two strong indications that this is not the case. First, there is no reason to suspect a symmetric universe to be unstable, which we saw in Chapter 2 to be a prerequisite for SSB. Second, if the symmetry would be spontaneously broken, we saw in Chapter 9 that we should expect both matter-dominated and antimatter-dominated domains in the universe. There is no observational indication that antimatter domains or matter–antimatter domain walls exist.

The commonly held view is therefore that instead of spontaneous symmetry breaking, there must exist a period of **baryogenesis** in the early universe dominated by processes that favour the creation of matter over antimatter. For an excess of matter to be created this way, the universe needs to satisfy the three so-called **Sakharov criteria** [173].

1. **Baryon number violation**. There need to exist processes that can change the net baryon number.
2. **$\mathcal{C}$ and $\mathcal{CP}$ violation**. Both charge conjugation and parity need to be broken because otherwise any baryon-number-violating processes would create left-handed matter at the same rate as right-handed antimatter.

3. **Out-of-equilibrium processes**. Because particles and antiparticles have the same mass, they will occur in equal numbers in any equilibrium setting. We therefore need a strong deviation from thermal equilibrium to allow the process that violates $\mathcal{C}$, $\mathcal{CP}$, and baryon number to create a net amount of matter.

Interestingly, all three conditions are satisfied by the electroweak interaction the electroweak theory. The Lagrangian of the Standard Model has $U(1)$ symmetry, with an associated baryon current:

$$j_B^{\mu} = \frac{1}{2}\overline{Q}\gamma^{\mu}Q. \tag{14.15}$$

Here, Q contains all flavours and colours of quarks. Since the Lagrangian has a symmetry, it seems like the baryon current should be a conserved Noether current, and hence imply the conservation of the net number of baryons. As we saw in Section 10.1.1, however, the symmetry of the Lagrangian may not be the same as that of the quantum action. In particular, the **chiral anomaly** and the fact that electroweak interactions are chiral themselves allows (non-perturbative) processes that create a net amount of left-handed baryons, so that the baryon current is in fact not conserved: $\partial_{\mu}j_B^{\mu} \neq 0$.

Besides the chiral anomaly, we also saw in Section 11.5 that the electroweak interaction manifestly violates both $\mathcal{C}$ and $\mathcal{CP}$ symmetries. Moreover, a departure from thermal equilibrium may be expect to occur at a discontinuous phase transition such as the one for the electroweak symmetry breaking in the early universe [195]. The electroweak interaction thus seems to simultaneously satisfy all three Sakharov criteria, leading to the hypothesis of **electroweak baryogenesis**.

Unfortunately, it does not work. The baryon number and $\mathcal{CP}$ violation in the Standard Model are too weak to explain the observed antimatter–matter ratio [195]. Moreover, we saw in Section 14.1.2 that the heavy Higgs mass implies a thermal crossover rather than a discontinuous electroweak transition in the early universe. To satisfy the Sakharov criteria and explain the baryon asymmetry, it is therefore commonly assumed we need physics beyond the Standard Model.

14.2 Inflation Era

The cosmic microwave background (CMB) radiation allows us to probe the universe on the grandest scale. One of its unexpected features is its smoothness. The puzzling aspect of this lies in the comparison of signals originating from regions of the universe that are separated by distances larger than ct_0, with c the speed of light and t_0 the age of the universe. Even though such regions seem causally disconnected, the relative variations in CMB radiation originating from them are less than $\Delta T/T \sim 10^{-4}$. It is surprising that the variations in energy density witnessed by the background radiation are so small, despite the time causally disconnected regions must have spent without any thermal contact.

A possible explanation for this observation is to assume there was a period of rapid expansion in the early universe, with the scale factor R growing exponentially in time: $R \sim e^{Ht}$. During such an epoch of **inflation**, the size of the universe is blown up enormously, allowing a near-equilibrium configuration of the very early universe to be stretched into causally disconnected regions without falling out of equilibrium. The homogeneity of the thermal state will then persist as long as there is a sufficiently smooth exit from the inflation era towards the current era of decelerated expansion.

14.2.1 Slow-Roll

Instead of asking why the universe is nearly homogeneous, we may then wonder how accelerated expansion may come about. For this, consider again the conservation of energy in an expanding universe, which through Einstein's equations gives rise to

$$\frac{\mathrm{d}^2R}{\mathrm{d}t^2} = -\frac{4\pi}{3}G\,(\rho + 3p)\,R. \tag{14.16}$$

Here, ρ is the energy density, and p is the pressure. To get accelerated expansion, we need $\mathrm{d}^2R/\mathrm{d}t^2 > 0$, and thus $p < -\rho/3$. In other words, inflation requires a medium exerting **negative pressure**. This can be accomplished by introducing a scalar field $\varphi(x)$ called the **inflaton** field. In general, the energy density and pressure of a homogeneous scalar field in a metric of the type of Eq. (14.1) are given by

$$\rho = \frac{1}{2}\,(\partial\varphi/\partial t)^2 + V(\varphi), \qquad p = \frac{1}{2}\,(\partial\varphi/\partial t)^2 - V(\varphi). \tag{14.17}$$

To obtain a negative pressure, the potential needs to be much larger than the rate of change of the field:

$$V(\varphi) \gg (\partial\varphi/\partial t)^2\,. \tag{14.18}$$

This condition can be satisfied, for example, if the field is stuck in a high-energy metastable minimum of the potential, which happens if you supercool through a first-order phase transition. This was the original proposal, known as the 'old inflation' theory, for describing rapid expansion in our universe, but it was abandoned because of the problematic latent heat associated with a first-order transition [117, 216].

So-called new inflation theories are instead constructed on the basis of second-order phase transitions. In these, the potential for the inflaton field has a local maximum at the symmetric point $\varphi = 0$ and a global minimum at $\varphi \neq 0$. To still ensure a negative pressure, the dynamics must be such that the field only slowly evolves from the symmetric state to the symmetry-broken phase. We can then substitute the expressions of Eq. (14.17) for energy density and pressure into the energy conservation equation $\partial\rho/\partial t = -3H(\rho + p)$ to find

$$\frac{\partial^2\varphi}{\partial t^2} + 3H\frac{\partial\varphi}{\partial t} + \frac{\partial V}{\partial\varphi} = 0. \tag{14.19}$$

Again neglecting the curvature term in the Friedmann equation of Eq. (14.2) and using the condition of Eq. (14.18) for obtaining negative pressure, the Hubble parameter H becomes $H \approx \sqrt{8\pi GV/3}$ and is approximately time-independent. Given that $\mathrm{d}R/\mathrm{d}t = HR$, this implies an exponential expansion of the scale factor, $R \sim e^{Ht}$, and thus an era of inflation.

In addition to Eq. (14.18), it is also commonly assumed that the second time-derivative of the inflaton field is small:

$$\left|\frac{\partial^2\varphi}{\partial t^2}\right| \ll \left|H\frac{\partial\varphi}{\partial t}\right|. \tag{14.20}$$

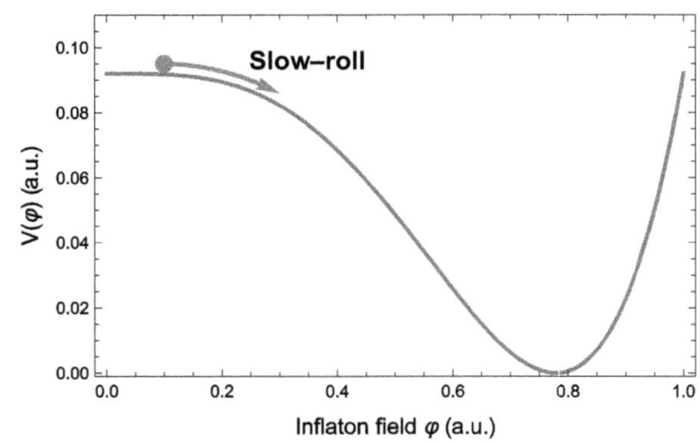

Figure 14.3 The Coleman–Weinberg potential of Eq. (14.22), which is one of the potentials proposed for the inflaton field. The potential satisfies the slow-roll conditions, which guarantee an era of exponential expansion in the early universe.

Equations (14.18) and (14.20) together are called the **slow-roll conditions**. Together, they guarantee that the dynamics of the inflaton field can generate a great expansion of the universe. Explicitly, the condition on the second time-derivative simplifies Eq. (14.19) to $3H\partial\varphi/\partial t \approx -\partial V/\partial\varphi$. This can be used to rewrite the integral of $(1/R)(\mathrm{d}R/\mathrm{d}t) = H$ in terms of the evolution of the inflaton field:

$$\frac{R(t_f)}{R(t_i)} \approx \exp\left[-\int_{\varphi_i}^{\varphi_f} \frac{3H^2}{\partial V/\partial\varphi} d\varphi\right] \approx \exp\left[-\int_{\varphi_i}^{\varphi_f} 8\pi G \frac{V(\varphi)}{\partial V/\partial\varphi} d\varphi\right]. \tag{14.21}$$

Here, the relation $H \approx \sqrt{8\pi G V/3}$ was used in the final step.

The scale factor can increase by several orders of magnitude during inflation if the potential $V(\varphi)$ is sufficiently flat. There are not enough observational constraints to determine the shape of the inflaton potential, and therefore there are many possible theoretical models. One candidate is the so-called Coleman–Weinberg order-by-disorder potential:

$$V(\varphi) = \lambda\phi^4 \log\frac{\phi}{M} + \frac{\lambda M^4}{4e}. \tag{14.22}$$

This potential has a global minimum at $\phi_c = Me^{-1/4}$, but is extremely flat for $\phi \ll M$, as shown in Figure 14.3. If the field is initially at a small value, the flatness of the potential guarantees a 'slow roll' towards the minimum, and thus an era of inflation. Regardless of the precise shape of the inflaton potential, the idea that a scalar field slowly undergoes a symmetry-breaking transition is central in all theories of inflation.

14.2.2 Broken Time-Translation Symmetry

We argued in Section 14.1 that the observed expansion of the universe necessarily implies broken time-translation symmetry. In making such a claim in the context of general relativity, we should be careful what exactly is meant by time-translation symmetry. For example, consider a so-called De Sitter universe with perfect exponential expansion, described by the metric

$$ds^2 = -c^2dt^2 + e^{2Ht}(dx^2 + dy^2 + dz^2). \tag{14.23}$$

Even though translations in the coordinate t are definitely not a symmetry of this metric, it is invariant under the so-called **dilation isometry** given by $t \to t + \Delta t$ and $x \to e^{-H\Delta t}x$ and similar for y and z. This ensures that it is always possible to define a local time-translation operation (corresponding to a time-like Killing vector) that leaves the metric invariant across a sufficiently large local patch. The observation of the CMB combined with the necessity of a smooth exit from the inflation era, however, imply that the expansion of our universe cannot be described by a pure De Sitter spacetime, and that any kind of time-translation symmetry is definitely broken.

Broken symmetries imply the existence of Nambu–Goldstone modes, which in the case of gravity should be massive because it is a gauge theory, just like the theories discussed in Chapter 8. Coincidentally, the theory of inflation requires a new scalar field, the inflaton, which is not in the Standard Model. It is then natural to wonder whether the inflaton field could in fact be the Nambu–Goldstone mode associated with time-translation symmetry breaking. This constitutes an active field of research [161].

14.2.3 Primordial Fluctuations

We mentioned in Section 14.1 that the CMB radiation is surprisingly uniform and isotropic. It is not completely uniform, however, and shows small variations in temperature of the order of $\Delta T/T \sim 10^{-4}$. This makes sense because if it were truly uniform on all scales, then each region of space would be indistinguishable from any other, and cosmic structure such as stars and galaxies could not emerge. For this reason, the fluctuations are called **primordial fluctuations**.

Going into a bit more detail, the observed CMB fluctuations in different directions can be expanded into spherical harmonics $\Delta T(\theta, \phi)/T = \sum_{l,m} a_{lm} Y_{lm}(\theta, \phi)$, which defines the coefficients a_{lm}. The associated power spectrum $P(l) = |a_{lm}|^2$ of the CMB anisotropies is found to be independent of m [117, 216]. A large part of the field of cosmology is dedicated to explaining how the universe we observe today can arise from the inhomogeneities in the early universe witnessed by the observed variations of the CMB and their power spectrum. Beyond that, however, the question remains what the origin of the primordial fluctuations is.

An argument commonly put forward starts from a scalar field $\varphi(x)$, which is usually taken to be the inflaton field, but could be any field that dominated the energy density in the early universe. Its action in curved spacetime is given by

$$S = -\int \mathrm{d}^4x\, \sqrt{-g} g^{\mu\nu} \partial_\mu \varphi \partial_\nu \varphi = \int \mathrm{d}t \mathrm{d}^3x\, a^3 \left(\frac{1}{c^2} (\partial_t \varphi)^2 - \frac{1}{a^2} (\nabla \varphi)^2 \right). \tag{14.24}$$

Here we used the metric Eq. (14.1), with dimensionless parameter $a = R/R_0$, and R_0 the scale factor at a reference time. We then switch to conformal time τ defined through $\mathrm{d}t = \mathrm{d}\tau a(\tau)$, where the distant past is given by $\tau \to -\infty$ and the distant future by $\tau \to 0$. The action becomes

$$S = \int \mathrm{d}\tau \mathrm{d}^3x\, a^2 \left(\frac{1}{c^2} (\partial_\tau \varphi)^2 - (\nabla \varphi)^2 \right). \tag{14.25}$$

Next, introduce a rescaled field $\nu = a\varphi$. The action in terms of this field is

$$S = \int \mathrm{d}\tau \mathrm{d}^3x \left(\frac{1}{c^2} (\partial_\tau \nu)^2 - (\nabla \nu)^2 + \frac{1}{c^2} \frac{\partial_\tau^2 a}{a} \nu^2 \right). \tag{14.26}$$

Exercise 14.1 (Mukhanov–Sasaki action).
Derive Eq. (14.26). Use integration by parts and discard the total derivative term.

The equation of motion for the field ν is given by

$$\partial_\tau^2 \nu - c^2\nabla^2 \nu - \frac{\partial_\tau^2 a}{a}\nu = 0. \tag{14.27}$$

Expressing this in term of the Fourier components $\nu_{\mathbf{k}}$ of the field yields

$$\partial_\tau^2 \nu_{\mathbf{k}} + \left(c^2k^2 - \frac{\partial_\tau^2 a}{a}\right)\nu_{\mathbf{k}} = 0. \tag{14.28}$$

This relation is known as the Mukhanov–Sasaki equation [117, 132]. Under the general condition that the energy density of the early universe was carried primarily by a scalar field, we find that its Fourier components $\nu_{\mathbf{k}}$ obey the equation of motion for a harmonic oscillator with time-dependent frequency $\omega_k = \sqrt{c^2k^2 - \partial_\tau^2 a/a}$.

From here, it is usually argued that the harmonic oscillator of Eq.(14.28) can be quantised, which gives an expression in terms of (time-independent) creation and annihilation operators $\hat{a}^\dagger(\mathbf{k})$ and $\hat{a}(\mathbf{k})$ [117]:

$$\hat{\nu}_{\mathbf{k}}(\tau) = \nu_{\mathbf{k}}(\tau)\hat{a}^\dagger(\mathbf{k}) + \nu_{\mathbf{k}}^*(\tau)\hat{a}(\mathbf{k}). \tag{14.29}$$

Here, $\nu_{\mathbf{k}}(\tau)$ are complex-valued mode functions. When the change of H is slow, we have $a(\tau) = -1/H\tau$, and $\partial_\tau^2 a/a = 2/\tau^2$. Then, the mode functions as a solution to Eq. (14.28) are

$$\nu_k(\tau) = \frac{\mathrm{e}^{-\mathrm{i}ck\tau}}{\sqrt{2k}}\frac{ck\tau - \mathrm{i}}{ck\tau}. \tag{14.30}$$

This form is sometimes called the Bunch–Davies mode function, and the vacuum for these modes, which is annihilated by all $\hat{a}_{\mathbf{k}}$, is called the **Bunch–Davies vacuum**. One of its important properties is that it is completely symmetric and uniform.

At early times $\tau \to -\infty$, the universe was small and all relevant dynamics took place well inside a Hubble radius cH^{-1}. Then, $c^2k^2 \gg \partial_\tau^2 a/a = 2/\tau^2$, and the modes are independent harmonic oscillators with frequency $\omega_k \approx k$. This is argued to give rise to Gaussian 'quantum vacuum fluctuations' at any $\mathbf{k}$, in the same sense that the static ground state of a quantum harmonic oscillator can be said to contain quantum fluctuations.

Due to the time dependence of the mode operator in Eq. (14.29), a configuration that is an eigenstate of the mode operator at any given time, will not be an eigenstate at later times. As the universe expands rapidly during inflation, however, the wavelength of any mode, $\lambda \sim k^{-1}$, grow with the expansion. At a point known as the horizon exit, $a\lambda$ exceeds the Hubble radius. Well beyond the horizon exit, $|ck\tau| \ll 1$, and the mode function of Eq. (14.30) becomes purely imaginary. This means that the mode operator can be expressed as $\hat{\nu}_{\mathbf{k}}(\tau) = \nu_{\mathbf{k}}(\tau)(\hat{a}^\dagger(\mathbf{k}) - \hat{a}(\mathbf{k}))$. The time dependence is now trivial, and an eigenvector of this operator remains an eigenvector for the remainder of time. This is said to imply that a 'fluctuation' with scale $\mathbf{k}$, upon traversing the horizon exit, will remain at that scale and 'no longer evolve'. This intuitive picture is also sometimes referred to as 'quantum fluctuations freezing in', or even 'quantum fluctuations becoming classical fluctuations'.

As it turns out, the CMB power spectrum $P(l)$ can be reproduced quite well from a collection of Gaussian fluctuations in the Bunch–Davies vacuum [117, 169]. But from the viewpoint of spontaneous symmetry breaking, there are some fundamental difficulties with the argumentation leading up to it.

First of all, as explained in Chapter 6, 'quantum fluctuations' are not fluctuations in time. They are corrections to a naive classical state that is not actually the ground state of the quantum system under consideration. Nevertheless, it could be that the quantum corrections in the state of the universe change as it expands. After all, the universe is not in equilibrium and time translation symmetry is broken, as discussed in Sections 14.1 and 14.2.2. This, however, leads to the problem of a **quantum-to-classical transition** yielding spatial density fluctuations. Apart from the question of how quantum corrections are transformed into classical fluctuations in the first place (see Appendix A.6), the classical quantities $v_{\mathbf{k}}$ that give rise to the power spectrum are identified with the variations of $\hat{v}_{\mathbf{k}}$, rather than their expectation values [169]. Indeed, the expectation value of any Gaussian fluctuation is exactly zero.

These problems with the quantum-to-classical transition are outlined in for instance in Refs. [162] and [190] and explained more elaborately in Ref. [169]. On the conceptual level, the central point is that the Bunch–Davies vacuum is completely symmetric. This symmetry cannot be broken, spontaneously or otherwise, without some external symmetry-breaking perturbation, as emphasised throughout this book. Since the initial vacuum state is supposed to encompass the entire universe, and nothing is external to the universe by definition, it seems from the symmetry-breaking viewpoint that primordial fluctuations cannot emerge from it.

14.3 Cosmological Defects

Given that the current configuration of the universe is in a symmetry-breaking state, it may contain both perturbative excitations (in this case given by the fundamental particles of the Standard Model), and topological defects of the type introduced in Chapter 7. The electroweak theory breaks $SU(2) \times U(1)$ down to $U(1)$, so that the associated coset space is $G/H \cong S^3$, the three-sphere. As shown in Table 7.2, the homotopy groups π_0, π_1, and π_2 are all trivial in this case, so that there are no domain walls, vortices, or monopoles in the electroweak model. Naively, we should therefore not expect any topological defects on cosmological scale associated with the electroweak symmetry breaking.

14.3.1 Semilocal Strings

The possibilities for topological defects to arise in the universe are richer than those suggested by the homotopy group analysis alone. In Section 8.4, we already found that while the energy of a $U(1)$ vortex diverges with system size, the energy of a $U(1)$ vortex coupled to a $U(1)$ gauge field remains finite. This argument can be extended to the situation with $G/H \cong S^3$. Consider a vortex line along the z-axis, so that in cylindrical coordinates we can write the Higgs field as

$$\Phi(\rho, \varphi, z) = \begin{pmatrix} 0 \\ v \end{pmatrix} f(\rho) e^{i\varphi}. \tag{14.31}$$

Here, $f(\rho)$ is a function that goes to zero at the vortex core and to one as $\rho \to \infty$. The factor $e^{i\varphi}$ implies a winding of the phase around the vortex core. This winding, however, can be removed without leaving the ground state manifold defined by the condition $\Phi^\dagger \Phi = v^2$. For example, the field configuration can be smoothly changed into the configuration $\Phi = \begin{pmatrix} v & 0 \end{pmatrix}^T$. This shows that vortices in the Higgs field are not topologically protected, as expected from the homotopy group analysis.

In the electroweak symmetry-broken phase, however, there is also a $U(1)$ gauge freedom associated with the massless photon. To understand its role, it is instructive to consider a simplified version of the Standard Model known as the **semilocal model**, in which a single $U(1)$ gauge field A_μ is coupled to the Higgs doublet:

$$\mathcal{L} = -\frac{1}{2} \left|\left(\partial_\mu - ieA_\mu\right) \Phi\right|^2 - \frac{1}{4} F_{\mu\nu} F^{\mu\nu} - \lambda \left(\Phi^\dagger \Phi - v^2\right)^2 . \tag{14.32}$$

Notice that this theory differs from the Ginzburg–Landau theory discussed in Chapter 8, because here the gauge field is coupled to a scalar doublet transforming under a global $SU(2)$ symmetry, rather than a single field. Nevertheless, a vortex configuration is still 'charged' under the $U(1)$ gauge freedom and may be in a configuration described by a non-zero gauge field pointing in the radial direction as well as a winding Higgs field:

$$\vec{A}(\rho, \varphi, z) = \frac{a(\rho)}{\rho} \hat{\varphi}. \tag{14.33}$$

Here, $a(\rho)$ goes from zero at $\rho = 0$ to one at $\rho \to \infty$. The corresponding magnetic field $\vec{B} = \vec{\nabla} \times \vec{A} = \frac{a'}{\rho} \hat{z}$ points along the vortex line and is constrained in a finite region around the vortex core.

In general, the energy of the vortex per line element contains three terms:

$$E/l_z = \int d^2x \left[\underbrace{\frac{1}{2} |(\vec{\nabla} - ie\vec{A})\Phi|^2}_{\text{Gradient}} + \underbrace{\frac{1}{2} B^2}_{\text{Magnetic}} + \underbrace{\lambda \left(\Phi^\dagger \Phi - v^2\right)^2}_{\text{Potential}} \right]. \tag{14.34}$$

Away from the vortex core, each term falls off quicker than $1/\rho$, which guarantees a finite contribution to the total energy. If we now again try to smoothly deform the Higgs field to remove the phase winding, however, we find that the gradient term diverges. For example, deforming Φ into $\begin{pmatrix} v & 0 \end{pmatrix}^T$ everywhere will not affect the magnetic or potential energy, but the gradient term at large distances goes to $\frac{1}{2} e^2 \vec{A}^2 |\Phi|^2 \to e^2 v^2 / (2\rho^2)$. As a result, the integral over all space diverges logarithmically.

Even though the configuration described by Eqs. (14.31) and (14.33) is not topologically protected, it will cost infinite energy to remove it, and it is therefore stable. A configuration like this is called a **semilocal string** [1]. The combination of global $SU(2)$ symmetry and a $U(1)$ gauge freedom allows for semilocal strings, in contrast to the global $U(1)$ symmetry and $U(1)$ gauge freedom of Section 8.4, in which the vortex string is a topological excitation.

The semilocal model is a simplified version of the Standard Model, which actually has three more gauge fields. In particular, the $U(1)$ gauge field is mixed with the Z gauge boson because the electroweak mixing angle of Section 11.3.2 is nontrivial. This gives the corresponding semilocal string a richer structure as well. The vortex core in that case contains a Z boson and the structure is therefore called a **Z-string**. A Z-string of finite length will have magnetic monopoles at its ends. Short, finite, rotating Z-strings are called **dumbbells**.

Unlike the pure semilocal string, a Z-string is not stable. But even though Z-strings have a finite lifetime, these nontrivial Higgs field configurations will likely have been formed during the electroweak transition, owing to the Kibble–Zurek mechanism of Section 9.3. If a Z-string decays because the Higgs field unwinds, the magnetic field of the string core cannot be removed. If indeed Z-strings were generated during the electroweak transition, this should then lead to a cosmic background magnetic field that ought to still be observable in the present time.

14.3.2 Sphalerons

The lack of conservation of baryon number discussed in Section 14.1 is closely relation to the only true topological defect that the electroweak theory allows. This is an **instanton**, characterised by the homotopy group $\pi_3(SU(2) \times U(1)/U(1)) = \pi_3(S^3) = \mathbb{Z}$. As discussed in Section 7.2.2, an instanton is a topological defect localised at a single spacetime event, which connects distinct ground states of the theory and thus allows transitions between them as a function of time.

To describe the instanton configurations associated with the electroweak theory, it is convenient to write the Higgs field not in its usual scalar doublet form, $\begin{pmatrix}\phi_1\\ \phi_2\end{pmatrix}$, but rather as a matrix:

$$\breve{\Phi} = \begin{pmatrix} \phi_2^* & \phi_1 \\ -\phi_1^* & \phi_2 \end{pmatrix} \tag{14.35}$$

Here, $\breve{\Phi}$ is the same $U(2)$ form of the field as the one we used in Exercise 11.9 to describe chiral symmetry breaking. Similarly, we can write the three $SU(2)$ gauge fields of the electroweak theory in matrix form, $\breve{A}_\mu(x) = A^a_\mu(x)\sigma_a$, with σ_a the Pauli matrices. In the gauge fix $\breve{A}_0 = 0$, any ground state can be written as a time-independent $SU(2)$ matrix field $\breve{U}(\mathbf{x})$ with the property that it approaches the unit matrix as $\mathbf{x} \to \infty$, and that

$$\breve{\Phi}(\mathbf{x}, t) = v\breve{U}(\mathbf{x}), \tag{14.36}$$
$$\breve{A}_i(\mathbf{x}, t) = i\breve{U}(\mathbf{x})\partial_i\breve{U}^\dagger(\mathbf{x}). \tag{14.37}$$

Here, the state $\breve{\Phi} = \mathrm{diag}(v, v)$ is the vacuum configuration identified in Exercise 11.9 for the Higgs field before coupling to gauge fields.

The field $\breve{U}(\mathbf{x})$ is a map from $\mathbb{R}^3$ to $SU(2)$ with all the points at infinity identified, so that it is classified by $\pi_3(S^3) = \mathbb{Z}$. This means there exist infinitely many topologically distinct ground states of the electroweak theory, which can be distinguished by an integer-valued topological invariant. It turns out the invariant can be written as

$$N_H = \frac{1}{24\pi^2}\int d^3x \epsilon^{ijk}\mathrm{Tr}\left[(\breve{U}^\dagger\partial_i\breve{U})(\breve{U}^\dagger\partial_j\breve{U})(\breve{U}^\dagger\partial_k\breve{U})\right]. \tag{14.38}$$

This is called the **Higgs winding number**. Topologically, distinct ground states with different winding numbers are separated by an energy barrier, and the unstable classical field configuration at the top of that energy barrier is called a **sphaleron**.

Instanton configurations connect ground states with different winding numbers. Similar to the nucleation described in Section 9.4, the rate Γ at which instantons appear is determined by the height of the energy barrier separating neighbouring states: $\Gamma \sim e^{-E_{\mathrm{sphaleron}}/T}$.

Within the Standard Model the energy of a sphaleron is approximately 10 TeV, which makes it highly unlikely for one to appear in the current universe. In the younger and hotter universe, however, sphalerons would have been accessible. This is important for the theory of baryogenesis discussed in Section 14.1.3, since ground states with different Higgs winding number turn out to contain different numbers of baryons, and sphalerons thus provide a way of realising baryon number violations.

14.3.3 Cosmic Strings

There are no topological defects in the Standard Model beyond Z-strings and sphalerons. The observation of any cosmological defects other than these two would therefore provide a clear window on physics beyond the Standard Model. One possible such extension would be the existence of a complex scalar field that spontaneously breaks a global $U(1)$ symmetry. Such a field could give rise to regular $U(1)$ vortices, dubbed **cosmic strings** in this context.

There are several proposed ways in which the Standard Model can be augmented with an additional complex scalar field. For example, as discussed in Section 11.5, the electroweak symmetry violates $\mathcal{CP}$ symmetry. There is then no theoretical reason for the strong force to respect $\mathcal{CP}$ symmetry, but experiments suggest it does. This unexpected experimental fact is sometimes called the 'strong $\mathcal{CP}$ problem'. One of the proposed mechanisms for explaining the unexpected presence of $\mathcal{CP}$ symmetry is the so-called Peccei–Quinn theory [152]. It introduces a new scalar field called the **axion** field, which spontaneously breaks a new $U(1)$ symmetry, and simultaneously guarantees the $\mathcal{CP}$ symmetry of the strong force.

Another possible hint of physics beyond the Standard Model comes from observations of gravitational pull in our universe beyond what is expected from the visible matter it contains. This suggests there may be an overabundance of matter that does not interact with visible light, but does carry mass. Such cold dark matter could perhaps be a new scalar field with a mass m greater than the Hubble constant, $H^2 \ll m^2$. That would result in a Bose condensate of such cold heavy particles, breaking $U(1)$ symmetry.

Finally, we already discussed the inflaton in Section 14.2.1, which actually also is a new scalar field beyond the Standard Model. The theory of inflation does not constrain the number of components of the inflaton field, but if it happens to be a complex scalar field it will possess a global $U(1)$ symmetry. The slow breaking of this symmetry that generates the inflation era would then also leave a broken $U(1)$ symmetry in the present universe.

In the end, it may even be possible that the axion, the inflaton, and cold dark matter are all the same. In fact, the axion field in Peccei–Quinn theory has been shown not to couple to the electromagnetic field and is thus a form of dark matter. Similarly, the equations proposed for describing cold dark matter scalar fields are the same as those used to describe the era of inflation. Whether or not there actually are new fields beyond the Standard Model should eventually be determined by experiment, and the detection of cosmic strings gives one possible inroad to exploring the workings of our universe by leveraging the physics of spontaneous symmetry breaking.

PART IV

PROJECTS

15 Gauging Rulers and Protractors

Box 15.1 Prerequisites

Read Section 1.6.
This is also a useful project to revisit after reading Chapter 8.

Box 15.2 Goal of this project

◇ Identify the relation between choosing coordinate systems and gauge degrees of freedom.
◇ Distinguish physical from unphysical degrees of freedom.

15.1 The Random Ruler

In Section 1.6, we introduced gauge freedoms and argued that the most common form of gauge freedom that we encounter regularly in daily life is the choice of units for measuring things in. For example, when measuring the extension of a spring, different people may reach different numerical values depending on whether they express their answers in centimetres or inches. But all of them will agree on the dimensionless ratio between the extended and equilibrium lengths of the spring. Moreover, as long as they keep in mind the units being used, different observers can translate between their own measurements and those of others. Switching from one choice of units to another changes the numerical value assigned to any particular 'length', but not any physical predictions following from those values. The choice of units is thus a gauge freedom: changing units affects the way in which we express our predictions, but it does not alter the physical observations that we expect to make.

To make the distinction between choosing how to measure something (a gauge freedom) and predicting physically different measurement outcomes (states connected by a physical transformation) more precise, consider two people who will measure the length of a spring. One, say Alice, will use a regular ruler, with equally spaced tick marks demarcating intervals along its edge. The other, Bob, uses the more exotic ruler shown in Figure 15.1, with randomly separated tick marks. Calling the positions of Alice's tick marks x_j with $j \in [0, 1, 2, 3, \dots]$, and those of Bob y_j, we can imagine that Bob's tick marks are displaced versions of Alice's, with the displacements defined as $u_j \equiv x_j - y_j$.

First, consider a situation in which both Alice and Bob align the left end of their spring with their leftmost tick mark at $j = 0$, and both happen to find the right end of the spring

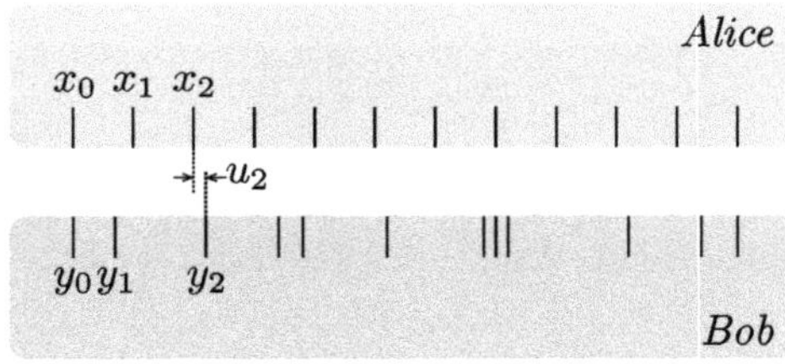

Figure 15.1 Rulers used by Alice (top) and Bob (bottom).

precisely coinciding with the tick mark at position $j = N$ on their respective rulers. Alice declares that the length of the spring is $L = x_N - x_0 = \sum_{j=0}^{N-1} x_{j+1} - x_j$.

Exercise 15.1 Express the length measured by Bob as a sum over intervals between his tick marks, and use this expression to show that $\sum_{j=0}^{N-1} u_{j+1} - u_j = 0$.

Already in this situation, it is clear that moving around the tick marks between the initial and final points does not influence the observed length of the spring. Their positions are gauge degrees of freedom, while the total length is gauge invariant.

In this case, the invariance of the total length is simply imposed, by the condition that the tick marks at $j = 0$ and $j = N$ stay fixed. We can lift that constraint by considering a different situation, in which Alice and Bob again both align the $j = 0$ tick marks of their rules with the left end of the spring, but now Alice finds the right end to coincide with her tick mark at $j = N_A$, while Bob registers it at $j = N_B$ on his ruler. Rather than simply disagreeing on the length of the spring expressed in the number of tick marks, however, Alice notices that she can express both results as a sum over N_A intervals, as long as she takes into account the shifts between her and Bob's tick marks.

Exercise 15.2 Show that the length observed by Alice can be expressed in terms of the intervals measured by Bob as

$$L = \sum_{j=0}^{N_A-1} y_{j+1} - y_j + w_j. \tag{15.1}$$

Define w_j in terms of the displacements u_j.

Notice that this expression for L remains the same even if Bob pulls out a different random ruler, with tick marks displaced in yet different ways. For any given set of displacements with respect to Alice's ruler, Eq. (15.1) predicts the same length of the spring as measured by Alice's ruler. Alice will thus regard the displacement of ticks on Bob's ruler as a gauge freedom, and in fact a local gauge freedom, since each tick can be moved independently. The value for L reported by Alice using Eq. (15.1) is invariant under any changes of the local displacements describing movement of Bob's tick marks. In this case, the field w_j is the gauge field ensuring this gauge invariance.

The gauge-invariant expression constructed by Alice in Eq. (15.1) is comparable to, for example, the gauge-invariant combination $\nabla\phi - \vec{A}$ appearing in the Noether current of Eq. (8.3). In both expressions, a gauge transformation involves a change in field values

(ϕ or y) accompanied by a corresponding change in the values of a gauge field ($\vec{A}$ or w), in such a way that physical predictions are unaffected.

Exercise 15.3

- Explicitly show the effect of changing $u_j \to u_j + \Delta_j$ for any $0 < j < N_B$ on the values of y_j and w_j. That is, express how the fields in Eq. (15.1) transform under gauge transformations.
- Show that these gauge transformations leave the value of L in Eq. (15.1) invariant.
- We can even consider displacements that move tick marks past one another, so that Bob ends up with a disordered sequence of the labels $j = 1, 2, 3, \ldots$. Show that this still results in the same gauge-invariant value for L.

Although the gauge invariant expression for length measured on a random ruler involves a reference ruler and is thus somewhat artificial, it brings to the fore the role of the gauge field as a means of keeping track of local changes in the way we describe physical observables such as the size of an interval. As long as we keep track of all the changes we make to the positions of any tick marks, we can always work back to a reference ruler and agree on the physical length of any object compared to the size of its unit intervals.

To see the true extent of the relation between local changes in coordinate systems and gauge freedoms, however, we need to move beyond the single ruler and instead consider measurements of field values at different locations being performed with different measurement devices.

15.2 Random Protractors

Consider the situation sketched in Figure 15.2, in which different people scattered throughout a meadow measure the angle $\theta(\vec{x})$ that the local wind velocity $\vec{v}(\vec{x})$ makes with respect to a reference direction $\gamma(\vec{x})$, which was determined by the local wind direction at noon.

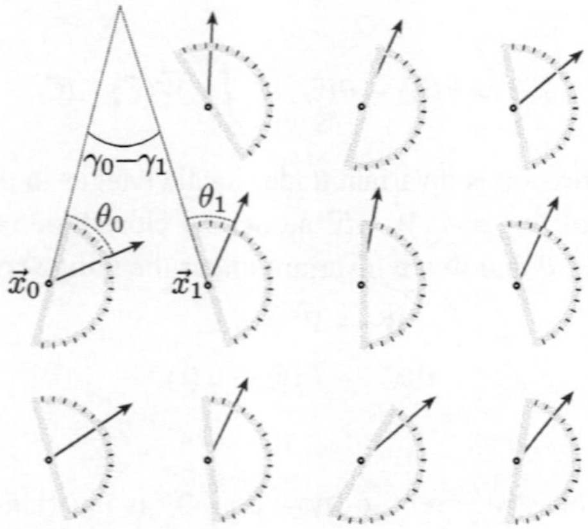

Figure 15.2

A field of protractors measuring local wind directions.

Exercise 15.4

- Explain why specifiying $\theta(\vec{x}) = 0$ does not necessarily imply a uniform wind profile.
- Give at least two distinct specifications of $\theta(\vec{x})$ and $\gamma(\vec{x})$ that correspond to a uniform wind profile.

Because wind is just a current in air, fluid dynamics demands that there is an energy cost proportional to the gradient of the wind direction: $E \propto \nabla \cdot \vec{v}$. Putting all their local readings together, the people in the meadow would like to determine this gradient contribution to the wind energy. They realise this is possible only if they calibrate (or 'gauge') their individual protractors with respect to an agreed reference direction. They patiently wait for the wind to become as uniform as possible and then quickly all align the zero of their protractors with the local wind direction. These directions will be called $\gamma(\vec{x})$ from here on. Because the wind might not have been perfectly uniform, $\gamma(\vec{x})$ may not be identically zero everywhere.

Exercise 15.5

- Explain why the gradient contribution to the energy is proportional to $E \propto \sum_{\langle i,j\rangle} \cos(\theta_i - \theta_j + W_{i,j})$, where $\theta_j \equiv \theta(\vec{x}_j)$ with $\vec{x}_j$ the position of one of the people in the meadow. Define $W_{i,j}$ in terms of local values of $\gamma(\vec{x})$.
- Show that the difference in local wind direction between any two people in the meadow, located at $\vec{x}_i$ and $\vec{x}_f$ can be written as $\Delta\theta \equiv \theta_i - \theta_f + \sum_{j=i}^{f-1} W_{j,j+1}$, where the sum runs over a series of nearest-neighbour positions connecting the two people.
- Show that any of the people between $\vec{x}_i$ and $\vec{x}_f$ recalibrating (reorienting) their protractors do not affect the value of $\Delta\theta$, as long as they dutifully report their updated values of $\gamma(\vec{x})$.
- Moreover, show that even the people at $\vec{x}_i$ and $\vec{x}_f$ are free to recalibrate their protractors, as long as they update their reported values of both $\gamma(\vec{x})$ and $\theta(\vec{x})$.
- Finally, show that a sum of $W_{j,j+1}$ along any closed path beginning and ending at the same position is unaffected by local recalibrations.

The field $W_{i,j}$ can be written as a vector field in the continuum limit, using $\vec{W}(\vec{r}) = (W_{r,r+\delta x}, W_{r,r+\delta y})$. The expression for the difference between two locally observed angles then becomes

$$\Delta\theta = \theta(\vec{x}_i) - \theta(\vec{x}_f) + \int_{\vec{x}_i}^{\vec{x}_f} \vec{W}(\vec{x}') \cdot d\vec{x}'. \tag{15.2}$$

As you showed, this expression is invariant under local changes in the orientation of protractors, and so is the value of $\Phi_C \equiv \oint_C \vec{W} \cdot d\vec{x}'$ along any closed loop C. In terms of the vector field $\vec{W}$, this implies that $\Delta\theta$ and Φ are invariant under the **gauge transformation**:

$$\begin{aligned} \vec{W} &\to \vec{W} + \nabla\alpha, \\ \theta(\vec{x}) &\to \theta(\vec{x}) + \alpha(\vec{x}). \end{aligned} \tag{15.3}$$

Exercise 15.6 Use Stokes' theorem to show that Φ_C is invariant under the transformation of Eq. (15.3) for any closed loop C.

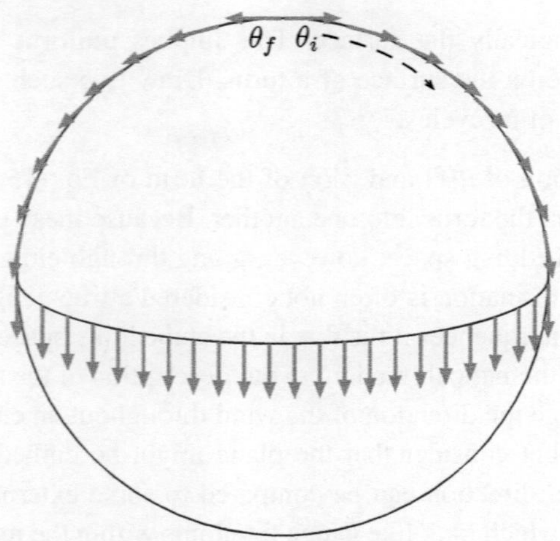

Figure 15.3

Curvature of a sphere evidenced by the Hannay angle of a closed curve.

The gauge invariant form of Eq. (15.2) and the gauge transformation of Eq. (15.3) can be compared directly to the expressions for gauge transformations of the electromagnetic field coupled to electrically charged particles in Eqs. (1.46) and (1.47) in Chapter 1, or Eq. (8.4) in Chapter 8. In that case, the gauge freedom can be interpreted as recalibrating the way we measure the quantum mechanical phase of charged particles. Local recalibrations lead to local shifts of the reported phase, as well as a corresponding addition of a gradient to the electromagnetic vector potential, in just the same way that these operations appear in Eq. (15.3) for measurements of the local wind direction.

The fact that the line integral of the electromagnetic vector potential around a closed loop is a gauge invariant quantity in the case of electromagnetism is familiar from the fact that it corresponds to the value of the magnetic flux piercing the loop. Because this is a physical, observable quantity, it must necessarily be gauge invariant. In fact, the gauge invariant value of Φ_C in the case of the protractors on a plane also has a physical meaning. You can easily convince yourself that $\Phi_C = 0$ for any closed loop on a flat plane. However, this is no longer true for a field of protractors on a curved surface. For example, taking $\theta(\vec{x})$ around a closed curve on the surface of a sphere, as indicated in Figure 15.3, you can keep every pair of neighbouring wind directions as parallel as possible and still end up with a non-zero rotation at the end. The so-called Hannay angle of rotation obtained by parallel transporting a vector along a closed loop is determined by the curvature of the surface enclosed within the curve. Equivalently, it can be expressed in terms of a flux piercing the curve.

Exercise 15.7

- Given that $\Phi_C = \pi$ along a ring surrounding the meadow, give at least two distinct specifications of $\theta(\vec{x})$ and $\vec{W}(\vec{x})$ that correspond to the same wind profile.
- Minimising the gradient contribution to the energy of the wind profile on earth (a sphere) as much as possible, pictorially show there are always two vortices that cannot be removed. The same is not true on a flat plane.

- A torus is a topologically flat surface. This implies uniform wind profiles without vortices are possible on the surface of a torus. Draw two such profiles, one for wind blowing along each of its cycles.

There is a transformation of $\theta(\vec{x})$ and $\vec{W}(\vec{x})$ of the form of Eq. (15.3) that transforms the two uniform profiles on the torus into one another. Because these two profiles imply different fluxes in its embedding space, however, going through either the hole of the torus or its inside, this transformation is often not considered a true gauge freedom. That is, it could have physical consequences if the flux in the embedding space can be measured, and hence is not a mere mathematical freedom in our description of the system. A similar situation arises if we change the direction of the wind throughout an entire plane everywhere by the same amount, but consider that the plane might be embedded in a larger space where the overall wind direction can be compared to some external frame of reference. Such transformations, which look like gauge freedoms within the manifold of interest, but which have physical consequences in its embedding space are called **large gauge transformations**. Within the embedding space, they really are physical transformations and not gauge freedoms.

The main difference between electromagnetism and random protractors on a surface is that the value of the magnetic field can be changed. The vector potential is thus associated with a dynamical degree of freedom of its own, unrelated to the presence of any charged particles. For the curvature measured by closed loops on surfaces this is often not the case, and gauge invariant curvature instead represents a constraint on the effect that recalibrations can have on the fields $\theta(\vec{x})$ and $\vec{W}(\vec{x})$. In other words, the fact that these fields are not independent degrees of freedom (recalibrating γ affects both θ and W) is signalled by the gauge freedom affecting both of them simultaneously. Notice that it is possible to imagine situations in which the local curvature of a manifold can be changed by physical degrees of freedom residing on the manifold itself. In that case, the flux Φ_C measured in a field of protractors will represent a gauge-invariant dynamical degree of freedom in addition to imposing a constraint, in full analogy to the gauge-invariant description of electromagnetism.

16 Simulating the BKT Transition

Box 16.1 Prerequisites

Read Chapter 7.

Box 16.2 Goal of this project

- ◇ Learn how to write your own Monte Carlo simulation code.
- ◇ Calculate the BKT transition temperature using the divergence of the susceptibility and correlation length.
- ◇ Distinguish phases with short-range correlations from algebraic long-range order.

16.1 The *XY*-Model

In Section 7.4, we introduced the Berezinskii–Kosterlitz–Thouless (BKT) phase transition. Using an entropic argument, we showed that at a critical temperature, the system goes from having short-range correlations (at high T) to algebraic correlations (at low T). In this project, we will simulate this transition on the square lattice XY-model.

In the XY-model, there are in-plane spins on each lattice site, described by $s_i = (\cos\theta_i, \sin\theta_i)$. The energy of a spin configuration $\mathcal{C}$ is given by

$$E[\mathcal{C}] = -\sum_{\langle ij\rangle} s_i \cdot s_j = -\sum_{\langle ij\rangle} \cos(\theta_i - \theta_j). \tag{16.1}$$

Here, $\langle ij\rangle$ is a sum over all nearest neighbours in the square lattice, and the lattice contains $N = L \times L$ sites with periodic boundary conditions. The dimensionless energy E defined here corresponds to expressing the energy of Section 7.4 in units of J. Some typical spin configurations are shown in Figure 16.1.

As usual, the partition function at temperature T is given by

$$Z = \sum_{\mathcal{C}} \mathrm{e}^{-E[\mathcal{C}]/T}. \tag{16.2}$$

Similarly, the expectation value of an observable A is defined as

$$\langle A\rangle = Z^{-1} \sum_{\mathcal{C}} A \mathrm{e}^{-E[\mathcal{C}]/T}. \tag{16.3}$$

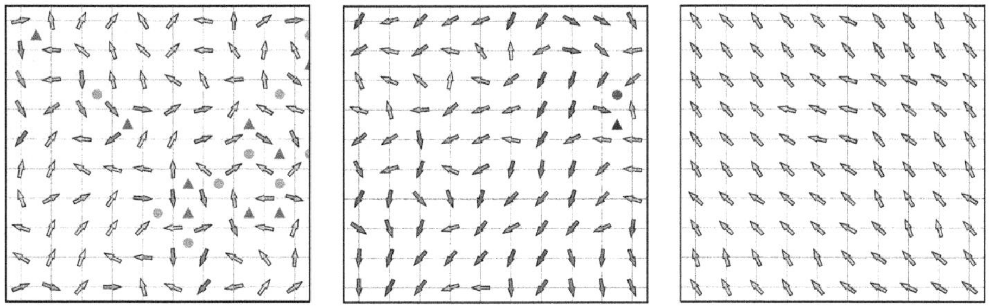

Figure 16.1 Typical XY-spin configurations at different temperatures. (a) At high temperature, $T = 2$, the spins are almost randomly oriented and many plaquettes contains a vortex (circle) or antivortex (triangle). (b) Close to the BKT transition at $T = 1$, only close vortex–antivortex pairs can exist. (c) At low temperature, $T = 0.1$, we find algebraic long-range order, which appears like spontaneous symmetry breaking due to the finite system size.

Using this, we can calculate the average energy $\langle E[\mathcal{C}]\rangle$, as well as the magnetisation m, the susceptibility χ, and the correlation functions $C(i,j)$ given by

$$\begin{aligned} m &= \frac{1}{N}\left\langle \sum_i s_i \right\rangle \equiv \frac{1}{N}\langle s_{\text{tot}}\rangle, \\ \chi &= \frac{1}{N}\left(\langle s_{\text{tot}}^2\rangle - \langle s_{\text{tot}}\rangle^2\right), \\ C(i,j) &= \langle s_i \cdot s_j\rangle. \end{aligned} \tag{16.4}$$

We can also calculate whether a given spin configuration contains vortices. On each bond, we calculate $\delta\theta_{\langle ij\rangle} = \theta_i - \theta_j$ such that it lies in the interval $[-\pi, \pi)$. The sum over $\delta\theta$ over all four sides of a plaquette, divided by 2π, is either 0 (no vortex) or ± 1 (for a vortex/antivortex).

16.2 Monte Carlo Simulation

For large system sizes, it becomes impossible to calculate the partition sum exactly. Therefore, we will use an approximation scheme called **Monte Carlo simulation** [193]. In this type of simulation, rather than summing over all possible configurations, we randomly generate a smaller number N_{MC} of spin configurations $\{\mathcal{C}\}_{\text{MC}}$. These spin configurations are created with a probability distribution given by the Boltzmann weights $P[\mathcal{C}] = Z^{-1}\mathrm{e}^{-E[\mathcal{C}]/T}$. The average magnetisation, for example, is then approximated by simply averaging over the generated random configurations, $m = \frac{1}{N_{\text{MC}}}\sum_{\{\mathcal{C}\}_{\text{MC}}}\left(\frac{1}{N}\sum_i s_i\right)$.

A very good way to generate the random spin configuration is using a **Markov chain**. The idea is to start with an initial configuration $\mathcal{C}$. Then, a new configuration $\mathcal{C}'$ is constructed by changing only a small part of the old configuration $\mathcal{C}$ depending on a conditional probability $P_M[\mathcal{C}'|\mathcal{C}]$. This conditional probability should satisfy the so-called detailed balance condition:

$$\frac{P_{\text{M}}[\mathcal{C}|\mathcal{C}']}{P_{\text{M}}[\mathcal{C}'|\mathcal{C}]} = \frac{P[\mathcal{C}']}{P[\mathcal{C}]}. \tag{16.5}$$

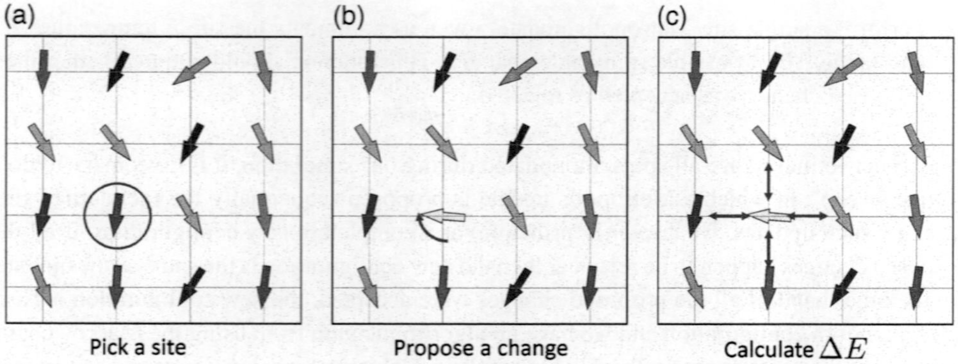

Figure 16.2

The steps of a single spin Metropolis algorithm update. First, a site is chosen (a). Then, we propose a change to the spin on that site following Eq. 16.6 (b). The change in energy ΔE is calculated, which determines whether the proposed change is accepted or rejected using Eq. 16.7 (c).

Using this conditional probability, a sequence of configurations $\mathcal{C}_1 \to \mathcal{C}_2 \to \mathcal{C}_3 \to \ldots$ is created.

Exercise 16.1 Show that if Eq. (16.5) is satisfied, then the probability of finding any given configuration $\mathcal{C}$ in the random sequence is given by its Boltzmann weight $P[\mathcal{C}]$.

For our purposes, we will use the **single spin Metropolis** algorithm. This procedure to create a new configuration for the *XY*-model consists of three steps. First, we choose a site i. Second, a change of spin on site i is proposed:

$$\theta_i \to \theta_i + \delta\theta. \tag{16.6}$$

Here, $\delta\theta$ is a random number on the interval $[0, \Delta\theta]$, with an adjustable parameter $\Delta\theta$ signifying the range of possible changes. The rotation of θ_i changes the energy of the configuration by an amount ΔE. In the final step, the change is accepted with probability p given by

$$p = \min\left(\mathrm{e}^{-\Delta E/T}, 1\right). \tag{16.7}$$

If the change is not accepted, it is rejected, and the angle θ_i is restored to its previous value. The sequence of steps is visualised in Figure 16.2.

Exercise 16.2 Show that the single spin Metropolis algorithm satisfies detailed balance, Eq. (16.5).

Exercise 16.3 Write a code with the following functions:

- Initialise a random spin configuration on an $L \times L$ square lattice;
- Compute the energy E of a spin configuration;

- Perform a single site Metropolis update, given as arguments the site i, temperature T and range $\Delta\theta$ of possible spin angle changes. This function should output whether the proposed change was accepted or rejected.

In order to make sure all spins are updated during our simulation, it is customary to do a 'lattice sweep', in which a Metropolis update is proposed sequentially for each lattice site. After L^2 such updates, we arrive (in principle) at a completely new configuration. If all the proposed changes happen to be rejected, then the new configuration is the same as the old one. On the other hand, if all the proposed changes were accepted, the new configuration is basically a random configuration and we have no algorithmic gain from using the Markov chain. It turns out the most numerically reliable results are obtained when we tune the acceptance rate to about 50%.

For the XY-model, the acceptance rate can be altered by changing the range $\Delta\theta$ of proposed angle changes. This is typically done during an 'equilibration sweep': a repeated number N_{eq} of lattice sweeps. The goals of such an equilibration run are to forget about the initial random configuration and to adjust $\Delta\theta$ in order to bring the acceptance rate r close to one half.

Exercise 16.4 Write the following functions:

- Perform a lattice sweep where you run through all lattice sites and do a Metropolis update on each. The function should depend on temperature, and it should output the acceptance rate r;
- An equilibration routine of N_{eq} lattice sweeps. After each sweep adjust $\Delta\theta$, such that at the end of equilibration the acceptance rate approaches $r = 0.5$.

16.3 Numerical Measurements

After equilibration, we are ready to actually compute the physically relevant quantities. For that purpose, we need a set of measurement functions. We already have the energy function, but we need four others as well.

Exercise 16.5 Write the functions:

- Susceptibility for a given spin configuration;
- Density of vortex pairs for a given spin configuration;
- Correlation function for a given spin configuration. To reduce the computational cost, only calculate the correlation function along one direction on the lattice, $C(n\hat{\mathbf{x}} + \mathbf{r}, \mathbf{r})$, but average this quantity over all possible sites $\mathbf{r}$.

We now have functions that can approximate physically meaningful quantities for each spin configuration. The final step is to average over a long sequence of random configurations generated by lattice sweeps. We call each time a physical quantity is calculated for a spin configuration a 'measurement'.

Exercise 16.6 Write a function that performs `nr_measurements` measurements of all physical quantities, with `nr_sweeps_per_measurement` lattice sweeps between each measurement.

As with experimental measurements, we should estimate the errors we make in our numerical measurements of physical observables. A traditional approach is to not only calculate the average but also the variance of all the measurements. However, doing this naively does not give you an error estimate but a different physical quantity. For example, measuring the variance of the energy yields the specific heat, while the variance of magnetisation is the susceptibility.

To avoid this, we can do 'binning'. The idea is to split the total number of measurements of A into N_{bins} bins, so there are N_m/N_{bins} measurements per bin. The measurements are averaged per bin, so there are now N_{bins} independent measurements A_i. The error estimate of our physical quantity A is then obtained using

$$\sigma^2 = \frac{1}{N_{\text{bins}} - 1} \sum_{i=1}^{N_{\text{bins}}} \left(A_i^2 - \langle A \rangle^2 \right). \tag{16.8}$$

Exercise 16.7 Write a measurement function that uses binning to produce error estimates. (Note: if you are familiar with parallel programming, this is a good place to parallelise by putting each bin on a separate CPU.)

Exercise 16.8 Make a so-called annealing function, in which we start at a high temperature (e.g., $T = 2$) with a random configuration. Slowly reduce the temperature with small steps $\Delta T = 0.05$ and perform an equilibration followed by a measurement run at each temperature. Output all the measured quantities (energy, magnetisation, susceptibility, vortex-pair density, and the correlation function) with their error estimates to a file.

Exercise 16.9 Run your code for $L = 10, 20, 40, 60, 80$, and 100, for $N_{\text{eq}} = 1{,}000$ and 10 bins, with 40 measurements per bin, with 5 lattice sweeps per measurement.

16.4 Analysis of Results

We now have a range of 'raw' results for system sizes up to $L = 100$, as shown in Figure 16.3. The final step of this project is to extract physical conclusions from these raw results. This means finding the critical temperature T_{BKT}, finding how the correlation length and the susceptibility diverge, and how the vortex pair density vanishes at the transition.

Exercise 16.10 Check that the energy is a smooth function for all temperatures and can thus not be used to detect the transition temperature.

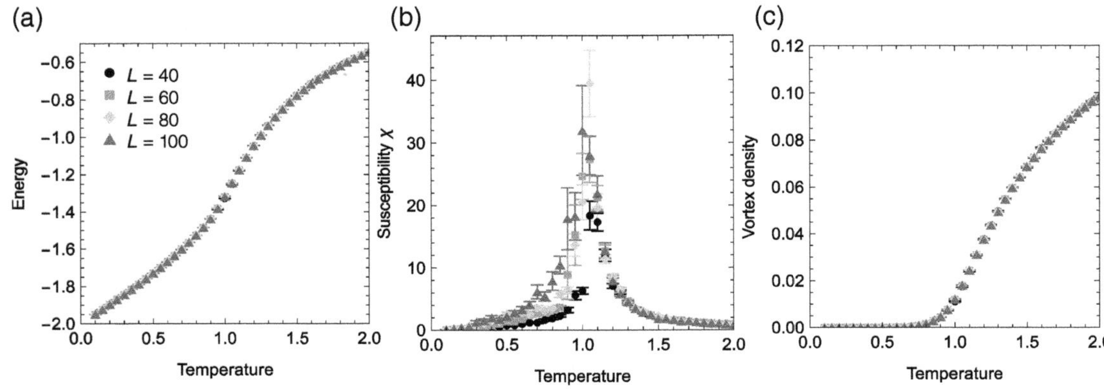

Figure 16.3 Typical raw results obtained for a Monte Carlo simulation of the *XY*-model. The energy per site (a) remains a smooth function at all temperatures. The susceptibility (b), however, seems to diverge, and the vortex pair density (c) vanishes.

To extract the correlation length for the correlation function, we need to anticipate the shape of $C(\mathbf{r}) = \langle s_{\mathbf{r}} \cdot s_0 \rangle$. At very low temperatures, in the phase with algebraic long-range order, the angle difference $\theta_i - \theta_j$ on a bond connecting neighbouring sites is typically small. The energy can then be expanded using $\cos(\theta_i - \theta_j) \approx 1 - \frac{1}{2}(\theta_i - \theta_j)^2$. Coarse-grained, this implies that in the algebraic phase the correlations are approximated by

$$C_0(\mathbf{r}) = \exp\left[TG(\mathbf{r})\right], \tag{16.9}$$

$$G(\mathbf{r}) = \frac{1}{N_k} \sum_{\mathbf{k}} \frac{e^{i\mathbf{k}\cdot\mathbf{r}} - 1}{4 - 2\cos k_x - 2\cos k_y}. \tag{16.10}$$

This should be the form of the correlation function when we approach $T = 0$.

Exercise 16.11 Show that in the limit $r \to \infty$, the correlation function Eq. (16.9) describes algebraic decay of the form $C_0(r) \sim r^{-(T/2\pi)}$.

At higher temperatures, corrections beyond the quadratic approximations become significant, and we write the general form:

$$C(n\hat{\mathbf{x}}) = e^{2\pi\eta G(n\hat{\mathbf{x}})} \left(e^{-n/\xi} + e^{-(L-n)/\xi} \right). \tag{16.11}$$

This function has two fitting parameters: the correlation length ξ and the algebraic exponent η.

Exercise 16.12 Write a code to fit the measured correlation function to the form of Eq. (16.11) to extract ξ and η. Apply this code to the Monte Carlo results obtained in the previous section to find ξ and η as a function of temperature T and system size L.

An example of the fitting results is shown in Figure 16.4. We are now in a position to extract the critical behaviour of the model. Defining the reduced temperature as $t = (T - T_{\mathrm{BKT}})/T_{\mathrm{BKT}}$, the correlation length and susceptibility will diverge exponentially [96, 97]:

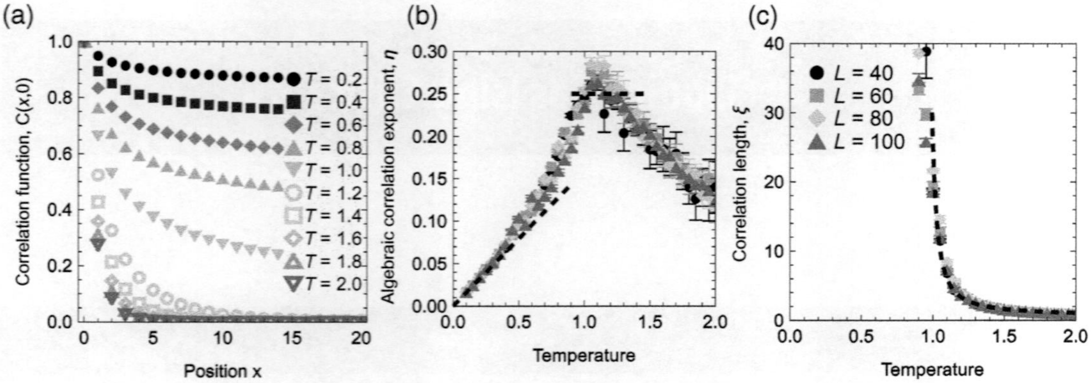

Figure 16.4 The correlation function (a) as a function of temperature goes from rapidly decaying at high temperatures to slowly decaying at low temperatures. We can fit these correlation functions using Eq. (16.11). We find that at low T, indeed the correlation function is algebraically decaying with $\eta = T/2\pi$, slowly increasing to $\eta = 0.25$ at the BKT transition (b). The correlation length, for fits above the BKT transition, diverges around $T_{\text{BKT}} = 0.89$ with critical exponent $\nu = 0.70$ (c).

$$\xi \sim e^{b/t^\nu}, \quad \chi \sim e^{b'/t^\nu}. \tag{16.12}$$

Here, b and b' are fitting parameters. Note, however, that due to the finite system sizes in our calculations, we can never see a full divergence. It is therefore important to discard any results for the correlation length ξ when the measured ξ is larger than L.

Exercise 16.13 Write a code to fit the ξ and η, extracting both the critical temperature T_{BKT} and the critical exponent ν. How do your results compare to the result in Ref. [193]? Can you find more recent results in the literature for the critical temperature and the exponent ν?

17 Liquid–Gas Ising Criticality

Box 17.1 Prerequisites

Read Chapter 5.

Box 17.2 Goal of this project

◇ Learn to interpret the liquid–gas transition as a lattice gas model.
◇ Learn how to calculate the mean field critical exponents of the liquid–gas critical point.

17.1 Liquid–Gas Transition

It might come as a surprise that a book on spontaneous symmetry breaking deals with the phase transition from liquid to gas. After all, unlike the liquid–solid transition, there is no symmetry difference between a gas and a liquid. Both are translationally invariant, and the only difference between them is their average particle density.

However, the power of universality (see Section 5.5) enables us to use concepts of SSB also when studying the liquid–gas transition. Because this transition is described by a scalar field (the liquid or gas particle density ρ) in $d = 3$ dimensions, the associated critical point must have exactly the same critical exponents as the $d = 3$ Ising model. This is because the order parameter in an Ising system, the magnetisation m, is also a scalar. Through an appropriate mapping of ρ onto m we should thus be able to understand the critical point of the liquid–gas transition.

Our starting point is the ideal gas law, expressed as

$$pv = k_B T. \tag{17.1}$$

Here, p is the pressure, $v = V/N$ is the volume per particle, T is the temperature, and k_B Boltzmann's constant. Note that the volume per particle is the inverse of the particle density $v = 1/\rho$. Of course, an ideal gas of non-interacting point-like particles cannot undergo any form of phase transition.

To allow for a liquid phase, we need to include both the fact that particles are not point-like, and interactions between them. Because particles then occupy a small volume, we will replace v with a reduced volume, $v - b$; and because the interactions and thus pressure increase when particles are close to one another, we replace p with $p + a/v^2$. The result is the **Van der Waals equation of state**:

$$\left(p + \frac{a}{v^2}\right)(v - b) = k_B T. \tag{17.2}$$

The parameters a and b are the **Van der Waals constants**, and their values differs from substance to substance.

Exercise 17.1 Plot the curves of p vs v for various temperatures T. At low temperatures, you will find not one, but three possible volumes for each pair of temperature and pressure values. What is the critical temperature T_c where you go from having one possible solution to three? Also calculate the critical values of pressure and volume.
Hint: At the critical temperature, pressure and volume, we have $\frac{\partial p}{\partial v} = \frac{\partial^2 p}{\partial v^2} = 0$.
Answer: $v_c = 3b$, $p_c = \frac{a}{27b^2}$ and $k_B T_c = \frac{8a}{27b}$.

Exercise 17.2 Given the values of the critical temperature, volume, and pressure, show that the Van der Waals equation can be simplified using reduced parameters $\overline{T} = k_B T/T_c$, $\overline{v} = v/v_c$, and $\overline{p} = p/p_c$. Also rewrite it in terms of the density $\rho = 1/v$.
Answer: $\left(\overline{p} + \frac{3}{\overline{v}^2}\right)\left(\overline{v} - \frac{1}{3}\right) = \frac{8}{3}\overline{T}$ or $\overline{p} = \frac{8\rho\overline{T}}{3-\overline{\rho}} - 3\overline{\rho}^2$.

From now on, for convenience, we will only use the reduced parameters introduced in Exercise 17.2, but without writing the bar on top all the time. At high temperatures, $T > T_c$, there is only one phase defined by the Van der Waals equation. Upon approaching the critical point several quantities diverge, and in particular, the **isothermal compressibility**:

$$\kappa = \frac{1}{\rho}\left.\frac{d\rho}{dp}\right|_T. \tag{17.3}$$

This quantity tells us how the density changes when you change the pressure. The fact that this quantity diverges suggests that something radical will happen with the density of the gas as we cross the critical point.

Exercise 17.3 Calculate the isothermal compressibility by taking the pressure-derivative of the Van der Waals equation of state. Show that at the critical density $\rho = \rho_c$ and $T > T_c$ the compressibility diverges as

$$\kappa \sim (T - T_c)^{-\gamma}. \tag{17.4}$$

In this case, the critical exponent is given by $\gamma = 1$.

The diverging compressibility at the critical temperature, $T = T_c$, implies that the pressure–density relation at that point is not linear, but is instead described by a higher-order power law.

Exercise 17.4 Use the Van der Waals equation to show that at $T = T_c$, the relation between pressure and density scales as

$$(p - p_c) \sim (\rho - \rho_c)^{\delta}. \tag{17.5}$$

Here, the critical exponent is $\delta = 3$.

At temperatures below T_c, we saw that the Van der Waals equation gives three possible values of the density ρ for a fixed pressure p. The relation between ρ and p is a cubic polynomial equation, so in principle, it is exactly solvable, but the tediousness of cubic roots obscures the physics in its solutions. Instead, we will expand the Van der Waals equation for density, pressure and temperatures close to the critical point. Of the three possible densities given a fixed pressure and temperature, we will be able to tell which one corresponds to the thermodynamically stable phase by looking which one has the lowest **free energy**.

Exercise 17.5 First, calculate the Helmholtz free energy $F[v, T]$ by integrating the Van der Waals equation, using the relation $p = -\left.\frac{\partial F[v,T]}{dv}\right|_T$. Also show that solutions of the Van der Waals equation correspond to a local extremum of the Gibbs free energy $G[v, p, T] = F[v, T] + pv$, with respect to volume v. Equivalently, the Van der Waals equation corresponds to a local extremum with respect to density ρ if we write the Gibbs free energy as a function of density instead of volume.
Answer: $G[\rho, p, T] = -\frac{8}{3}T \log\left[\frac{3}{\rho} - 1\right] - 3\rho + \frac{p}{\rho}$.

Write a Taylor expansion of the Gibbs free energy close to the critical point. You can define $\delta\rho = \rho - \rho_c$, and similarly for the other parameters. Keep terms up to $\mathcal{O}(\delta\rho^4)$. Terms independent of pressure and volume can be neglected.
Finally, we simplify this equation a bit. Show that around the critical point the prefactor of the fourth order term is always positive, so we can replace this with a constant. Use a shift $\delta\rho \to \delta\rho - \frac{2}{3}(\delta p - \delta T)$ to remove the term cubic in $\delta\rho$ from the free energy, after which you can discard any terms quadratic or higher in δp or δT. You should find the following free energy functional:

$$G[\rho, p, T] = (-\delta p + 4\delta T)\delta\rho + (\delta p - \delta T)\delta\rho^2 + \frac{3}{8}\delta\rho^4. \tag{17.6}$$

The free energy found in Eq. (17.6) has a familiar shape: it looks exactly like the Landau functional encountered in Section 5.2, supplemented with a source term linear in $\delta\rho$. When $\delta p < \delta T < 0$, there are indeed three local extrema of the free energy. Note that if $\delta p = 4\delta T$, the free energy is an even function of $\delta\rho$, which simplifies finding a solution.

Exercise 17.6 Along the line $\delta p = 4\delta T$, find the three solutions that minimise the free energy for $T < T_c$.
The trivial solution $\delta\rho = 0$ has a peculiar property: it seems to be a configuration with negative isothermal compressibility κ. This unphysical result is a consequence of the configuration $\delta\rho = 0$ not being a state of stable thermodynamic equilibrium below the critical temperature.
Show that the other two solutions have equal magnitude $\delta\rho$ and equal Gibbs free energy. Also show that one of them has a much higher compressibility than the other. As expected, the low-density solution with $\delta\rho < 0$ corresponds to a compressible gas, whereas the high-density solution can be interpreted as a liquid that is hard to compress.
Show that close to the critical point, the density of the liquid and gas scale as

$$|\delta\rho| \sim |T - T_c|^\beta. \tag{17.7}$$

In this case, the critical exponent is $\beta = 1/2$.

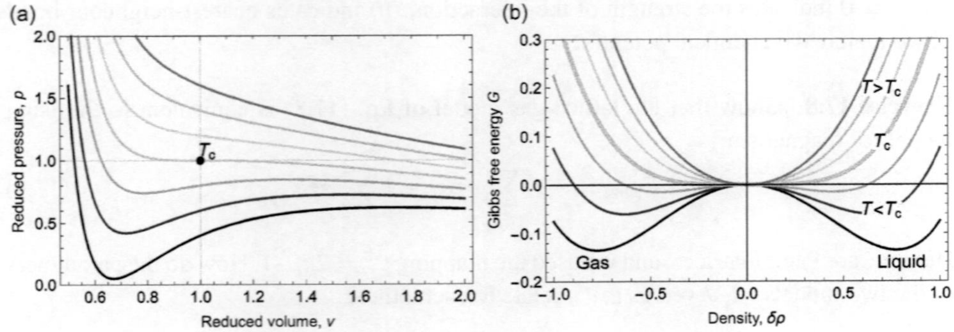

Figure 17.1

(a) The $p - v$ curves given by the Van der Waals equation of state, Eq. (17.2), for a range of temperatures. Above T_c, a fixed pressure p yields only one possible solution for the volume per particle v. Below T_c, however, three possible solutions for v exist. (b) The Gibbs free energy following Eq. (17.6) along the coexistence curve $\delta p = 4\delta T$ shows the phase transition from a unique state to two possible phases. On the coexistence curve, the gas and liquid phases have the same free energy.

As it happens, the mathematically convenient line $\delta p = 4\delta T$ is exactly the place in the phase diagram where the liquid and gas solutions have the same free energy. This line is known as the liquid–gas **coexistence curve**. It is where the phase transition occurs between the gas and the liquid. Moreover, given the fact that the free energy is symmetric under $\delta\rho \to -\delta\rho$, the system has an **emergent $\mathbb{Z}_2$ symmetry** at this line of transitions. Along the coexistence curve, for $T < T_c$, this emergent symmetry between the gas and liquid will be spontaneously broken. Indeed, the free energy along this curve in Figure 17.1 shows a familiar form in the context of spontaneous symmetry breaking.

Exercise 17.7 Draw a phase diagram in the (p, T) plane close to the critical point. Clearly mark the critical endpoint as well as the coexistence curve. Indicate where the liquid and the gas phases are. After working through the next section, compare the results of this exercise with Figure 17.2.

17.2 Ising and the Lattice Gas Model

Having described the liquid–gas transition on a purely thermodynamic level, we will now construct a statistical (approximate) description of interacting molecules. One particular, greatly simplified model is known as the **lattice gas** model. Consider a lattice, which for convenience we take to be the cubic lattice in d dimensions, and say that each lattice site may either contain a particle or not, indicated by $n_i = 0$ (no particle) or $n_i = 1$ (one particle). To mimic the short-range attraction of the Van der Waals potential, we say that having two particles on neighbouring sites lowers the overall energy as compared to them being further apart. The energy of a given particle configuration can then be described by

$$E_{LG} = -J \sum_{\langle ij \rangle} n_i n_j - \mu \sum_i n_i. \tag{17.8}$$

Here, $J > 0$ indicates the strength of the interaction, $\langle ij \rangle$ indicates nearest-neighbour bonds, and we included a chemical potential μ.

Exercise 17.8 Show that the lattice gas model of Eq. (17.8) is equivalent to the Ising model for magnetism:

$$E_{\text{Ising}} = -J' \sum_{\langle ij \rangle} \sigma_i^z \sigma_j^z - h \sum_i \sigma_i^z. \tag{17.9}$$

Here, σ^z are Pauli matrices, and we used the mapping $\sigma_i^z = 2n_i - 1$. How do the parameters of the two models (μ, J versus h, J') relate to each other?

Both the lattice gas model and the Ising model have a symmetry that can be spontaneously broken. We discussed the $\mathbb{Z}_2$ symmetry of the spins, given by $\sigma_i^z \to -\sigma_i^z$, in various places in this book. Note, however, that the magnetic field term $h \sum_i \sigma_i^z$ is not invariant under this transformation. The Ising model can therefore only exhibit spontaneous symmetry breaking when $h = 0$. The same applies to the lattice gas model: only along a particular line in phase space there is a symmetry that can be spontaneously broken.

Exercise 17.9 Show that the lattice gas model has **particle–hole** symmetry, given by $n_i \to 1 - n_i$, for a specific choice of parameters μ, J.

Having established that the lattice gas model has a symmetry, it makes sense to explore the possibility of spontaneous symmetry breaking. In particular, we know the Ising model will spontaneously order at low temperatures. Here, we will use a mean-field theory to extract the free energy and phase diagram of the lattice gas model.

Exercise 17.10 First, introduce the average particle density $n = \frac{1}{N} \sum_i n_i$, with N the number of lattice sites. By splitting the particle number into an average value and fluctuations, $n_i = n + \delta n_i$, rewrite the energy of Eq. (17.8) in terms of n and δn_i.
Assuming the fluctuations are small, ignore terms of second order in δn, and write the energy back in terms of n and n_i. The result is known as the mean-field energy:

$$E_{\text{LG}}^{\text{MF}}[n] = Jn^2 dN - (\mu + 2Jnd) \sum_i n_i. \tag{17.10}$$

Exercise 17.11 The partition sum $Z = \sum_{\{n_i\}} \mathrm{e}^{-E/k_{\text{B}}T}$ is a sum over all possible particle configurations, from which the free energy $F = -k_{\text{B}} T \log Z$ can be calculated. Use the fact that the mean-field energy does not have any inter-site correlations to exactly calculate the mean-field free energy:

$$F_{\text{LG}}/N[n, T] = Jn^2 d - k_{\text{B}} T \log \left[1 + \mathrm{e}^{(\mu + 2Jnd)/k_{\text{B}}T} \right]. \tag{17.11}$$

Exercise 17.12 Show by expanding the free energy of Eq. (17.11) up to fourth order in n, that there is a critical point below which there are two possible solutions for n. What are the critical values of μ, n, and T? Show that $\delta n \sim |T - T_{\text{c}}|^{\beta}$ with $\beta = 1/2$. We can identify the two solutions as a low-density gas and a high-density liquid phase.

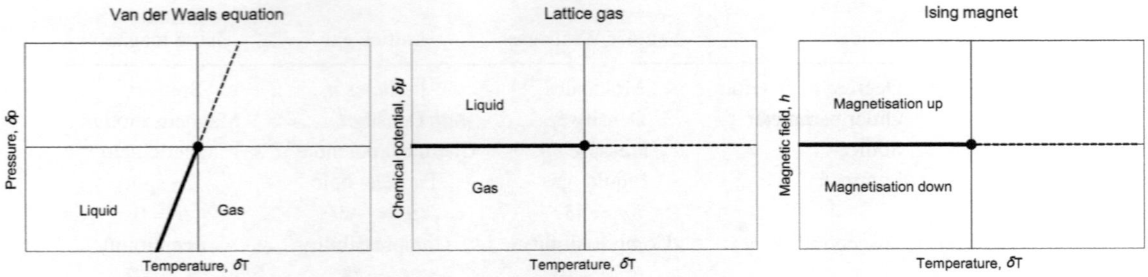

Figure 17.2 The phase diagrams close to the critical point for the Van der Waals equation of state, the lattice gas model and the Ising model. All models are equivalent, and contain a line along which the system is symmetric (dashed line), a coexistence curve signalling a first-order transition (solid line), and a critical endpoint (big dot).

Exercise 17.13 Give a self-consistent expression for n by finding the minimum of the free energy, $\frac{\partial F}{\partial n} = 0$. Show that upon approaching the critical temperature T_c from high T along the symmetry line $\mu = -dJ$, the particle compressibility $\kappa = \frac{\partial n}{\partial \mu}$ diverges as $\kappa \sim (T - T_c)^{-\gamma}$ with $\gamma = 1$.
Hint: Use the same trick as for the liquid–gas transition.

Exercise 17.14 At the critical temperature T_c, we can expand the self-consistent equation in terms of $\delta\mu = \mu - \mu_c$ and $\delta n = n - n_c$. Show that in this case $\delta\mu \sim \delta n^{\delta}$ with $\delta = 3$.

Using mean-field theory, we have seen that the lattice gas model has a critical point, below which it has both a gas and a liquid phase. This is just like in the Van der Waals theory. Because the lattice gas model is also equivalent to the Ising model, however, we can also interpret the two phases as ferromagnetic states with magnetisation pointing up or down.

Exercise 17.15 Use the mapping from the lattice gas to the Ising model to develop the mean-field theory of the Ising ferromagnet with order parameter $m = \frac{1}{N}\sum_i \sigma_i^z$. This means deriving the mean-field energy E^{MF}, the mean-field free energy $F[m]$, and the corresponding self-consistent equation. Instead of the compressibility, calculate the divergence of the magnetic susceptibility $\chi = \frac{\partial m}{\partial h}$.

Exercise 17.16 Draw the phase diagrams close to the critical point of the lattice gas model (with chemical potential vs. temperature) and the Ising model (with field vs. temperature). Compare your results to Figure 17.2.

17.3 From Liquid–Gas Transition to the Ising Model

We have seen that the liquid–gas transition as described by the Van der Waals equation, the lattice gas model, and the Ising model have equivalent free energies, phase diagrams

	Van der Waals gas	Lattice gas	Ising magnet
Degrees of freedom	Molecules	Particles n_i	Spins σ_i^z
Order parameter	Density $\delta\rho$	Density δn	Magnetisation m
Source	Pressure δp	Chemical potential μ	Magnetic field h
Symmetry	Liquid–gas	Particle–hole	$\mathbb{Z}_2$ spin
	$\delta p = 4\delta T$	$\mu = -dJ$	$h = 0$
Susceptibility	Compressibility	Compressibility	Susceptibility
	$\kappa = \frac{1}{\rho}\frac{\partial \rho}{\partial p}$	$\kappa = \frac{\partial n}{\partial \mu}$	$\chi = \frac{\partial m}{\partial h}$

Table 17.1 Liquid–gas to Ising mapping.

(see Figure 17.2), and critical exponents. Even though we typically only associate the Ising model ferromagnetism with spontaneous symmetry breaking, the other two models also have emergent symmetries along a line in their phase diagrams. The relations between the main ingredients in the three models are shown in Table 17.1.

Exercise 17.17 Find the relations that map the Van der Waals liquid–gas model with parameters (ρ, p, T, G) to the mean-field models of the lattice gas model with parameters (n, μ, J, T, F) and the Ising model with parameters (m, h, J, T, F).

Note that all three theories discussed here are mean-field theories. That means any statistical, spatial or temporal fluctuations of the order parameter fields ρ, n, and m have been completely ignored. They can be included by promoting the free energy functionals to a full-fledged field theory, and using renormalisation group theory to calculate the critical exponents. In $d = 3$, this would yield the more accurate exponents $\beta \approx 0.33$, $\gamma \approx 1.24$, and $\delta \approx 4.79$. The mapping between the models remains unchanged when fluctuations are included, so these exponents still apply to the Ising ferromagnet as well as the liquid–gas critical point. The agreement of their critical exponents, a hallmark of universality, has even been confirmed experimentally.

The Tower of States from Exact Diagonalisation 18

Prerequisites Box 18.1

Read Sections 2.4 and 3.6.

Goal of this project Box 18.2

◇ Learn how to write exact diagonalisation code.
◇ Calculate the scaling of the tower of states in systems that break a continuous symmetry.

18.1 Ground State of the Heisenberg Model

The ground state of a finite-size system is typically unique, even when the system will exhibit spontaneous symmetry breaking. In this case, a **tower of states** exists (see Section 3.6), whose excitation gap from the ground state scales as $1/N$, with N the system size. The goal of this project is to explicitly calculate these states without any assumptions whatsoever. The technique we will use is called **exact diagonalisation**. It consists of constructing the Hamiltonian as a matrix on a finite-size system and calculating some of its eigenvalues exactly using a diagonalisation procedure.

Consider the Heisenberg antiferromagnet on the two-dimensional square lattice:

$$\hat{H} = J \sum_{\langle ij \rangle} \hat{\vec{S}}_i \cdot \hat{\vec{S}}_j. \tag{18.1}$$

Here, $\hat{\vec{S}}_i$ are spin-$\frac{1}{2}$ operators acting on site i of a square lattice. The sum runs over nearest neighbours, and the coupling $J > 0$ is positive. As discussed in Section 2.4, this model will spontaneously order antiferromagnetically. It can be proven, however, that any finite-size system will have indeed a unique spin singlet (total $S = 0$) ground state [122].

Exercise 18.1 Show that the two-site Heisenberg model indeed has a spin singlet (total $S = 0$) ground state.

Exercise 18.2 Rederive the proof in Ref. [122], which shows that the ground state of the $S = \frac{1}{2}$ Heisenberg model on a square lattice is a total spin singlet.

A good overview of the properties of the square-lattice Heisenberg model, both obtained via spin wave theory and using numerical methods like exact diagonalisation, can be found in Ref. [120].

18.2 Exact Diagonalisation

On an intuitive level, exact diagonalisation (ED) is the most straightforward numerical tool a quantum physicist can employ. The Hamiltonian of Eq. (18.1), when put on a finite lattice of $N = L_x \times L_y$ sites, is a matrix operator. In order to explicitly calculate its eigenvalues and eigenvectors, we express it in a basis. The most convenient basis is those of product states in terms of the eigenstates of individual $\hat{S}_i^z$ operators. For example, for a $N = 2 \times 2$ lattice, that basis will consist of the 16 states shown in Figure 18.1.

Exercise 18.3 For an $N = L_x \times L_y$ lattice, how many states are there in the Hilbert space?

Exercise 18.4 Using the spin-z product states as a basis set, write a code that constructs the Heisenberg Hamiltonian as a matrix for a system with linear dimensions L_x and L_y, using periodic boundary conditions. Use standard packages to compute the whole spectrum of eigenvalues and eigenvectors.
What is the largest system size you can reasonably calculate on your home computer? To find this out, monitor both the memory usage and duration of your program, as a function of L_x and L_y.

You probably found that both the memory usage and the duration of your code rapidly get out of hand, already for relatively small system sizes. The challenge of exact diagonalisation is thus to use various tricks to speed up the code, enabling us to go to larger system sizes.

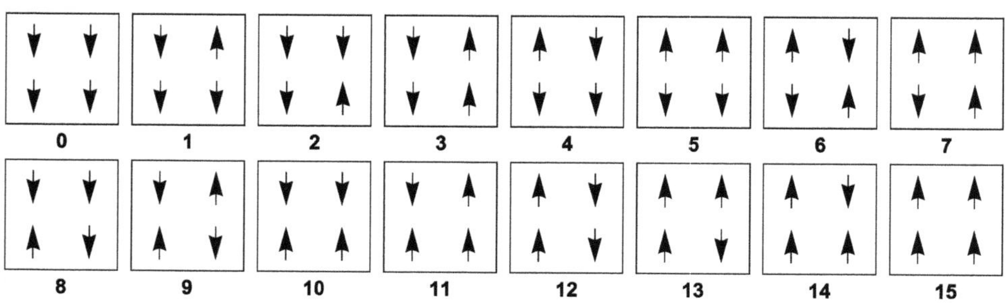

Figure 18.1 The 16 possible spin states on a $N = 2 \times 2$ lattice.

The first trick we will use is that of using 'sparse matrices'. In general, an $N \times N$ matrix has N^2 elements. However, in most matrices we are interested in, including the Hamiltonian, spin, and translation operators, the vast majority of matrix elements are zero. In these cases, it makes sense to only store the $\mathcal{O}(N)$ non-zero matrix elements rather than the whole $\mathcal{O}(N^2)$ matrix.

Exercise 18.5 How many non-zero matrix elements does the Hamiltonian have, as a function of L_x and L_y? How many operations do you need in order to multiply the Hamiltonian matrix with a vector? How much speed-up can you get compared to matrices that have the same size but are not sparse?

Exercise 18.6 Write a code that constructs the Hamiltonian as a sparse matrix.
Hint: The Python package SciPy has special features to deal with sparse matrices.

The second trick is to employ symmetry. We know that the Heisenberg model has global $SU(2)$ symmetry, as well as translation symmetry. Both will be broken in the antiferromagnetic state. However, any eigenstate of a finite system will reflect those symmetries. As a result, we can restrict our calculations to fixed values for the eigenvalue of the total-spin operator $\hat{S}^z$. Similarly, we can focus on fixed eigenvalues of the translation operators $\hat{T}_x$ and $\hat{T}_y$. Note that finding a basis with fixed eigenvalue of the total-spin operator $\hat{S}^2$ is not so trivial, which is why we will use only the $\hat{S}^z$ and translation symmetries.

Exercise 18.7 Write a code that constructs the matrices corresponding to operators for total spin, $\hat{\vec{S}} = \sum_i \hat{\vec{S}}_i$, and total spin squared $\hat{\vec{S}}^2$, given L_x and L_y. Verify that these operators commute with the Hamiltonian matrix. Do the same for the translation operators $\hat{T}_x$ and $\hat{T}_y$, which shift the whole system one site in the x- or y-direction, respectively. Show that the eigenvalues of the translation operators on a finite system have the form $\mathrm{e}^{\mathrm{i}k_{x,y}}$ with momentum $k_{x,y} = \frac{2\pi n}{L_{x,y}}$ and n an integer.

Exercise 18.8 Write a code that constructs the Hamiltonian in a restricted Hilbert space with fixed eigenvalues for $\hat{S}^z$, $\hat{T}_x$, and $\hat{T}_y$. Notice that the resulting matrices are much smaller than tose in the full Hilbert space. For example, a system with $L_x = L_y = 4$ has $2^{16} = 65\,536$ states in its Hilbert space, but the largest restricted Hilbert space with eigenvalue zero for $\hat{S}^z$ and eigenvalues one for $\hat{T}_x$ and $\hat{T}_y$ contains only 812 states.

The third and final trick has to do with finding solutions to the eigenvalue problem. We are after the tower of states, which is found in the low-lying excitation spectrum. This means we do not want to know all of the eigenvalues, but only those closest to the ground state energy. These can be found by the so-called **Lanczos algorithm**, which uses the fact that repeated application of a particular matrix operator onto any state will zoom in on the eigenstates with the largest (absolute) eigenvalues. Every self-respecting programming language or tool will have a Lanczos package available. For example, in Python one can use `scipy.sparse.linalg`. Alternatively, in scientific software such as Mathematica (`eigensystem`) and MATLAB (`eigs`) the Lanczos functionality is built in.

Exercise 18.9 Write a code that finds the states closest to the ground state using the Lanczos algorithm.
Hint: The spectrum of the Heisenberg matrix $\hat{H}$ is approximately symmetric, so the Lanczos algorithm will provide states close to the ground state as well as the most highly excited states. To ensure the algorithm only provides states close to the ground state, shift the Hamiltonian by a constant energy: $\hat{H} \to \hat{H} - \lambda\hat{\mathbb{I}}$.

18.3 Calculating the Tower of States

We are now in a position to actually calculate the tower of states.

Exercise 18.10 For a range of system sizes from $L_x = L_y = 2$ up to at least $L_x = L_y = 4$, calculate the low-energy spectrum, up to a few hundred eigenvalues for each of a few fixed eigenvalue of $\hat{S}^z$ close to zero. For each energy eigenvalue, store the quantum number for total spin S, the polarisation S^z, and the translation eigenvalues characterised by momentum k_x and k_y. You will need to be careful with degenerate eigenvalues, although you can always find unique states once you simultaneously diagonalise the symmetry operators. Visualise the spectrum as in Figure 18.2.

The results you now have should look like Figure 18.2. At first, you may notice something strange: we claimed before that the global spectrum ($k = 0$) contains the tower of states, but

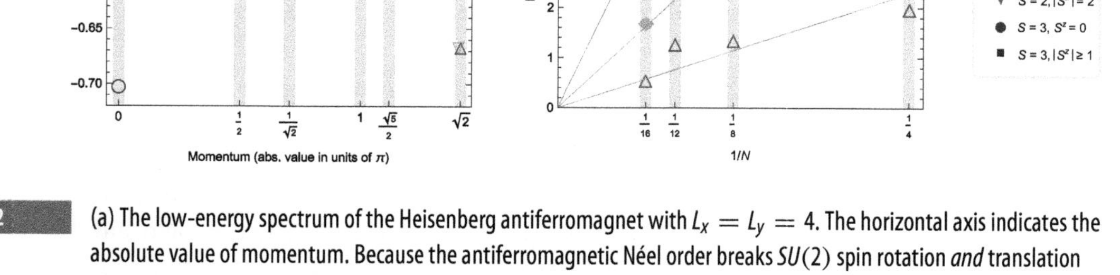

Figure 18.2 (a) The low-energy spectrum of the Heisenberg antiferromagnet with $L_x = L_y = 4$. The horizontal axis indicates the absolute value of momentum. Because the antiferromagnetic Néel order breaks $SU(2)$ spin rotation *and* translation symmetry, the first excited state of the tower of states has $S = 1$ and $T_x = T_y = -1$ (meaning $k_x = k_y = \pi$, which is indicated on the plot as $|k| = \sqrt{2}$). The $S = 1$ states at finite momenta represent the Nambu–Goldstone modes. (b) The excitation gap of the first three states in the tower of states vanishes as $S(S + 1)/N$ (dashed lines).

the first excited $k = 0$ state actually has $S = 2$, and there is an excitation with lower energy and $S = 1$ at momentum $\vec{k} = (\pi, \pi)$. In order to see whether this state actually belongs to the tower of states, we need to see how the excitation energy scales with the system size N.

Exercise 18.11 Calculate how the gap from the ground state energy scales with inverse system size $1/N$ for the first few excited states. Which states correspond to the tower of states? What is the degeneracy of each member of the tower of states (remember that there can be states with different S^z but the same energy)? Plot the energy scaling of the first few excited states, as is done in Figure 18.2.

As it turns out, the states with momentum $\vec{k} = (\pi, \pi)$ actually belong to the tower of states. Specifically, those tower-states with odd-integer spin ($S = 1, 3$, etc.) will have non-zero momentum, whereas all even-integer spin states are at zero momentum. In general, the fact that the tower of states contains states with a non-zero momentum reveals that the resulting symmetry-broken state not only breaks the $SU(2)$ spin symmetry but also translational symmetry. That is, in the degenerate manifold that is formed as N tends to infinity, linear combinations of states with different momentum as well as different total spin quantum numbers can be formed. This way, the antiferromagnetic Néel state can break sublattice as well as spin-rotation symmetry.

In the antiferromagnetic state the unit cell is enlarged, resulting in a smaller Brillouin zone, such that the original non-zero momentum $\vec{k} = (\pi, \pi)$ is folded back onto $\vec{k} = (0, 0)$. There exists a systematic way to determine which non-zero momenta will play a role in the tower of states for which symmetry-breaking pattern, using group representation theory. See Ref. [221] for an introduction. Without going into further detail, the ED results found here are consistent with antiferromagnetic order.

Now that we have an explicit form for the ground state and the tower of states, we can do some further analysis on the states. First, you will show that the symmetric singlet ground state is quite unstable. Second, we can see that a staggered magnetic field $h(-1)^i\hat{S}^z_i$ couples the different parts of the tower of states. And finally, the symmetry-broken state, which is close to the classical Néel state is a superposition of states in the tower of states.

Exercise 18.12 For the singlet ground state, calculate the spin correlation function $C(i,j) = \langle\psi_0|\hat{S}^z_i\hat{S}^z_j|\psi_0\rangle$ and show it does not satisfy cluster decomposition if you make your system large enough.

Exercise 18.13 Show that a staggered magnetic field $h(-1)^i\hat{S}^z_i$ couples the different states within the tower of states.

Exercise 18.14 Calculate the overlap of the perfect Néel state with the different states of the tower of states.

19 Entanglement Entropy of Symmetry Breaking

Box 19.1 Prerequisites

Read Chapter 3 (in particular Section 3.6), Chapter 4, and Section A.2.

Box 19.2 Goal of this project

◇ Learn about bipartite many-body quantum entanglement.
◇ Calculate the size dependence of the entanglement entropy of a system that exhibits spontaneous symmetry breaking.

19.1 Ground State Wave Function

We learned in Chapter 3 that most finite-size systems prone to spontaneous symmetry breaking have a unique, symmetric, ground state, and that the long-range entanglement present in such a symmetric state is an indicator for its instability and tendency towards symmetry breaking. The goal of this project is to analytically compute the associated entanglement entropy, following Ref. [127].

The first step is to find an expression for the ground state wave function on a volume V. From this, we can construct the density matrix, followed by the reduced density matrix ρ_A on a restricted region A with volume $V_A \subset V$. From the reduced density matrix we can calculate the entanglement entropy. This entanglement entropy will depend on the linear size ℓ of region A (such that $V_A \sim \ell^d$).

In this project, we will consider the 'type A' breaking of $O(N)$ symmetry in d dimensions. The order parameter is a vector $\vec{n}(\mathbf{x})$ that takes values on the $(N-1)$-sphere $S^{(N-1)}$ of directions for the $O(N)$ degrees of freedom. This includes, for example, Heisenberg antiferromagnetic order whose sublattice magnetisation has a direction on S^2, and XY magnetic order or superfluids whose angle or phase takes values on S^1 (the circle). We will describe our system with a field theory, so that the components of the ground state wave function are given by a functional $\psi\left[\vec{n}(\mathbf{x})\right]$ of the order parameter field.

As discussed in Section 3.6, the low-energy spectrum of such a system consists of a tower of states and the Nambu–Goldstone modes. The tower of states describes the global properties, and its ground state is a symmetric singlet. This means the global, averaged order parameter $\vec{n}_0 \equiv \frac{1}{V}\int \mathrm{d}^d x\, \vec{n}(\mathbf{x})$ cannot have any particular direction in the ground state configuration. On the other hand, the spatial correlations of the order parameter field in the ground state may be nontrivial and are generated by the Nambu–Goldstone modes.

Exercise 19.1 Write a quadratic Hamiltonian for the low-energy Nambu–Goldstone modes $\hat{b}_{\mathbf{k}\alpha}$ using their expected number and dispersion relation.
Answer: There are $(N-1)$ bosons with a linear dispersion $\omega_{\mathbf{k}} = c|\mathbf{k}|$, hence $\hat{H}_{\mathrm{NG}} = \sum_{\alpha=1}^{N-1}\sum_{\mathbf{k}\neq 0} c|\mathbf{k}|\, \hat{b}^{\dagger}_{\mathbf{k}\alpha}\hat{b}_{\mathbf{k}\alpha}$. Here, c is the mode velocity.

Exercise 19.2 Write out the bosonic field operators $\hat{\phi}_\alpha(\mathbf{x})$ in terms of the creation and annihilation operators $\hat{b}^{\dagger}_{\mathbf{k}\alpha}$ and $\hat{b}_{\mathbf{k}\alpha}$. What is the correlation function $\langle\hat{\phi}_\alpha(\mathbf{x})\hat{\phi}_\beta(\mathbf{y})\rangle$ in the ground state?
Answer: You can find from the canonical quantisation, in standard theory textbooks, that $\hat{\phi}_\alpha(\mathbf{x}) = \frac{1}{\sqrt{V}}\sum_{\mathbf{k}\neq 0}\frac{1}{\sqrt{2\omega_{\mathbf{k}}}}\left(\hat{b}_{\mathbf{k}\alpha} + \hat{b}^{\dagger}_{-\mathbf{k}\alpha}\right)\mathrm{e}^{\mathrm{i}\mathbf{k}\cdot\mathbf{x}}$, see, for example, chapter 2 of Ref. [159]. The correlation function in the ground state is given by $\langle\hat{\phi}_\alpha(\mathbf{x})\hat{\phi}_\beta(\mathbf{y})\rangle = \delta_{\alpha\beta}\frac{1}{V}\sum_{\mathbf{k}\neq 0}\frac{1}{2\omega_{\mathbf{k}}}\mathrm{e}^{\mathrm{i}\mathbf{k}\cdot(\mathbf{x}-\mathbf{y})}$.

Exercise 19.3 Given the components $\psi[\phi]$ of the ground state wave function, the correlation functions can be calculated by

$$\langle\hat{\phi}_\alpha(\mathbf{x})\hat{\phi}_\beta(\mathbf{y})\rangle = \int \mathcal{D}\phi\; (\psi[\phi])^*\, \phi_\alpha(\mathbf{x})\phi_\beta(\mathbf{y})\; \psi[\phi]. \tag{19.1}$$

Here, ϕ_α are eigenvalues of the operator $\hat{\phi}_\alpha$. Consider the Gaussian ansatz:

$$\psi[\phi] \propto \exp\left(-\frac{1}{4}\sum_\alpha \int \mathrm{d}^d x \mathrm{d}^d y\; \phi_\alpha(\mathbf{x}) Q(\mathbf{x},\mathbf{y})\phi_\alpha(\mathbf{y})\right). \tag{19.2}$$

Find the function $Q(\mathbf{x},\mathbf{y})$ that correctly reproduces the ground state wave function correlations identified in the previous exercise. You do not need to worry about normalisation of the wave function.
Answer: In order to match the previous correlation function, we need to have $Q(\mathbf{x},\mathbf{y}) = \sum_{\mathbf{k}\neq 0} 2\omega_{\mathbf{k}}\mathrm{e}^{\mathrm{i}\mathbf{k}\cdot(\mathbf{x}-\mathbf{y})}$.

Now that we found a wave function in terms of the Nambu–Goldstone modes, the final step is to symmetrise the wave function. We can relate the overall order parameter field $\vec{n}(\mathbf{x})$ to the Nambu–Goldstone fields $\phi_\alpha(\mathbf{x})$ via

$$\vec{n}(\mathbf{x}) = \vec{n}_0\sqrt{1 - \frac{\sum_\alpha \phi_\alpha^2(\mathbf{x})}{\rho}} + \sum_\alpha \frac{\vec{e}_\alpha \phi_\alpha(\mathbf{x})}{\sqrt{\rho}}. \tag{19.3}$$

The vectors $\vec{e}_\alpha$ with $\alpha = 1, \ldots, N-1$ are unit vectors orthogonal to the average $\vec{n}_0$, while ρ is the so-called order parameter stiffness.

Exercise 19.4 Use the relation of Eq. (19.3) to express the components of the ground state wave function in terms of the order parameter field $\vec{n}(\mathbf{x})$, up to second order in the field. There is no need to normalise the wave function.
Answer: $\psi[\vec{n}] \propto \exp\left(-\frac{\rho}{4}\sum_\alpha \int \mathrm{d}^d x \mathrm{d}^d y\, Q(\mathbf{x},\mathbf{y})\vec{n}(\mathbf{x})\cdot\vec{n}(\mathbf{y})\right)$.

Exercise 19.5 Verify explicitly that the wave function components $\psi[\vec{n}]$ do not depend on the average $\vec{n}_0$.

19.2 Reduced Density Matrix

Just as the components $\psi[\vec{n}]$ of the ground state wave function are given by a functional of the order parameter field $\vec{n}(\mathbf{x})$, the components of the density matrix can be expressed as a functional of two-order parameter fields:

$$\rho[\vec{n},\vec{n}'] = \psi[\vec{n}]\,\psi[\vec{n}']^*. \tag{19.4}$$

To obtain the **reduced density matrix**, we split our system into two subregions A and B, with volumes V_A and V_B, respectively. We thus also split the order parameter field in two, $\vec{n}(\mathbf{x}) = \vec{n}_A(\mathbf{x}) + \vec{n}_B(\mathbf{x})$, with $\vec{n}_A$ being non-zero only in region A, and $\vec{n}_B$ only in region B. The reduced density matrix is given by the partial trace of the full density matrix over region B, which can be done by writing

$$\rho_A[\vec{n}_A,\vec{n}'_A] = \int \mathcal{D}\vec{n}_B\; \rho[\vec{n}_A+\vec{n}_B,\vec{n}'_A+\vec{n}_B]. \tag{19.5}$$

To perform this integration, we replace $\vec{n}$ with $\vec{n}_0$ and the Nambu–Goldstone fields ϕ_α following Eq. (19.3).

Exercise 19.6 Calculate $\rho_A[\vec{n}_A,\vec{n}'_A]$ by first rewriting the wave function components in terms of $\vec{n}_0$ and ϕ_α. Then perform the integration over ϕ_α in region B. Finally, write the result back in terms of $\vec{n}_A$ and $\vec{n}'_A$.
Hint: The reduced density matrix can be written in the form

$$\rho_A[\vec{n}_A,\vec{n}'_A] \propto \exp\left(-(\vec{n}_A-\vec{n}'_A)\tilde{Q}(\vec{n}_A-\vec{n}'_A) - (\vec{n}_A+\vec{n}'_A)\tilde{R}(\vec{n}_A+\vec{n}'_A)\right). \tag{19.6}$$

Here, $\vec{n}\tilde{X}\vec{m}$ is a short-hand notation for $\int \mathrm{d}^d x \mathrm{d}^d y\, \vec{n}(\mathbf{x})\tilde{X}(\mathbf{x},\mathbf{y})\vec{m}(\mathbf{y})$. What are $\tilde{Q}$ and $\tilde{R}$?

Answer: Consider $Q(\mathbf{x},\mathbf{y})$ from the previous exercises and define Q_{AB} to be restricted so that $\mathbf{x}$ is in A and $\mathbf{y}$ in B. Then, $\tilde{Q} = Q_{AA}$ and $\tilde{R} = Q_{AB}Q_{BB}^{-1}Q_{BA}$. Use this to complete the remainder of the exercise.

Interestingly, even though the wave function components of the whole system did not depend on the average order parameter, the components of the reduced density matrix do depend on the average order parameter in subregion A.

Exercise 19.7 Show that the reduced density matrix for a constant order parameter field $\vec{n}_A(\mathbf{x}) = \vec{N}_A$ equals

$$\rho_A(\vec{N}_A, \vec{N}'_A) \propto \exp\left(-\frac{\chi}{2}(\vec{N}_A - \vec{N}'_A)^2\right). \tag{19.7}$$

Express χ in terms of an integral over the region A.
Also calculate the value of this integral for a region A that is a d-dimensional sphere with radius ℓ, to leading order in ℓ, using a short-distance (UV) cut-off a.
Answer: The integral is $\chi = \frac{\rho}{4}\int_A \mathrm{d}^d x \mathrm{d}^d y\, Q(\mathbf{x}, \mathbf{y}) = \frac{\rho}{2\pi}\ell^{d-1}\log \ell/a$.

As it turns out, the interesting leading order (in ℓ) contribution to the entanglement entropy beyond the usual area law comes from the part of the reduced density matrix that only depends on the average (constant) order parameter $\vec{N}_A$. From now on, we therefore ignore the rest, although it can give rise to interesting subtle subleading contributions to the entanglement entropy.

19.3 Rényi Entropy

It is notoriously difficult to directly compute the Von Neumann entanglement entropy given by

$$S_{\mathrm{VN}} = -\mathrm{Tr}\left[\rho_A \log \rho_A\right]. \tag{19.8}$$

The problem is the calculation of the logarithm of the density matrix. A common trick is to calculate the **Renyi entropy** instead, which is defined by

$$S_n = -\frac{1}{n-1}\log \mathrm{Tr}\left[\rho_A^n\right]. \tag{19.9}$$

This only requires calculating the nth power of the reduced density matrix.

Exercise 19.8 Show that the Von Neumann entropy equals $S_{\mathrm{VN}} = \lim_{n\to 1} S_n$.

The powers of the reduced density matrix are obtained through integration. For example, ρ_A^2 is given by

$$\rho_A^2(\vec{N}_A, \vec{N}''_A) = \int \mathcal{D}N'_A\, \rho_A(\vec{N}_A, \vec{N}'_A)\rho_A(\vec{N}'_A, \vec{N}''_A). \tag{19.10}$$

Exercise 19.9 Calculate ρ_A^2. You can ignore overall multiplicative factors but you should keep track of how ρ_A^2 depends on $\vec{N}_A, \vec{N}''_A$ and the parameter χ.
Hint: The integral over $\vec{N}'_A$ is approximately Gaussian. Be careful with how many components you need to integrate over. The ρ^2 you'll find looks very similar to Eq. (19.7) with a χ-dependent prefactor.
Answer: The integral is only non-zero when $\vec{N}_A$, $\vec{N}'_A$, and $\vec{N}''_A$ are all close to each other. Without loss of generality, consider them all close to the north pole. Then there are

$(N-1)$ orthogonal directions, meaning the integral over $\vec{N}'_A$ is just a $(N-1)$-dimensional Gaussian integral. As a result, completing the square and performing the integration yields $\rho_A^2(\vec{N}_A, \vec{N}''_A) \propto \chi^{-(N-1)/2} \exp\left(-\frac{\chi}{4}(\vec{N}_A - \vec{N}''_A)^2\right)^2$.

Exercise 19.10 Use induction to calculate $\rho_A^n(\vec{N}_A, \vec{N}''_A)$.
Answer: Repeated integration yields

$$\rho_A^n(\vec{N}_A, \vec{N}''_A) \propto \chi^{-(n-1)(N-1)/2} \exp\left(-\frac{\chi}{2n}(\vec{N}_A - \vec{N}''_A)^2\right)^2. \tag{19.11}$$

We are now in a position to calculate the entanglement entropy. The final step consists of taking the trace of ρ_A^n to get the Renyi entropy.

Exercise 19.11 Calculate the Renyi entropy of Eq. (19.9) to leading order in χ. You can safely ignore all constant contributions.
Answer: The trace means setting $\vec{N}_A = \vec{N}''_A$. As a result, $\mathrm{Tr}\rho_A^n = \chi^{-(n-1)(N-1)/2}$, and thus the Renyi entropy is $S_n = \frac{N-1}{2} \log \chi$.

The Renyi entropy is, to leading order in χ, independent of n. Taking the limit $n \to 1$ to get the Von Neumann entanglement entropy is thus trivial.

Exercise 19.12 Write down the Von Neumann entanglement entropy to leading order in the subsystem size ℓ.
Answer: $S_{\mathrm{VN}} = \frac{N-1}{2} \log \ell^{d-1}$.

You have found that the entanglement entropy has a logarithmic term whose prefactor depends on the number of Nambu–Goldstone modes. This result may come as a surprise, given that the ground state wave function we started with was symmetric, and Nambu–Goldstone modes are typically thought of as excitations of the symmetry-broken configuration. In fact, the result can be turned upside down. If one can somehow calculate the ground state wave function of a finite-size system, the appearance of a logarithmic term in the bipartite entanglement entropy acts as an indication that spontaneous symmetry breaking giving rise to Nambu–Goldstone modes will occur. This was done in Refs. [86, 100, 116], where Quantum Monte Carlo calculations where used for the XY model ($N = 2$) and the Heisenberg antiferromagnet ($N = 3$).

Note that even though we derived this result for a specific model, it is more general and applies also to more complicated coset spaces, such as $SO(3)$ [166]. Further details of the entanglement entropy, such as subleading terms, will reveal even more information about the symmetry-breaking pattern.

20 Dynamical Symmetry Breaking

Prerequisites — Box 20.1

Read Chapter 2. For the second half of the project, familiarity with methods of quantum field theory is assumed. Chapters 10 and 11 provide useful background knowledge for this part.

Goal of this project — Box 20.2

◇ Understand the concept of dynamical symmetry breaking arising from interactions.
◇ Be able to analyse instances of dynamical symmetry breaking in both condensed matter and high-energy settings.

In Chapter 2, the focus was on isolated systems, such as crystals, antiferromagnets, and Bose–Einstein condensates, in which the state of spontaneously broken symmetry minimised the potential energy. In many cases, however, an interaction between several fields or even self-interactions of a single field rather than a potential energy lead to symmetry-breaking states, and these interactions typically obey the symmetry. How symmetric interactions cause symmetry-broken states to arise becomes clear after integrating out one of the fields, much like the appearance of the Cooper pair bound state after integrating out the phonon field in Section 12.2.

This type of mutual spontaneous symmetry breaking among interacting fields is known as **dynamical symmetry breaking**. It occurs throughout both low- and high-energy physics, for example in the emergence of electron mass discussed in Section 11.3, or the pair creation in BCS superconductors and other fermionic condensates. In this project, we consider two simplified models to highlight the mechanism of dynamical symmetry breaking and how it relates to spontaneous symmetry breaking triggered by an external influence.

20.1 Charge Density Wave

A charge density wave is a state of matter in which the electron density in an otherwise homogeneous material is modulated with a wavelength λ larger than the lattice spacing a of the underlying atomic lattice. As an elementary model for charge density wave formation, consider a one-dimensional chain of atoms hosting spinless electrons, interacting through a repulsive nearest-neighbour (Coulomb) interaction:

$$\hat{H} = \sum_j \left\{ -t(\hat{c}_j^\dagger \hat{c}_{j+1} + \hat{c}_{j+1}^\dagger \hat{c}_j) + V \hat{c}_j^\dagger \hat{c}_j \hat{c}_{j+1}^\dagger \hat{c}_{j+1} \right\}. \tag{20.1}$$

Here, $\hat{c}_j^\dagger$ creates an electron at position $x = ja$. The constants t and V are both positive and we focus on the half-filled chain, with N atomic sites and $N/2$ electrons.

Exercise 20.1

- What is the ground state of $\hat{H}$ in the absence of interactions (i.e., $V = 0$)?
- Which symmetry of $\hat{H}$ is broken by a state in which the local electron density $\langle \hat{c}_j^\dagger \hat{c}_j \rangle$ is larger on sites with even values of j than on odd sites?

Anticipating the breakdown of translational symmetry, we may write the Hamiltonian in terms of the operators:

$$\begin{aligned} \hat{a}_m^\dagger &\equiv \hat{c}_{2m-1}^\dagger, \\ \hat{b}_m^\dagger &\equiv \hat{c}_{2m}^\dagger. \end{aligned} \tag{20.2}$$

Here, we used $m \in \{1, 2, \ldots, N/2\}$, while the index j of the original $\hat{c}_j^\dagger$ operators is an integer between 1 and N.

Exercise 20.2 Show that the Hamiltonian written in terms of $\hat{a}_m^\dagger$ and $\hat{b}_m^\dagger$ operators is invariant under the exchange of a and b particles. That is, it has a symmetry $\hat{U}$ defined such that $\hat{U} \hat{a}_m^\dagger \hat{U}^\dagger = \hat{b}_m^\dagger$.

Notice that the interaction term in the Hamiltonian does not break the symmetry of exchanging a and b particles, and that we will not introduce any potential energy terms, infinitesimal or otherwise, that do break the symmetry. Nevertheless, the symmetry will be spontaneously broken in the thermodynamic limit, due to the interactions. We will consider the ordering at the mean-field level and neglect the effects of quantum corrections and thermal fluctuations discussed in Section 6.2, which would prevent true long-range order from arising in a strictly one-dimensional situation.

Exercise 20.3

- Introduce the mean field Ansatz $\langle \hat{a}_m^\dagger \hat{a}_m \rangle = n_0 - n$ and $\langle \hat{b}_m^\dagger \hat{b}_m \rangle = n_0 + n$, and show that the mean field Hamiltonian may be written as

$$\begin{aligned} \hat{H}_{\text{MF}} = \sum_k \Big\{ &(2Vn_0 + 2Vn) \hat{a}_k^\dagger \hat{a}_k + (2Vn_0 - 2Vn) \hat{b}_k^\dagger \hat{b}_k \\ &- t\left(1 + e^{\mathrm{i}2ka}\right) \hat{a}_k^\dagger \hat{b}_k - t\left(1 + e^{-\mathrm{i}2ka}\right) \hat{b}_k^\dagger \hat{a}_k - 2V(n_0^2 - n^2) \Big\}. \end{aligned} \tag{20.3}$$

- Write the mean field Hamiltonian in matrix form, and show that the mean field dispersions are given by

$$E_{\text{MF}}^\pm(k) = -2V(n_0^2 - n^2) + 2Vn_0 - \mu \pm 2\sqrt{t^2 \cos^2(ka) + V^2 n^2}. \tag{20.4}$$

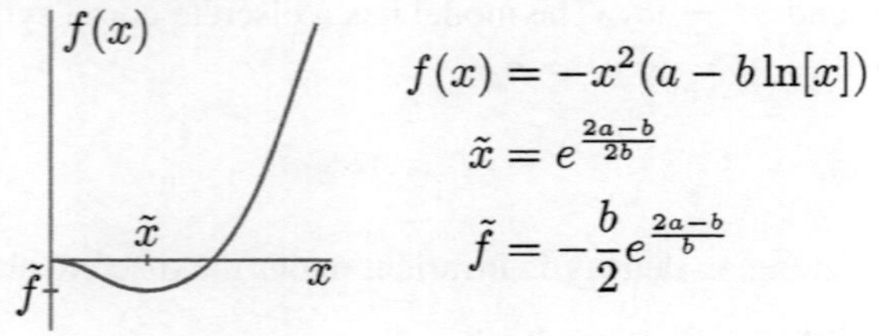

Figure 20.1

A sketch of the function $f(x) = -x^2(a - b\ln(x))$.

Notice that a non-zero value for the order parameter n indicates that the electron density on even sites differs from that on odd sites. That is, a state with non-zero n breaks the exchange symmetry of a and b particles, or equivalently, lowers the discrete translational symmetry of the Hamiltonian as written in Eq. (20.1). The resulting electronic configuration is known as a **charge density wave** with wavelength $\lambda = 2a$, doubling the periodicity of the underlying atomic lattice.

Exercise 20.4 Calculate the total ground state energy for the half-filled chain in the continuum limit and show that the mean field value of n is non-zero (indicating that the mean-field ground state is indeed a charge density wave).
Hint 1: You may want to use the following identity for the elliptic integral:

$$\int_{-\pi/2}^{\pi/2} \mathrm{d}x\sqrt{A^2\cos^2(x) + B^2} \approx 2A + \frac{B^2}{4A}\left(2 - \ln\left[\frac{|B|}{4A}\right]\right). \tag{20.5}$$

Hint 2: See Figure 20.1 for a sketch of the function $f(x) = -x^2(a - b\ln(x))$.

The self-consistent mean-field value for the charge density modulation behaves as $|n| \sim (t/V)\mathrm{e}^{-t/V}$. We thus find that n will be non-zero for any non-zero value of V. That the critical value for obtaining an ordered ground state lies at vanishing interaction strength ($V = 0$) is a peculiarity of the chain being defined in one spatial dimension, but the fact that interaction can lead to the formation of charge density wave order without perturbations in the electronic potential is general.

20.2 The Gross–Neveu Model

The Gross–Neveu model describes (1, 1)-dimensional fermionic fields with local density–density interactions. For N interacting species of fermions, it is given by the Lagrangian density:

$$\mathcal{L} = \sum_{i=1}^{N} \bar{\psi}_i \mathrm{i}\partial\!\!\!/ \psi_i + \sum_{i\neq j} g_0^2 \bar{\psi}_i\psi_i\bar{\psi}_j\psi_j. \tag{20.6}$$

Here, the fields $\psi_i(x)$ are two-component spinors, occurring in N different flavours. The dashed partial derivative indicates $\partial\!\!\!/ = \gamma^\mu\partial_\mu$ as usual, while the barred fields are defined $\bar{\psi}_i = \psi_i^\dagger\gamma_0$, and the two-dimensional gamma matrices can be written in terms of Dirac matrices as $\gamma^0 = \sigma_y$ and $\gamma^1 = \mathrm{i}\sigma_x$. The model has a discrete chiral symmetry generated by the γ^5 matrix, defined as $\gamma^5 = \gamma^0\gamma^1 = \sigma_z$.

Exercise 20.5

- Show that the Lagrangian density is invariant under the discrete chiral transformation $\psi_i \to \psi_i' = \gamma^5\psi_i$.
- Show that theories with this discrete chiral symmetry cannot contain a mass term of the form $m\bar{\psi}_i\psi_i$. This implies that as long as the chiral symmetry remains unbroken, the fields ψ_i are necessarily massless.

To see the effect of the interaction between fields of different flavours, we can use a Hubbard–Stratonovich transformation by introducing a real, scalar auxiliary field ρ to decouple the interaction term. Classical solutions for ρ, solving the saddle point equation for the action, then correspond to mean-field solutions for the Gross–Neveu model with ρ the mean field order parameter.

Exercise 20.6 The partition function can be written in terms of the Lagrangian density in the usual way:

$$Z = \int \mathcal{D}[\bar{\psi}_i, \psi_i]\, \mathrm{e}^{\mathrm{i}\int \mathrm{d}^2x\, \mathcal{L}}. \tag{20.7}$$

- Introduce a real scalar field ρ and use the identity $\int \mathcal{D}[\rho]\mathrm{e}^{\int \mathrm{d}^2x\,(\rho+A)B(\rho+A)} = 1$ to write the partition function for the Gross–Neveu model in the form

$$Z = \int \mathcal{D}[\bar{\psi}_i, \psi_i, \rho]\, \mathrm{e}^{\mathrm{i}\int \mathrm{d}^2x\, \sum_i \bar{\psi}_i \mathrm{i}\partial\!\!\!/ \psi_i - \rho\bar{\psi}_i\psi_i - \frac{1}{2g_0^2}\rho^2}. \tag{20.8}$$

Notice that by writing the decoupling in terms of a single field ρ, we anticipate finding a mean-field value for $\rho \sim \langle\bar{\psi}_i\psi_i\rangle$ that is independent of fermion flavour. This may be natural because all flavours couple with the same strength g_0, but it is not the most general solution.

- Using the fact that $\mathcal{L}$ is invariant under the discrete chiral transformation at every space-time point, argue what the effect of the transformation on ρ should be. Confirm that your result is consistent with the interpretation of $\rho \sim \langle\bar{\psi}_i\psi_i\rangle$ being proportional to the local fermion density.

The partition function can be written as a theory of only the auxiliary field ρ, by integrating out the fermions, which is made possible by the action now being quadratic in terms of fermion fields. The result will be partition function of the form $Z = \int \mathcal{D}[\rho]\, \mathrm{e}^{\mathrm{i}V[\rho]}$, where V is the effective potential for the auxiliary field. The equilibrium configuration of ρ will minimise this effective potential.

Exercise 20.7 Integrating out the fermions to obtain an expression for the effective potential involves several steps.

- First, isolate the integrals over fermionic fields, by writing

$$Z = \int \mathcal{D}[\rho]\, B[\rho] \left(\int \mathcal{D}[\bar{\psi}_i, \psi_i]\, A[\bar{\psi}_i, \psi_i, \rho] \right)^N . \tag{20.9}$$

Use a Fourier transform to show that A can be expressed in terms of momentum-space fields $\psi_i(p)$ as

$$A[\bar{\psi}_i, \psi_i, \rho] = e^{i \int d^2p\, \sum_i \bar{\psi}_i \not{p} \psi_i - \rho \bar{\psi}_i \psi_i} . \tag{20.10}$$

You may assume that the field ρ is constant in both space and time.

- Employing a standard Gaussian integral and an identity for the determinant, show that $\int \mathcal{D}[\bar{\psi}_i, \psi_i] A = e^{\mathrm{Tr}[\ln(\not{p}-\rho)]}$. Here, the trace indicates an integration over 2-momentum as well as a sum over Dirac or spin indices.
- Using the same identity for the determinant again, show that $\int \mathcal{D}[\bar{\psi}_i, \psi_i] A = e^{\int d^2p/(2\pi)^2 \ln(\det[\not{p}-\rho])}$. Notice that in this expression, the determinant is taken only of a 2×2 matrix in spin indices. It can therefore be computed explicitly by writing out the matrix components.
- Noting the implicit multiplication of ρ by an identity matrix in spin indices, show that $\int \mathcal{D}[\bar{\psi}_i, \psi_i] A = e^{\int d^2p/(2\pi)^2 \ln(\rho^2 - p^2)}$, where $p^2 = p^\mu p_\mu$.

The remaining integral in the final expression obtained so far is divergent. It can be evaluated using dimensional regularisation by writing it as the limit $\epsilon \to 0$ of the expression $\int d^{2-\epsilon}p \ln(\rho^2 - p^2)$. This is a standard integral that can be expressed in terms of the Γ-function, which in turn has an asymptotic expansion in ϵ (as shown in many textbooks on quantum field theory, such as Refs. [159, 215]) that allows us to finally write

$$\int \mathcal{D}[\bar{\psi}_i, \psi_i] A = e^{i \frac{\rho^2}{4\pi}[1 - \ln(\rho^2/M^2)]} \lim_{\epsilon \to 0} e^{i \frac{\rho^2}{4\pi} \frac{1}{\epsilon}} . \tag{20.11}$$

Here, we defined $M^2 = e^\gamma/(4\pi)$ with γ the Euler–Mascheroni constant. The final diverging exponential factor can be argued to be cancelled exactly by an opposing divergent factor in $B = e^{-i \int d^2p\, \rho^2/(2g_0^2)}$. That is, if the interaction strength g_0 also depends on the dimensional cut-off ϵ, the divergences in the factors A and B may cancel, leaving only the non-divergent, renormalised expression:

$$Z = \int \mathcal{D}[\rho]\, e^{i\left\{ -\frac{\rho^2}{2g^2} + N \frac{\rho^2}{4\pi}[1 - \ln(\rho^2/M^2)] \right\}} . \tag{20.12}$$

Here, g is the renormalised, physical value of the interaction after cancelling the divergent terms.

Exercise 20.8

- Show that the effective potential is minimised for $\rho = \pm M e^{-\pi/(Ng^2)}$. These are its possible classical, mean field values.
- Explain why the non-zero value of ρ implies that the discrete chiral symmetry is broken in the classical equilibrium configuration and use the form of the Lagrangian density in Eq. (20.8) to explain that interactions rather than an external perturbation render the fields ψ_i massive in the symmetry-broken state.

Going beyond the mean-field description, it can be shown that all quantum corrections are at most of order $1/N$, so that they disappear in the limit of infinite flavours. Similar to the thermodynamic limit, in the limit of an infinite number of degrees of freedom, in this case flavours, the theory reduces to its classical, mean-field result. In this limit, the symmetry breaking occurs for any non-zero value of the interaction strength.

Finally, notice that the Gross–Neveu model discussed here can be straightforwardly related to the charge density wave model we studied before, by considering it with a fixed, non-zero density of fermions. In that case, the interaction causes a local attraction or repulsion (depending on the sign of g_0) between fermions of different flavours. Allowing for multiple mean fields ρ_i as well as spatial dependence of the mean fields, different situations can then arise. If the density of each flavour is fixed independently, the local interactions may cause the formation of density waves, such that concentrations of one flavour either avoid or coincide with concentrations of others. Alternatively, if only the overall density is fixed, it may cause flavour symmetry breaking, in which the occupation of one type renders all others massive and unoccupied.

20.3 Discussion

Having seen two concrete examples of dynamical symmetry breaking, some comments are in order.

First, in both the charge density wave and Gross–Neveu models, the energy gained by realising one of the symmetry-broken states, rather than the symmetric configuration with vanishing mean field, scales extensively with system size. This implies that in the thermodynamic limit even an infinitesimal dip in the effective potential will suffice to break the symmetry, and the symmetry breaking may be considered 'spontaneous'. In fact, one may construct the equilibrium mean-field configuration proceeding from a situation with the value of the mean-field order parameter being infinitesimally small, causing a small but finite energetic advantage for a state with a slightly larger value of the order parameter, which in turn favours further enhancement of the order parameter, and so on. It can be shown that such a calculation converges to a self-consistent value of the order parameter that agrees with a mean-field equilibrium value defined by extremising the effective potential.

Imagining the self-consistent process to occur in time, gradually building up the ordered state starting from an instantaneous infinitesimal perturbation, shows why this type of spontaneous symmetry breaking has been dubbed 'dynamical'. Notice, however, that this is misleading. The mean-field solution is for an equilibrium configuration, and does not indicate that it can be reached by a self-enforcing dynamical cycle. In fact, it is straightforwardly checked that such a self-enforcing cycle is never generated by the time evolution of the charge density wave Hamiltonian or Gross–Neveu Lagrangian, simply because their kinetic terms do not favour symmetry-breaking states, and the interaction terms conserve the value of the order parameter. Just like other forms of spontaneous symmetry breaking discussed in Chapter 2, dynamical symmetry breaking refers to symmetry-broken configurations being stable equilibrium states in the thermodynamic limit. It does not refer to how one obtains these symmetry-broken states dynamically.

Second, as discussed at length in Chapters 1, 2, and 8, symmetry breaking always refers to a relative observable, contrasting the two systems that conspire to break each other's symmetries. In the examples of dynamical symmetry breaking above, this can be seen from the fact that the mean-field order parameter can be either positive or negative in both the charge density wave and the Gross–Neveu model. Without an external reference, it is impossible to tell whether the order parameter is positive or negative. To do so would require a convention of what sites to call even or odd in the charge density wave, or a convention of what to call particles and what holes in the Gross–Neveu model. Even without an external reference making such conventions possible, however, we can say with confidence that in the charge density wave the density on any site is different from that on neighbouring sites, and likewise in the Gross–Neveu model the occupation of any particular flavour of particles has the same sign as that of all other flavours. In this sense, the relative occupation between flavours, or between even and odd sites, is broken even in the isolated system.

Finally, notice that actually, all forms of spontaneous symmetry breaking may be argued to be dynamical symmetry breaking. The distinction between the minima of the potential being either immediately apparent in the form of the Lagrangian that one chooses to describe an isolated system with, or arising only in the effective Lagrangian after considering interactions between fields, is subjective. Any potential energy may always be considered to be an effective description of interactions between fields. Regardless of the description one chooses, spontaneous symmetry breaking encompasses a symmetric action giving rise to stable low-energy states that do not obey the symmetry.

21 Scattering Amplitudes

Box 21.1 Prerequisites

This project builds on the content of Chapter 10. Familiarity with methods of quantum field theory is assumed.

Box 21.2 Goal of this project

◇ Understand the origin and implications of Adler zeroes in the scattering amplitudes of Nambu–Goldstone modes.
◇ Be able to formulate a convenient parametrisation of Nambu–Goldstone fields, and calculate their scattering amplitudes.

In Chapter 10, the coset-space construction was used to argue that Nambu–Goldstone fields generically appear in the effective field theory accompanied by derivatives, which suppress their coupling constants at low energy. The resulting vanishing of scattering amplitudes for processes involving Nambu–Goldstone modes in the zero-momentum limit is known as an Adler zero. In this project, we will explore the appearance of Adler zeroes in the prototypical linear sigma model, which appears in different settings in both high-energy and condensed matter physics. You will see that identifying Adler zeroes is straightforward in an opportune parametrisation of the Nambu–Goldstone fields, but requires a non-obvious exact cancellation of diagrams in more naive parametrisations.

21.1 NG Modes in the Linear Sigma Model

Consider the so-called linear sigma model described by the Lagrangian density:

$$\mathcal{L} = \sum_i \partial_\mu \phi_i \partial^\mu \phi_i + M^2 \sum_i \phi_i^2 - \frac{\lambda}{2}\left(\sum_i \phi_i^2\right)^2. \tag{21.1}$$

Here, $i \in \{1, 2, \ldots, N\}$ is an index labelling N real-valued scalar fields ϕ_i. The coupling constants M and λ are both real and positive. Spacetime indices μ are implicitly summed over, but we keep the sums over the index i explicit for later convenience.

Exercise 21.1

- Thinking of ϕ_i as components of an N-dimensional vector, explain why any field configuration can be written as $\phi_i = \rho \sum_j R_{ij} n_j$, with ρ a real and positive constant, R_{ij} an orthogonal matrix, and n_j the components of an (arbitrary) unit vector.
- Using this parametrisation, explain why the group of global symmetry transformations under which $\mathcal{L}$ is invariant is $O(N)$, the group of orthogonal matrices describing the symmetries of an N-dimensional sphere.
- What requirement on the amplitude ρ will ensure the potential is minimised?

The ground state of the linear sigma model will spontaneously break the $O(N)$ symmetry and pick a direction for the vector ϕ. The residual symmetry will be $O(N-1)$, the group of orthogonal matrices describing rotations in N dimensions keeping a single axis fixed.

Exercise 21.2

- Explain why the residual symmetry group should be $O(N-1)$, and what the residual symmetry transformations do to the ground state vector.
- Describe what the broken symmetry transformations do to the ground state vector, and why these leave the Lagrangian invariant.
- Given that $O(N)$ has $N(N-1)/2$ generators, and that the Lagrangian is invariant under Lorentz transformations, how many Goldstone modes do you expect in this system?

Next, consider local variations of the amplitude and orientation of the ground state vector: $\phi_i' = (1 + A(x)) \sum_{jk} R^{\text{broken}}_{ij}[\theta_a(x)] R^{\text{residual}}_{jk}[\eta_r(x)] \phi_k$, with ϕ_k the components of a symmetry-broken ground state. Here, we separated the $O(N)$ rotation matrices into rotations $R^{\text{residual}} \in O(N-1)$ in the residual symmetry group, and rotations R^{broken} in the set of broken symmetries. The sets of spatially dependent rotation angles θ_a and η_r are defined by writing the rotation matrices in terms of their generators: $R^{\text{residual}}[\eta_r(x)] = \mathrm{e}^{\mathrm{i}\eta_r(x)L^r}$ and $R^{\text{broken}}[\theta_a(x)] = \mathrm{e}^{\mathrm{i}\theta_a(x)L^a}$, with L^r and L^a the unbroken and broken generators of $O(N)$. As shown by Eq. (4.4) in the proof of Goldstone's theorem, local versions of the broken symmetry transformation excite Nambu–Goldstone modes. We can therefore say that acting with $\mathrm{e}^{\mathrm{i}\theta_a(x)L^a}$ on the symmetry-broken ground state creates the Nambu–Goldstone field $\theta_a(x)$. The cost of exciting these modes, and also the massive excitations parametrised by $A(x)$ and $\eta_r(x)$, is described by comparing the action with that of the symmetry-broken ground state:

$$S[\theta_a(x), \eta_r, A] = S[\phi'] - S[\phi]. \tag{21.2}$$

Exercise 21.3

- Explain why the potential in Eq. (21.1) is not affected by the creation of (local) Nambu–Goldstone modes, and why in general, the potential in a system undergoing spontaneous symmetry breaking is unaffected by local versions of the broken symmetry.
- Show that the only term in the action $S[\theta_a(x), A]$ containing the Nambu–Goldstone fields is proportional to $\partial_\mu \theta^a \partial^\mu \theta_a$.

The fact that the action for the Nambu–Goldstone modes contains only a kinetic energy term, indicates that the Nambu–Goldstone modes are massless (because there are no quadratic terms in the potential) and non-interacting (because there are no higher-order terms). In this particular model, they therefore do not scatter off one another at all. More generally, the potential being invariant under the excitation of Nambu–Goldstone modes directly leads to only derivatives of these modes appearing in the action, which suppresses scattering amplitudes at low wave numbers. This is a general feature as long as these modes are parametrised by modulations of the order parameter field in the broken symmetry direction.

21.2 Linear Parametrisation of Excitations

Consider again the linear sigma model of Eq. (21.1), focussing now on the symmetry-breaking ground state with $\phi_i = 0$ for all $i \in \{1 \ldots N-1\}$, and $\phi_N = \sqrt{M^2/\lambda}$. This time, we write excitations on top of the symmetry-broken state in terms of the fields $\delta\phi_i$ defined by $\phi'_i = \phi_i + \delta\phi_i$, with ϕ the symmetry-broken ground state and ϕ' the state containing excitations. The field $\delta\phi_N$ is purely a (massive) amplitude mode. The excitations $\delta\phi_i$ with $i \neq N$, however, are *not* pure Goldstone modes.

Exercise 21.4

- Explain why the excitations $\delta\phi_i$ with $i \neq N$ are not by themselves precisely Goldstone modes.
- Describe the excitation that is purely a Goldstone mode and identify an expansion parameter for which $\delta\phi_i$ with $i \neq N$ coincides with the Goldstone mode to linear order in the expansion.
  Hint: Consider $N = 2$ and draw the ground state field configurations in the $\phi_1 - \phi_2$ plane.

The fact that the excitations $\delta\phi_i$ coincide with Goldstone modes to linear order is reflected in their mass spectrum and scattering amplitudes. The mass term should disappear in accordance with Goldstone's theorem, while the presence of Adler zeroes may be expected to suppress the scattering amplitudes.

Exercise 21.5 Show that the Lagrangian density for the linear excitation modes can be written in the form

$$\begin{aligned}\mathcal{L} = \frac{M^4}{2\lambda} &+ \sum_i \partial_\mu \delta\phi_i \partial^\mu \delta\phi_i - \frac{\lambda}{2}\left(\sum_i \delta\phi_i^2\right)^2 \\ &+ \partial_\mu \delta\phi_N \partial^\mu \delta\phi_N - 2M^2 \delta\phi_N^2 - 2\sqrt{M^2\lambda}\,\delta\phi_N^3 - \frac{\lambda}{2}\delta\phi_N^4 \\ &- 2\sqrt{M^2\lambda}\left(\sum_i \delta\phi_i^2\right)\delta\phi_N - \lambda\left(\sum_i \delta\phi_i^2\right)\delta\phi_N^2. \end{aligned} \tag{21.3}$$

Here, the index i runs from 1 through $N-1$. Notice that the fields $\delta\phi_i$ do not have a (quadratic) mass term, but do interact. These interactions could lead to effective mass terms.

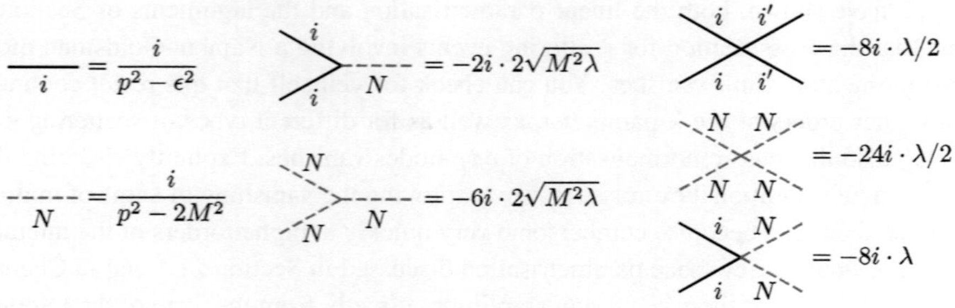

Figure 21.1

Feynman rules and diagrams for the Lagrangian density of Eq. (21.3).

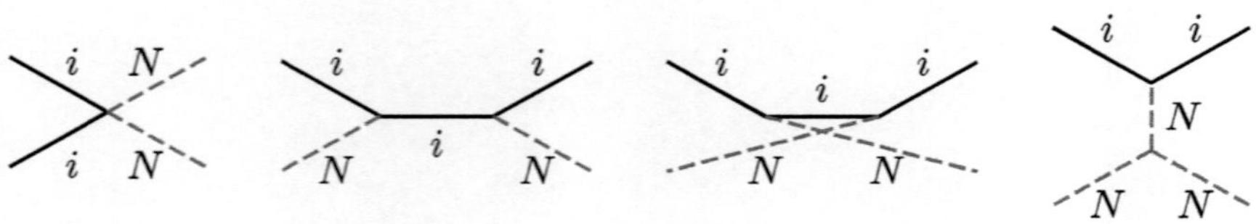

Figure 21.2

Feynman diagrams for the scattering between $\delta\phi_i$ and $\delta\phi_N$ to first order in λ.

To calculate the scattering cross sections involving Nambu–Goldstone modes, or the effect of the interactions on their mass, we first introduce the Feynman rules, in Figure 21.1. Notice that in these Feynman rules, we give the propagator for the fields $\delta\phi_i$ an infinitesimal mass ϵ, to avoid formal divergences. We will take the limit $\epsilon \to 0$ at the end of all calculations.

Exercise 21.6 Derive the Feynman rules in Figure 21.1 from the Lagrangian density in Eq. (21.3). Pay special attention to the combinatorial prefactors.

As an example of a scattering event involving Nambu–Goldstone modes in the zero-momentum limit, consider the scattering between one $\delta\phi_i$ excitation and one $\delta\phi_N$ mode. To first order in λ, the four diagrams contributing to the scattering cross section are shown in Figure 21.2.

Exercise 21.7

- Show that the total cross section resulting from the sum of diagrams in Figure 21.2 can be written as

$$-i\lambda\left(8 + 4\frac{4M^2}{(k+p)^2 - \epsilon^2} + 4\frac{4M^2}{(k+q)^2 - \epsilon^2} - 12\frac{4M^2}{(p+q)^2 - 2M^2}\right). \tag{21.4}$$

Here, the squared momenta are all relativistic four-vectors. Indicate which momenta belong to which propagators in the Feynman diagrams.
- Now consider the limits $\epsilon \to 0$, $p \to 0$, and $k^2 \to 2M^2$, corresponding to the scattering of a zero-momentum $\delta\phi_i$ mode with an (on-shell) $\delta\phi_N$ mode of momentum k. Explain why these limits imply that we must also take $q \to 0$, and show that the sum of diagrams in Figure 21.2 vanishes in this limit.

As you have shown, both the linear parametrisation and the arguments of Section 21.1 indicate that the cross section for scattering events involving a Nambu–Goldstone mode in the zero-momentum limit vanishes. You can check for yourself that this result continues to hold at higher orders of the λ parameter, as well as for different types of scattering events. Moreover, also the mass renormalisation of $\delta\phi_i$ modes vanishes. Explicitly checking this to order λ^2 is a nice additional exercise. However, proving the vanishing of sums of multitudes of Feynman diagrams becomes cumbersome very quickly at higher orders of the interaction. That the non-linear coset-space parametrisation discussed in Section 21.1 and in Chapter 10 reveals the vanishing of these scattering amplitudes directly from the form of the Lagrangian density is one example of the power of using a parametrisation tailored to both broken and residual symmetries.

A Superposition of Symmetry-Broken States

22

Prerequisites Box 22.1

Read Chapter 2, in particular Section 2.4.

Goal of this project Box 22.2

◇ Show the relation between the symmetric ground state and the symmetry-broken states of the Lieb–Mattis Hamiltonian.

In this project, you will show that for the Lieb–Mattis Hamiltonian, the symmetric ground state is an equal-weight superposition of all possible symmetry-broken states. In the derivation, we will follow the result of Ref. [167].

22.1 Introduction: The Lieb–Mattis Model

In Section 2.4, we introduced the **Lieb–Mattis Hamiltonian**, describing the collective properties of the Heisenberg antiferromagnet:

$$\hat{H}_{\mathrm{LM}} = \frac{J}{N} \hat{\mathbf{S}}_A \cdot \hat{\mathbf{S}}_B. \tag{22.1}$$

Here, J is positive, $\hat{\mathbf{S}}_{A/B}$ are the operators for components of the total spin on the A/B sublattice, N is the total number of sites, and on each site there is a spin-S. This model is a perfect toy model for describing how a symmetry-broken state can arise even though the ground state of the model for any finite size is symmetric.

Exercise 22.1 Prove that the Lieb–Mattis Hamiltonian of Eq. (22.1) commutes with the operator for total spin squared on each sublattice $\hat{\mathbf{S}}^2_{A/B}$ and with the total spin squared $\hat{\mathbf{S}}^2_{\mathrm{tot}} \equiv (\hat{\mathbf{S}}_A + \hat{\mathbf{S}}_B)^2$ of the entire lattice. Use this to calculate the energy of the ground state, in terms of the spin value S, the coupling J, and the system size N.
Answer: You can rewrite the Lieb–Mattis Hamiltonian as $\hat{H}_{\mathrm{LM}} = \frac{J}{2N}\left(\hat{\mathbf{S}}^2_{\mathrm{tot}} - \hat{\mathbf{S}}^2_A - \hat{\mathbf{S}}^2_B\right)$, which directly implies that $\hat{H}_{\mathrm{LM}}$ commutes with the total spin and the total spin on each

sublattice. All eigenstates are therefore labelled by a total spin and total spin per sublattice, $|S_{\text{tot}}, S_A, S_B\rangle$. The ground state is obtained by minimising the total spin ($S_{\text{tot}} = 0$) and maximising the total spin per sublattice ($S_{A/B} = NS/2$). This gives the ground state energy $E_0 = -\frac{JS}{2}\left(\frac{NS}{2}+1\right)$.

You have found that the ground state of the Lieb–Mattis Hamiltonian is a total spin singlet state. This state is symmetric under any rotation of the total spin. On the other hand, we learned that a sufficiently large system will invariably become antiferromagnetic in the presence of small perturbations . The quintessential antiferromagnetic state is the Néel state. This is a product state with all spins pointing one way on sublattice A and the opposite way on the B sublattice,

$$|\psi_{\text{N}}\rangle = \prod_{i\in A} |S\rangle_i \prod_{j\in B} |-S\rangle_j, \tag{22.2}$$

where $|S\rangle_i$ is the maximally polarised state of the spin-S on site i. For example, for $S = 1/2$, the Néel state would read $|\psi_{\text{N}}\rangle = \prod_{i\in A} |\uparrow\rangle_i \prod_{j\in B} |\downarrow\rangle_j$.

Exercise 22.2 Is the Néel state an eigenstate of the Lieb–Mattis Hamiltonian? What is the energy (either eigenenergy or energy expectation value) of the Néel state? How does the energy density E/N of the Néel state compare to that of the ground state for large N? *Answer*: the Néel state is not an eigenstate because $\hat{\mathbf{S}}_A \cdot \hat{\mathbf{S}}_B = \hat{S}^z_A\hat{S}^z_B + \frac{1}{2}(\hat{S}^+_A\hat{S}^-_B + \hat{S}^-_A\hat{S}^+_B)$ and the $\hat{S}^+\hat{S}^-$ terms change the Néel state. The energy expectation value is $E_N = -\frac{1}{4}JNS^2$. Therefore, the energy density in the thermodynamic is the same in the Néel state and the ground state: $E/N = -\frac{JS^2}{4}$.

Each possible orientation of the spins on the A-sublattice defines a distinct Néel state. All these Néel states are related to each other through rotations. A common way to quantify this rotation is by using polar coordinates:

$$|\theta, \phi\rangle = \mathrm{e}^{-\mathrm{i}\phi\hat{S}^z}\mathrm{e}^{-\mathrm{i}\theta\hat{S}^y}|S\rangle. \tag{22.3}$$

Here, $|S\rangle$ is the maximally polarised state in the z-direction. The general Néel state can be obtained by applying this rotation to every site:

$$|\psi_{\text{N}}(\theta, \phi)\rangle = \mathrm{e}^{-\mathrm{i}\phi\hat{S}^z_{\text{tot}}}\mathrm{e}^{-\mathrm{i}\theta\hat{S}^y_{\text{tot}}}|\psi_{\text{N}}\rangle. \tag{22.4}$$

In this case, ψ_{N} is given by Eq. (22.2).

We can create a symmetric superposition of all possible Néel states by integrating over all directions with the Haar measure:

$$|\psi_{\text{S}}\rangle = \int_0^{2\pi} \mathrm{d}\phi \int_0^{\pi} \sin\theta \mathrm{d}\theta \;\; |\psi_{\text{N}}(\theta, \phi)\rangle. \tag{22.5}$$

Exercise 22.3 Show that the symmetrised Néel state $|\psi_{\text{S}}\rangle$ is invariant under spin rotations.

22.2 Explicit Calculation for $S = 1/2$

We will now show that the symmetrised Néel state of Eq. (22.5) is identical to the singlet ground state of the Lieb–Mattis model. There are various ways to do this. One of the methods is to show that $|\psi_S\rangle$ is an eigenstate of $\hat{H}_{LM}$ with energy equal to the ground state energy. For simplicity, we consider $S = 1/2$.

The first step is to explicitly calculate the symmetrised wave function $|\psi_S\rangle$ in the S^z basis of the individual spins.

Exercise 22.4 For a single spin, calculate $|\theta, \phi\rangle = e^{-i\phi\hat{S}^z} e^{-i\theta\hat{S}^y} |\uparrow\rangle$ in the basis of $|\uparrow\rangle$, $|\downarrow\rangle$.
Answer: $|\theta, \phi\rangle = e^{-i\phi/2}\cos(\theta/2)\,|\uparrow\rangle + e^{i\phi/2}\sin(\theta/2)\,|\downarrow\rangle$.

Exercise 22.5 Any many-spin wave function can be written in a basis of spin configurations C, defined by products of $\hat{S}^z$-eigenstates on each site, as $|\psi\rangle = \sum_C a_C |\{C\}\rangle$. Write out the coefficients a_C for the general Néel state $|\psi_N(\theta, \phi)\rangle$.
Hint: The coefficient a_C depends only on the total number of up/down spins on the A/B sublattice in configuration C.
Answer: $a_C = \int \sin(\theta) d\theta \ \int d\phi \ \left(e^{-i\phi/2}\cos(\theta/2)\right)^{N_A^\uparrow} \left(e^{i\phi/2}\sin(\theta/2)\right)^{N_A^\downarrow} \times \left(-e^{-i\phi/2}\sin(\theta/2)\right)^{N_B^\uparrow} \left(e^{i\phi/2}\cos(\theta/2)\right)^{N_B^\downarrow}$.

Exercise 22.6 Given the coefficients a_C for the Néel state in a general direction, what are the coefficients for the symmetrised state $|\psi_S\rangle$? Can you normalise the coefficients such that $\sum_C |a_C|^2 = 1$?
Hint: The coefficients a_C can be expressed purely in terms of the number of up spins on the B sublattice $N_B^\uparrow$.
Answer: $a_C = (-1)^{N_B^\uparrow} \Big/ \left(\sqrt{N/2+1} \binom{N/2}{N_B^\uparrow} \right)$.

We now have an explicit expression for the symmetrised wave function. This allows us to explicitly verify that $|\psi_S\rangle$ is an eigenstate of the Lieb–Mattis Hamiltonian. For this, we need to find the coefficients a'_C such that

$$\hat{H}_{LM}|\psi_S\rangle = \sum_C a'_C |C\rangle. \tag{22.6}$$

The Lieb–Mattis Hamiltonian itself can be written as a sum over all pairs of sites i, j with i on sublattice A and j on sublattice B:

$$\hat{H}_{LM} = \frac{J}{N} \sum_{i\in A, j\in B} \left(\hat{S}_i^z \hat{S}_j^z + \frac{1}{2}\left(\hat{S}_i^+ \hat{S}_j^- + \hat{S}_i^- \hat{S}_j^+ \right) \right). \tag{22.7}$$

Exercise 22.7 Write, using the Lieb–Mattis Hamiltonian in the form of Eq. (22.7), the new coefficients a'_C in terms of the coefficients a_C.
Hint: Also the coefficients a'_C can be written in terms of the number of up-spins on the B sublattice.

You will have found that the coefficients $a'_C = -\frac{J}{4}\left(\frac{N}{4}+1\right)a_C$, thus proving that the symmetrised state is an eigenstate of H_{LM} with the ground state energy as its eigenvalue:

$$\hat{H}_{\mathrm{LM}}|\psi_{\mathrm{S}}\rangle = E_0|\psi_{\mathrm{S}}\rangle. \tag{22.8}$$

Consequently, $|\psi_{\mathrm{S}}\rangle$ is the ground state of the Lieb–Mattis model.

While we proved here that the symmetrised superposition of Néel states equals the total-spin singlet ground state for the special case of $S = 1/2$, it is a more general phenomenon that finite-size symmetric ground states can be constructed from an infinite superposition of symmetry-broken states. This can only work when the symmetric ground state has some non-zero overlap with the symmetry-broken state, and if the symmetry-broken state has the same energy density as the ground state in the thermodynamic limit. It means that the symmetric ground state already 'knows' about many properties of the symmetry-broken states, as we found also in the exact diagonalisation project of Chapter 18.

Appendix A: Other Aspects of Spontaneous Symmetry Breaking

Symmetry and symmetry breaking affects almost all areas and aspects of physics. In this book, we focussed on an important, but nevertheless limited, scope within the realm of spontaneous symmetry breaking, and necessarily leave out many interesting physical effects and observations. Although this is unavoidable in general, it is particularly regrettable for a few topics that are especially closely related to the main content of this book, or that are the focus of especially intense current interest in the literature. We therefore include brief introductions to the role of symmetry in some selected topics as short appendices.

A.1 Glasses

The rigidity of collective states of matter is governed by the spontaneous breaking of symmetry. This is most apparent in crystalline solids, where the framework of gapless Nambu–Goldstone modes (in this case, phonons) assures that crystals are both rigid and solid. However, most solid states surrounding us in everyday life are not crystalline: they are amorphous solids that one can collectively call **glasses**.

A glass is a solid when it comes to short-time response functions, and indeed glasses are in many macroscopic ways indistinguishable from crystalline solids: they feel rigid, carry phonons, can break, and, most importantly, they break the translational symmetry of space. On the microscopic level, however, glasses are disordered. For example, the SiO_2 molecules in an ordinary windowpane sit, immobile, at seemingly random positions and there is no long-range order in any conventional correlation function. Notice that, in this respect, glasses differ from polycrystalline systems, in which crystalline order exists at some intermediate length scale.

From the perspective of symmetry breaking, the fact that glasses spontaneously break translational symmetry directly implies their rigidity and the existence of phonons (they are the Nambu–Goldstone modes). The main difference between glasses and crystalline solids is that in the latter there is a residual discrete translational symmetry, whereas in glasses the translation is completely broken.

A detailed theoretical description of the glass phase remains elusive to this day. In fact, one of the major open questions is whether the glass phase is even a genuine phase of matter, or if it should perhaps be understood as an extremely slow, viscous liquid. A closely related question is whether there exists a true thermodynamic **glass transition**, separating a high-temperature liquid from a low-temperature stable glass phase. Both of these questions are difficult to answer because most glasses are in practice made by supercooling a liquid to avoid crystallisation, and noticing that upon further cooling the viscosity increases exponentially (see also the related discussion in Section 9.4 on supercooling).

In theoretical simulations, the word ‘glasses’ is often used for systems with so-called **quenched disorder**. These are described by Hamiltonians in which the values of parameters are chosen at random from some probability distribution. A famous example is the Sherrington–Kirkpatrick model for disordered magnets, in which each Ising spin has a random interaction with every other spin. This model has a low-temperature rigid phase without long-range order, akin to a glass. However, the states observed in these types of simulations do not spontaneously break translational symmetry because the quenched disorder already explicitly broke it to begin with. Nonetheless, within an ensemble of Hamiltonians with different realisations of the randomly chosen parameter values, quenched disorder glasses do break the symmetry between different disorder realisations, which is known as **replica symmetry breaking**. Some good books on such quenched disorder glasses are Refs. [128, 135, 188]. A detailed introduction of glasses without quenched disorder, which can, for example, be made using the supercooling of liquids, can be found in Refs. [44, 224].

Finally, a second class of solid materials without long-range order consists of **quasicrystals**. In a quasicrystal, there is no periodicity in the atomic positions. The Fourier transform of the atomic positions, however, does consist of perfectly sharp peaks. In fact, there are infinitely many of them arranged in a fractal pattern. A quasicrystal can therefore be discovered by having an unusual but sharp electron diffraction pattern. From the perspective of symmetry breaking, it is interesting to note that a quasicrystal completely breaks translational symmetry (unlike normal crystals, which break it into a discrete subgroup), but that they can break rotational symmetry down to just a discrete subgroup. The remaining rotational symmetries in quasicrystals can be types that are not allowed in crystalline matter, such as fivefold or tenfold, which gives another window for discovering them.

A.2 Many-Body Entanglement

A relatively recent development in quantum physics is the study of **many-body entanglement**. This topic can be introduced most easily by first reviewing some properties of two-particle entanglement, of which many good discussions can be found in elementary textbooks on quantum mechanics, such as Refs. [65, 109, 174].

First then, consider a system with two spin-$\frac{1}{2}$ degrees of freedom. The full Hilbert space is spanned by four states:

$$\mathcal{H} = \{|\uparrow\rangle_1 \otimes |\uparrow\rangle_2, |\downarrow\rangle_1 \otimes |\uparrow\rangle_2, |\uparrow\rangle_1 \otimes |\downarrow\rangle_2, |\downarrow\rangle_1 \otimes |\downarrow\rangle_2\}. \tag{A.1}$$

Any state can be written as a linear superposition of these four states. A particularly interesting one to consider is the singlet state:

$$|0\rangle = \frac{1}{\sqrt{2}} \left(|\downarrow\rangle_1 \otimes |\uparrow\rangle_2 - |\uparrow\rangle_1 \otimes |\downarrow\rangle_2\right). \tag{A.2}$$

This state is certainly entangled, but just from its definition, it is not easy to quantify just how entangled it is.

One way to approach this question is to consider the properties of the **reduced density matrix**. To begin with, we (arbitrarily) write the full Hilbert space as a tensor product of two parts: $\mathcal{H} = \{|\uparrow\rangle_1, |\downarrow\rangle_1\} \otimes \{|\uparrow\rangle_2, |\downarrow\rangle_2\}$. The reduced density matrix on spin 1 is then obtained by 'tracing' the full density matrix over the degrees of freedom of spin 2. For the pure singlet state of Eq. (A.2), the full density matrix is just the projection operator $\rho = |0\rangle\langle 0|$, and the trace operation can be carried out explicitly:

$$\begin{aligned}\rho_1 &= \mathrm{Tr}_2\,\rho \\ &= \langle\uparrow_2|\rho|\uparrow_2\rangle + \langle\downarrow_2|\rho|\downarrow_2\rangle \\ &= \frac{1}{2}\left(|\uparrow_1\rangle\langle\uparrow_1| + |\downarrow_1\rangle\langle\downarrow_1|\right).\end{aligned} \tag{A.3}$$

The fact that spins 1 and 2 were entangled is reflected in the fact that the reduced density matrix ρ_1 no longer represents a pure state. To quantify this, Von Neumann proposed to compute the **entanglement entropy** of the reduced density matrix, defined as

$$S_{\mathrm{vN}} = -\mathrm{Tr}\rho_1 \log \rho_1. \tag{A.4}$$

In the case of the singlet state $|0\rangle$, the entanglement entropy equals $S_{\mathrm{vN}} = \log 2$.

The notion of entanglement entropy can be extended to systems with many degrees of freedom, like a many-spin system. In such cases, one can still talk about **bipartite entanglement** in the same way as we showed for two spins. One needs to (again arbitrarily) split the system into two parts (commonly called A and B), compute the reduced density matrix on one subsystem and use that to compute the entanglement entropy defined by Eq. (A.4).

Typically, it is then interesting to explore how the entanglement entropy scales with the size of subsystem. There are a few special cases to consider:

1. The first is a so-called product state, like the Néel antiferromagnet, in which there is no entanglement at all between any two spins. In this case, the entanglement entropy is identically zero.
2. The second case is when spins are only entangled with a few spins that are close by. This is then reflected in the entanglement entropy by the fact that it scales with the size of the boundary of region A, in what is known as area law entanglement. All ground states of locally interacting systems obey the area law.
3. A final, extreme case is when each spin is entangled with every other spin, in which case the entanglement entropy scales with the volume of subsystem A. Such a 'volume law' typically applies to sufficiently highly excited eigenstates of interacting Hamiltonians or their thermal states at sufficiently high temperature.

In Section A.4, we will introduce topologically ordered states, which are considered to be the most entangled states possible, though a thorough discussion of their entanglement properties goes beyond our present scope (see, e.g., Ref. [175]). The possible connection between entanglement entropy and the classical, statistical entropy at any given temperature is an active field of research.

To see how many-body entanglement is related to spontaneous symmetry breaking, recall the discussion of Section 3.7, in which we showed that exact ground states of symmetric Hamiltonians displaying long-range correlations are an indication that the symmetry in these models may be spontaneously broken. This implies that unstable ground states are always highly entangled, while stable symmetry-breaking states are unentangled product states that satisfy cluster decomposition. One must be careful, however, not to directly equate the absence of bipartite entanglement with stability. Take, for example, the unstable ground state of the Ising model, defined in Eq. (3.22). In quantum information theory, this state is known as the **Greenberger–Horne–Zeilinger state** (GHZ state) and is considered a maximally entangled state. The entanglement entropy, however, is exactly $S_{\mathrm{vN}} = \log 2$, regardless of how the system is divided in two. Clearly, entanglement entropy by itself is not a sufficient tool for distinguishing between stable (unentangled) and unstable (entangled) ground states, and the search for alternative measures of entanglement is ongoing.

Nevertheless, the entanglement entropy does capture some subtle properties related to spontaneous symmetry breaking. In symmetry-breaking systems with a symmetric, finite-size ground state, the entanglement entropy has logarithmic corrections to its generic area law. The coefficient in front of the logarithm counts the number of Nambu–Goldstone modes N_{NG}, [127, 166]:

$$S_{\mathrm{vN}}(L^d) = aL^{d-1} + \frac{N_{\mathrm{NG}}(d-1)}{2} \log L + \dots . \tag{A.5}$$

This result is derived in the project of Chapter 19. Furthermore, the spectrum of the so-called **entanglement Hamiltonian** $H \equiv -\log \rho_A$, with ρ_A the reduced density matrix, has the same structure as the tower of states of Section 3.6. These are just two examples of how information about the symmetry-broken state and its excitations is hidden in the entanglement structure of the symmetric ground state.

A.3 Quantum Phase Transitions

Our discussion of phase transitions in Chapter 5 covered ordinary, thermal phase transitions. The change from one phase of matter to another in those transitions occurs as a result of a change in temperature. The transition from the ordered to the disordered phase (low to high temperature, or lower to higher symmetry) can then be said to be caused by the increase of thermal fluctuations.

It is also possible to induce a phase transition at fixed temperature as some other parameter is varied. Examples of such parameters are the density, an external magnetic field, pressure, or an applied current. These transitions, however, can always be viewed as thermal transitions by tuning to the critical value of the tuning parameter and then changing temperature. This is particularly obvious when considering the phase diagram with temperature on one axis and the alternative parameter on the other, as shown in Figure A.1. Whichever way you cross a transition line in such a diagram, the crossing point, and hence the critical behaviour, can also be reached in a purely thermal transition.

This is not true for transitions that occur at precisely zero temperature, in which the system undergoes a transition between two distinct stable phases as some parameter p is tuned. Such a transition is called a **quantum phase transition**. If the phase transition is continuous (see Section 5.1), then we speak of a **quantum critical point** at the critical value $p = p_{\mathrm{c}}$. The difference between thermal and quantum phase transitions is shown in the cartoon phase diagram of Figure A.1.

At zero temperature, the system should be in a (possibly symmetry-broken) ground state at any value of the tuning parameter. Thermal excitations or fluctuations do not play any role. It is sometimes said that, in analogy to thermal transitions, the quantum phase transition is due

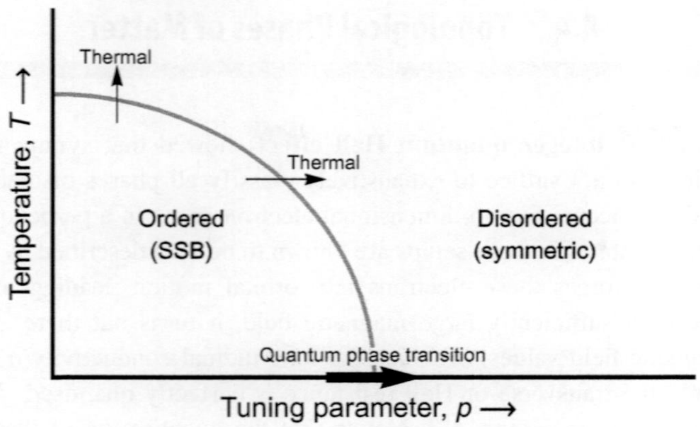

Figure A.1

A typical phase diagram with temperature T (vertical axis) and another tuning parameter p (horizontal axis). In this simple system, there are two phases: a symmetry-broken ordered phase on the left, and a symmetric 'disordered' state on the right. Any transition at non-zero temperature, whether it is induced by tuning temperature or p, can be interpreted as a thermal transition. By contrast, the phase transition at zero temperature is a quantum phase transition.

to quantum fluctuations, but as explained in Chapter 6 that is a misnomer. The system is in a ground state of its Hamiltonian throughout the transition, and nothing fluctuates or changes as a function of time in any state encountered. Instead, a quantum phase transition can be described in the language of **quantum corrections**. In the symmetry-broken phase, far away from the quantum critical point, the system at zero temperature resembles the classical state in the sense of Section 3.3. With respect to this classical state, quantum corrections become more and more important as p approaches p_c, and they completely overwhelm the classical correlations at the quantum critical point. Far across on the other side of the transition, the system (often) approaches a different classical state.

Just as in the thermal phase transitions discussed in Section 5.5, the quantum critical point is characterised by scale invariance and universality, dictated by the symmetry. It turns out that the universality class of a quantum critical point is linked to that of a thermal critical point with the same symmetry-breaking pattern, but in one dimension higher. This can be understood by recalling the calculation of the thermal and quantum corrections in Section 6.2. At zero temperature, all Matsubara frequencies must be taken into account, and these effectively become an additional dimension. At non-zero temperature, on the other hand, a finite number of Matsubara frequencies can always be interpreted as thermal corrections to the static equilibrium state.

Exactly zero temperature is beyond the reach of any experiment in the real world, so the discussion of quantum phase transitions may seem purely academic. However, it turns out that the existence of the quantum critical point has large influence on the physics near the critical value p_c even at non-zero temperatures, sometimes up to hundreds of Kelvin. This is referred to as **quantum criticality** and is one of the strongest manifestations of non-classical physics in condensed matter. In fact, much more so than, for example, superconductors, which are described by classical states just as much as any crystal. A good resource on this topic is Ref. [170].

A.4 Topological Phases of Matter

The discovery of the **integer quantum Hall effect** showed that symmetry and symmetry breaking alone do not suffice to exhaustively classify all phases of matter. The integer quantum Hall effect occurs in two-dimensional electron gases in a perpendicular magnetic field. The electronic states in these setups are known to be well described by non-interacting electrons. The field forces these electrons into orbital motion, leading to the formation of Landau levels. In sufficiently large magnetic field, it turns out there are **plateaus**, or regimes of magnetic field values, in which the longitudinal conductivity σ_{xx} is completely suppressed while the transverse or Hall resistance is perfectly quantised: $R_{xy} = \frac{h}{\nu e^2}$ with ν an integer, as shown in Figure A.2. Notice that the quantisation of the Hall resistance is in terms of natural constants only. It does not depend on any property of the material hosting the two-dimensional electron gas, nor on the size or shape of the sample used to measure it. Interesting introductions to the physics of the quantum Hall effect are given in Refs. [118, 194].

Quantum Hall phases with a different **filling fraction** ν cannot be distinguished based on symmetry: no non-trivial symmetry breaking occurs upon going from one to the other, and there is no local order parameter for any value of the filling fraction. The only physical quantity that distinguishes different quantum Hall states is the quantised Hall conductivity itself. The integer part of the conductivity is also known as a Chern number, and this is historically the first example of a **topological invariant** being recognised as characterising a phase of matter. The word 'topology' here is not directly related to the topological defects of Chapter 7. Instead, invariants like the Chern number demarcate classes of gapped systems. They are topological in the sense that all states within a given class (with one particular value of the invariant) can be smoothly deformed into one another, while states in distinct classes cannot. Consequently, topological invariants are unaffected by local deformations of the state or Hamiltonian.

The physics of quantum Hall states can be generalised to include **topological insulators** and **topological superconductors**. These too are characterised by topological invariants. Symmetry does make an important comeback here because the types of topological invariants that are relevant to any given system depend strongly on its symmetries. For example,

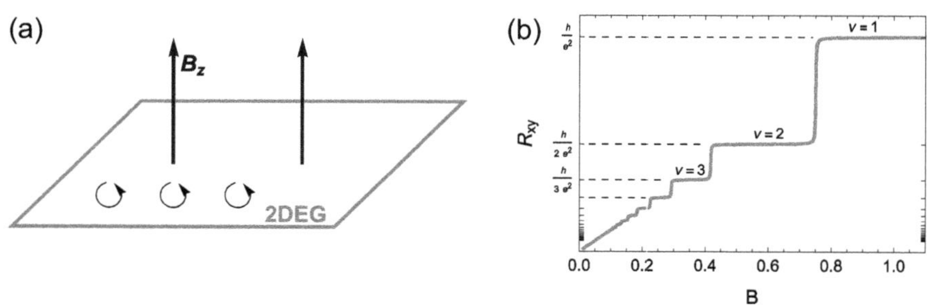

Figure A.2 (a) The integer quantum Hall effect occurs in two-dimensional electron gases (2DEG) subject to a large perpendicular magnetic field. (b) In such a large magnetic field, electron energies are quantised into Landau levels. Consequently, the transverse or Hall resistance R_{xy} is quantised and described by a topological invariant.

topological insulators often occur in crystalline materials and the symmetries of the crystal lattice play an intricate role in determining the number and types of available topological invariants [99]. The overarching name for these crystalline states is **symmetry-protected topological phases** of matter.

Returning to the quantum Hall effect, it was found that states with quantised Hall conductance may also emerge at magnetic field values corresponding to certain rational, non-integer filling fractions, in what was soon dubbed the **fractional quantum Hall effect**. In this case, there are strong interactions between electrons and they are essential in establishing the fractional quantum Hall state. The notion of **topological order** was pioneered to distinguish between different fractional quantum Hall states as well as related states in other interacting quantum liquids [218].

Topologically ordered materials have a non-zero ground state degeneracy, with distinct ground states labelled by a topological invariant. They are long-range entangled (as defined in Section A.2), as opposed to the short-range entanglement of symmetry-protected topological phases. Topologically ordered systems also typically have fractionalised excitations, such as the quasiparticles in the fractional quantum Hall states, whose electric charge is a rational fraction of the elementary electron charge. Reviews of this relatively new notion of order are given in Refs. [218], [219], and [220], as well as in some textbooks [129, 171].

In recent years, it has been suggested that symmetry-protected topological phases can be characterised by anomalies (see Section 10.1) and topological order can perhaps be cast into the language of spontaneous symmetry breaking if one includes **higher-form symmetries**, which are discussed in Section A.5 [124].

A.5 Higher-Form Symmetries

To describe the concept of **higher-form symmetries**, we need to recast our understanding of symmetries from Chapter 1 into the language of **differential forms**. Recall the Noether current J_μ with its conservation law $\partial^\mu J_\mu = 0$. This Noether current can be viewed as a differential 1-form, and in that language the conservation law is written as

$$d \star J = 0. \tag{A.6}$$

Here, the $\star$ is the Hodge duality operation and d is the exterior derivative. A generalised or **higher-form symmetry** has the same form of conservation law as Eq. (A.6), but with J a differential p-form [55, 61, 148]. For example, if the conserved current is a 2-form, it can be expressed by an antisymmetric tensor $J_{\mu\nu}$ that satisfies $\partial^\mu J_{\mu\nu} = 0$.

Similarly, the definition of the Noether charge, which was given in Eq. (1.13) as an integral over all space $Q = \int_V \mathrm{d}^d x\, j^0(x)$, can be generalised. For a p-form current in a $D = (d+1)$-dimensional theory, there exists a conserved charge when integrating over a subspace Σ_{D-p} of dimension $D - p$:

$$Q_\Sigma = \int_{\Sigma_{D-p}} \star J. \tag{A.7}$$

This charge is a topological quantity in the sense that it only depends on the state of the system and the topology of the subspace Σ_{D-p}. Just like the 'regular' Noether charge Q, the generalised charges Q_Σ act as generators of unitary symmetry operators, $U(\Sigma) = e^{i\alpha Q_\Sigma}$.

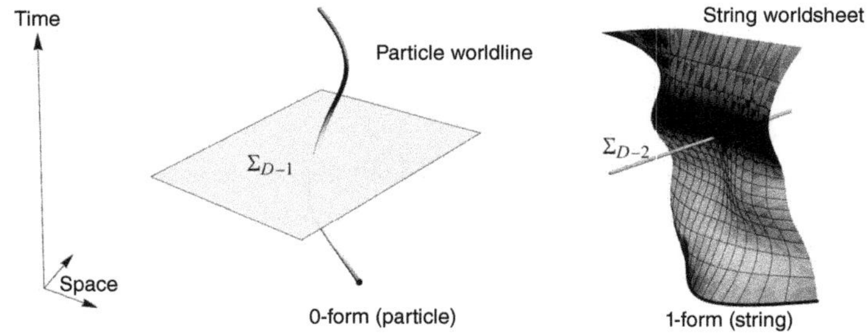

Figure A.3 In a zero-form symmetry, charges are carried by pointlike particles, and the total conserved charge can be found by integrating over a $D-1$-dimensional surface Σ_{D-1}. The total charge is given by how many particle worldlines cross the integration surface. This can be generalised to higher-form symmetries, such as the 1-form symmetry where charges are carried by strings. The total charge is integrated over a $D-2$-dimensional subspace Σ_{D-2}. The total charge is now given by how many string worldsheets intersect with the charge surface Σ_{D-2}.

The generalisation to higher forms also affects the type of objects that carry the charges Q_Σ. In the case of a regular symmetry, the total charge Q counts how many particle worldlines cross the $d=(D-1)$-dimensional volume of integration. Since particles are pointlike 0-forms, the regular symmetries introduced in Chapter 1 are known as **zero-form symmetries**. For higher-form symmetries, the charges are carried by extended objects such as strings and surfaces, as shown in Figure A.3. For example, to calculate the conserved charge corresponding to a 1-form symmetry, we need to integrate the 2-form Noether current J over a $(D-2)$-dimensional surface, and this counts how many times a string worldsheet intersects the charge surface Σ_{D-2}. In general, p-form symmetries in $D=(d+1)$ dimensions have charges carried by p-dimensional objects, a conserved current J that is a $(p+1)$-form, and conserved charge operators that are defined on $(d-p)$-dimensional subspaces.

While higher-form symmetries may seem exotic, they can be found in mundane settings such as ordinary Maxwell electromagnetism in $D=(3+1)$ dimensions. To see this, we rewrite the electromagnetic field tensor as

$$J^{\mu\nu}_{(m)} = \epsilon^{\mu\nu\rho\sigma} F_{\rho\sigma}. \tag{A.8}$$

The Bianchi identity guarantees that $\partial_\mu J^{\mu\nu}_{(m)} = 0$. In other words, the absence of magnetic monopoles in electromagnetism may be interpreted as a conservation law for the 2-form J, which must be associated with a 1-form symmetry. Introducing the **dual gauge field** $\tilde{A}_\mu$ via the relation $J^{\mu\nu}_{(m)} = \partial_\mu \tilde{A}_\nu - \partial_\nu \tilde{A}_\mu$ (or equivalently $F_{\mu\nu} = \partial_\kappa \epsilon_{\kappa\mu\nu\lambda}\tilde{A}_\lambda$), we find that the charged string operators corresponding to the 1-form symmetry are so-called **'t Hooft loops**:

$$H_C = \mathrm{e}^{\mathrm{i}\oint \mathrm{d}x^\mu \tilde{A}_\mu}. \tag{A.9}$$

The associated charges equal the flux going through a two-dimensional surface:

$$Q_\mathrm{M} = \int_\Sigma \mathrm{d}S_{mn}\, F_{mn} = \int_\Sigma \mathrm{d}S_{mn}\, \epsilon_{mnk} B_k. \tag{A.10}$$

Interestingly, in the absence of any electric charges, the equation for the Maxwell field strength tensor, $\partial_\mu F^{\mu\nu} = 0$, also represents a 1-form symmetry. The associated string operators are known as **Wilson loops**:

$$W_{\mathcal{C}} = \mathrm{e}^{\mathrm{i}\oint \mathrm{d}x^\mu A_\mu}, \qquad \text{Wilson loop.} \tag{A.11}$$

In this case, the associated electric charges are given by the electric field lines threading a two-dimensional surface:

$$Q_\mathrm{E} = \int_\Sigma \mathrm{d}S_{tm}\, F_{tm} = \int_\Sigma \mathrm{d}S_{tm}\, E_m. \tag{A.12}$$

We thus find that Maxwell's theory of electromagnetism in empty space can be described as having both electric and a magnetic 1-form symmetries.

Just like regular symmetries, the higher-form symmetries can be broken. In the case of 1-forms, line operators akin to the 't Hooft and Wilson loops will act as (non-local) order parameters. Clearly, the expectation value of a line operator depends on its integration contour $\mathcal{C}$. If it satisfies an area law, such that $W_{\mathcal{C}} \propto \mathrm{e}^{-|\mathcal{S}|}$ with $\mathcal{S}$ the area enclosed by $\mathcal{C}$ and $|\mathcal{S}|$ its areal size, then the magnitude of the line operator expectation value falls off quickly with increasing contour size, and the symmetry is said to be unbroken. If, on the other hand, it satisfies a perimeter law, such that $W_{\mathcal{C}} \propto \mathrm{e}^{-|\mathcal{C}|}$ with $|\mathcal{C}|$ the length of $\mathcal{C}$, its magnitude falls off slowly with contour size, and the symmetry is said to be broken. This also generalises to higher forms with the appropriate integration domains. The language of 1-form symmetries allows us to rephrase the **confinement transition** introduced in Section 11.4 as a symmetry breaking transition.

Going back to the Maxwell's electromagnetism, the magnetic symmetry Q_M is spontaneously broken in the Maxwell vacuum, as the 't Hooft line $H_{\mathcal{C}}$ satisfies a perimeter law. The associated Nambu–Goldstone mode is the photon A_μ itself. This higher-form symmetry is restored is a superconductor, where the photon becomes gapped by the Anderson–Higgs mechanism, as described in Section 8.3. This is, in a sense, a dual description of the superconducting phase transition. One way to understand how it comes about, is to realise that Q_M counts magnetic flux enclosed in its contour. In the vacuum, magnetic fields are free, or unconfined, and cost little energy to create. They are analogous, for example, to bosons in a Bose–Einstein condensate. On the other hand, magnetic flux in a superconductor is confined into vortex lines and is expensive to create, analogous to ordinary gapped bosonic excitations. The p-form generalisation of the Goldstone theorem is given in Refs. [76, 102].

Finally, it has been suggested that the topological orders discussed in Section A.4 may be interpreted as phases with spontaneously broken higher-form symmetries [124, 148].

A.6 Decoherence and Spontaneous Unitarity Violation

One of the hallmarks of quantum physics is the ability to create superpositions of any two states of a system. If the system is completely isolated, the phases of superposed energy eigenstates evolve, but the superposition never decays. In reality, however, no system is completely isolated, and in practice it is hard to maintain coherent superpositions of macroscopically distinct states for any extended period of time. The mechanism by which macroscopic objects interacting with an external environment seemingly lose the ability of being superposed, is known as **decoherence**, and was pioneered in the 1970s [36, 82, 165, 228, 231, 234].

The relevance of decoherence to symmetry breaking lies in the observation that we typically encounter everyday objects with a broken symmetry (such as macroscopic solid objects breaking translations), but we hardly ever see superpositions of symmetry-breaking configurations (such as macroscopic solid objects superposed over macroscopic distances). The explanation offered by decoherence for why such extended superpositions do not seem to occur in reality, is the identification of a particular type of preferred basis, known as the **pointer basis**. For macroscopic objects with a degenerate manifold of symmetry-breaking states, the pointer basis is defined simply as the eigenbasis of the part of the Hamiltonian describing its interactions with the environment [231].

Starting from any pure state $|\psi\rangle$ of the combined system of macroscopic object and environment, we may write its density matrix $\rho = |\psi\rangle\langle\psi|$ in the pointer basis. Since we do not typically observe the microscopic particles making up the environment of a macroscopic, symmetry-breaking object, all information about quantities that we are likely to make physical predictions about is contained in the **reduced density matrix**, which is obtained by tracing ρ over all uncontrollable environmental degrees of freedom (as introduced in Section A.2). The central observation of the decoherence programme is the fact that generically, off-diagonal elements of the reduced density matrix obtained from ρ written in the pointer basis decay exponentially in time [177]. This renders the reduced density matrix diagonal after a short time, and because the reduced density matrix contains all accessible information about the system, no quantum coherence between distinct pointer states can be observed after decoherence.

For systems undergoing symmetry breaking, the pointer basis of macroscopic objects is precisely the basis of broken-symmetry states [133, 199]. Therefore, if one were to create a superposition of distinct symmetry-breaking states, quantum interference between its components would be unobservable after a decoherence time. Typically, this decoherence time scales inversely with the size of the system. The symmetry-broken states by themselves are stable against interactions with the environment, in close analogy to the stability discussed in Section 3.7.

Although decoherence prevents the observation of quantum interference between superposed pointer states, it does not preclude the formation of such states. Starting from a symmetric equilibrium state and tuning the parameters of its Hamiltonian through a phase transition, unitary quantum dynamics will always predict the formation of a superposition of distinct ordered states [199]. This is unavoidable because unitary time evolution is invertible, so information about the initial state cannot be lost. Adding small symmetry-breaking fields, as we did to explain the stability of the ordered state in equilibrium in Chapter 2, cannot resolve the question of how one ends up in just a single symmetry-breaking state under any type of unitary time evolution [198, 199], as you showed yourself in Exercise 9.1.

In Chapter 9, we considered how symmetry-broken states can arise from noisy time evolution, but in fact all derivations in that chapter were classical. For example, the noisy evolution discussed in Section 9.1.2, describes the expectation value of the order parameter operator, which cannot be brought into superposition by construction. The linear response theory of Section 9.1.3 is likewise formulated in terms of expectation values, and not individual quantum states. These theories therefore do not explain how a symmetric quantum state can evolved into a symmetry-broken state under unitary quantum dynamics.

The unavoidable environmental decoherence that is also present during phase transitions, renders the reduced density matrix diagonal, but it does not indicate which of the symmetry-breaking states will be produced in a single realisation of the phase transition. This is not because of our ignorance of the state of the environment, but rather because the reduced density matrix can only describe ensembles of experiments, averaged over all possible states of the environment [3, 198]. For any given, specific state of the environment, unitary dynamics will produce a superposition of ordered states.

It has been suggested that the emergence of a single ordered state as a system goes through a phase transition may be a manifestation of the unitarity of quantum mechanical time evolution itself being broken spontaneously [197, 198, 199]. Such **spontaneous unitarity violation** would be different from the spontaneous symmetry breaking discussed in this book, as it would need to be caused by a small non-unitary perturbation to the time translation operator, and hence a small perturbation to Schrödinger's equation that is beyond quantum mechanics. Nevertheless, it is has been shown that the physical implications of such a perturbation, if it exists, are very similar to those of equilibrium spontaneous symmetry breaking. Most importantly, there is a singular limit in the quantum evolution of macroscopic objects: in the presence of a non-unitary perturbation of strength ϵ, a symmetric N-state superposition of ordered ground states will evolve into just one of the symmetry-breaking configurations in a time $\tau \sim 1/(N\epsilon)$. In the thermodynamic limit, even an infinitesimal perturbation thus suffices to reduce the initial superposition to a single pointer state. The pointer states themselves are stable under time evolution, in the sense that under the influence of a non-unitary perturbation, they evolve over times $\tau' \sim 1/\epsilon$, without the dependence on system size. It has been shown that any system susceptible to breaking a symmetry in equilibrium will also be susceptible to this type of spontaneous unitarity violation [126, 197, 198].

Finally, notice that a single macroscopic, ordered state emerging from a phase transition is similar to a single macroscopic configuration of a measurement machine emerging from quantum measurement. The peculiar parallels between the **measurement problem** and spontaneous symmetry breaking have been noted several times [33, 63, 104, 131, 197], and in particular it has been pointed out that both require a breakdown of unitary time evolution [3, 157, 198]. This suggests the intriguing possibility that quantum measurement and the emergence of order during phase transitions are both consequences of spontaneous unitarity violation [126, 133, 134, 197, 198].

Appendix B: Further Reading

We can recommend several textbooks and review articles to continue your study of spontaneous symmetry breaking.

Symmetry and Noether's Theorem

- Goldstein, H., Poole, C., and Safko, J. 2002. *Classical Mechanics*, Addison Wesley – a solid standard reference.
- Kleinert, H. 2008. *Multivalued Fields*, World Scientific, Chapter 3 – two side-by-side derivations of the theorem, both in mechanics and field theory.
- Butterfield, J. 2005. *On Symmetry and Conserved Quantities in Classical Mechanics* – in-depth review of the theorem from Hamiltonian and Lagrangian perspectives.
- Bradin, K. A. 2002. Which Symmetry? Noether, Weyl, and Conservation of Electric Charge. *Stud. Hist. Phil. Sci.*, B33(1), 3 – lucid treatment of Noether's second theorem.
- Banados, M., and Reyes, I. 2016. A Short Review on Noether's Theorems, Gauge Symmetries and Boundary Terms. *Int. J. Mod. Phys. D*, 25(10), 1630021 – very similar to our treatment.

Symmetry Breaking

- Weinberg, S. 1996. *Quantum Theory of Fields, Volume II*. Cambridge University Press. Chapter 19 'Spontaneously Broken Global Symmetries' – an excellent introduction of SSB from the high energy perspective, including a discussion of pions and quark mass terms.
- Guralnik, G., Hagen, C., and Kibble, T. 1967. Broken Symmetries and the Goldstone Theorem. Pages of 567–708 of: *Advances in Particle Physics*, Vol 2. Interscience, New York – one of the earliest reviews on symmetry breaking and the Anderson–Higgs mechanism.
- Anderson, P. W. 1984. *Basic Notions of Condensed Matter Physics*. Benjamin Cummings. Chapter 1 – perspective of condensed matter physics, discussing Bose condensation, crystals, magnets, and so forth.
- Landsman, K. 2017. *Foundations of Quantum Theory*. Springer. Chapter 10 – spontaneous symmetry breaking from a mathematical physics perspective.

Goldstone Theorem

- Guralnik, Hagen & Kibble, *op. cit.* – contains several proofs of the theorem.

- Burgess, C. P. 2020. *Introduction to Effective Field Theory: Thinking Effectively about Hierarchies of Scale*. Cambridge University Press – a modern review.
- Brauner, T. 2010. Spontaneous Symmetry Breaking and Nambu–Goldstone Bosons in Quantum Many-body Systems. *Symmetry*, 2(2), 609 – first review to consider type-B NG modes.

Mermin–Wagner–Hohenberg–Coleman Theorem

- Nagaosa, N. 1999. *Quantum Field Theory in Condensed Matter Physics, Texts and Monographs in Physics*. Springer. Berlin – our derivation in Section 6.2 is based on this one.

Phase Transitions

- Chaikin, P., and Lubensky, T. 2000. *Principles of Condensed Matter Physics*. Cambridge University Press – good review of phase transition physics, also information on topological defects.
- Yeomans, J. 1992. *Statistical Mechanics of Phase Transitions.* Clarendon Press; Goldenfeld, N. 1992. *Lectures on Phase Transitions and the Renormalization Group*, No. 85 in Frontiers in Physics. Perseus Books – two standard textbooks on phase transitions and criticality.
- Herbut, I. 2007. *A Modern Approach to Critical Phenomena.* Cambridge University Press – another solid textbook with many explicit calculations.

Topological Defects

- Mermin, N. 1979. The Topological Theory of Defects in Ordered Media. *Rev. Mod. Phys.*, 51, 591 – essential and accessible review.

Mathematics

- Jones, H. 1998. *Groups, Representations, and Physics*. Institute of Physics Publishing.
- Georgi, H. M. 1999. *Lie Algebras in Particle Physics*, 2nd ed., Frontiers in Physics. Perseus. – two solid introductions to group theory and Lie algebras in physics.
- Nakahara, M. 2003. *Geometry, Topology and Physics*, 2nd ed., Graduate Student Series in Physics. Taylor & Francis – advanced physics textbook on topology, including homotopy theory.

Particle Physics

- Weinberg, S. 1995. *The Quantum Theory of Fields*. Cambridge University Press – volume I and II of this masterwork discuss everything you need to know about particle physics.
- Peskin, M. E., and Schroeder, D. V. 2018. *An Introduction To Quantum Field Theory*. CRC Press – a pedagogical introduction to quantum field theory, including quantum electrodynamics and the Standard Model.

Cosmology

- Mukhanov, V. 2005. *Physical Foundations of Cosmology*. Cambridge University Press – a very good introduction to cosmology that covers the history of the universe, including inflation and cosmological perturbations.
- Weinberg, S. 2008. *Cosmology*. Oxford University Press – a very deep book that includes a lot of details on the structure of the universe.
- Boyanovsky, D., de Vega, H. J., and Schwarz, D. J. 2006. Phase Transitions in the Early and Present Universe. *Annu. Rev. Nucl. Part. Sci.,* 56, 441–500.
- Hindmarsh, M., Lüben, M., Lumma, J., and Pauly, M. Phase Transitions in the Early Universe. *SciPost Phys. Lect. Notes*, 024.
- Mazumdar, A., and White, G. 2019. Review of Cosmic Phase Transitions: Their Significance and Experimental Signatures. *Rep. Prog. Phys.*, 82, 076901 – the field of early universe phase transitions is advancing rapidly, here are three excellent recent reviews.
- Achúcarro, A., and Vachaspati, T. 2000. Semilocal and Electroweak Strings. *Phys. Rep.,* 327, 347–426 – a very clear introduction into the field of topological defects of the electroweak theory.

References

[1] Achúcarro, A., and Vachaspati, T. 2000. Semilocal and Electroweak Strings. *Phys. Rep.*, **327**(6), 347–426.

[2] Adler, S. L. 1969. Axial-Vector Vertex in Spinor Electrodynamics. *Phys. Rev.*, **177**(5), 2426–2438.

[3] Adler, S. L. 2003. Why Decoherence Has Not Solved the Measurement Problem: A Response to P. W. Anderson. *Stud. Hist. Phil. Mod. Phys.*, **34**(1), 135–142.

[4] Adzhemyan, L. Ts., Evdokimov, D. A., Hnatič, M., Ivanova, E. V., Kompaniets, M. V., Kudlis, A., and Zakharov, D. V. 2022. The Dynamic Critical Exponent z for 2D and 3D Ising Models from Five-Loop ϵ Expansion. *Phys. Lett. A*, **425**, 127870.

[5] Alberte, L., and Nicolis, A. 2020. Spontaneously Broken Boosts and the Goldstone Continuum. *JHEP*, **2020**(7), 76.

[6] Anderson, P. W. 1952. An Approximate Quantum Theory of the Antiferromagnetic Ground State. *Phys. Rev.*, **86**, 694–701.

[7] Anderson, P. W. 1963. Plasmons, Gauge Invariance, and Mass. *Phys. Rev.*, **130**, 439–442.

[8] Anderson, P. W. 1966. Considerations on the Flow of Superfluid Helium. *Rev. Mod. Phys.*, **38**, 298–310.

[9] Anderson, P. W. 1984. *Basic Notions of Condensed Matter Physics*. Frontiers in Physics, Vol. 55. Benjamin Cummings.

[10] Annett, J. F. 2004. *Superconductivity, Superfluids and Condensates*. Oxford Master Series in Physics. Oxford University Press.

[11] Aoki, T., Homma, S., and Nakano, H. 1980. Dynamical Properties of the Three-Dimensional XY Model at Low Temperatures. *Prog. Theor. Phys.*, **64**, 448.

[12] Arovas, D., Schrieffer, J. R., and Wilczek, F. 1984. Fractional Statistics and the Quantum Hall Effect. *Phys. Rev. Lett.*, **53**, 722–723.

[13] Ashcroft, N. W., and Mermin, N. D. 1976. *Solid State Physics*. Saunders College.

[14] Baker, M., and Glashow, S. L. 1962. Spontaneous Breakdown of Elementary Particle Symmetries. *Phys. Rev.*, **128**, 2462–2471.

[15] Banados, M., and Reyes, I. 2016. A Short Review on Noether's Theorems, Gauge Symmetries and Boundary Terms. *Int. J. Mod. Phys. D*, **25**(10), 1630021.

[16] Barceló, C., Carballo-Rubio, R., Di Filippo, F., and Garay, L. J. 2016. From Physical Symmetries to Emergent Gauge Symmetries. *JHEP*, **2016**(10), 84.

[17] Bardeen, J., Cooper, L. N., and Schrieffer, J. R. 1957. Theory of Superconductivity. *Phys. Rev.*, **108**, 1175–1204.

[18] Beekman, A. J. 2015. Criteria for the Absence of Quantum Fluctuations after Spontaneous Symmetry Breaking. *Ann. Phys. (N.Y.)*, **361**, 461–489.

[19] Beekman, A. J. 2020. Theory of Generalized Josephson Effects. *Prog. Theor. Exp. Phys.*, **2020**(7), 073B09.

[20] Beekman, A. J., Wu, K., Cvetkovic, V., and Zaanen, J. 2013. Deconfining the Rotational Goldstone Mode: The Superconducting Quantum Liquid Crystal in (2+1) Dimensions. *Phys. Rev. B*, **88**, 204121.

[21] Beekman, A. J., Nissinen, J., Wu, K., and Zaanen, J. 2017a. Dual Gauge Field Theory of Quantum Liquid Crystals in Three Dimensions. *Phys. Rev. B*, **96**, 165115.

[22] Beekman, A. J., Nissinen, J., Wu, K., Liu, K., Slager, R.-J., Nussinov, Z., Cvetkovic, V., and Zaanen, J. 2017b. Dual Gauge Field Theory of Quantum Liquid Crystals in Two Dimensions. *Phys. Rep.*, **683**, 1–110.

[23] Beekman, A. J., Rademaker, L., and van Wezel, J. 2019. An Introduction to Spontaneous Symmetry Breaking. *SciPost Phys. Lect. Notes*, 11.

[24] Bell, J. S., and Jackiw, R. 1969. A PCAC Puzzle: $\pi 0 \rightarrow \gamma\gamma$ in the σ-Model. *Il Nuovo Cim. A (1965–1970)*, **60**(1), 47–61.

[25] Berezinskii, V. L. 1972. Destruction of Long-Range Order in One-Dimensional and Two-Dimensional Systems Possessing a Continuous Symmetry Group II: Quantum Systems. *Sov. Phys. JETP*, **34**, 610.

[26] Bernevig, B. A., and Hughes, T. L. 2013. *Topological Insulators and Topological Superconductors*. Princeton University Press.

[27] Berry, M. 2002. Singular Limits. *Phys. Today*, **55**(5), 10–11.

[28] Borsboom, S. 2024. *Spontaneous Breaking of Global Gauge Symmetries in the Higgs Mechanism*. M.Phil. thesis, University of Amsterdam.

[29] Boyanovsky, D., de Vega, H. J., and Schwarz, D. J. 2006. Phase Transitions in the Early and Present Universe. *Annu. Rev. Nucl. Part. Sci.*, **56**(1), 441–500.

[30] Brading, K. A. 2002. Which Symmetry? Noether, Weyl, and Conservation of Electric Charge. *Stud. Hist. Phil. Sci.*, **B33**(1), 3–22.

[31] Brauner, T. 2010. Spontaneous Symmetry Breaking and Nambu–Goldstone Bosons in Quantum Many-Body Systems. *Symmetry*, **2**(2), 609–657.

[32] Brauner, T. 2024. *Effective Field Theory for Spontaneously Broken Symmetry*. Lecture Notes in Physics. Springer.

[33] Brox, H., Olaussen, K., and Nguyen, A. K. 2008. Collapse of the Quantum Wavefunction. *arXiv*:0809.1575.

[34] Bruno, P. 2013. Impossibility of Spontaneously Rotating Time Crystals: A No-Go Theorem. *Phys. Rev. Lett.*, **111**, 070402.

[35] Burgess, C. P. 2020. *Introduction to Effective Field Theory: Thinking Effectively about Hierarchies of Scale*. Cambridge University Press.

[36] Caldeira, A. O., and Leggett, A. J. 1983. Path Integral Approach to Quantum Brownian Motion. *Physica A*, **121**, 587–616.

[37] Callan, C. G., Coleman, S., Wess, J., and Zumino, B. 1969. Structure of Phenomenological Lagrangians. II. *Phys. Rev.*, **177**, 2247–2250.

[38] Carroll, S. 2003. *Spacetime and Geometry: An Introduction to General Relativity*. Benjamin Cummings.

[39] Chaikin, P. M., and Lubensky, T. C. 2000. *Principles of Condensed Matter Physics*. Cambridge University Press.

[40] Choi, S., Choi, J., Landig, R., Kucsko, G., Zhou, H., Isoya, J., Jelezko, F., Onoda, S., Sumiya, H., Khemani, V., et al. 2017. Observation of Discrete Time-Crystalline Order in a Disordered Dipolar Many-Body System. *Nature*, **543**(7644), 221.

[41] Coleman, S. 1985. *Aspects of Symmetry: Selected Erice Lectures*. Cambridge University Press.

[42] Coleman, S., Wess, J., and Zumino, B. 1969. Structure of Phenomenological Lagrangians. I. *Phys. Rev.*, **177**, 2239–2247.

[43] de Gennes, P. G. 1999. *Superconductivity of Metals and Alloys*. Advanced book classics. Perseus.

[44] Debenedetti, P. G. 1996. *Metastable Liquids: Concepts and Principles*. Physical Chemistry: Science and Engineering. Princeton University Press.

[45] Delacrétaz, L. V., Endlich, S., Monin, A., Penco, R., and Riva, F. 2014. (Re)Inventing the Relativistic Wheel: Gravity, Cosets, and Spinning Objects. *JHEP*, **8**, 008.

[46] Desai, S. 2024. Astrophysical and Cosmological Searches for Lorentz Invariance Violation. Pages 433–463 of: Bambi, Cosimo, and Cárdenas-Avendaño, Alejandro (eds.), *Recent Progress on Gravity Tests: Challenges and Future Perspectives*. Springer Series in Astrophysics and Cosmology. Springer.

[47] DeWitt, B. S. 1965. *Dynamical Theory of Groups and Fields*. Gordon and Breach.

[48] Dirac, P. A. M. 1955. Gauge-Invariant Formulation of Quantum Electrodynamics. *Can. J. Phys.*, **33**(11), 650–660.

[49] Dirac, P. A. M. 2001. *Lectures on Quantum Mechanics*. Dover.

[50] Elitzur, S. 1975. Impossibility of Spontaneously Breaking Local Symmetries. *Phys. Rev. D*, **12**, 3978–3982.

[51] Else, D. V., Monroe, C., Nayak, C., and Yao, N. Y. 2020. Discrete Time Crystals. *Annu. Rev. Condens. Matter Phys.*, **11**(1), 467–499.

[52] Feynman, R. P., Leighton, R. B., and Sands, M. 1965. *The Feynman Lectures on Physics. Vol. 3: Quantum Mechanics*. Addison-Wesley Publishing Co., Inc.

[53] Fradkin, E., and Shenker, S. H. 1979. Phase Diagrams of Lattice Gauge Theories with Higgs Fields. *Phys. Rev. D*, **19**(12), 3682–3697.

[54] Fröhlich, J. 2023. Gauge Invariance and Anomalies in Condensed Matter Physics. *J. Math. Phys.*, **64**(3), 031903.

[55] Gaiotto, D., Kapustin, A., Seiberg, N., and Willett, B. 2015. Generalized Global Symmetries. *JHEP*, **2015**(2), 172.

[56] Georgi, H. M. 1999. *Lie Algebras in Particle Physics*, 2nd ed. Frontiers in Physics. Perseus.

[57] Ginzburg, V. L., and Landau, L. D. 2009. On the theory of superconductivity. Pages 113–137 of: *On Superconductivity and Superfluidity*. Springer.

[58] Goldenfeld, N. 1992. *Lectures on Phase Transitions and the Renormalization Group*. Frontiers in Physics, no. 85. Perseus Books.

[59] Goldman, A. M. 2013. The Berezinskii–Kosterlitz–Thouless Transition in Superconductors. In: Jose, J. V. (ed), *40 years of Berezinskii–Kosterlitz–Thouless theory*. World Scientific.

[60] Goldman, V. J., and Su, B. 1995. Resonant Tunneling in the Quantum Hall Regime: Measurement of Fractional Charge. *Science*, **267**, 1010–1012.

[61] Gomes, P. R. S. 2023. An Introduction to Higher-form Symmetries. *SciPost Phys. Lect. Notes*, 074.

[62] Gould, H., and Tobochnik, J. 2010. *Statistical and Thermal Physics: With Computer Applications*. Princeton University Press.

[63] Grady, M. 1994. Spontaneous Symmetry Breaking as the Mechanism of Quantum Measurement. *arXiv*, hep–th/9409049.

[64] Greensite, J. 2011. *An Introduction to the Confinement Problem*. Lecture Notes in Physics. Springer Berlin Heidelberg.

[65] Griffiths, D. J. 2016. *Introduction to Quantum Mechanics*. Cambridge University Press.

[66] Halmos, P. R. 1995. *Linear Algebra Problem Handbook*. Mathematical Association of America.

[67] Halperin, B. I., and Nelson, D. R. 1978. Theory of Two-Dimensional Melting. *Phys. Rev. Lett.*, **41**(2), 121–124.

[68] Halperin, B. I., Hohenberg, P. C., and Ma, S.-K. 1972. Calculation of Dynamic Critical Properties Using Wilson's Expansion Methods. *Phys. Rev. Lett.*, **29**(23), 1548–1551.

[69] Halperin, B. I., Lubensky, T. C., and Ma, S.-K. 1974. First-Order Phase Transitions in Superconductors and Smectic-*A* Liquid Crystals. *Phys. Rev. Lett.*, **32**, 292–295.

[70] Hayata, T., and Hidaka, Y. 2015. Dispersion Relations of Nambu-Goldstone Modes at Finite Temperature and Density. *Phys. Rev. D*, **91**, 056006.

[71] Henneaux, M., and Teitelboim, C. 1994. *Quantization of Gauge Systems*. Princeton University Press.

[72] Herbut, I. 2007. *A Modern Approach to Critical Phenomena*. Cambridge University Press.

[73] Hidaka, Y. 2013. Counting Rule for Nambu-Goldstone Modes in Nonrelativistic Systems. *Phys. Rev. Lett.*, **110**, 091601.

[74] Higgs, P. W. 1964. Broken Symmetries and the Masses of Gauge Bosons. *Phys. Rev. Lett.*, **13**, 508–509.

[75] Hindmarsh, M., Lüben, M., Lumma, J., and Pauly, M. 2021. Phase Transitions in the Early Universe. *SciPost Phys. Lect. Notes*, 024.

[76] Hofman, D., and Iqbal, N. 2019. Goldstone Modes and Photonization for Higher Form Symmetries. *SciPost Phys.*, **6**(1), 006.

[77] Hohenberg, P. C., and Halperin, B. I. 1977. Theory of dynamic critical phenomena. *Rev. Mod. Phys.*, **49**(3), 435.

[78] Hohenberg, P. C., and Krekhov, A. P. 2015. An introduction to the Ginzburg–Landau theory of phase transitions and nonequilibrium patterns. *Phys. Rep.*, **572**, 1–42.

[79] Ivanov, E. A., and Ogievetsky, V. I. 1975. The Inverse Higgs Phenomenon in Nonlinear Realizations. *Teor. Mat. Fiz.*, **25**, 164–177.

[80] Jackson, J. D., and Okun, L. B. 2001. Historical Roots of Gauge Invariance. *Rev. Mod. Phys.*, **73**, 663–680.

[81] Jones, H. F. 1998. *Groups, Representations, and Physics*. Institute of Physics Pub.

[82] Joos, E., and Zeh, H. D. 1985. The Emergence of Classical Properties Through Interaction with the Environment. *Z. Phys. B: Condens. Matter*, **59**, 223–243.

[83] Josephson, B. D. 1962. Possible New Effects in Superconductive Tunnelling. *Phys. Lett.*, **1**(7), 251–253.

[84] Kajantie, K., Laine, M., Rummukainen, K., and Shaposhnikov, M. 1996. Is There a Hot Electroweak Phase Transition at mH $\gtrsim$ mW? *Phys. Rev. Lett.*, **77**(14), 2887–2890.

[85] Kajita, T. 2016. Nobel Lecture: Discovery of Atmospheric Neutrino Oscillations. *Rev. Mod. Phys.*, **88**(3), 030501.

[86] Kallin, A. B., Hastings, M. B., Melko, R. G., and Singh, R. R. P. 2011. Anomalies in the Entanglement Properties of the Square-Lattice Heisenberg Model. *Phys. Rev. B*, **84**(16), 165134.

[87] Kapustin, A. 2012. Remarks on Nonrelativistic Goldstone Bosons. *arXiv*, 1207.0457.

[88] Karch, A., and Tong, D. 2016. Particle-Vortex Duality from 3D Bosonization. *Phys. Rev. X*, **6**, 031043.

[89] Kibble, T. W. B. 1976. Topology of Cosmic Domains and Strings. *J. Phys. A: Math. Gen.*, **9**(8), 1387–1398.

[90] Kibble, T. W. B. 1980. Some Implications of a Cosmological Phase Transition. *Phys. Rep.*, **67**, 183.

[91] Kibble, T. W. B. 2007. Phase-Transition Dynamics in the Lab and the Universe. *Physics Today*, **60**, 47.

[92] Kleinert, H. 1983. Limitations to the Coleman-Weinberg Mechanism of Spontaneous Mass Generation. *Phys. Lett. B*, **128**(1), 69–72.

[93] Kleinert, H. 2009. *Path Integrals in Quantum Mechanics, Statistics, Polymer Physics, and Financial Markets*. EBL-Schweitzer. World Scientific.

[94] Kobayashi, M., and Nitta, M. 2015. Interpolating Relativistic and Nonrelativistic Nambu-Goldstone and Higgs Modes. *Phys. Rev. D*, **92**, 045028.

[95] Kogut, J. B. 1979. An Introduction to Lattice Gauge Theory and Spin Systems. *Rev. Mod. Phys.*, **51**(4), 659–713.

[96] Kosterlitz, J. M., and Thouless, D. J. 1972. Long Range Order and Metastability in Two Dimensional Solids and Superfluids. (Application of dislocation theory). *J. Phys. C*, **5**(11), L124.

[97] Kosterlitz, J. M., and Thouless, D. J. 1973. Ordering, Metastability and Phase Transitions in Two-Dimensional Systems. *J. Phys. C*, **6**(7), 1181.

[98] Kramers, H. A., and Wannier, G. H. 1941. Statistics of the Two-Dimensional Ferromagnet Part I. *Phys. Rev.*, **60**(3), 252–262.

[99] Kruthoff, J., de Boer, J., van Wezel, J., Kane, C. L., and Slager, R.-J. 2017. Topological Classification of Crystalline Insulators through Band Structure Combinatorics. *Phys. Rev. X*, **7**, 041069.

[100] Kulchytskyy, B., Herdman, C. M., Inglis, S., and Melko, R. G. 2015. Detecting Goldstone Modes with Entanglement Entropy. *Phys. Rev. B*, **92**(11), 115146.

[101] Kuno, Y., and Okada, Y. 2001. Muon Decay and Physics beyond the Standard Model. *Rev. Mod. Phys.*, **73**(1), 151–202.

[102] Lake, E. 2018. Higher-Form Symmetries and Spontaneous Symmetry Breaking. *arXiv*, 1802.07747.

[103] Landau, L. D. 1937. On the Theory of Phase Transitions. *Zh. Eksp. Teor. Fiz.*, **7**, 19–32.

[104] Landsman, K. 2017. *Foundations of Quantum Theory: From Classical Concepts to Operator Algebras*. Fundamental Theories of Physics. Springer International Publishing.

[105] Landsman, N. P. 2013. Spontaneous Symmetry Breaking in Quantum Systems – Emergence or Reduction? *Stud. Hist. Phil. Mod. Phys.*, **44**(4), 379–394.

[106] Langacker, P. 1981. Grand Unified Theories and Proton Decay. *Phys. Rep.*, **72**(4), 185–385.

[107] Lange, R. V. 1966. Nonrelativistic Theorem Analogous to the Goldstone Theorem. *Phys. Rev.*, **146**, 301–303.

[108] Le Bellac, M. 1996. *Thermal Field Theory*. Cambridge University Press.

[109] Le Bellac, M., and de Forcrand-Millard, P. 2011. *Quantum Physics*. Cambridge University Press.

[110] Lee, Y.-L., and Lee, Y.-W. 2003. Quantum Dynamics of Tunneling between Ferromagnets. *Phys. Rev. B*, **68**, 184413.

[111] Leggett, A. J. 1975. A Theoretical Description of the New Phases of Liquid ^{3}He. *Rev. Mod. Phys.*, **47**, 331–414.

[112] Leggett, A. J., and Sols, F. 1991. On the Concept of Spontaneously Broken Gauge Symmetry in Condensed Matter Physics. *Found. Phys.*, **21**(3), 353–364.

[113] Leutwyler, H. 1994. Nonrelativistic Effective Lagrangians. *Phys. Rev. D*, **49**, 3033–3043.

[114] Li, Z. B., Schülke, L., and Zheng, B. 1994. Dynamic Monte Carlo Measurement of Critical Exponents. *Phys. Rev. Lett.*, **74**(17), 3396–3398.

[115] Low, I., and Manohar, A. V. 2002. Spontaneously Broken Spacetime Symmetries and Goldstone's Theorem. *Phys. Rev. Lett.*, **88**, 101602.

[116] Luitz, D. J., Plat, X., Alet, F., and Laflorencie, N. 2015. Universal Logarithmic Corrections to Entanglement Entropies in Two Dimensions with Spontaneously Broken Continuous Symmetries. *Phys. Rev. B*, **91**(15), 155145–155146.

[117] Lyth, D. H., and Liddle, A. R. 2009. *The Primordial Density Perturbation: Cosmology, Inflation and the Origin of Structure*. Cambridge University Press.

[118] MacDonald, A. H. 1994. Introduction to the Physics of the Quantum Hall Regime. *arXiv*, cond–mat/9410047.

[119] Mahan, G. D. 2000. *Many-Particle Physics*. Physics of Solids and Liquids. Springer US.

[120] Manousakis, E. 1991. The spin-$\frac{1}{2}$ Heisenberg Antiferromagnet on a Square Lattice and Its Application to the Cuprous Oxides. *Rev. Mod. Phys.*, **63**, 1–62.

[121] Maris, H. J., and Kadanoff, L. P. 1978. Teaching the Renormalization Group. *Am. J. Phys.*, **46**, 652.

[122] Marshall, W. 1955. Antiferromagnetism. *Proc. R. Soc. Lond.*, **A232**, 48–68.

[123] Mazumdar, A., and White, G. 2019. Review of Cosmic Phase Transitions: Their Significance and Experimental Signatures. *Rep. Prog. Phys.*, **82**(7), 076901.

[124] McGreevy, J. 2022. Generalized Symmetries in Condensed Matter. *Ann. Rev. Cond. Matt. Phys.*, **14**(1), 57–82.

[125] Mermin, N. D. 1979. The Topological Theory of Defects in Ordered Media. *Rev. Mod. Phys.*, **51**, 591–648.

[126] Mertens, L., Wesseling, M., Vercauteren, N., Corrales-Salazar, S., and van Wezel, J. 2021. Inconsistency of Linear Dynamics and Born's Rule. *Phys. Rev. A*, **104**, 052224.

[127] Metlitski, M. A., and Grover, T. 2011. Entanglement Entropy of Systems with Spontaneously Broken Continuous Symmetry. *arXiv*, 1112.5166.

[128] Mézard., M., Parisi., G., and Virasoro, M. 1987. *Spin Glass Theory and beyond: An Introduction to the Replica Method and Its Applications*. World Scientific Lecture Notes in Physics. World Scientific Publishing Company.

[129] Moessner, R., and Moore, J. E. 2021. *Topological Phases of Matter*. Cambridge University Press.

[130] Mong, R. S. K., Clarke, D. J., Alicea, J., Lindner, N. H., Fendley, P., Nayak, C., Oreg, Y., Stern, A., Berg, E., Shtengel, K., and Fisher, M. P. A. 2014. Universal

Topological Quantum Computation from a Superconductor-Abelian Quantum Hall Heterostructure. *Phys. Rev. X*, **4**, 011036.

[131] Morikawa, M., and Nakamichi, A. 2006. Quantum Measurement Driven by Spontaneous Symmetry Breaking. *Prog. Theor. Phys.*, **116**(4), 679–698.

[132] Mukhanov, V. 2005. *Physical Foundations of Cosmology*. Cambridge University Press.

[133] Mukherjee, A., and van Wezel, J. 2024. Colored-Noise-Driven Unitarity Violation Causing Dynamical Quantum State Reduction. *Phys. Rev. A*, **109**, 032214.

[134] Mukherjee, A., Gotur, S., Aalberts, J., van den Ende, R., Mertens, L., and van Wezel, J. 2024. Quantum State Reduction of General Initial States through Spontaneous Unitarity Violation. *Entropy*, **26**, 131.

[135] Mydosh, J. A. 1993. *Spin Glasses: An Experimental Introduction*. Taylor & Francis.

[136] Naegels, D. 2022. Goldstone Boson Physics and Effective Field Theories. *Proceedings of Science, Modave2021*, 004.

[137] Nagaosa, N. 1999. *Quantum Field Theory in Condensed Matter Physics*. Texts and Monographs in Physics. Springer.

[138] Nakahara, M. 2003. *Geometry, Topology and Physics, 2nd ed., Graduate Student Series in Physics*. Taylor & Francis.

[139] Nambu, Y. 2004. Spontaneous Breaking of Lie and Current Algebras. *J. Stat. Phys.*, **115**(1–2), 7–17.

[140] Nelson, D. R., and Halperin, B. I. 1979. Dislocation-Mediated Melting in Two Dimensions. *Phys. Rev. B*, **19**(5), 2457–2484.

[141] Nicolis, A., and Piazza, F. 2012. Spontaneous Symmetry Probing. *J. High Energ. Phys.*, **2012**, 25.

[142] Nicolis, A., and Piazza, F. 2013. Implications of Relativity on Nonrelativistic Goldstone Theorems: Gapped Excitations at Finite Charge Density. *Phys. Rev. Lett.*, **110**, 011602.

[143] Nicolis, A., Penco, R., Piazza, F., and Rosen, R. A. 2013. More on Gapped Goldstones at Finite Density: More Gapped Goldstones. *JHEP*, **2013**, 55.

[144] Nicolis, A., Penco, R., and Rosen, R. A. 2014. Relativistic Fluids, Superfluids, Solids, and Supersolids from a Coset Construction. *Phys. Rev. D*, **89**, 045002.

[145] Nicolis, A., Penco, R., Piazza, F., and Rattazzi, R. 2015. Zoology of Condensed Matter: Framids, Ordinary Stuff, Extra-Ordinary Stuff. *JHEP*, **06**, 155.

[146] Nielsen, H. B., and Chadha, S. 1976. On How to Count Goldstone Bosons. *Nucl. Phys. B*, **105**, 445.

[147] Nightingale, M. P., and Blöte, H. W. J. 1996. Dynamic Exponent of the Two-Dimensional Ising Model and Monte Carlo Computation of the Subdominant Eigenvalue of the Stochastic Matrix. *Phys. Rev. Lett.*, **76**(24), 4548.

[148] Nussinov, Z., and Ortiz, G. 2009. A Symmetry Principle for Topological Quantum Order. *Ann. Phys.*, **324**(5), 977–1057.

[149] Ogievetsky, V. I. 1974. Nonlinear Realizations of Internal and Space-Time Symmetries. In: *10th Karpacz Winter School of Theoretical Physics*.

[150] Onsager, L. 1944. Crystal Statistics I A Two-Dimensional Model with an Order-Disorder Transition. *Phys. Rev.*, **65**(3/4), 117–149.

[151] Pauli, W. 1940. The Connection between Spin and Statistics. *Phys. Rev.*, **58**, 716–722.

[152] Peccei, R. D., and Quinn, H. R. 1977. CP Conservation in the Presence of Pseudoparticles. *Phys. Rev. Lett.*, **38**(25), 1440–1443.

[153] Pekker, D., and Varma, C. M. 2015. Amplitude/Higgs Modes in Condensed Matter Physics. *Ann. Rev. Cond. Mat. Phys.*, **6**(1), 269–297.

[154] Penco, R. 2020. An Introduction to Effective Field Theories. *arXiv*, hep–th/2006.16285.

[155] Penrose, O. 1951. CXXXVI. On the Quantum Mechanics of Helium II. *Phil. Mag.*, **7**, 1373.

[156] Penrose, O., and Onsager, L. 1956. Bose-Einstein Condensation and Liquid Helium. *Phys. Rev.*, **104**(3), 576–584.

[157] Penrose, R. 2009. Black Holes, Quantum Theory and Cosmology. *J. Phys. Conf. Ser.*, **174**, 012001.

[158] Peres, A. 1995. *Quantum Theory: Concepts and Methods*. Fundamental Theories of Physics. Springer Netherlands.

[159] Peskin, M. E., and Schroeder, D. V. 1995. *An Introduction to Quantum Field Theory*. Westview Press.

[160] Phillips, P. 2012. *Advanced Solid State Physics*. 2nd ed. Cambridge University Press. Chap. 8, page 109–114.

[161] Piazza, F., and Vernizzi, F. 2013. Effective Field Theory of Cosmological Perturbations. *Classical and Quantum Gravity*, **30**(21), 214007.

[162] Pinto-Neto, N., Santos, G., and Struyve, W. 2012. Quantum-to-Classical Transition of Primordial Cosmological Perturbations in de Broglie–Bohm Quantum Theory. *Phys. Rev. D*, **85**(8), 083506.

[163] Polchinski, J. 1992. Effective Field Theory and the Fermi Surface. Pages 235–276 of: *Theoretical Advanced Study Institute (TASI 92): From Black Holes and Strings to Particles*.

[164] Poole, C. P., Farach, H. A., Creswick, R. J., and Prozorov, R. 2010. *Superconductivity*. Elsevier Science.

[165] Prokof'ev, N. V., and Stamp, P. C. E. 2000. Theory of the Spin Bath. *Rep. Prog. Phys.*, **63**, 669.

[166] Rademaker, L. 2015. Tower of States and the Entanglement Spectrum in a Coplanar Antiferromagnet. *Phys. Rev. B*, **92**(14), 144419 – 8.

[167] Rademaker, L. 2019. Exact Ground State of the Lieb-Mattis Hamiltonian as a Superposition of Néel States. *Phys. Rev. Res.*, **1**, 032018.

[168] Ryder, L. H. 1996. *Quantum Field Theory*. Cambridge University Press.

[169] Ryssens, W. 2012. *On the Quantum-to-Classical Transition of Primordial Perturbations*. M.Phil. thesis, KU Leuven.

[170] Sachdev, S. 2011. *Quantum Phase Transitions*. 2nd ed. Cambridge University Press.

[171] Sachdev, S. 2023. *Quantum Phases of Matter*. Cambridge University Press.

[172] Sachdev, S., and Young, A. P. 1997. Low Temperature Relaxational Dynamics of the Ising Chain in a Transverse Field. *Phys. Rev. Lett.*, **78**, 2220–2223.

[173] Sakharov, A. D. 1991. Violation of CP Invariance, C Asymmetry, and Baryon Asymmetry of the Universe. *Usp. Fiz. Nauk*, **161**, 61–64.

[174] Sakurai, J. J., and Napolitano, J. 2017. *Modern Quantum Mechanics*. Cambridge University Press.

[175] Savary, L., and Balents, L. 2016. Quantum Spin Liquids: A Review. *Rep. Prog. Phys.*, **80**(1), 016502–56.

[176] Schäfer, T., Son, D. T, Stephanov, M. A., Toublan, D., and Verbaarschot, J. J. M. 2001. Kaon Condensation and Goldstone's Theorem. *Phys. Lett. B*, **522**(1–2), 67–75.

[177] Schlosshauer, M. A. 2007. *Decoherence: And the Quantum-To-Classical Transition*. The Frontiers Collection. Springer.

[178] Schmitt, A. 2014. *Introduction to Superfluidity. Field-theoretical Approach and Applications*. Lecture Notes in Physics, Vol. 188. Springer Cham.

[179] Seiberg, N., Senthil, T., Wang, C., and Witten, W. 2016. A Duality Web in 2+1 Dimensions and Condensed Matter Physics. *Ann. Phys.*, **374**, 395–433.

[180] Shifman, M. 2012. *Advanced Topics in Quantum Field Theory: A Lecture Course*. Cambridge University Press.

[181] Sinova, J., Hanna, C. B., and MacDonald, A. H. 2002. Quantum Melting and Absence of Bose–Einstein Condensation in Two-Dimensional Vortex Matter. *Phys. Rev. Lett.*, **89**, 030403.

[182] Skyrme, T. H. R. 1962. A Unified Field Theory of Mesons and Baryons. *Nucl. Phys.*, **31**, 556–569.

[183] Son, D. T. 2001. Hydrodynamics of Relativistic Systems with Broken Continuous Symmetries. *Int. J. Mod. Phys. A*, **16**, 1284–1286.

[184] Son, D. T. 2002. Low-Energy Quantum Effective Action for Relativistic Superfluids. *arXiv*, hep–ph/0204199.

[185] Son, D. T. 2005. Effective Lagrangian and Topological Interactions in Supersolids. *Phys. Rev. Lett.*, **94**(17).

[186] Srednicki, M. 2007. *Quantum Field Theory*. Cambridge University Press.

[187] Steigman, G. 1976. Observational Tests of Antimatter Cosmologies. *Ann. Rev. of Astronomy and Astrophysics*, **14**, 339–372.

[188] Stein, D. L., and Newman, C. M. 2013. *Spin Glasses and Complexity*. Primers in Complex Systems. Princeton University Press.

[189] Stewart, G. R. 1984. Heavy-fermion Systems. *Rev. Mod. Phys.*, **56**, 755–787.

[190] Sudarsky, D. 2011. Shortcomings in the Understanding of Why Cosmological Perturbations Look Classical. *Int. J. Mod. Phys. D*, **20**(04), 509–552.

[191] Tasaki, H. 2020. *Physics and Mathematics of Quantum Many-Body Systems*. Springer Graduate Texts in Physics.

[192] Tinkham, M. 1996. *Introduction to Superconductivity*. McGraw Hill.

[193] Tobochnik, J., and Chester, G. V. 1979. Monte Carlo study of the planar spin model. *Phys. Rev. B*, **20**, 3761–3769.

[194] Tong, D. 2016. Lectures on the Quantum Hall Effect. *arXiv*, 1606.06687.

[195] Trodden, M. 1999. Electroweak Baryogenesis. *Rev. Mod. Phys.*, **71**(5), 1463–1500.

[196] van der Waals, J. D. 1873. *Over de Continuïteit van den Gas en Vloeistoftoestand*. Ph.D. thesis, Leiden University.

[197] van Wezel, J. 2008. Quantum Dynamics in the Thermodynamic Limit. *Phys. Rev. B*, **78**, 054301.

[198] van Wezel, J. 2010. Broken Time Translation Symmetry as a model for Quantum State Reduction. *Symmetry*, **2**, 582–608.

[199] van Wezel, J. 2022. Phase Transitions as a Manifestation of Spontaneous Unitarity Violation. *J. Phys. A Math. Theor.*, **55**, 401001.

[200] van Wezel, J., and van den Brink, J. 2008. Spontaneous Symmetry Breaking and Decoherence in Superconductors. *Phys. Rev. B*, **77**, 064523.

[201] van Wezel, J., Zaanen, J., and van den Brink, J. 2006. Relation between Decoherence and Spontaneous Symmetry Breaking in Many-Particle Qubits. *Phys. Rev. B*, **74**, 094430.

[202] Veenstra, J., Binysh, J., Seinen, V., Naber, R., Poisson, D. R., Hunt, A., van Saarloos, W., Souslov, A., and Coulais, C. 2025. Wave Coarsening Drives Time Crystallization in Active Solids. *arXiv*, 2508.20052.

[203] Volkov, D. V. 1973. Phenomenological Lagrangians. *Fiz. Elem. Chast. Atom. Yadra*, **4**, 3–41.

[204] Volovik, G. E. 1992. *Exotic Properties of Superfluid Helium 3*. World Scientific.

[205] Watanabe, H. 2019. Counting Rules of Nambu–Goldstone Modes. *Ann. Rev. Cond. Matt. Phys.*, **11**(1), 1–19.

[206] Watanabe, H., and Brauner, T. 2011. Number of Nambu-Goldstone Bosons and Its Relation to Charge Densities. *Phys. Rev. D*, **84**, 125013.

[207] Watanabe, H., and Murayama, H. 2012. Unified Description of Nambu-Goldstone Bosons without Lorentz Invariance. *Phys. Rev. Lett.*, **108**, 251602.

[208] Watanabe, H., and Murayama, H. 2013. Redundancies in Nambu-Goldstone Bosons. *Phys. Rev. Lett.*, **110**, 181601.

[209] Watanabe, H., and Murayama, H. 2014a. Effective Lagrangian for Nonrelativistic Systems. *Phys. Rev. X*, **4**, 031057.

[210] Watanabe, H., and Murayama, H. 2014b. Nambu-Goldstone Bosons with Fractional-Power Dispersion Relations. *Phys. Rev. D*, **89**, 101701(R).

[211] Watanabe, H., and Oshikawa, M. 2015. Absence of Quantum Time Crystals. *Phys. Rev. Lett.*, **114**, 251603.

[212] Watanabe, H., Brauner, T., and Murayama, H. 2013. Massive Nambu-Goldstone Bosons. *Phys. Rev. Lett.*, **111**, 021601.

[213] Weinberg, S. 1994. *Dreams of a Final Theory: The Scientist's Search for the Ultimate Laws of Nature*. Vintage Books. Knopf Doubleday Publishing Group.

[214] Weinberg, S. 1995. *The Quantum Theory of Fields*. Vol. 1. Cambridge University Press.

[215] Weinberg, S. 1996. *The Quantum Theory of Fields*. Vol. 2. Cambridge University Press.

[216] Weinberg, S. 2008. *Cosmology*. OUP Oxford.

[217] Weinberg, S. 2015. *Lectures on Quantum Mechanics*. 2nd ed. Cambridge University Press.

[218] Wen, X.-G. 2002. *An Introduction of Topological Orders*. https://web.archive.org /web/20170829201814/http://dao.mit.edu/ wen/topartS3.pdf.

[219] Wen, X.-G. 2004. *Quantum Field Theory of Many-Body Systems: From the Origin of Sound to an Origin of Light and Electrons*. OUP Oxford.

[220] Wen, X.-G. 2017. Colloquium: Zoo of Quantum-Topological Phases of Matter. *Rev. Mod. Phys.*, **89**, 041004.

[221] Wietek, A., Schuler, M., and Läuchli, A. M. 2017. Studying Continuous Symmetry Breaking Using Energy Level Spectroscopy. *arXiv*, 1704.08622.

[222] Wilczek, F. 2012. Quantum Time Crystals. *Phys. Rev. Lett.*, **109**, 160401.

[223] Wilson, K. G. 1971. Renormalization Group and Critical Phenomena. I. Renormalization Group and the Kadanoff Scaling Picture. *Phys. Rev. B*, **4**, 3174–3183.

[224] Wolynes, P. G., and Lubchenko, V. 2012. *Structural Glasses and Supercooled Liquids: Theory, Experiment, and Applications*. Wiley.

[225] Yeomans, J. M. 1992. *Statistical Mechanics of Phase Transitions*. Clarendon Press.

[226] Young, A. P. 1979. Melting and the Vector Coulomb Gas in Two Dimensions. *Phys. Rev. B*, **19**(4), 1855–1866.

[227] Zaanen, J., Nussinov, Z., and Mukhin, S. I. 2004. Duality in 2+1D Quantum Elasticity: Superconductivity and Quantum Nematic Order. *Ann. Phys.*, **310**(1), 181–260.

[228] Zeh, H. D. 1970. On the Interpretation of Measurement in Quantum Theory. *Found. Phys.*, **1**(1), 69–76.

[229] Zhang, J., Hess, P. W., Kyprianidis, A., Becker, P., Lee, A., Smith, J., Pagano, G., Potirniche, I.-D., Potter, A. C., Vishwanath, A., et al. 2017. Observation of a Discrete Time Crystal. *Nature*, **543**(7644), 217.

[230] Zinn-Justin, J. 2021. *Quantum Field Theory and Critical Phenomena, 5th ed.,* Oxford University Press.

[231] Zurek, W. H. 1981. Pointer Basis of Quantum Apparatus: Into What Mixture Does the Wave Packet Collapse? *Phys. Rev. D*, **24**, 1516–1525.

[232] Zurek, W. H. 1985. Cosmological Experiments in Superfluid Helium? *Nature*, **317**(6037), 505–508.

[233] Zurek, W. H. 1996. Cosmological Experiments in Condensed Matter Systems. *Phys. Rep.*, **276**, 177.

[234] Zurek, W. H. 2003. Decoherence, Einselection, and the Quantum Origins of the Classical. *Rev. Mod. Phys.*, **75**, 715–775.

Index

For EU product safety concerns, contact us at Calle de José Abascal, 56–1°,
28003 Madrid, Spain or eugpsr@cambridge.org.

www.ingramcontent.com/pod-product-compliance
Ingram Content Group UK Ltd.
Pitfield, Milton Keynes, MK11 3LW, UK
UKHW052355290926
475916UK00019B/329